P9-AZX-457

DISCARDED
FROM
UNH LIBRARY

Handbook of

ENVIRONMENTAL ANALYSIS

Chemical Pollutants in Air, Water, Soil, and Solid Wastes

Pradyot Patnaik, Ph.D.

LEWIS PUBLISHERS

Boca Raton New York London Tokyo

Acquiring Editor: Neil Levine
Project Editor: Carole Sweatman
Marketing Manager: Greg Daurelle
Direct Marketing Manager: Arline Massey
Cover Design: Dawn Boyd
Prepress: Kevin Luong
Manufacturing: Sheri Schwartz

Library of Congress Cataloging-in-Publication Data

Patnaik, Pradyot.
Handbook of environmental analysis : chemical pollutants in air, water, soil, and soild wastes / Pradyot Patnaik.
p. cm.
Includes bibliographical references and index.
ISBN 0-87371-989-1 (alk. paper)
1. Pollutants--Analysis. I. Title.
TD193.P38 1996
628.5′2--dc20 96-32647
CIP

Direct all inquiries to CRC Press, Inc., 2000 Corporate Blvd., N.W., Boca Raton, Florida 33431.

Lewis Publishers is an imprint of CRC Press

International Standard Book Number 0-87371-989-1
Library of Congress Card Number 96-32647
Printed in the United States of America 1 2 3 4 5 6 7 8 9 0
Printed on acid-free paper

TABLE OF CONTENTS

Dedication

Preface

The Author

Acknowledgments

Glossary of Terms: Units, Conversions, and Abbreviations

List of Tables

Part 1 Analytical Techniques

1.1 Introduction 3
1.2 Precision and Accuracy of Analysis 5
1.3 Analysis of Organic Pollutants by Gas Chromatography 17
1.4 Analysis of Organic Pollutants by Gas Chromatography/Mass Spectrometry 25
1.5 Extraction of Organic Pollutants and Sample Cleanup 33
1.6 Titrimetric Analysis 39
1.7 Colorimetric Analysis 65
1.8 Analysis of Metals by Atomic Absorption and Emission Spectroscopy 69
1.9 Ion-Selective Electrode Analysis 77
1.10 Application of High Performance Liquid Chromatography in Environmental Analysis 83
1.11 Ion Chromatography 87
1.12 Air Analysis 91
1.13 Application of Immunoassay Techniques in Environmental Analysis 101

Part 2 Specific Classes of Substances and Aggregate Properties

2.1 Aldehydes and Ketones 107
2.2 Alkalinity 115
2.3 Bromide 117
2.4 Chloride 121
2.5 Cyanate 127
2.6 Cyanide, Total 129
2.7 Cyanide Amenable to Chlorination 137
2.8 Fluoride 141

2.9 Halogenated Hydrocarbons 143
2.10 Hardness 151
2.11 Herbicides: Chlorophenoxy Acid 153
2.12 Hydrocarbons 159
2.13 Hydrocarbons, Polynuclear Aromatic 165
2.14 Nitrogen (Ammonia) 171
2.15 Nitrogen (Nitrate) 179
2.16 Nitrosamines 183
2.17 Oxygen Demand, Biochemical 187
2.18 Oxygen Demand, Chemical 193
2.19 Pesticides: Carbamate, Urea and Triazine 199
2.20 Pesticides: Organochlorine 207
2.21 Pesticides: Organophosphorus 215
2.22 pH and Eh 221
2.23 Phenols 227
2.24 Phosphorus 233
2.25 Phthalate Esters 239
2.26 Polychlorinated Biphenyls (PCBs) 243
2.27 Polychlorinated Dioxins and Dibenzofurans 249
2.28 Silica 253
2.29 Sulfate 257
2.30 Sulfide 261
2.31 Sulfite 269
2.32 Surfactant: Anionic 275
2.33 Thiocyanate 279

Part 3 Selected Individual Compounds

3.1 Acetaldehyde 285
3.2 Acetone 287
3.3 Acetonitrile 289
3.4 Acrolein 291
3.5 Acrylonitrile 293
3.6 Aniline 295
3.7 Arsine 297
3.8 Asbestos 299
3.9 Benzene 301
3.10 Benzidine 303
3.11 Benzyl chloride 305
3.12 1,3-Butadiene 307
3.13 Carbon Disulfide 309
3.14 Carbon Monoxide 311
3.15 Carbon Tetrachloride 313
3.16 Captan 315
3.17 Chloroacetic acid 317
3.18 Chlorobenzene 319

3.19 Chloroform 321
3.20 2-Chlorophenol 323
3.21 Cumene 325
3.22 Cyanogen 327
3.23 Cyanuric Acid 329
3.24 Diazomethane 331
3.25 Diborane 333
3.26 Dichlorobenzene 335
3.27 1,1-Dichlorethylene 337
3.28 2,4-Dichlorophenol 339
3.29 Diethyl Ether 341
3.30 2,4-Dinitrotoluene 343
3.31 2,6-Dinitrotoluene 345
3.32 Epichlorohydrin 347
3.33 Ethyl Benzene 349
3.34 Ethyl Chloride 351
3.35 Ethylene Chlorohydrin 353
3.36 Ethylene Dibromide 355
3.37 Ethylene Glycol 357
3.38 Ethylene Oxide 359
3.39 Formaldehyde 361
3.40 Freon-113 363
3.41 Hydrogen Cyanide 365
3.42 Hydroquinone 367
3.43 Hydrogen Sulfide 369
3.44 Isophorone 371
3.45 Methane 373
3.46 Methyl Bromide 375
3.47 Methyl Chloride 377
3.48 Methylene Chloride 379
3.49 Methyl Iodide 381
3.50 Methyl Isobutyl Ketone 383
3.51 Methyl Isocyanate 385
3.52 Methyl Methacrylate 387
3.53 Nitrobenzene 389
3.54 Nitrogen Dioxide 391
3.55 Pentachlorophenol 393
3.56 Phosgene 395
3.57 Pyridine 397
3.58 Pyrocatechol 399
3.59 Pyrogallol 401
3.60 Resorcinol 403
3.61 Stibine 405
3.62 Strychnine 407
3.63 Styrene 409
3.64 Sulfur Dioxide 411

3.65 Tetrachloroethylene 413
3.66 Tetraethyllead 415
3.67 Tetraethyl Pyrophosphate 417
3.68 Tetrahydrofuran 419
3.69 Toluene 421
3.70 Toluene-2,4-diisocyanate 423
3.71 *o*-Toluidine 425
3.72 1,1,1-Trichloroethylene 427
3.73 Trichloroethylene 429
3.74 2,4,6-Trichlorophenol 431
3.75 Vinylchloride 433
3.76 Xylene 435

Bibliography 437

Appendices

Appendix A Some Common QC Formulas and Statistics 443
Appendix B Sample Containers, Preservations, and Holding Times 449
Appendix C Preparation of Molar and Normal Solutions of Some Common Reagents 453
Appendix D Total Dissolved Solids and Specific Conductance: Theoretical Calculations 455
Appendix E Characteristic Masses for Identification of Additional Organic Pollutants (Not Listed in the Text) by GC/MS 463
Appendix F Volatility of Some Additional Organic Substances (Not Listed in Text) for Purge and Trap Analysis 467
Appendix G Analysis of Elements by Atomic Spectroscopy: An Overview 471
Appendix H Analysis of Trace Elements by Inductively Coupled Plasma Mass Spectrometry 473
Appendix I Oil and Grease Analysis: An Overview 477
Appendix J NIOSH Methods for Air Analysis 481
Appendix K U.S. EPA Methods for Air Analysis 489
Appendix L Inorganic Test Procedures for Analysis of Aqueous Samples: EPA, SM, and ASTM Reference Methods 493
Appendix M U.S. EPA Analytical Methods for Organic Pollutants in Aqueous and Solid Matrices 499

Indices

Chemical Compounds 537
CAS Registry Numbers 561

DEDICATION

To

Manisha and Chirag

PREFACE

The subject on environmental analysis has expanded in recent years into a fully grown scientific field on its own merit. There is, however, a dearth of a single volume on the subject which adequately covers all aspects of environmental analysis. This book attempts to combine the features of both a reference handbook as well as a textbook.

This book presents a brief discussion on the analytical techniques and the methods of determination of chemical pollutants in aqueous, solid, and air samples at trace concentrations.

The topics in this book are presented under three broad sections. Part 1 highlights different analytical techniques including instrumentations, sample preparations, wet methods, air analysis, and immunoassay. Instrumental methods primarily include gas chromatography, mass spectrometry, high performance liquid chromatography and spectrophotometric methods. Part 2 presents analytical methodologies for different classes of organic and inorganic pollutants in aqueous, solid, and air matrices. Substances of similiar structures, or functional groups, or similar properties have been grouped together. This should guide users of this book to select analytical procedures to analyze compounds not listed in this text. In Part 3, analytical methodologies and physical properties are presented individually for some selected compounds.

Most of the analytical methods in Parts 2 and 3 of this text are abstracted from the methodologies of the U.S. Environmental Protection Agency, National Institute for Occupational Safety and Health, Standard Methods for the Examination of Water and Wastewater, and the American Society for Testing and Materials. Some methods are abstracted from other reliable publications or journal articles. The few suggested methods of analysis as found in this handbook are either based on the structural features of the compounds or the author's own experimental work and personal communications.

Appendices at the end of the volume provide instant information on a wide array of topics ranging from sample preservation to statistical calculation. Chemical equations, structures, problems, and examples of their solutions are presented wherever necessary.

The author would greatly appreciate any comments and suggestions that would serve to improve this book in its future editions.

THE AUTHOR

Pradyot Patnaik, Ph.D., is currently the Director of the Environmental Laboratory of Interstate Sanitation Commission at Staten Island, New York and is a researcher at the Center for Environmental Science at City University of New York at Staten Island. Dr. Patnaik teaches as an Adjunct Professor at the New Jersey Institute of Technology in Newark, New Jersey and at the Community College of Philadelphia. He was formerly the Director of the Special Research Project at the Environmental Testing and Technologies in Westmont, New Jersey, and after that, the Director of Rancocas Environmental Laboratory in Delanco, New Jersey.

Earlier Dr. Patnaik was a post-doctoral research scientist at Cornell University, Ithaca, New York. He had obtained a B.S. and M.S. in chemistry from Utkal University in India and a Ph.D. from the Indian Institute of Technology, Bombay.

Dr. Patnaik is the author of the book, *A Comprehensive Guide to the Hazardous Properties of Chemical Substances*.

ACKNOWLEDGMENTS

I wish to thank the following organizations for their research literature: Dionex Corporation, Hach Company, J&W Scientific, Millipore Corporation, Ohmicro Corporation, Orion Research, Restek Corporation, and Supelco Incorporated. I wish to express my thanks to Mr. Neil Levine, acquisition editor; Ms. Joann Fazzi, editorial assistant; Mrs. Carole Sweatman, project editor; and all others in the staff of Lewis Publishers/CRC Press for the production of this book. Also, I would like to thank Mr. Skip DeWall, former editor at Lewis Publishers, for initiating this project.

I gratefully acknowledge the help received from Mrs. Mary Ann Richardson for typing the manuscript and drawing some of the chemical structures and Mrs. Lalitha Subramanian for preparing the indices.

Special words of thanks go to my wife, Sanjukta, for her unusual patience and moral support during this long ordeal.

GLOSSARY OF TERMS

UNITS, CONVERSION AND ABBREVIATIONS

Metric and English Units

1 liter (L) = 1000 milliliter (mL)
1 gallon (gal) = 3.784 L
1 L = 0.264 gal
1 quart (qt) = 0.9464 L = 946.4 mL
1 L = 1.057 qt
1 fluid ounce = 29.6 mL
1 mL = 1000 microliters (μL)
1 kg = 1000 grams (g)
1 g = 1000 milligrams (mg)
1 mg = 1000 micrograms (μg)
1 μg = 1000 nanograms (ng)
1 ng = 1000 picrogram (pg)
1 kg = 2.205 pounds (lb)
1 lb = 453.6 g
1 ounce (oz) = 28.35 g
1 cubic meter (m^3) = 1,000,000 cubic centimeters (cc) or cm^3
1 cc = 1 mL
1 m^3 = 1000 L

Conversions

- Concentration of a solution as weight percent (w/w)

$$= \frac{\text{mass of solute (g)}}{\text{100 g solution}} \times 100\%$$

- Concentration of a solution as volume percent (v/v)

$$= \frac{\text{mL substance}}{\text{100 mL solution}} \times 100\%$$

- Concentration of a solution as weight per volume percent (w/v)

$$= \frac{\text{mass of substance (g)}}{\text{100 mL solution}} \times 100\%$$

- Concentration, as part per million (ppm) = mg substance/L solution; that is, ppm = mg/L or μg/mL or ng/μL; part per billion (ppb) = μg/L or ng/mL.

- 1 Atmosphere (atm) = 760 torr
  = 760 mm mercury
  = 14.6 pounds per square inch (psi)
  = 101.306 kPa

- °F = (1.8 × °C) + 32

- $°C = \frac{(°F - 32)}{1.8}$

- K (Kelvin) = °C + 273

- $\text{Density} = \frac{\text{mass}}{\text{volume}}$ (The units commonly used for density are g/mL for liquids, g/cm^3 for solids, and g/L for gases.)

- 1 mole (mol) = molecular or formula weight in grams (i.e., 1 mol NaOH = 40 g NaOH)

- $\text{Molarity } (M) = \frac{\text{mol of substance}}{\text{liter solution}}$

- $\text{Normality } (N) = \frac{\text{gram equivalent weight of substance}}{\text{liter solution}}$ or milligram equivalent/mL solution

- STP = Standard Temperature and Pressure, which is 0°C and 1 atm

- NTP = Normal Temperature and Pressure, which is 25°C and 1 atm
  (At STP volume of 1 mol of any gas = 22.4 L, while at NTP 1 mol of any gas would occupy 24.45 L.)

- Ideal gas equation: $PV = nRT$
  where P = pressure,
  V = volume,
  T = temperature (K),
  R = 0.082 L.atm/mol.K (when pressure and volume are expressed in atm and L, respectively.)

- Vapor pressure is the pressure of a vapor in equilibrium with its liquid or solid form. It is temperature dependent, and expressed in mm Hg or torr. It is a characteristic of the volatility of a substance. The higher the vapor pressure of a substance, the more volatile it is. Vapor pressure data in this text are presented at the temperature 20°C.

- $\text{Density of a gas at STP} = \frac{\text{gram molecular weight (mole)}}{22.4\ \text{L}}$

- $\text{Density of a gas at NTP} = \frac{\text{gram molecular weight (mole)}}{24.45\ \text{L}}$

- To determine how heavy a gas is with respect to air, divide the molecular weight of the gas by 29 (i.e., $\frac{\text{gas } x}{\text{air}} = \frac{\text{mol. wt. of } x}{29}$).

Abbreviations

AA	Atomic Absorption (Spectrophotometry)
ACS	American Chemical Society
ASTM	American Society for Testing and Materials
CAS	Chemical Abstract Service
ECD	Electron Capture Detector
EPA (U.S. EPA)	The United States Environmental Protection Agency
FID	Flame Ionization Detector
FPD	Flame Photometric Detector
GC	Gas Chromatography
GC/MS	Gas Chromatography/Mass Spectrometry
HECD	Hall Electrolytic Conductivity Detector
HPLC	High Performance Liquid Chromatography
ICP	Inductively Coupled Plasma (Emission Spectrometry)
ICP-MS	Inductively Coupled Plasma (Mass Spectrometry)
IR	Infrared (Spectrophotometry)
LLE	Liquid-Liquid Extraction
NIOSH	National Institute for Occupational Safety and Health
NPD	Nitrogen-Phosphorus Detector
NTP	Normal Temperature and Pressure
OSHA	Occupational Safety and Health Administration
SFE	Supercritical Fluid Extraction
SM	Standard Methods (for the Examination of Water and Wastewater)
SPE	Solid Phase Extraction
TCD	Thermal Conductivity Detector
UV	Ultraviolet (Spectrophotometry)

LIST OF TABLES

Table 1.3.1
Polarity of Stationary Phases 19

Table 1.3.2
Separation Efficiency and Sample Capacity of GC Columns
of Varying Internal Diameters (ID) 20

Table 1.4.1
BFB Tuning Requirement for Volatile Organic Analysis..................................28

Table 1.4.2
DFTPP Tuning Requirement for Semivolatile Organic Analysis 29

Table 1.5.1
Stationary Phases for Solid Phase Extraction 35

Table 1.5.2
Cleanup Methods for Organic Extracts 37

Table 1.6.1
Titrimetric Procedures Applied in Environmental Analysis..............................40

Table 1.6.2
Some Common Acid-Base Indicators..41

Table 1.6.3
Common Oxidizing Agents for Redox Titrations 47

Table 1.6.4
A Set of Potentiometric Titration Data as an Example.....................................64

Table 1.7.1
Common Pollutants...67

Table 1.8.1
Acid Combination Suggested for Sample Preparation.......................................70

Table 1.8.2
Recommended Wavelength, Flame Type, and Technique
for Flame Atomic Absorption Analysis 71

Table 1.8.3
Substances Added to the Sample for the Removal of Interference in Graphite Furnace Atomic Absorption Method 72

Table 1.8.4
Recommended Wavelength and the Instrument Detection Level in Inductively Coupled Plasma Emission Spectrometry 76

Table 1.9.1
Application of Ion-Selective Electrodes in Environmental Analysis 81

Table 1.10.1
Determination of Organic Analytes by High Performance Liquid Chromatography 85

Table 1.11.1
Application of Ion Chromatography in Environmental Analysis 88

Table 1.11.2
Some Common Eluants Used in Ion Chromatography 89

Table 2.1
Some Commercially Used Aldehydes and Ketones 108

Table 2.9.1
Purgeable Volatile Halogenated Hydrocarbons and Their Characteristic Masses 146

Table 2.9.2
Some Solvent Extractable Halogenated Hydrocarbon Pollutants and Their Characteristic Masses 147

Table 2.9.3
NIOSH Methods for Air Analysis for Halogenated Hydrocarbons 149

Table 2.9.4
U.S. EPA Methods for the Air Analysis of Halogenated Hydrocarbons 150

Table 2.11.1
Some Common Chlorophenoxy Acid Herbicides 153

Table 2.13.1
Common Polynuclear Aromatic Hydrocarbons 166

Table 2.13.2
Characteristic Ions for Polynuclear Aromatic Hydrocarbons 168

Table 2.16.1
Nitrosamines Classified as Priority Pollutants by U.S. EPA
under the Resource Conservation and Recovery Act 183

Table 2.16.2
Characteristic Masses for Some Common Nitrosamine Pollutants 185

Table 2.19.1
Some Common Carbamate Pesticides 201

Table 2.19.2
Some Common Pesticides Containing Urea Functional Group.......................203

Table 2.19.3
DFTPPO Tuning Criteria for LC/MS System Performance Check 204

Table 2.19.4
Some Common Triazine Pesticides 205

Table 2.20.1
Some Common Chlorinated Pesticides and Their Degradation Products 208

Table 2.20.2
Elution Patterns for Pesticides in Florisil Column Cleanup 209

Table 2.20.3
Characteristic Masses for Chlorinated Pesticides..212

Table 2.21.1
Common Organophosphorus Pesticides 217

Table 2.21.2
Organophosphorus Pesticides Containing Halogen Atoms 218

Table 2.21.3
Characteristic Ions for Identification of
Some Common Organophosphorus Pesticides by GC/MS 219

Table 2.22.1
Potentials of Redox Standard Solutions for Selected
Reference Electrodes..225

Table 2.23.1
Phenols Classified as U.S. EPA Priority Pollutants 228

Table 2.23.2
Elution Pattern of Phenol Derivatives Using Silica Gel 230

Table 2.23.3
Characteristic Mass Ions for GC/MS Determination
of Some Common Phenol Pollutants..231

Table 2.25.1
Characteristic Ions for Phthalate Esters..241

Table 2.25.2
EPA and NIOSH Methods for the Analysis of Phthalate Esters 241

Table 2.26.1
PCBs and Their Chlorine Contents 244

Table 2.26.2
Common Analytical Columns and Conditions for GC Analysis of PCBs 244

Table 2.31.1
Sulfite Standards...273

Table 2.32.1
Series of Calibration Standard Solutions..276

Table 3.8.1
Asbestos Types and Their Compositions..299

Appendix A
t Values at the 99% Confidence Level...447

Appendix B
Sample Containers, Preservations, and Holding Times 449

Appendix C
Common Reagents in Wet Analysis 454

Appendix D
Table 1
Equivalent Conductance, $\lambda°$ for Common Ions in Water at 25°C...................459

Appendix E
Characteristic Masses of Miscellaneous Organic Pollutants
(Not Listed in Text) for GC/MS Identification 463

Appendix F
Volatility of Some Additional Organic Substances
(Not Listed in Text) for Purge and Trap Analysis....................467

Appendix G
Analysis of Elements by Atomic Spectroscopy: An Overview 471

Appendix H
Table 1
Recommended Analytical Masses for Element Detection....................474

Table 2
Mass Ions for Internal Standards....................474

Table 3
Interference from Isobars and Matrix Molecular Ions....................475

Table 4
Recommended Elemental Equations for Calculations 475

Appendix I
Table 1
Extraction Solvents for Oil and Grease....................478

Appendix J
NIOSH Methods for Air Analysis 481

Appendix K
Table 1
Description of U.S. EPA Methods for Air Analysis....................489

Table 2
Individual Organic Pollutants in Air: U.S. EPA Methods....................490

Appendix L
Inorganic Test Procedures for Analysis of Aqueous Samples:
EPA, SM, and ASTM Reference Methods....................493

Appendix M
The U.S. EPA's Analytical Methods for Organic Pollutants
in Aqueous and Solid Matrices....................500

1 ANALYTICAL TECHNIQUES

1.1 INTRODUCTION

The analysis of chemical pollutants in the environmental matrices has entered a new phase in the last decade. Modifications in instrumentation, sampling, and sample preparation techniques have become essential to keep up with the requirements of achieving ppt to low ppb detection levels, as well as to achieve a faster speed of analysis. In addition, more stringent quality-control (QC) requirements in analytical methods have become necessary to obtain high data quality. This has led to the many new methodologies that are different from the conventional macro and semicmicro analytical approach.

Environmental analysis today — like any other scientific field — relies heavily on instrumentation. Organic pollutants are primarily determined by gas chromatography (GC), gas chromatography/mass spectrometry (GC/MS), and high performance liquid chromatography (HPLC), methods. There is, however, also a growing interest in alternative techniques, such as Fourier transform infrared spectroscopy (FTIR). Specially designed capillary columns have come up for GC analysis to achieve high resolution and better separation of many closely eluting isomers. Another major development in organic analysis is HPLC determination using postcolumn derivatization. Many classes of substances such as aldehydes, ketones, and carboxylic acids may be accurately determined by using such techniques.

Most organic compounds may be best analyzed by GC/MS. Such GC/MS or GC analysis, however, is preceded by either a "purge and trap" concentration step or a liquid or solid phase extraction step using a suitable organic solvent. The purge and trap method for aqueous samples is applicable for volatile substances that have lower solubility in water. A mass spectrometer should be used wherever possible to identify the compounds more correctly. Although it has a lower sensitivity than other GC detectors, mass spectrometry is, by far, the most confirmatory test for compound identification.

Methodologies for inorganic anions and metals have undergone rapid growth similar to chromatographic techniques. Notable among these technologies are the atomic absorption and emission spectroscopy and ion chromatography (IC). The latter is a rapid method to determine several anions, simultaneously. The IC approach may be modified further to measure such weak anions as carboxylates and cyanide.

Sample preparation is a key step in all environmental analyses. Two major areas of development in this area have been solid phase extraction and supercritical fluid extraction. Both techniques have made the extraction of pollutants from aqueous and nonaqueous matrices relatively simple, fast, and less expensive. These processes, along with gel permeable chromatography, provide efficient methods of removing interferences.

The methods of analysis of pollutants in ambient air has developed tremendously in recent years. Although these methods employ the same analytical instrumentation (i.e., GC, GC/MS, HPLC, IR, atomic absorption, ion chromatography, and the electrode methods), the air sampling technique is probably the most important component of such analysis. The use of cryogenic traps and high pressure pumps has supplemented the impinger and sorbent tube sampling techniques.

The number of pollutants that are currently regulated constitute only a fraction of those found in the environment. In addition, their chemical characteristics and concentrations may vary widely. New and alternative methodologies that are simple, rapid, and reliable need to be developed. Enzyme immunoassay and portable GC and IR techniques need greater attention.

This book presents a detailed discussion of various analytical methodologies, including sample preparations, cleanup, and instrumentation to identify different classes of substances and selected individual compounds and also to derive a method to analyze uncommon pollutants based on their physical and chemical properties. Only the chemical testings are discussed in this book; microbial and radiological testings are excluded in this edition.

1.2 PRECISION AND ACCURACY OF ANALYSIS

QUALITY ASSURANCE AND QUALITY CONTROL IN ENVIRONMENTAL ANALYSIS

Quality assurance and quality control programs mandate that every laboratory follow a set of well-defined guidelines so as to achieve analytical results at a high degree of accuracy. The term quality assurance refers to a set of principles that are defined, documented, and strictly observed such that the accuracy of results of analysis may be stated with a high level of confidence and legally defensible. The quality assurance plan includes documentation of sampling events, receipt of samples in the laboratory, and their relinquish to respective individuals to perform analysis; all of which are recorded on chain-of-custody forms with dates and times, as well as the names and signatures of individuals responsible to perform the tasks. The plan, in a broader sense of the term, also includes quality control.

The laboratory quality control program has several components: documentation of standard operating procedures for all analytical methods, periodic determination of method detection levels for the analytes, preparation of standard calibration curves and daily check of calibration standards, analysis of reagent blank, instrument performance check, determination of precision and accuracy of analysis, and preparation of control charts. Determination of precision and accuracy of analysis and method detection limits are described under separate subheadings in the following sections. The other components of the quality control plan are briefly discussed below.

The preparation of a standard calibration curve is required for many colorimetric and gas chromatography analyses. A fresh calibration check standard at any selected concentration should be prepared daily and analyzed prior to sample analysis. If the response for the check standard falls outside of ± 15% standard deviation for the same concentration in the standard calibration curve, then a new calibration curve should be prepared.

PRECISION AND ACCURACY

The determination of precision and accuracy is an important part of environmental analysis because it indicates the degree of bias or any error in the measurement.

Precision determines the reproducibility or repeatability of the analytical data. It measures how closely multiple analysis of a given sample agree with each other. If a sample is repeatedly analyzed under identical conditions, the results of each measurement, x, may vary from each other due to experimental error or causes beyond control. These results will be distributed randomly about a mean value which is the arithmetic average of all measurements. If the frequency is plotted against the results of each measurement, a bell-shaped curve known as normal distribution curve or gaussian curve, as shown below, will be obtained (Figure 1.2.1). (In many highly dirty environmental samples, the results of multiple analysis may show skewed distribution and not normal distribution.)

The mean $\bar{x}$ of all distributions is equal to $\frac{\sum x}{n}$ (i.e., the sum of all measurements divided by the number of measurements). Standard deviation, which fixes the width of the normal distribution, consists of a fixed fraction of the values making up the curve. An estimate of standard deviation, s, can be calculated as follows:

$$s = \sqrt{\frac{\sum (x - \bar{x})^2}{n - 1}}$$

In a normal distribution curve 68.27% area lies between $\bar{x} \pm 1s$, 95.45% area lies between $\bar{x} \pm 2s$, and 99.70% area falls between $\bar{x} \pm 3s$. In other words, 99.70% of replicate measurement should give values that should theoretically fall within three standard deviations about the arithmetic average of all measurements. Therefore, $3s$ about the mean is taken as the upper and lower control limits in control charts. Any value outside $\bar{x} \pm 3s$ should be considered unusual, thus indicating that there is some problem in the analysis which must be addressed immediately.

Standard deviation can be calculated alternatively from the following equation, which does not differ much from the one shown above.

$$s = \sqrt{\frac{\sum x^2 - \frac{(\sum x)^2}{n}}{n - 1}}$$

where $\sum x^2$ = sum of the squares of individual measurements,
$\sum x$ = sum of individual measurements, and
n = number of individual measurements.

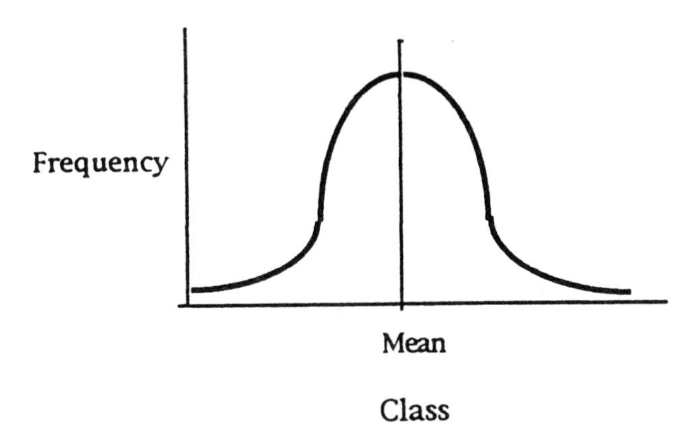

Figure 1.2.1 Normal distribution curve.

Although precision or reproducibility of analysis can be expressed in terms of standard deviation, the magnitude of the analyte in the sample may alter the standard deviation quite significantly. This is shown below in the following two examples.

Example 1

The total petroleum hydrocarbon (TPH) in an effluent sample in six replicate analyses was found to be as follows: 5.3, 4.9, 5.1, 5.5., 4.7, and 5.0 mg/L.

Determine the standard deviation.

x	x^2
5.3	28.09
4.9	24.01
5.1	26.01
5.5	30.25
4.7	22.09
5.0	25.00
30.5	155.45

$$\sum x = 30.5; \sum x^2 = 155.45$$

$$\left(\sum x\right)^2 = (30.5)^2 = 930.25; \ n = 6$$

$$s = \sqrt{\frac{155.45 - \frac{930.25}{6}}{6-1}} = \sqrt{\frac{0.41}{5}} = 0.29 \text{ mg}$$

Example 2

If the results of six replicate analyses for TPH in an influent sample were ten times greater (i.e., 53, 49, 51, 55, 47, and 50 mg/L), the standard deviation would be as follows:

$$\sum x = 305;\ \sum x^2 = 15545$$

$$\left(\sum x\right)^2 = 93025;\ n = 6$$

$$s = \sqrt{\frac{15545 - \frac{93025}{6}}{5}}\ \text{mg/L} = 2.86\ \text{mg/L}$$

A further increase in the magnitude of analyte concentrations as 530, 490, 510, 550, 470, and 500 mg/L on replicate analysis would give a standard deviation of 28.6 mg/L. Thus, standard deviation, which varies with the magnitude or size of the measurements, has no meaning unless the magnitude of analyte concentrations is stated.

In other words, precision of analysis will always be very low for any influent sample relative to the corresponding effluent sample. This may cause some confusion which would not arise if precision is expressed in other scales for which analyte size need not be stated. One such scale is relative standard deviation (RSD) or the coefficient of variance (CV) which is a ratio of standard deviation to the arithmetic mean of the replicate analyses and expressed as a percent.

$$\text{RSD} = \frac{s}{\bar{x}} \times 100\%$$

In Examples 1 and 2 above, the RSDs are as follows:

$$\frac{0.29\ \text{mg/L}}{5.3\ \text{mg/L}} \times 100\% = 5.4\%$$

$$\frac{2.86\ \text{mg/L}}{53\ \text{mg/L}} \times 100\% = 5.4\%$$

Thus, the RSDs in the replicate analyses of influent and effluent samples with ten times variation in magnitude remain the same at 5.4%; while their standard deviations are 0.29 and 2.8 mg/L, respectively.

Another scale of measurement of precision is standard error of mean (M) which is the ratio of the standard deviation to the square root of number of measurements (n).

$$M = \frac{s}{\sqrt{n}}$$

The above scale, however, would vary in the same proportion as standard deviation with the size of the analyte in the sample.

In routine tests of environmental samples, many repeat analyses of a sample aliquots are not possible. Therefore, the precision of the test required is measured by a scale known as relative percent difference (RPD). This is determined from the duplicate analysis performed under identical conditions on two aliquots of one of the samples in a batch. It is calculated by dividing the difference of test results by the average of test results and expressing as percent. Thus,

$$\mathrm{RPD} = \frac{(a_1 - a_2) \text{ or } (a_2 - a_1)}{\frac{(a_1 + a_2)}{2}} \times 100\%$$

where a_1 and a_2 are the results of duplicate analysis of a sample.

Example 3

Concentrations of chloride in two aliquots of a sample were found to be 9.7 and 11.1 mg/L. Determine the precision of analysis as RPD.

$$\mathrm{RPD} = \frac{(11.1\ \mathrm{mg/L} - 9.7\ \mathrm{mg/L})}{\frac{(11.1\ \mathrm{mg/L} + 9.7\ \mathrm{mg/L})}{2}} \times 100\%$$

$$= \frac{(1.4\ \mathrm{mg/L})}{10.4\ \mathrm{mg/L}} \times 100\% = 13.5$$

Accuracy determines the closeness of the analytical data to the true value. It is estimated from the recovery of a known standard spiked into the sample. Based on the percent spike recovery, a correction for the bias may be made. Routine environmental analyses generally do not require such corrections in the results. However in specific types of analysis, correction for bias may be required when the percent spike recovery for a QC batch sample is greater than 0 and less than 100. In the wastewater analyses for certain organics, U.S. EPA has set forth the range for percent recovery. If the spike recovery for any analyte falls outside the range, the QC criteria for that analyte is not met.

The matrix spike recovery may be defined in two different ways: (1) one method determines the percent recovery only for the standard added to the spiked sample, as followed by U.S. EPA, and (2) the other method calculates the percent recovery for the combined unknown sample and standard. Spike recovery calculated by both these methods would give different values.

Definition 1 (U.S. EPA Percent Recovery Formula)

$$\%\text{Recovery} = 100\left(X_s - X_u\right)/K$$

where X_s = measured value for the spiked sample,
X_u = measured value for the unspiked sample adjusted for dilution of the spike, and
K = known value of the spike in the sample.

Definition 2

$$\%\text{Recovery} = \frac{\text{Measured concentration}}{\text{Theoretical concentration}} \times 100\%$$

Theoretical concentration can be calculated as:

$$\frac{\left(C_u \times V_u\right)}{\left(V_u + V_s\right)} + \frac{\left(C_s \times V_s\right)}{\left(V_u + V_s\right)}$$

where C_u = measured concentration of the unknown sample, and
C_s = concentration of the standard

while V_u and V_s are the volumes of the unknown sample and standard, respectively.

The percent spike recovery as per either definition may be calculated by taking either the concentration or the mass of the analyte into consideration. This is shown below in the following examples.

Example 4

A wastewater sample was found to contain 3.8 mg/L cyanide. A 100 mL aliquot of this sample was spiked with 10 mL of 50 mg/L cyanide standard solution. The concentration of this spiked solution was measured to be 8.1 mg/L.

(a) Calculate the percent spike recovery as per Definition 1.

1. Calculation based on concentration:

$$X_s = 8.10 \text{ mg/L}$$

$$X_u = \frac{(3.8 \text{ mg/L})(100 \text{ mL})}{110 \text{ mL}} = 3.454 \text{ mg/L}$$

$$K = \frac{(50 \text{ mg/L})(10 \text{ mL})}{110 \text{ mL}} = 4.545 \text{ mg/L}$$

$$\%\text{Recovery} = \frac{8.10 \text{ mg/L} - 3.454 \text{ mg/L}}{4.545 \text{ mg/L}} \times 100\%$$

$$= 102.2\%$$

2. Calculation based on mass (alternative method):
   The total mass of CN^- ions in 110 mL of sample and spike solution

$$= \frac{8.1 \text{ mg}}{\text{L}} \times 0.110 \text{ L} = 0.891 \text{ mg}.$$

The mass of CN^- ions present in 100 mL of sample aliquot before spiking

$$= \frac{3.8 \text{ mg}}{\text{L}} \times 0.100 \text{ L} = 0.380 \text{ mg}.$$

The mass of CN^- ions in 10 mL standard solution spiked

$$= \frac{50 \text{ mg}}{\text{L}} \times 0.01 \text{ L} = 0.500 \text{ mg}$$

$$\%\text{Recovery} = \frac{(0.891 \text{ mg} - 0.380 \text{ mg})}{0.500 \text{ mg}} \times 100$$

$$= 102.2\%.$$

(b) Calculate the percent spike recovery as per Definition 2.

1. Calculation based on concentration.
   Measured concentration of CN^- after spiking = 8.1 mg/L. Actual concentration of CN^- expected after the sample aliquot was spiked

= Initial concentration of CN^- in the aliquot + concentration of CN^- in the spike standard

$$= \frac{(3.8 \text{ mg/L})(100 \text{ mL})}{(110 \text{ mL})} + \frac{(50 \text{ mg/L})(10 \text{ mL})}{(110 \text{ mL})}$$

$$= 8.00 \text{ mg/L}$$

$$\%\text{Recovery} = \frac{8.10 \text{ mg/L}}{8.00 \text{ mg/L}} \times 100$$

$$= 101.2$$

2. Calculation based on mass (alternative method).
   Measured amount of cyanide in a total volume of 110 mL (sample + spike)

$$= \frac{8.1 \text{ mg}}{\text{L}} \times 0.110 \text{ L} \equiv 0.891 \text{ mg}$$

The actual amount of cyanide to be expected in 110 mL of sample plus spike solution

$$= \left(\frac{3.8\ \text{mg}}{\text{L}} \times 0.100\ \text{L}\right) + \left(\frac{50\ \text{mg}}{\text{L}} \times 0.01\ \text{L}\right)$$

$$= 0.880\ \text{mg}$$

$$\%\text{Recovery} = \frac{0.891}{0.880} \times 100 = 101.2$$

Example 5

A sample measured 11.7 mg/L. A 50 mL portion of the sample was spiked with 5 mL of 100 mg/L standard. The result was 18.8 mg/L. Calculate the spike recovery as per Definitions 1 and 2. (Calculations are based on concentrations.)

Definition 1

$$X_s = 18.8\ \text{mg/L}$$

$$X_u = \frac{(11.7\ \text{mg/L})(50\ \text{mL})}{55\ \text{mL}} = 10.64\ \text{mg/L}$$

$$K = \frac{(100\ \text{mg/L})(5\ \text{mL})}{55\ \text{mL}} = 9.09\ \text{mg/L}$$

$$\%\text{Recovery} = \frac{18.80\ \text{mg/L} - 10.64\ \text{mg/L}}{9.09\ \text{mg/L}} \times 100 = 89.8\%$$

Definition 2

$$\text{Measured concentration} = 18.8\ \text{mg/L}$$

$$\text{Actual concentration} = \frac{(11.7\ \text{mg/L})(50\ \text{mL})}{55\ \text{mL}} + \frac{(100\ \text{mg/L})(5\ \text{mL})}{55\ \text{mL}} = 19.7\ \text{mg/L}$$

$$\%\text{Recovery} = \frac{18.8\ \text{mg/L}}{19.7\ \text{mg/L}} \times 100 = 95.3\%$$

When the spike recovery is less than 100%, U.S. EPA formula (Definition 1) gives a *lower* value than that calculated as per Definition 2, as we see in Example 5. However, if the recovery is above 100%, Definition 1 formula gives a *higher* value (Example 4).

Unlike aqueous samples, spike recovery for soil and solid wastes often does not require any correction to be made in the volume of spike solution. Because the analysis of all soil and solid matrice requires that the analyte in the solid sample be extracted into a definite volume of solvent, there is no need to make any volume or mass correction for the spike in solution added. This is shown in Example 6.

Example 6

A soil sample was soxhlet extracted with freon and the extract was analyzed for petroleum hydrocarbons (PHC) by IR spectrometry. The concentration of PHC in the sample was found to be 285 mg/kg. A 40 g portion of this sample was spiked with 2 mL of 1000 mg/L PHC standard. The concentration of the spiked sample was measured as 326 mg/kg. Determine the accuracy of the analysis as the percent recovery of the amount spiked.

The mass of PHC in the spiked sample

$$= 40 \text{ g} \times \frac{326 \text{ mg}}{1000 \text{ g}} = 13.04 \text{ mg}.$$

The mass of PHC in the sample before spiking

$$= 40 \text{ g} \times \frac{285 \text{ mg}}{1000 \text{ g}} = 11.40 \text{ mg}.$$

$$\text{Amount Spiked} = 2 \text{ mL} \times \frac{1000 \text{ mg}}{1000 \text{ mL}} = 2.00 \text{ mg}$$

$$\%\text{Recovery} = \frac{13.04 \text{ mg} - 2.00 \text{ mg}}{11.40 \text{ mg}} = 96.8\%$$

The mass of the spiked sample is considered as 40 g and not 43 g (to include the mass of 2 mL of spiking solution of approximate density 1.5 g/mL) in the above calculation. This is true because this 2 mL of solvent which is added onto the soil as spike standard readily mixes into the soxhlet extract. Therefore, the mass of the sample after extraction (i.e., the mass of the solid residue) almost remains the same as it was before its extraction.

The percent spike recovery in the above example may alternatively be determined by U.S. EPA formula as shown below:

$$X_s = 326 \text{ mg/kg}$$

$$X_u = 285 \text{ mg/kg}$$

$$K = \frac{\left(2 \text{ mL} \times \frac{1000 \text{ mg}}{1000 \text{ mL}}\right)}{40 \text{ g}} \times \frac{1000 \text{ g}}{1 \text{ kg}} = 50 \text{ mg/kg}$$

$$\%\text{Recovery} = \frac{(326 \text{ mg/kg} - 285 \text{ mg/kg})}{50 \text{ mg/kg}} \times 100 = 82.0\%$$

No mass or volume correction is made in the above calculation involving solid samples. Thus, X_u was considered as 285 mg/kg.

The percent spike recovery in the above example employing Definition 2 would then be

$$\frac{(326 \text{ mg/kg})}{(285 \text{ mg/kg}) + (50 \text{ mg/kg})} \times 100 = 97.3\%$$

In the organic analysis, in addition to the matrix spike (which means adding a measured amount of one or more of the same substances that are analyzed for), a few surrogate substances are also spiked onto the sample. Surrogates are compounds that have chemical properties similar to those of analytes but are not found in the environmental matrices. Deuterated or fluoro substituted analogs of the analytes, such as phenol-d_6, nitrobenzene-d_5, or pentafluorobenzene, etc. are examples of surrogates. Surrogate spike recovery indicates extraction efficiency and measures the accuracy of the method as a whole. Surrogates for specific classes of analytes are listed in Part 2 under their respective group headings.

CONTROL CHARTS

There are two types of control charts: accuracy chart and precision control chart. Accuracy control charts are prepared from the percent spike recoveries data obtained from multiple routine analysis. Precision control charts may be prepared from the relative percent difference (RPD) of analyte concentrations in the samples and their duplicate analytical data. Alternatively, RPDs are calculated for percent recoveries of the analytes in the matrix spike and matrix spike duplicate in each batch and twenty (or any reasonable number of data points) are plotted against the frequency or number of analysis. If the samples are clean and the analytes are not found, the aliquots of samples must be spiked with the standard solutions of the analytes and the RPD should be determined for the matrix spike recoveries. Ongoing data quality thus can be checked against the background information of the control chart. Sudden onset of any major problem in the analysis can readily be determined from the substantial deviation of the data from the average.

Thus, control charts measure both the precision and accuracy of the test method. A control chart is prepared by spiking a known amount of the analyte of interest into 4 to 6 portions of reagent grade water. The recoveries are measured and the average recovery and standard deviation are calculated. In routine analysis, one sample in a batch is spiked with a known concentration of a standard and the percent spike recovery is measured. An average of 10 to 20 such recoveries are calculated and the standard deviation about this mean value is determined. The spike recoveries are plotted against the frequency of analysis or the number of days. A typical control chart is shown below in Figure 1.2.2.

The upper and lower warning limits (UWL and LWL) are drawn at $2s$ above and below, respectively, of the mean recovery. The upper and lower control limits (UCL and LCL) are defined at $3s$ value about the mean. If any data point falls outside UCL or LCL, an error in analysis is inferred that must be determined and corrected. The recoveries should fall between both the warning limits (UWL

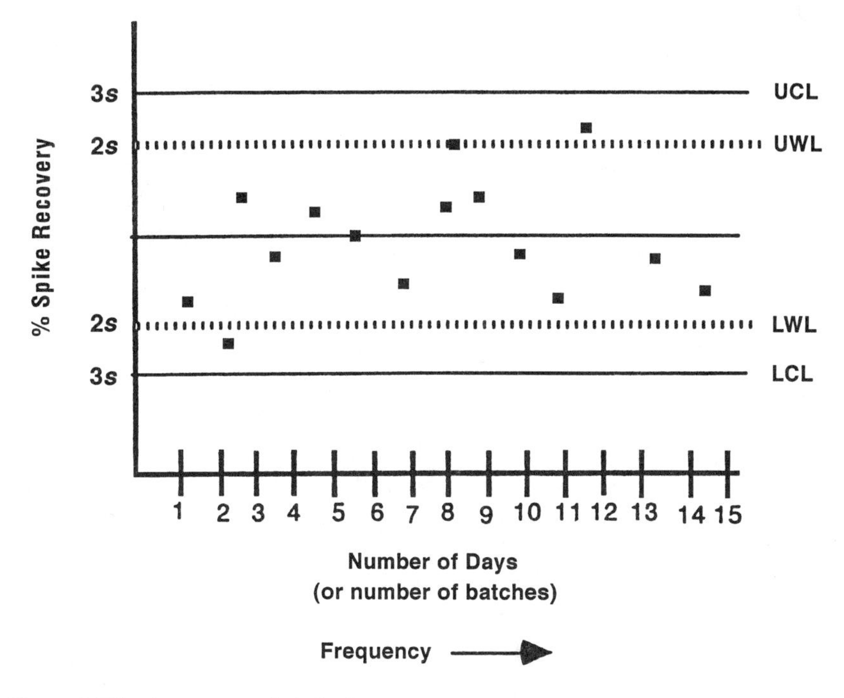

Figure 1.2.2 Accuracy control chart.

and LWL). Similarly, seven data points falling consecutively above or below the mean should indicate analytical error. Also, two thirds of the number of points should fall within one standard deviation of the mean.

The control charts of the type discussed above measure both the accuracy and the precision of analysis. The accuracy, however, is not well defined. For example, if the calibration standards were mistakenly prepared from a wrong stock solution (of wrongly labeled concentration), the true value of spike recoveries for the whole set of data would be different. Theoretically, another drawback of this concept is that the standard deviation windows (i.e., $2s$ and $3s$ about the mean) cannot be fixed and would continuously change as the mean would change with the addition of each data point. Also, there is an element of bias in our interpretation favoring the initial results over the latter results, i.e., the quality of data rests on assuming a high accuracy of initial data in the control chart. Despite these minor drawbacks, control charts are very useful in assessing data quality in environmental analysis.

Precision control charts may, alternatively, be constructed by plotting the RPDs of duplicate analysis measured in each analytical batch against frequency of analysis (or number of days). The mean and the standard deviation of an appropriate number (e.g., 20) of RPDs are determined. The upper and lower warning limits and the uppper and lower control limits are defined at 2 and $3s$, respectively. Such a control chart, however, would measure only the quality of precision in the analysis. This may be done as an additional precision check in conjunction with the spike recovery control chart.

Control charts are key ingredients of QC programs in environmental analysis.

1.3 ANALYSIS OF ORGANIC POLLUTANTS BY GAS CHROMATOGRAPHY

Gas chromatography (GC) is the most common analytical technique for the quantitative determination of organic pollutants in aqueous and nonaqueous samples. In environmental analysis, a very low detection limit is required to determine the pollutants at trace levels. Such low detection can be achieved by sample concentration followed by cleanup of the extract to remove interfering substances. Sample extractions and cleanup procedures are described in detail in Chapter 5 of Part 1 of this text.

Aqueous samples containing volatile organics can be directly analyzed by GC (without any separate sample extraction steps) interfaced with a purge and trap setup. The analytes in the sample are concentrated by the purge and trap technique, (as discussed in the following section), prior to their analysis by GC or GC/MS. The volatile organics in soils, sediments, and solid wastes may be analyzed in a similar way by subjecting an aqueous extract of the sample to purge and trap concentration. Alternatively, the analytes may be thermally desorbed from the solid matrices and transported onto the GC column by a carrier gas.

At the outset, one must understand certain principles of GC to assess if it is a proper analytical tool for the purpose. If so, how to achieve the best separation and identification of component mixtures in the sample with reasonable precision, accuracy, and speed? And what kind of detector and column should be selected for the purpose? It is, therefore, important to examine the type of compounds that are to be analyzed and certain physical and chemical properties of these compounds. Information regarding the structure and the functional groups, elemental composition, the polarity in the molecule, its molecular weight, boiling point, and thermal stability are very helpful for achieving the best analysis. After we know these properties, it is very simple to perform the GC analysis of component mixtures. To achieve this, just use an appropriate column and a proper detector. Properties of columns and detectors are highlighted below in the following sections.

Efficiency of the chromatographic system can be determined from the number of theoretical plates per meter. Although this term primarily describes the property and resolution efficiency of a column, other extra column variables, such as the detector, inlet, injection technique, and the carrier gas velocity can also affect the theoretical plates. It is calculated from the following equation:

$$\text{Theoretical plates/m} = \frac{5.54\left(\frac{t}{w}\right)^2}{L}$$

where t = retention time of the test compound,
w = width of the peak at half height, and
L = length of the column (m).

The number of theoretical plates also depend on the partition ratio k of the test compound and its solubility in the liquid phase. Substances that have higher k values have lower plate numbers. Greater plate numbers indicate greater resolution or better separation of the component mixture.

SELECTION OF COLUMN

A variety of GC columns are commercially available to meet the specific purpose. Selection of columns, their stationary phases, inside diameters, lengths, and the film thickness are briefly discussed below.

A capillary column is usually preferred over packed column for better resolution and lower detection limit. Efficiency of a column to separate organic compounds depends on the stationary phase and the polarity of the analyte molecules. The polarity of the phase, therefore, is the most important characteristic in selecting a column. While a nonpolar phase most effectively separates nonpolar molecules, a polar phase is required to achieve the separation of polar compounds. Aliphatic hydrocarbons containing C-H single bonds are all nonpolar compounds while those containing carbon-carbon double bonds, such as olefins and aromatics, are polarizable compounds. On the other hand, organic compounds containing oxygen, nitrogen, sulfur, phosphorus, or halogen atoms should exhibit greater polarity. Examples of such polar compounds include carboxylic acids, ketones, aldehydes, esters, alcohols, thiols, ethers, amines, nitroaromatics, nitrosamines, nitriles, halocarbons, PCBs, and organic phosphates. An increase in polarity reduces the thermal stability of the stationary phase. Therefore, a phase of least polarity should be selected, whenever possible, to enhance the life of the column. Table 1.3.1 lists the polarity of various stationary phases of common capillary GC columns.

Capillary columns are composed of fused silica which has shown remarkable properties of inertness and efficiency in chromatographic analysis. Glass capillaries and stainless steel are sometimes used too; but their applications are nowadays limited.

The inside diameter of the capillary column is another major factor that often dictates the separation of components. The narrowbore columns with internal diameter (ID) 0.20, 0.25, and 0.32 mm provide the best separation for closely eluting

Table 1.3.1 Polarity of Stationary Phases

Polarity	Stationary phase	Examples
Nonpolar	Polydimethylsiloxane	AT-1, BP-1, DB-1, DC-200, HP-1, OV-1, OV-101, RSL-160, Rtx-1, SF-96, SP-2100, SPB-1, ULTRA-1
	Polyphenylmethylsiloxane	AT-5, BP-5, DB-5, HP-5, MPS-5, OV-73, RSL-200, Rtx-5, SE-52, SPB-5, ULTRA-2
Intermediate	Polyphenylmethylsiloxane	AT-20, BP-10, DB-17, HP-17, MPS-50, OV-17, RSL-300, Rtx-20, SP-2250, SPB-20
	Polycyanopropylphenyldimethylpolysiloxane	AT-1301, DB-1301, Rtx-1301
	Polycyanopropylphenylmethylsiloxane	AT-1701, DB-1701, GB-1701, OV-1701, Rtx-1701, SPB-1701
Polar	Polytrifluoropropylsiloxane	AT-210, DB-210, OV-210, QF-1, RSL-400, SP-2401
	Polyphenylcyanopropylmethylsiloxane	AT-225, DB-225, HP-225, OV-225, RSL-500, Rtx-225
	Polyethyleneglycol	AT-WAX, BP-20, Carbowax 20M, CP/WAX 51, DB-WAX, HP-20M, Stabilwax, Supelcowax 10, Superox II
Very polar (acidic)	Polyethyleneglycol ester	AT-1000, FFAP, Nukol, OV-351, SP-1000, Superox-FA

components and isomers. The smaller the ID, the greater is the resolution. On the other hand, a major disadvantage of such a narrowbore column, however, is its low sample capacity [i.e., the quantity of sample that can be applied without causing the peak(s) to overload]. Wider bore columns of 0.53 and 0.75 mm ID do not have this problem. These wider columns have greater sample capacity than narrowbore columns but relatively lower resolution capacity. Such widebore columns, however, are better suited for environmental samples that often contain pollutants at high concentrations. In addition, widebore columns of 0.53 and 0.75 mm ID provide sufficient sensitivity for minor components peaks without being overloaded with the major components. The sample capacity of a column may further be increased by temperature programming. It also depends on the polarity of the components and the phase—the polar phase has a high capacity for polar components, while the nonpolar phase has a high capacity for nonpolar components.

The stationary phase can be bound to the tubing either as a physical coating on the wall, or can be chemically immobilized. The former type phases are called nonbonded phases, while the chemically bound phases, cross-linked within the tubing are known as bonded phases. The latter is preferred because it can be used at high temperatures with less bleeding and can be rinsed with solvents to remove nonvolatile substances that accumulate on the column.

The film thickness of the stationary phase is another major factor that should be taken into account for column selection. A thicker film increases the resolution

Table 1.3.2 Separation Efficiency and Sample Capacity of GC Columns of Varying Internal Diameters (ID)

Column ID (mm)	Sample capacity (ng)	Separation efficiency (theoretical plates/m)[a]	Carrier gas flow rate, optimum (cc/min)
		Capillary	
0.20	10–30	5000	0.4
0.25	50–100	4200	0.7
0.32	400–500	3300	1.4
0.53	1000–2000	1700	2.5
0.75	10,000–15,000	1200	5
		Packed	
2.0	20,000	2000	20

Note: The data presented are for a 60-m capillary column and a 2-m packed column.

[a] Numbers are rounded off. The higher this number, the greater is the resolution efficiency.

on a nonpolar column, but decreases the same on a polar column. It also increases the sample capacity, retention of sample components, and therefore, the retention time and the temperature at which the components would elute from the column—and the column bleed. Thus, it has both advantages and disadvantages. Use a thicker film (>1 μm) to analyze gases or substances with low boiling points or to analyze highly concentrated samples. On the other hand, a thin film (<0.25 μm) should be used to analyze compounds with high boiling points (>300°C) and should be employed with a shorter column (10 to 15 m length). Film thickness of the stationary phase and the column ID are interrelated, as follows:

$$\text{Phase ratio, } \beta = \frac{\text{column radius } (\mu\text{m})}{2 \times \text{phase thickness } (\mu\text{m})}$$

Columns with equal beta value (β) will provide similar separations under the same analytical conditions. For example, a capillary column with 0.32 μm ID and 0.8 μm phase film thickness could be substituted with a column of same phase with 0.53 μm ID and 1.3 μm film thickness to produce very similar separation. Standard film thickness (0.25 to 0.8 μm) should, however, work for most chemical analyses.

Separation of closely eluting components can be efficiently achieved on a longer column. The greater the length of the capillary column, the higher is its resolution efficiency. On the other hand, the long column enhances the time of analysis, increasing the retention times of the components. As mentioned earlier, high resolution can also be attained with narrowbore columns. Therefore, optimizing the column length and ID can provide good separation in the desired analysis time.

DETECTORS

Selection of GC detectors is very crucial in chemical analysis. Flame ionization detector (FID) and thermal conductivity detector (TCD) can be used for all general purposes. The detection limits for analytes, however, are high, especially for the TCD. The latter is commonly used for gas analysis.

When using FID, aqueous samples can be directly injected onto the GC without any sample extraction. The detection limit of an analyte, however, in such a case would be much higher (low ppm level) than what is desired in environmental analysis. When appropriate sample concentration steps are adopted, organic compounds in aqueous and solid matrices and air can be effectively determined at a much lower detection level. Carbon disulfide is commonly used in the air analysis of many organics by GC-FID.

Halogen specific detectors, such as electron capture detector (ECD) and Hall electrolytic conductivity detector (HECD) show the best response to compounds that contain halogen atoms. Most nitrogen-containing organics can be determined by nitrogen-phosphorus detector (NPD) in nitrogen mode while organophosphorus compounds can be analyzed by the same detector in phosphorus specific mode. Flame photometric detector (FPD) is also equally efficient for determining phosphorus compounds. FPD, however, is primarily used to analyze sulfur-containing organics. Photoionization detector (PID) is sensitive to substances that contain the carbon–carbon double bond such as aromatics and olefins, as well as their substitution products.

CALIBRATION

Prior to the analysis of the unknown, a calibration standard curve is prepared by running at least four standards. There are two ways in which calibration is performed: external standard method and internal standard method. External standard method involves preparation of a calibration curve by plotting area or height response against concentrations of the analyte(s) in the standards. The calibration factor is then calculated as the ratio of concentrations to area/height response and should be constant over a wide range of concentrations. To determine the concentration of the analyte in the unknown sample (extract), the response for the unknown should be compared with that of the standards within the linear range of the curve. Alternatively, an average of response ratios may be calculated which is compared with the response of the analyte. A single point calibration may be used if the area/height response of the analyte is within ±20% of the response of the standard.

The internal standard method is more reliable than the external standard method. Equal amounts of one or more internal standards are added onto equal volumes of sample extracts and the calibration standards. The response factor (RF) is then calculated as follows:

$$\mathrm{RF} = \frac{A_s \times C_{is}}{A_{is} \times C_s}$$

where A_s and A_{is} are the area (or height) response for the analyte and the internal standard, respectively; while C_s and C_{is} are their concentrations. Thus, RF for analytes may be determined by running standard solutions of the analytes containing internal standards. If the RF values over the working range of concentrations fall within ± 20% relative standard deviation, an average RF value should be used in the above equation to determine the concentration of the analytes in the sample. Alternatively, a calibration curve may be plotted between response ratio (A_s/A_{is}) vs. RF.

The concentration of the analyte in the sample = $\dfrac{A_s \times C_{is} \times D}{A_{is} \times \mathrm{RF}}$ where D is the dilution factor. For aqueous samples, the concentration of the analyte is usually expressed in µg/L. All concentration terms including those of the calibration standards and internal standards must be in the same unit.

Calculations

The concentration of an analyte in an aqueous or nonaqueous sample may be calculated by one of the following methods:

External Standard Calibration

The area/height response for the analyte peak is compared with that of the standards from the calibration curve or from the calibration factor.

$$\text{Concentration, µg/L} = \frac{A_{unk} \times Q_{std} \times V_{tot} \times D}{A_{std} \times V_{inj} \times V_{sample}}$$

where A_{unk} = Area count or peak height of the analyte
Q_{std} = Amount of standard injected or purged in ng
V_{tot} = Volume of total extract in µL
D = Dilution factor, dimensionless
A_{std} = Area or peak response for the standard
V_{inj} = Volume of extract injected in µL
V_{sample} = Volume of sample extracted or purged in mL

For nonaqueous samples, the concentration of the analyte is calculated in the same way except that weight of the sample W is substituted for the volume of the sample, V_{sample}. Thus,

$$\text{Concentration, µg/kg} = \frac{A_{unk} \times Q_{std} \times V_{tot} \times D}{A_{std} \times V_{inj} \times W}$$

The concentration calculated above is on the sample "as is" and not as dry weight corrected. Concentration on a dry weight basis may be calculated by dividing the above result with the percent total solid expressed in decimal.

Example

A 500-mL sample aliquot was extracted with hexane to a final volume of 2 mL. The volume of sample extract and the standard injected were 4 μL. The concentration of the analyte in the standard was 50 μg/L. The area response of the analyte in the sample extract and the standard solutions were 28,500 and 24,800, respectively. Determine the concentration of the analyte in the sample.

$$A_{unk} = 28,500$$

$$A_{std} = 24,800$$

$$Q_{std} = 4\ \mu\text{L} \times \frac{50\ \mu\text{g}}{1\ \text{L}} \times \frac{1\ \text{L}}{1000000\ \mu\text{L}} \times \frac{1000\ \text{ng}}{1\ \mu\text{g}} = 0.2\ \text{ng}$$

$$V_{tot} = 2\ \text{mL} \times \frac{1000\ \mu\text{L}}{1\ \text{mL}} = 2000\ \mu\text{L}$$

$$V_{inj} = 4\ \mu\text{L}$$

$$V_{sample} = 500\ \text{mL}$$

$$D = 1,\ \text{the extract was not diluted}$$

Concentration of the analyte in the sample, μg/L

$$= \frac{28,500 \times 0.2\ \text{ng} \times 2000\ \mu\text{L} \times 1}{24,800 \times 4\ \mu\text{L} \times 500\ \text{mL}}$$

$$= 0.23\ \text{ng/mL}$$

$$= 0.23\ \mu\text{g/L}$$

Alternative Calculation

When the aliquots of the sample extract and the standard injected into the column are the same (i.e., 4 μL), the concentration of the analyte in the sample may be calculated in a simpler way as shown below:

Concentration of analyte, μg/L = Analyte concn. (in the extract) determined from the calibration curve,

$$\mu\text{g/L} \times \frac{\text{mL extract}}{\text{mL sample}} \times D$$

In the above problem the concentration of the analyte in the *extract* is

$$\frac{28,500}{24,800} \times 50\ \mu\text{g/L} = 57.5\ \mu\text{g/L}$$

(taking single point calibration) which can also be determined from the external standard calibration curve. Therefore, the concentration of the analyte in the *sample*

$$= 57.5\ \mu g/L \times \frac{2\ mL}{500\ mL} \times 1$$

$$= 0.23\ \mu g/L$$

Internal Standard Method

The concentration of the analyte can be determined from the RF using the equation

$$C_s = \frac{A_s \times C_{is}}{A_{is} \times RF}$$

For this, the internal standard eluting nearest to the analyte should be considered.

ROUTINE ANALYSIS

Routine GC analysis for environmental samples involve running one of the calibration check standards before sample analysis to determine if the area or height response is constant (i.e., within 15% standard deviation of the response factor or calibration factor, and to check if there is a shift in the retention times of the analytes' peaks. The latter can occur to a significant degree due to any variation in conditions, such as temperature or the flow rate of the carrier gas. Therefore, an internal standard should be used if possible in order to determine the retention time shift or to compensate for any change in the peak response. If an analyte is detected in the sample, its presence must be ascertained and then confirmed as follows:

1. Peak matching of the unknown with the known should be done, additionally, at a different temperature and/or flow rate conditions.
2. The sample extract should be spiked with the standard analyte solution at a concentration to produce a response which is two-to-three times the response of the unknown peak.
3. The identification of the peak must be finally confirmed on a second GC column. This may be done either after performing steps 1 and 2 or by injecting the extract straight onto the second column (confirmatory GC column) without going through steps 1 and 2.

In addition to determining the presence or absence of pollutants of interest in the sample, the routine analysis must include quality control/quality assurance tests to determine the precision and accuracy of the test results and any possible source of errors such as sample contamination, absence of preservative, or exceeding of sample holding time. The QA/QC is discussed at length in Chapter 1.2.

1.4 ANALYSIS OF ORGANIC POLLUTANTS BY GAS CHROMATOGRAPHY/MASS SPECTROMETRY

Gas chromatography/mass spectroscopy (GC/MS) is probably the best technique to identify a wide array of unknown organic substances in sample matrices. It is also the most positive confirmatory test to determine the presence of pollutants in the sample. Its application in environmental analysis has grown up enormously in the last decade.

The method is based on the principle of chromatographic separation of components of a mixture on a GC column, followed by their identification from their mass spectra. The compounds are separated on a suitable GC column, following which, the components eluted from the column are subjected to electron-impact or chemical ionization. The fragmented and molecular ions are identified from their characteristic mass spectra. Thus, the substances present in the sample are determined from their characteristic primary and secondary ions and also from their retention times.

For the analysis of organic pollutants in environmental samples, U.S. EPA has classified them into two categories: volatile organics and semivolatile organics (Figure 1.4.1). The substances in the former category are those that are more or less volatile at ambient temperature and pressure. This classification, however, is based on the analytical technique used rather than the chemical structures of pollutants. For example, chloroform and *p*-xylene are very different in their structures and chemical properties. The only property that groups them together is that both are volatile substances. Although the volatile pollutants are expected to have low boiling points, there are a few compounds such as ethylbenzene, xylenes, or dichlorobenzenes that have boiling points higher than that of water.

VOLATILE ORGANICS BY PURGE AND TRAP METHOD

Two techniques may be applied to transfer the volatile analytes from the sample matrices onto the GC column: purge and trap technique and thermal desorption. In the purge and trap method, an aliquot of aqueous sample (usually 5 mL for wastewaters and 25 mL for drinking waters) is bubbled through by helium or nitrogen for 11 min. The analytes are purged out from the sample and carried over with the purging gas onto a trap consisting of activated charcoal, tenax, and silica gel

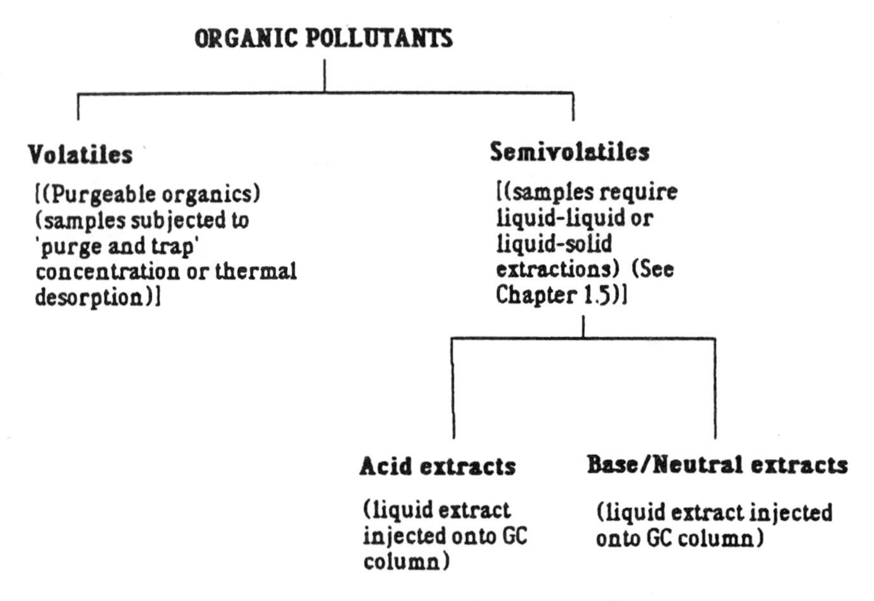

Figure 1.4.1 Classification of organic pollutants based on sample extraction technique.

adsorbents where they are adsorbed. The trap is then heated at 180°C for 4 min. The analytes are desorbed and transported by the carrier gas onto the GC column.

For soil or solid waste samples, an accurately weighted amount of sample is treated with methanol. A small portion of the methanol extract (usually between 10 and 100 mL, depending on the expected level of analytes concentrations in the sample and the presence of matrix interference substances) is injected into 5 mL of laboratory reagent-grade water taken in the purging vessel which is then purged with an inert gas. Analytes adsorbed over the trap are desorbed by heating and transported to the GC column.

Calculation involving analyte concentration in the solid matrix is shown in the following example.

Example

Two g soil was treated with 10 mL methanol. A 100-μL aliquot of methanol extract was injected into 5 mL reagent grade water for purge and trap analysis of a volatile organic compound. The concentration of the compound was found to be 22 mg/L. Determine its concentration in the soil.

This problem can be solved by either of the following two simple ways:

1. A mass calculation:

$$22 \text{ mg compound/L} \equiv 22 \text{ mg} \times \frac{5 \text{ mL water}}{1000 \text{ mL water}}$$

$$\equiv 0.11 \text{ mg compound in 5 mL water}$$

In other words, 0.11 mg compound was found to be present in 5 mL water in the purging vessel spiked with methanol extract. This amount must have come from 100 μL methanol extract injected into this 5 mL water in the purging vessel; i.e., 0.11 mg compound/100 mL methanol extract.

$$\text{extract} \equiv 0.11\ \text{mg} \times \frac{10\ \text{mL (or } 10{,}000\ \mu\text{L) extract}}{100\ \mu\text{L extract}}$$

$$\equiv 11\ \text{mg compound in 10 mL methanol extract,}$$

which would have come from 2 g sample.

Thus, 2 g sample contains 11 mg compound or 1000 g sample would contain $11\ \text{mg} \times \frac{1000\ \text{g}}{2\ \text{g}}$ or 5500 mg of compound, i.e., the concentration of the compound in the soil = 5500 mg/kg.

2. Dimensional analysis:

$$\frac{22\ \text{mg compound}}{1\ \text{L water}} \times \frac{1\ \text{L water}}{1000\ \text{mL water}} \times \frac{5\ \text{mL water}}{100\ \mu\text{L methanol}} \times \frac{1000\ \mu\text{L methanol}}{1\ \text{mL methanol}}$$

$$\times \frac{10\ \text{mL methanol}}{2\ \text{g soil}} \times \frac{1000\ \text{g soil}}{1\ \text{kg soil}}$$

$$= \frac{22 \times 5 \times 1000 \times 10 \times 1000}{1000 \times 1000 \times 2}\ \mu\text{g compound/kg soil}$$

$$= 5500\ \mu\text{g compound per kg soil}$$

SEMIVOLATILE ORGANICS

Organic substances that are not volatile are grouped under semivolatiles. The latter class also includes substances of very low volatility such as chlorinated biphenyls and polynuclear aromatics. As far as the GC/MS technique goes, the principle of analysis of the semivolatile organics is not so distinctly different from that of the volatile organics. On the other hand, the method of extraction of analytes from the sample matrices and the sample concentration steps for these semivolatile organic compounds vastly differ from the volatile organics. Such extraction techniques and the sample cleanup methods are discussed more extensively in Chapter 1.5.

Tuning

Before beginning any analysis, an instrument must be checked for its performance. Every factory-built GC/MS instrument is equipped with a tuning substance, such as perfluorotributylamine (PFTBA). The characteristic mass ions of this compound (or any other tuning compound) and the relative area response corresponding

Table 1.4.1 BFB Tuning Requirement for Volatile Organic Analysis

Ion mass	Abundance criteria
50	15 to 40% of mass 95
75	30 to 60% of mass 95
95	Base peak, 100% relative abundance
96	5 to 9% of mass 95
173	Less than 2% of mass 174
174	Greater than 50% of mass 95
175	5 to 9% of mass 174
176	95 to 101% of mass 174
177	5 to 9% of mass 176

to these mass ions have been well established and set as guidelines to check the performance of the instrument. This is called tuning. All instruments must meet such tuning criteria to perform general organic analysis satisfactorily. In order to achieve highly reliable and accurate results, U.S. EPA has mandated the use of additional tuning substances and set the acceptable performance criteria for environmental analysis of organic pollutants by GC/MS. These substances are 4-bromofluorobenzene (BFB) and decafluorotriphenylphosphine (DFTPP) for volatile and semivolatile organic analysis, respectively. Decaflurotriphenylphosphine oxide (DFTPPO) is the tuning substance for LC/MS analysis of certain nitrogen-containing pesticides. The GC/MS system must be checked to see if the acceptable performance criteria are achieved. All the key m/z criteria for BFB and DFTPP tuning are listed in Table 1.4.1 and 1.4.2.

Inject 2 μL (50 ng) of BFB or DFTPP solution. Alternatively, 2 μL of BFB solution may be added to 5 mL reagent water and analyzed following purge and trap concentration. A background correction should be made prior to checking the m/z criteria of the mass spectra. If the instrument has passed the DFTPP test earlier in the day, it may not be necessary to perform a BFB test for volatiles when using the same instrument. It is only after the tuning criteria are met that standards, blanks, and samples are to be analyzed.

The following instrumental conditions are required for tuning and analysis:

Electron energy:	70 v (nominal)
Mass range:	Twenty to 260 amu for volatiles, and 35 to 450 amu for semivolatiles.
Scan time:	To give at least 5 scans per peak but not to exceed 7 seconds per scan.

For the analysis of semivolatile organics, a column performance test for base/neutral and acid fractions must be performed to test the efficiency of chromatographic column for separation of the analytes. For the base/neutral fraction, inject 100 ng of benzidine and determine the benzidine tailing factor which must be less than three. Similarly, for acid fraction, inject 50 ng of pentachlorophenol and calculate its tailing factor which must be less than five.

Table 1.4.2 DFTPP Tuning Requirement for Semivolatile Organic Analysis

Ion mass	Abundance criteria
51[a]	30 to 60% of mass 198
68	Less than 2% of mass 69
70[b]	Less than 2% of mass 69
127[c]	40 to 60% of mass 198
197[d]	Less than 1% of mass 198
198	Base peak, 100% relative abundance
199	5 to 9% of mass 198
275[e]	10 to 30% of mass 198
365[f]	Greater than 1% of mass 198
441	Present but less than mass 443
442[g]	Greater than 40% of mass 198
443[h]	17 to 23% of mass 442

[a] From 10 to 80% of mass 198 for Method 525 (Drinking Water).
[b] Must be present for CLP requirement.
[c] From 10 to 80% of mass 198 for Method 525 (Drinking Water) and 25 to 75% of mass 198 for CLP requirement.
[d] Less than 2% of mass 198 for Method 525 (Drinking Water).
[e] From 10 to 60% mass 198 for Method 525 (Drinking Water).
[f] Greater than 0.75% of mass 198 for CLP requirement.
[g] Greater than 50% of mass 198 for Method 525 (Drinking Water) and from 40 to 100% of mass 198 for CLP requirement.
[h] From 15 to 24% of mass 198 for Method 525 (Drinking water) and CLP requirement.

The calculation of tailing factor is shown below in Figure 1.4.2. Both benzidine and pentachlorophenol may be added onto the DFTPP standard and injected. If the column performance test fails, replace the column packing to achieve the tailing factor criteria.

COMPOUND IDENTIFICATION

The analytes are identified from their mass spectra and retention times. The retention time must fall within ±3 sec of the retention time of the known compound. The mass spectra of the unknown peak must have the primary and all secondary characteristic ions (and often the molecular ion).

Calibration

Quantitation of an analyte can be performed by either of the following three methods:

1. External standard method
2. Internal standard method
3. Isotope dilution method

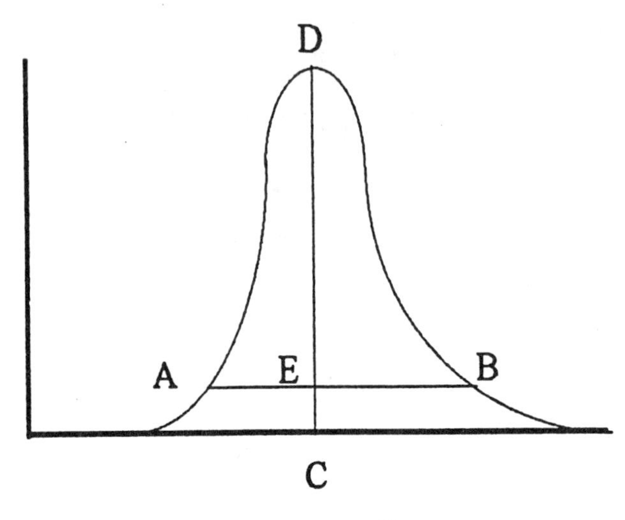

$$\text{Tailing factor} = \frac{EB}{AE}$$

Peak height, CD = 100 mm

10% Peak height, CE = 10 mm

Peak width AB at 10% peak height = 52 mm

$AE = 20$ mm

$EB = 32$ mm

$$\text{Tailing factor} = \frac{EB}{AE} = \frac{32}{20} = 1.6$$

Figure 1.4.2 Tailing factor calculation.

The first two methods have been discussed earlier under "Gas Chromatography" in Chapter 1.3. Isotope dilution method, which is not so frequently employed for quantitation because of cost, is somewhat similar to internal standard technique and is presented below in brief.

In the isotope dilution method, labeled analogs of analytes are first added onto the standard solutions of the analytes as well as to the samples/sample extracts. Before the analysis, a calibration curve is prepared plotting relative response (RR) against concentrations using a minimum of five data points. Relative response of a compound is measured from the isotope ratios of the compound and its labeled analog as follows:

$$\text{RR} = \frac{(R_l - R_m)(R_s + 1)}{(R_m - R_s)(R_l + 1)}$$

where R_s = isotope ratio in the pure standard,
R_l = isotope ratio in the pure labeled compound, and
R_m = isotope ratio in the mixture.

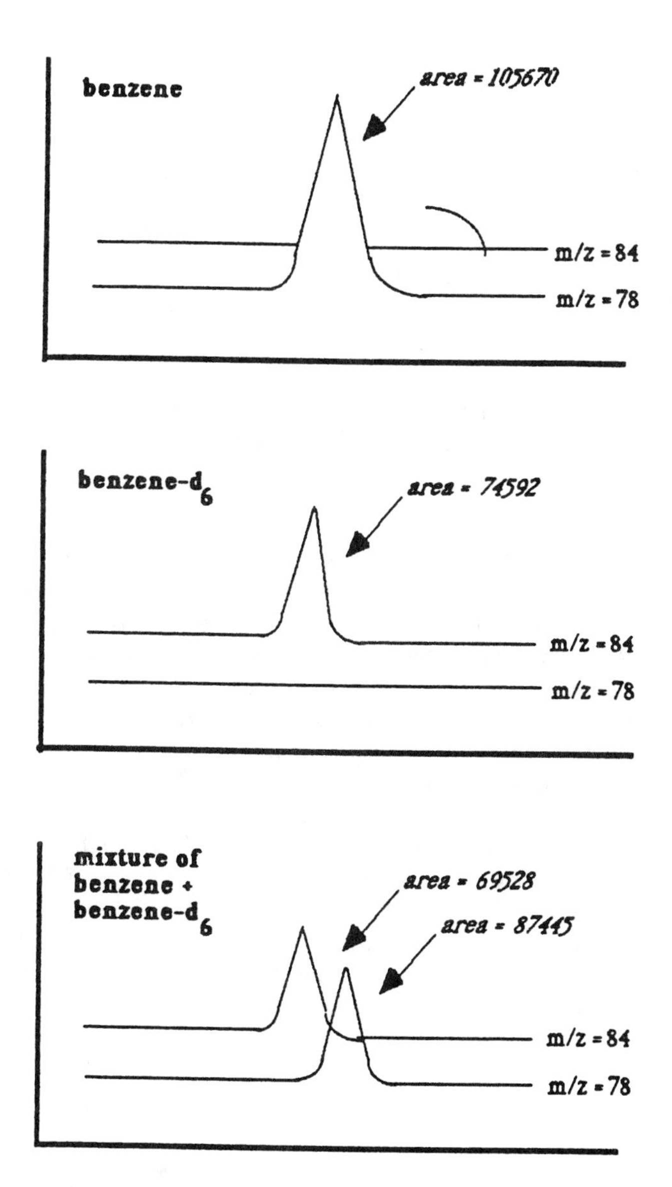

Figure 1.4.3 An example of extracted ion current profiles for benzene, benzene-d_6, and a mixture of benzene and benzene-d_6.

Isotope ratio is measured as the ratio of the area of the primary ion of the unlabeled compound to that of the labeled compound. When the area is zero, it is assigned a value of 1. The retention times of the analytes in most cases are the same as that of their labeled analogs. The isotope can be calculated from the extracted ion current profile (EICP) areas. An example of EICP for benzene, benzene-d_6, and a mixture of benzene and benzene-d_6 is presented in Figure 1.4.3. Calculation to determine the RR is given below:

$$R_l = \frac{1}{74592} = 0.000134$$

$$R_s = \frac{105670}{1} = 105670$$

$$R_m = \frac{m/z\,78}{m/z\,84} = \frac{87445}{69528} = 1.2577$$

Therefore:

$$\text{RR} = \frac{(0.0000134 - 1.2577)(105670 + 1)}{(1.2577 - 105670)(0.0000134 + 1)}$$

$$= 1.2577$$

As mentioned earlier, an appropriate constant amount of labeled compound is spiked to each of the calibration standards. The RR at each concentration is determined which is then plotted against the concentration to prepare a calibration curve. The RR of the analytes in the sample (or sample extract) are matched in the linearity range of the calibration curve to determine their concentrations. If the ratio of the RR to concentration is constant (<20% relative standard deviation) over the calibration range, an averaged ratio may alternatively be used instead of the calibration curve for that compound.

It may be noted that for calibration each analyte is needed to be spiked with its labeled analog. This enhances the cost of the analysis. High cost and often unavailability of labeled analogs are the major drawbacks of isotope dilution method as compared with external and internal standard calibration methods. The isotope dilution method should be, theoretically, more accurate than the internal standard method, as the chromatograpic response and the retention times of the labeled analogs are closest to the compounds.

1.5 EXTRACTION OF ORGANIC POLLUTANTS AND SAMPLE CLEANUP

SAMPLE EXTRACTION

Organic pollutants in potable or nonpotable waters, soils, sediments, sludges, solid wastes, and other matrices must be brought into an appropriate organic solvent for their injection into the gas chromatography (GC) column. Such extraction also enables the increase the concentration of analytes in samples by several order of magnitude for their detection at ppb or ppt level. Depending upon the nature of sample matrices, various extraction techniques may be effectively applied for accurate and low level detection of organics. These are outlined below in the following schematic diagram.

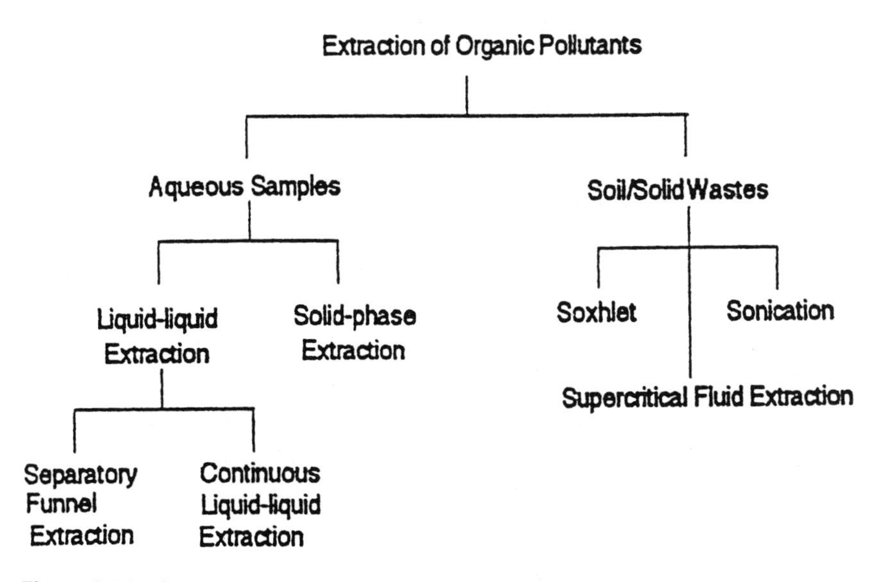

Figure 1.5.1 Schematic diagram of various extraction techniques.

Liquid-Liquid Extraction

Aqueous samples are commonly extracted by liquid-liquid extraction (LLE) technique. A measured volume of the liquid sample is repeatedly extracted with an immiscible organic solvent. The selection of organic solvent must meet the following criteria:

1. it must be immiscible with water,
2. the organic pollutants should be soluble in the solvent (their solubility must be greater in this solvent than in water), and
3. the density of the solvent should be greater than water when the extraction is carried out in a separatory funnel or a continuous liquid-liquid extractor; on the other hand, the solvent should be less dense than water, when microextraction is performed in a glass vial.

Upon mixing the aqueous sample with the solvent, the pollutants dissolve more in the latter because they are more soluble in the solvent. In other words, they partition or distribute in the aqueous and the solvent phases and at equilibrium the ratio of concentration of the solute in both the phases is constant.

The partition or distribution coefficient, P, is equal to the ratio of the concentration of the solute in the solvent to that in the water, i.e.,

$$P = \frac{C_{solvent}}{C_{water}}$$

Since P is independent of volume ratio but constant at any given temperature, increasing the volume of the solvent will cause more dissolution of the solute into the solvent. In other words, the greater the amount of solvent the more of the solute would dissolve in it. Again, for any given volume of solvent, repeated extractions using smaller portions in equal amounts will give greater extraction efficiency than a single-step extraction. For example, in the extraction of semivolatile organics, high-extraction efficiency is achieved by three successive extractions using 60 mL quantity of methylene chloride each time than one single step extraction with 180 mL solvent.

Certain widely used solvents such as diethyl ether or methylene chloride are highly volatile. Excess pressure buildup may cause rupture of the separatory funnel. Many accidents have been reported. It is important to vent out the excess pressure, especially after the first time shaking the sample with the solvent. Before extraction, rinse the separatory funnel with a few milliliters of the solvent. A glass container that has even a slight crack should not be used for extraction.

Solid Phase Extraction

Organic substances can be extracted from aqueous samples by solid-liquid (known as solid phase) extraction. The process is simple, fast, and cost effective in comparison to LLE. In addition, the analysis can be carried out using a smaller volume of sample. By using a suitable capillary column, a detection level comparable to LLE-packed column could be readily attained. The method requires a measured volume of the aqueous sample to be passed through a cartridge tube packed with a suitable solid adsorbent material. The organic pollutants in the sample are adsorbed onto the solid surface from which they are eluted by a properly selected solvent. The sample is applied at the top of the tube and drawn through the bed by a syringe or vacuum, maintaining a flow rate of 1 to 2 drops/sec. Alternatively, larger pore size particles may be used to allow fast flow rates for large volume samples. The tube is washed with a nonpolar solvent for polar analytes and polar solvent for nonpolar analytes. Finally, the analytes are eluted out of column by a suitable solvent. Polar solvents should be used for polar analytes and nonpolar solvents for nonpolar analytes. The sample extracts may be concentrated further by evaporation of solvent.

The selection of adsorbent packing material is based on the polarity of pollutants to be analyzed. The nonpolar hydrophobic adsorbents retain the nonpolar analytes and allow the polar substances to pass through the column. The hydrophilic adsorbents adsorb the polar components, allowing the nonpolar materials to pass through. Various stationary phases for solid phase extraction are listed below in Table 1.5.1.

Table 1.5.1 Stationary Phases for Solid Phase Extraction

Nonpolar compounds	Octadecyl (C-18) bonded silica, octyl (C-8) bonded silica
Moderately polar compounds	Phenyl-, ethyl-, and cyclohexyl bonded silica
Polar compounds	Silica, florisil, silica gel, cyano-, diol-, and amino groups bound to silica, silicates, and alumina

Soxhlett Extraction

Solid samples may be conveniently extracted for semivolatile and nonvolatile organic pollutants by Soxhlett extraction. The sample is placed in a porous extraction thimble and immersed in the solvent. The extraction comprises a series of batch processes involving distillation and condensation of the solvent along with periodic fill-in and siphoning of the solvent in and out of the extraction chamber. This causes an intimate mixing of the sample with the solvent. Soxhlett extraction using a fluorocarbon solvent is commonly employed to leach out petroleum hydrocarbons from the soil. Other than this use, its application in environmental analysis is limited, because it is slow, taking up several hours to complete. The extraction also requires a relatively large quantity of solvent and usually a preconcentration step is necessary.

Supercritical Fluid Extraction

A supercritical fluid is defined as a substance that is above its critical temperature and pressure. It exhibits remarkable liquid-like solvent properties and, therefore, high extraction efficiency. Such common gases as carbon dioxide and nitrous oxide have been successfully employed as supercritical fluids in the extraction of organics from solid matrices. The solid sample is placed in an extraction vessel into which the pressurized supercritical fluid is pumped. The organic analytes dissolve in the supercritical fluid and are swept out of the extraction chamber into a collection vessel. The pressure is released at the valve attached to the collection device where it drops down to atmospheric pressure. The supercritical fluid then returns to its gaseous state and escapes out, leaving analytes in the collection vessel in an appropriate solvent such as methylene chloride.

The extraction efficiency of supercritical fluids may be enhanced by mixing into it a small amount of a cosolvent such as acetone or methanol. Supercritical fluid extraction offers certain advantages over other extraction processes: (1) it is relatively a fast process with greater extraction efficiency; (2) sample concentration steps may be eliminated; and (3) unlike LLE or Soxhlett extraction, a large amount of organic solvents is not required.

CLEANUP

Sample extracts injected into the GC can cause loss of detector sensitivity, shorten the lifetime of a column, produce extraneous peaks, and deteriorate peak resolution and column efficiency. The sample extracts may be purified by one or more of the following techniques:

1. Partitioning between immiscible solvents
2. Adsorption chromatography
3. Gel permeation chromatography
4. Destruction of interfering substances with acids, alkalies, and oxidizing agents
5. Distillation

The cleanup procedures presented in Table 1.5.2 may be applied for different classes of organic substances.

Acid-Base Partitioning

This is applied to separate acidic or basic organics from neutral organics. The solvent extract is shaken with water that is highly basic. The acidic organics partition into the aqueous layer; whereas, the basic and neutral compounds stay in the organic solvent and separate out. After this, the aqueous layer is acidified to a pH below 2, and then extracted with methylene chloride. The organic layer now contains the acid fraction. Phenols, chlorophenoxy acid, herbicides, and semivolatile organic pollutants are cleaned up by the procedure described above.

Alumina Column Cleanup

Highly porous and granular aluminum oxide—available in three pH ranges (acidic, neutral and basic)—are used in column chromatography. Analytes are separated from the interfering compounds by virtue of their different chemical polarity.

The column is packed with alumina, and then covered under anhydrous Na_2SO_4. The extract is then loaded on it. A suitable solvent is selected to elute the analytes. The interfering compound is left adsorbed onto the column. The

Table 1.5.2 Cleanup Methods for Organic Extracts

Analyte group	Cleanup method(s)
Priority pollutants (semivolatiles)	GPC, acid-base, sulfur
Organochlorine pesticides	Florisil, GPC, sulfur
Polychlorinated biphenyls (PCBs)	Florisil, GPC, sulfur, $KMnO_4$-H_2SO_4
Organophosphorus pesticides	Florisil, GPC
Chlorinated herbicides	Acid-base
Chlorinated hydrocarbons	Florisil, GPC
Polynuclear aromatic hydrocarbons	Alumina, silica gel, GPC
Nitroaromatics	Florisil, GPC
Cyclic ketones	Florisil, GPC
Nitrosamines	Alumina, florisil, GPC
Phenols	GPC, acid-base, silica gel (on derivatized phenols)
Phthalate esters	Alumina, florisil, GPC
Petroleum waste	Alumina, acid-base

Note: GPC, Gel permeation chromatography

eluate is then concentrated further. Alumina can be prepared in various activity grades by adding water to Grade I (prepared by heating at >400°C until no more water is lost). Among the common pollutants, phthalate esters and nitrosamines are separated. Basic alumina (pH 9 to 10) is most active in separating basic and neutral compounds: alkali, alkaloids, steroids, alcohols, and pigments. Certain solvents such as acetone or ethyl acetate cannot be used. This form of alumina can cause polymerization, dehydration, and condensation reactions.

Neutral form is less active than the basic grade and is used to separate aldehydes, ketones, esters, and lactones, etc. Acidic form (pH 4 to 5) is used to separate strong acids and acidic pigments. Alumina column cleanup is also used to separate petroleum wastes.

Silca Gel Cleanup

Silica gel is a form of amorphous silica with weak acidic properties. It is made by treating H_2SO_4 with sodium silicate when used for cleanup purposes. Interfering compounds of different polarity are absorbed onto and retained on the column. There are two types of silica gel: activated and deactivated. The former is prepared by heating silica gel for several hours at 150°C. It is used to separate hydrocarbons. The deactivated form contains 10 to 20% water and is used to separate plasticizers, steroids, terpenoids, alkaloids, glycosides, dyes, lipids, sugar, esters, and alkali metal cations. In environmental analysis, silica gel is used to clean up sample extracts containing single component pesticides, PCBs, polynuclear aromatic hydrocarbons, and derivatized phenolic compounds. Methanol and ethanol decrease adsorbent activity.

Florisil Column Cleanup

Florisil is a form of magnesium silicate with acidic properties. It is used for clean up of sample extracts containing the following types of analytes: organochlorine pesticides, organophosphorus pesticides, phthalate esters, nitrosamines, haloethers, nitroaromatics, and chlorinated hydrocarbons. Florisil is also used to separate aromatic compounds from aliphatic-aromatic mixtures, as well as to separate esters, ketones, glycerides, steroids, alkaloids, and some carbohydrates. It also separates out nitrogen compounds from hydrocarbons.

The column is packed with florisil, topped with anhydrous sodium sulfate, and then loaded with the sample to be analyzed. Suitable solvent(s) are passed through the column. The eluate is concentrated for analysis, while the interfering compounds are retained on the column.

Gel-Permeation Cleanup

This separation is based on the size of the porous, hydrophobic gels. The pore size must be greater than the pore size of the molecules to be separated. Gel-permeation cleanup (GPC) is used for cleaning sample extracts from synthetic macromolecules, polymers, proteins, lipids, steroids, viruses, natural resins, and other high molecular weight compounds. Methylene chloride is used as the solvent for separation. A 5 mL aliquot of the extract is loaded onto the GPC column. Elution is carried out using a suitable solvent, and the eluate is concentrated for analysis.

Sulfur Cleanup

Sulfur is found in many industrial wastes, marine algae, and sediment samples. Sulfur may mask the region of chromatogram, overlapping with peaks of interest. For example, in pesticides analysis, sulfur can mask over many pesticides such as lindane, aldrin, and heptachlor. Sulfur has a solubility similar to the organochlorine and organophosphorus pesticides and it cannot be separated by Florisil cleanup method.

The removal of sulfur is achieved by treating the extract with one of the following three substances: copper, mercury, or tetrabutyl ammonium-sodium sulfite reagent.

The sample extract is vigorously shaken with one of the above reagents. The clean extract, free from sulfur, is then separated from the reagent.

Permanganate-Sulfuric Acid Cleanup

Interfering substances in the sample extract may often be destroyed by treating the extract with a strong oxidizing agent, such as $KMnO_4$ or a strong acid like conc. H_2SO_4, or a combination of both. In such a case, the analyte should be chemically stable to these reagents. For example, interfering substances in the sample extract for the analysis of polychlorinated biphenyls can be effectively destroyed by treatment with a small quantity of $KMnO_4$-H_2SO_4 mixture. PCBs are chemically stable under the condition of treatment, and do not react with the acid-permanganate mixture at such a short contact time.

1.6 TITRIMETRIC ANALYSIS

Titration is one of the most commonly employed techniques in wet analysis. Many routine analysis of wastewaters, potable waters, and aqueous extracts of sludges and soils can be effectively performed using various titrimetric techniques.

In general, any titrimetric procedure involves slow addition of a solution of accurately known concentraion (a standard solution) to a solution of unknown concentration (sample to be analyzed) until the reaction between both the solutions is complete. In other words, the standard titrant is added slowly up to the point known as *end point* at which the solute analyte in the sample is completely consumed by the solute in the standard solution. The completion of the reaction is usually monitored by using an indictor, which causes a color change at the end point.

Titrimetric methods generally employed in environmental analysis may be broadly classified into the following types:

1. Acid-base titration
2. General redox titration
3. Iodometric titration
4. Argentometric titration
5. Complexometric titration

The above classification highlights the common analytical methods. There is, however, a great deal of overlapping as far as the chemistry of the process is involved. For example, iodometric method involves an oxidation-reduction reaction between thiosulfate anion and iodine. It is, however, classified here under a separate heading because of its wide application in environmental analysis.

Table 1.6.1 highlights some of the aggregate properties and parameters that can be determined by various titrimetric methods.

ACID-BASE TITRATION

Acid-base titration involves a neutralization reaction between an acid and a soluble base. The reactants may be a strong acid and a strong base, a strong acid

Table 1.6.1 Tritimetric Procedures Applied in Environmental Analysis

Titrimetric methods	Aggregate properties/individual parameters that can be tested
Acid-base	Acidity, alkalinity, CO_2, ammonia, salinity
Iodometric	Chlorine (residual), chlorine dioxide, hypochlorite, chloramine, oxygen (dissolved), sulfide, sulfite
Oxidation-reduction (other than iodometric)	Specific oxidation states of metals, ferrocyanide, permanganate, oxalic acid, organic peroxides, chemical oxygen demand
Complexometric (EDTA type)	Hardness, most metals
Argentometric	Chloride, cyanide, thiocyanate

and a weak base, a weak acid and a strong base, or a weak acid and weak base, resulting in the formation of a soluble salt and water. This is shown below in the following examples:

$$HCl(aq) + NaOH(aq) \Rightarrow NaCl(aq) + H_2O(l) \tag{1}$$

$$H_2SO_4(aq) + Na_2CO_3(aq) \Rightarrow Na_2SO_4(aq) + H_2O(l) + CO_2(g) \tag{2}$$

In acid-base neutralization reactions, no change occurs in the oxidation number of the metal cation. In the above reactions, for example, there is no change in the oxidation number of sodium which remains in +1 oxidation state, both in the reactant and the product.

One of the most common acid-base indicators is phenolphthalein. In an aqueous solution of pH less than 8, it is colorless. As the pH approaches 10, the color turns red. Some of the common acid-base indicators are listed in Table 1.6.2. An acid-base titration may be graphically represented by a titration curve, which is a plot between the change of pH vs. the volume of acid or base added, causing such a pH change. Shapes of some of the titration curves are shown in Figures 1.6.1 to 1.6.3. Figure 1.6.1 illustrates the shape of a titration curve for a strong acid and strong base using the same concentrations of acid and base. Such a titration curve would have a long vertical section, typifying a strong acid-strong base titration. Near this vertical section, the addition of a very small amount of titrant causes a very rapid change in the pH. The midpoint of this vertical section is known as *equivalence point*, which theoretically should be equal to the end point of the titration. The equivalence point in an acid-base titraion involving equal concentrations of a strong acid and a strong base is 7. In other words, a strong acid would completely neutralize an equal volume of strong base of the same strength, or vice versa at pH 7. Figures 1.6.2 and 1.6.3 show the titration curves for acids and bases of different strengths. The vertical section is very short in the weak acid-weak base curve.

The end point in an acid-base titration can be determined either by an acid-base indicator or by means of a potentiometer (or a pH meter). An acid-base indicator is a weak organic acid or a weak organic base which can be written as

Table 1.6.2 Some Common Acid-Base Indicators

Indicators	Transition range pH[a]	Color change
Crystal violet	0.0 – 1.8	Yellow to blue
Methyl green	0.2 – 1.8	Yellow to blue
Quinaldine red	1.0 – 2.2	Colorless to red
Thymol blue	1.2 – 2.8	Red to yellow
	8.0 – 9.6	Yellow to blue
Benzopurpirine 48	2.2 – 4.2	Violet to red
Methyl yellow	2.9 – 4.0	Red to orange
Congo red	3.0 – 5.0	Blue to red
Methyl orange	3.2 – 4.4	Red to yellow
Resazurin	3.8 – 6.4	Orange to violet
Bromocresol purple	5.2 – 6.8	Yellow to purple
Phenol red	6.6 – 8.0	Yellow to red
Cresol purple	7.6 – 9.2	Yellow to purple
m-Nitrophenol	6.8 – 8.6	Colorless to yellow
Phenolphthalein	8.3 – 10.0	Colorless to red[b]
Thymolphthalein	9.4 – 10.6	Colorless to blue
Alizarin yellow R	10.1 – 12.0	Yellow to red
2,4,-6-Trinitrotoluene	11.5 – 13.0	Colorless to orange
Clayton yellow	12.2 – 13.0	Yellow to amber

[a] The pH range at which the indicator produces color change.
[b] Colorless at pH beyond 13.0.

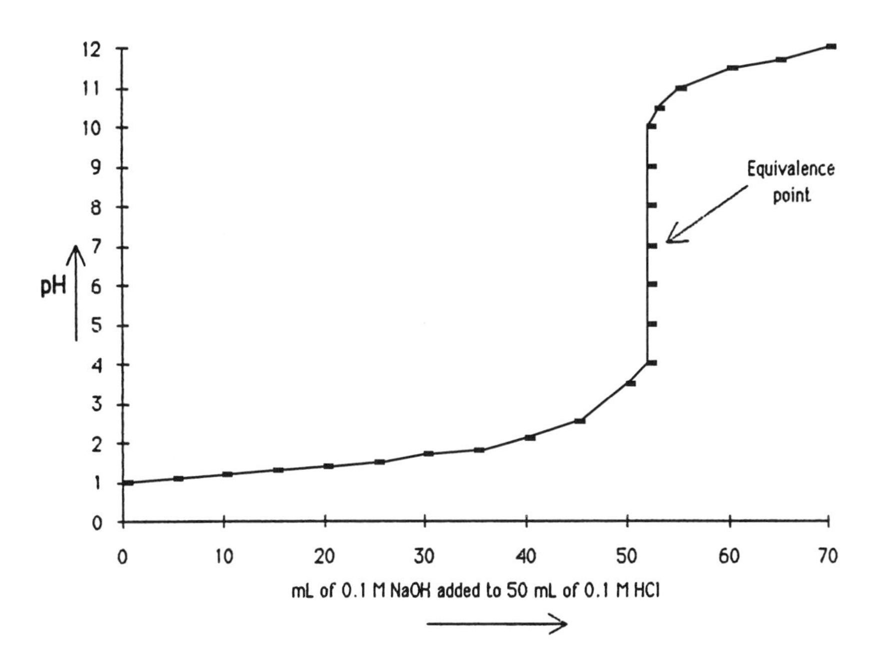

Figure 1.6.1 Titration curve for strong acid and strong base. Indicator producing color change in pH range 4 to 10 may be used.

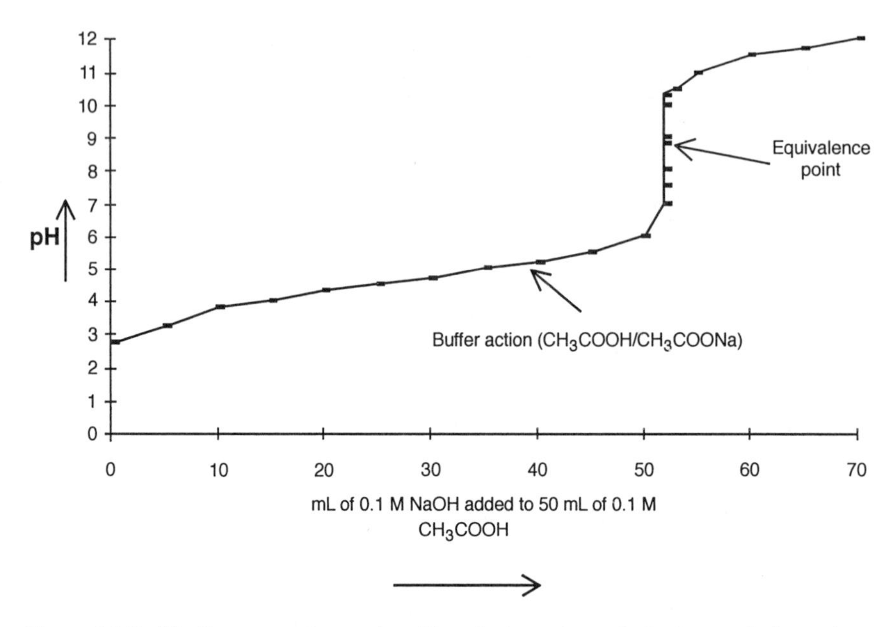

Figure 1.6.2 Titration curve for weak acid and strong base. Indicator producing color change in pH range 7 to 10 may be used.

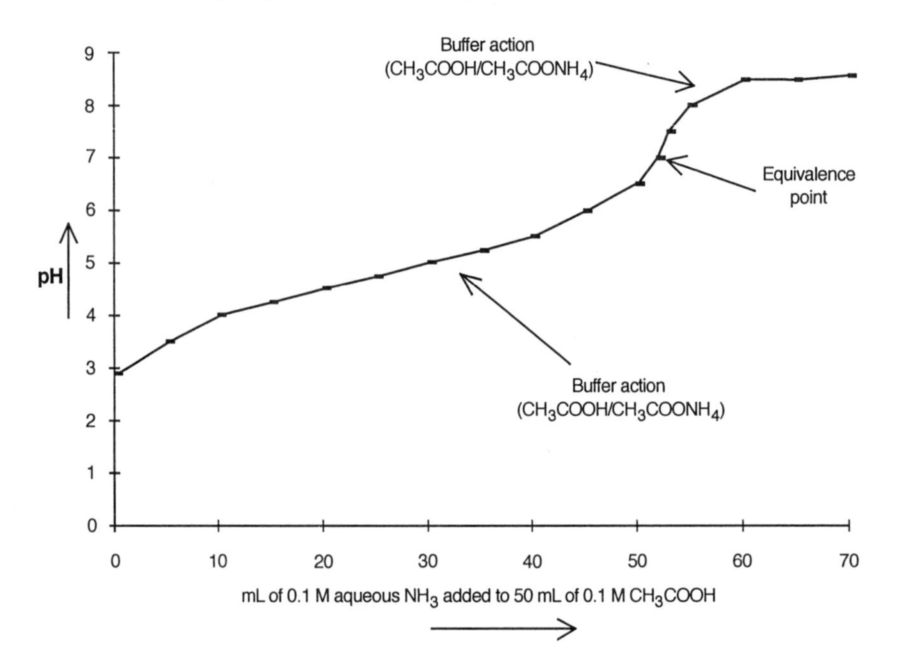

Figure 1.6.3 Titration curve for weak acid and weak base. No color indicator can be used.

HY or YOH, respectively, where Y is the complex organic group. In dilute aqueous solution, the indicator, say HY would dissociate as follows:

$$HY + H_2O \Leftrightarrow H_3O^+ + Y^-$$

color - 1 color - 2

In the above example, HY is a weak organic acid and Y^- is the anion or the conjugate base of this acid. HY and Y^- must have different colors. The addition of an acid or base would shift the equilibrium to left or right, respectively, in the above reaction, thus producing more of HY or Y^-. Generally, a concentration of Y^- that is ten times greater than HY should produce the color of Y^- (color-2). Similarly, a HY concentration ten times over Y^- should produce the color of HY. Selection of an indicator in acid-base titration can be made from looking at the titration curves. Any indicator that produces a color change at a pH range within the vertical section of the titration curve may be employed in the titration. For example, in a strong acid-strong base titration, indicators such as methyl red, methyl orange, phenolphthalein, or bromthymol blue that produce a color change in the pH range 5 to 10 may be used.

Calculation

The strength of an acid or a base may be determined by either the normality method or by mole calculation as shown below. When normality of one of the reactants is known, the strength of the other reactant can be determined from the following relationship:

$$\text{volume of acid} \times \text{its normality} = \text{volume of base} \times \text{its normality}$$

Therefore, the normality of acid

$$= \frac{\text{mL titrant (base)} \times \text{the normality of the titrant (base)}}{\text{mL acid of unknown strength taken in titration}}$$

Similarly, the normality of the base

$$= \frac{\text{mL acid titrant} \times \text{its normality}}{\text{mL base taken in titration}}$$

Normality of a solution is the number of gram equivalents (equivalent weight expressed in grams) of a substance in 1 L of solution or the number of milliequivalent (equivalent weight expressed in millgrams) in 1 mL of solution. Although equivalent weight of a substance may vary according to the reaction, in most cases, especially in acid-base titrations where neutralizations completely occur, it may be calculated as follows:

$$\text{Equivalent weight of acid} = \frac{\text{Formula weight}}{\text{Number of replaceable (dissociable) hydrogen atoms}}$$

For example, equivalent weight of HCl is 36.45/1, or 36.45, H_2SO_4 is 98/2 or 49, or CH_3COOH is 60/1 or 60. In acetic acid, CH_3COOH, although there are totally 5 H atoms, the acid has only one dissociable H atom which ionizes in water as follows:

$$CH_3COOH + H_2O \Rightarrow CH_3COO^- + H_3O^+$$

Similarly, the equivalent weight of a base can be determined by dividing its formula weight by the number of -OH group in its formula unit. For example, equivalent weight of NaOH is 40/1 or 40 and $Ca(OH)_2$ is 74/2 or 37. Many basic substances such as sodium carbonate or sodium bicarbonate do not contain any hydroxyl group. Gram equivalents of these bases are the amounts of these substances that would react with one gram equivalent of an acid. For example, the equivalent weights of sodium carbonate and sodium bicarbonate may simply be determined from the following reactions:

$$Na_2CO_3 + 2\ HCl \Rightarrow 2\ NaCl + CO_2 + H_2C$$

$$NaHCO_3 + HCl \Rightarrow NaCl + CO_2 + H_2O$$

In the above equations, 0.5 mol Na_2CO_3 and 1 mol $NaHCO_3$ react with 1 mol (or 1 g equivalent) of HCl, respectively. Therefore, the equivalent weights of Na_2CO_3 is one-half its formula weight, which is 0.5×106 or 53, and that of $NaHCO_3$ is the same as its formula weight, 84.

The strength or normality of an acid titrant prior to its use in the titration is determined by standardizing against a base such as sodium carbonate of known strength. Other primary standards for acid are tris(hydroxymethyl)aminomethane and sodium tetraborate decahydrate. Similarly, a base titrant is standardized against potassium hydrogen phthalate or potassium hydrogen iodate. The following example illustrates the normality based calculation to determine the strength of an acid or base from a titration experiment.

Example 1

23.5 mL of 0.025 *N* NaOH was required to titrate 20 mL of HCl. Determine the strength (normality) of the HCl solution.

$$\text{Normality of HCl} = \frac{23.5\ \text{mL} \times 0.025\ N}{20.0\ \text{mL}}$$

$$= 0.029$$

Sometimes, it may be more convenient to express the strength of an unknown acid or base in terms of molarity rather than normality. The titrimetric method remains the same. The calculation, however, is based on the molar ratios of reactants in the balanced equation, as cited below in the following two examples.

Example 2

56.8 mL of 0.01 M KOH was required to titrate 40 mL of a dilute H_2SO_4 solution. Determine the molarity of the acid solution.

$$2\ \text{KOH} + \text{H}_2\text{SO}_4 \Rightarrow \text{K}_2\text{SO}_4 + 2\ \text{H}_2\text{O}$$

2 mol 1 mol 1 mol 2 mol

Since molarity $(M) = \frac{\text{mole}}{\text{liter}}$ or $\frac{\text{millimole}}{\text{mL}}$, therefore, 56.8 mL of 0.001 M KOH

$\equiv (56.8 \times 0.01)$ or 0.568 mmol KOH.

In the above reaction, the molar ratio of KOH to H_2SO_4 is 2:1. Therefore, 0.568 mmol KOH would require 0.568/2 or 0.284 mmol of H_2SO_4.

The volume of H_2SO_4 taken in the titration was 40 mL. Therefore, the strength of the acid $= \frac{0.284\ \text{mmol}}{40\ \text{mL}} \equiv 0.0071\ M$.

Example 3

0.2872 g anhydrous Na_2CO_3 was dissolved in water and the volume was made up to 250 mL. 25 mL of this solution was required to titrate 11.2 mL of H_3PO_4. Determine the molarity of the acid. The balanced equation is as follows:

$$3\ \text{Na}_2\text{CO}_3 + 2\ \text{H}_3\text{PO}_4 \Rightarrow 2\ \text{Na}_3\text{PO}_4 + 3\ \text{H}_2\text{O} + 3\ \text{CO}_2$$

3 mol 2 mol 2 mol 3 mol 3 mol

0.2873 g Na_2CO_3 is equal to 0.2873 g$\times \frac{1\ \text{mol Na}_2\text{CO}_3}{106\ \text{g Na}_2\text{CO}_3}$ or 0.00271 mol Na_2CO_3.

0.00271 mol Na_2CO_3 would react with 0.00271 mol $Na_2CO_3 \times \frac{2\ \text{mol H}_3\text{PO}_4}{3\ \text{mol Na}_2\text{CO}_3}$

or 0.00181 mol H_3PO_4 or 1.81 mmol H_3PO_4.

Therefore, the molarity of phosphoric acid, $H_3PO_4 = \frac{1.81\ \text{mmol}}{11.2\ \text{mL}} = 0.16\ M$.

OXIDATION-REDUCTION TITRATIONS

An oxidation-reduction titration or redox titration is an oxidation-reduction reaction involving a transfer of electron(s) between two substances in solutions. A substance is said to be oxidized when it loses electron(s) and reduced when it gains electron(s). Examples of oxidation-reduction reactions are illustrated below:

$$Ce^{4+} + Fe^{2+} \Leftrightarrow Ce^{3+} + Fe^{3+}$$

$$H_3AsO_3 + 2\ Ce^{4+} + H_2O \Leftrightarrow H_3AsO_4 + 2\ Ce^{3+} + 2\ H^+$$

In the first reaction, Ce^{4+} is reduced to Ce^{3+}, while Fe^{2+} is oxidized to Fe^{3+}. In the second reaction shown above, arsenous acid is oxidized to arsenic acid by Ce^{4+}. The oxidation state of As changes from +3 in the reactant, H_3AsO_3, to +5 in the product, H_3AsO_4, while Ce^{4+} is reduced to Ce^{3+}. Ionic reactions like molecular reactions must be balanced. The net charge on the reactants must be equal to that of the products.

Redox titrations find numerous applications in environmental analysis. Iodometric titration involving the reaction of iodine with a reducing agent such as thiosulfate or phenylarsine oxide of known strength is a typical example of a redox titration. This method is discussed separately in the next section. Another example of redox titration is the determination of sulfite, (SO_3^{2-}) using ferric ammonium sulfate, [$NH_4Fe(SO_4)_2$].

Some common oxidants used as standard solutions in redox titrations are listed in Table 1.6.3. Potassium permanganate is a strong oxidizing agent that has a standard potential greater than other common oxidants. Its solution has a distinct purple color. One drop of 0.01 *M* $KMnO_4$ can impart color to 250 mL water. No indicator, therefore, is required in permanganate titration. The end point color, however, fades gradually within 30 sec. For dilute permanganate solution, a sharper end point may be obtained by using ferroin (or its derivatives) or diphenylamine sulfonic acid indicator.

Half-reaction for $KMnO_4$ is given below:

$$MnO_4^- + 8\ H^+ + 5e^- \Leftrightarrow Mn^{2+} + 4\ H_2O$$

where Mn is reduced from +7 to +2 oxidation state. The standard potential for this reaction is 1.51 V.

Aqueous solutions of permanganate are not so stable because MnO_4^- can form MnO_2 by reacting with water over time. Therefore, $KMnO_4$ solution needs to be standardized periodically, filtering out any MnO_2 formed, before standardization.

Cerium (IV) solution in 0.1 *M* H_2SO_4 is an ideal oxidant for many redox titrations. The half-reaction is as follows:

$$Ce^{4+} + e^- \overset{\text{(acid)}}{\Leftrightarrow} Ce^{3+} \qquad E° = 1.44\ V$$

Table 1.6.3 Common Oxidizing Agents for Redox Titrations

Oxidant	Standard potential (V)	Indicator
Potassium permanganate ($KMnO_4$)	1.51	None or ferroin
Cerium (IV)[a] (Ce^{4+})	1.44	Ferroin
Potassium bromate ($KBrO_3$)	1.42	α-Naphthoflavone
Potassium iodate (KIO_3)	1.42	α-Naphthoflavone
Potassium dichromate ($K_2Cr_2O_7$)	1.33	Diphenylamine sulfonic acid
Iodine (I_2)	0.536	Starch

[a] Cerium ammonium nitrate, $Ce(NO_3)_4 \cdot 2NH_4NO_3$; cerium ammonium sulfate, $Ce(SO_4)_2 \cdot 2(NH_4)_2SO_4 \cdot 2H_2O$; or cerium (IV) hydroxide, $Ce(OH)_4$ may be used to prepare Ce^{4+} standard solution.

The solution is remarkably stable, even at high temperatures. However, it has to be strongly acidic to prevent the precipitation of basic cerium salts.

Potassium dichromate is another common oxidizing reagent employed in redox titrations. The titration is carried out in the presence of 1 *M* HCl or H_2SO_4. The half-reaction is as follows:

$$Cr_2O_7^{2-} + 14\ H^+ + 6\ e^- \Leftrightarrow 2\ Cr^{3+} + 7\ H_2O \qquad E° = 1.33$$

The orange dichromate ion is reduced to green Cr (III) ion; the oxidation state of Cr changes from +6 to +3. The electrode potential of the reaction is lower than those of permanganate and Ce (IV). The dichromate reactions are relatively slow with many reducing agents. However, dichromate solutions are highly stable and can be boiled without decomposition. Diphenylamine sulfonic acid is commonly used as the indicator in dichromate titration. The indicator is colorless in reduced form and red-violet in oxidized form. Therefore, in a direct titration, the color changes from green (due to Cr^{3+}) to violet.

Periodic acid is another important oxidant. It can selectively oxidize many organic compounds. Compounds containing aldehyde, ketone or alcoholic groups on adjacent carbon atoms are rapidly oxidized. Similarly, hydroxylamines are oxidized to aldehydes. Some of the half-reactions are shown below:

$$\begin{array}{c}\text{OH}\quad\text{OH}\\ |\qquad |\\ CH_3-CH-CH-CH_3\end{array} \xrightarrow{H_5IO_6} 2\ \begin{array}{c}\text{O}\\ ||\\ CH_3-C-H\end{array} + 2\ H^+ + 2\ e^-$$

(Butylene Glycol)

$$\begin{array}{c}\text{OH}\quad\text{NH}_2\\ |\qquad |\\ CH_2-CH_2\end{array} \xrightarrow{H_5IO_6} 2\ \begin{array}{c}\text{O}\\ ||\\ H-C-H\end{array} + NH_3 + 2\ H^+ + 2\ e^-$$

(Ethanolamine)

The oxidation state of iodine in periodic acid is +7. In the presence of strong acid it is reduced to iodate ion. The half-reaction is as follows:

$$H_5IO_6 + H^+ + 2\,e^- \Rightarrow IO_3^- + 3\,H_2O \qquad E° = 1.6\ V$$

Sodium or potassium salt of periodic acid can be used as a primary standard to prepare periodate solutions.

The application of potassium bromate as oxidant in direct titration is very less. It is, however, used as a source of bromine for many bromination reactions in organic analysis. The analyte solution is acidified. An unmeasured quantity of KBr in excess is added onto it, followed by the addition of a measured volume of standard $KBrO_3$ solution. One mole BrO_3^- generates 3 mol stoichiometric quantity of bromine. The reaction is as follows:

$$BrO_3^- + 5\,Br^- + 6\,H^+ \Rightarrow 3\,Br_2 + 3\,H_2O$$

After the addition of bromate standard, let the reaction vessel stand for some time to allow the bromination to go to completion. The excess bromine is measured by adding excess KI and titrating the liberated iodine against a standard solution of sodium thiosulfate using starch indicator.

$$KI \Rightarrow K^+ + I^-$$

$$2\,I^- + Br_2 \Rightarrow I_2 + 2\,Br^-$$

Many other powerful oxidants are used in redox titrations. Often a metal ion may be present in more than one oxidation state which must be oxidized or reduced into the desired oxidation state. For example, a salt solution of iron may contain both Fe^{2+} and Fe^{3+} ions. Peroxydisulfates, bismuthates, and peroxides are often used as auxiliary oxidizing reagents to convert the ion of interest into the higher oxidation state. The half-reactions for these oxidants are as follows:

$$S_2O_8^{2-} + 2\,e^- \Leftrightarrow 2\,SO_4^{2-} \qquad E° = 2.01\ V$$

$$\text{(Peroxydisulfate)}$$

Ammonium peroxydisulfate in acidic medium can oxidize Mn^{2+} to permanganate, Ce^{3+} to Ce^{4+} and Cr^{3+} to Cr^{6+} (dichromate). The half-reaction for hydrogen peroxide in acid medium is as follows:

$$H_2O_2 + 2\,H^+ + 2\,e^- \Leftrightarrow 2\,H_2O \qquad E° = 1.78\ V$$

The bismuthate ion is reduced to BiO^+ in acid medium, the half-reaction being

$$NaBiO_3 + 4\ H^+ + 2\ e^- \Leftrightarrow BiO^+ + Na^+ + 2\ H_2O$$

Standardization of Oxidants

Oxidizing reagents such as tetravalent cerium or potassium permanganate solutions may be standardized by oxalic acid, sodium oxalate, or potassium iodide. The reactions of Ce^{4+} and permanganate ions with oxalic acid in acid medium are given below:

$$2\ Ce^{4+} + H_2C_2O_4 \Leftrightarrow 2\ Ce^{3+} + 2\ H^+ + 2\ CO_2$$

$$2\ MnO_4^- + 5\ H_2C_2O_4 + 6\ H^+ \Leftrightarrow 2\ Mn^{2+} + 10\ CO_2 + 8\ H_2O$$

Sodium oxalate in acid medium is converted to undissociated oxalic acid, thus undergoing the same reactions as shown above. Ce^{4+} is standardized against oxalic acid or sodium oxalate in HCl solution at 50°C in presence of iodine monochloride as a catalyst. Permanganate titration is carried out at elevated temperature (above 60°C). The reaction is initially slow. The Mn^{2+} produced causes autocatalysis. Thus, during the addition of the first few mL of permanganate solution, the violet color persists for a while, but later on, as the reduction progresses with the generation of Mn,$^{2+}$ the permanganate color decolorizes more rapidly. At the end point, the solution turns pink, which persists for about 30 sec.

Example

A solution of potassium permanganate was standardized against sodium oxalate ($Na_2C_2O_4$) primary standard. A 50.0 mL 0.01 *M* $Na_2C_2O_4$ standard solution required 37.8 mL $KMnO_4$ solution. Determine the strength (molarity) of $KMnO_4$ solution.

$$\text{Amount of } Na_2C_2O_4 = 0.05\ L\ Na_2C_2O_4 \times \frac{0.01\ mol\ Na_2C_2O_4}{1\ L\ Na_2C_2O_4}$$

$$= 0.0005\ mol\ Na_2C_2O_4$$

which would react with an amount of

$$KMnO_4 = 0.0005\ mol\ Na_2C_2O_4 \times \frac{2\ mol\ KMnO_4}{5\ mol\ Na_2C_2O_4}$$

$$= 0.0002\ mol\ KMnO_4$$

Thus the molarity of $KMnO_4$

$$= \frac{0.0002 \text{ mol } KMnO_4}{0.0378 \text{ L } KMnO_4}$$

$$= 0.0053\ M$$

Periodic acid or periodate solutions are standardized by adding an excess of potassium iodide and then titrating the liberated iodine against a standard solution of sodium thiosulfate, phenylarsine oxide, or sodium arsenite. One mole of periodate liberates an equimolar amount of iodine as per the following reaction:

$$H_4IO_6 + 2\ I^- \Rightarrow IO_3^- + I_2 + 2\ OH^- + H_2O$$

The solution must be maintained slightly alkaline by adding a buffer such as sodium bicarbonate or borax. This prevents any further reduction of iodate.

Among the other oxidizing agents, potassium dichromate and potassium bromate may be used as primary standards without any standardization. The solids should be dried at 150°C before weighing. Potassium metaperiodate K_5IO_6 may also be used as a primary standard. It is stable if prepared in alkaline solution.

Reducing Agent

Redox titrations are often performed for metal analysis. Metals in their lower oxidation states are common reducing agents. This includes Fe^{2+}, Sn^{2+}, Mo^{3+}, W^{3+}, Ti^{3+}, Co^{2+}, U^{4+}, and VO^{2+}. Sodium thiosulfate, ($Na_2S_2O_3$) is one of the most widely used reductants in iodometric titrations. Other reducing agents include sodium arsenite and phenylarsine oxide. Iodometric titration is discussed separately in the next section.

Oxalic acid is another common reducing agent. Many metals may be analyzed by reactions with oxalic acid. The metals are converted to their sparingly soluble metaloxalates, which are then filtered, washed, and dissolved in acid. Acid treatment liberates oxalic acid which is titrated with a standard permangante or Ce^{4+} solution. Nitrous acid (HNO_2) is another suitable reducing agent. It may be analyzed by treatment with a measured excess of permanganate solution. Nitrous acid oxidizes to nitrate and the excess permanganate is back titrated.

Often a metal ion may be present in a solution in more than one oxidation state. To determine its total content, it should be converted into one specific oxidation state by treatment with an auxiliary oxidizing or reducing agent. Some of the preoxidants were mentioned earlier in this section. Among the auxiliary reducing agents, amalgamated zinc (Zn/Hg) and granular metallic silver are common. These are used in Jones and Walden reductors, respectively. Other metals, such as Al, Cd, Ni and Cu, are also used for prereduction of analytes.

Oxidation-Reduction Indicators

Redox indicators are substances that change color when oxidized or reduced. These substances have different colors in their oxidized and reduced states. The end point in the titration is detected from the change in the color. As in acid-base titration, color change occurs at a specific pH range. In redox titrations, the change in color depends on the change in electrode potential of the system as the titration progresses. Many indicators are reduced by protons (H^+). Therefore, the transition potential and, hence, the color change is pH dependent too.

Ferroin is one of the most commonly used oxidation-reduction indicators. It is a 1,10-phenanthroline complex of iron having the following structure:

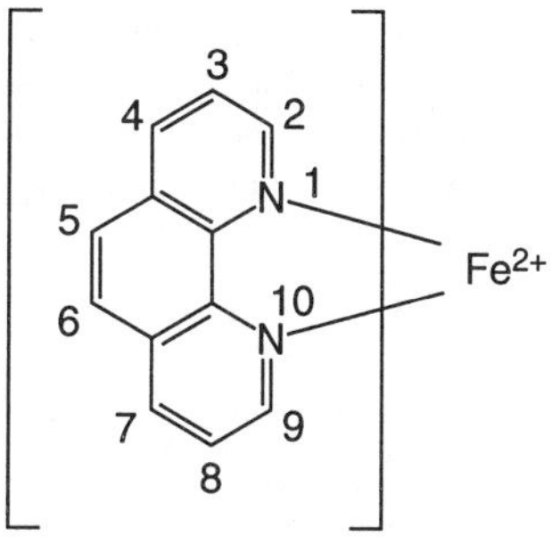

3 4 2 N 1 5 Fe^{2+} 10 6 N 7 9 8

tris(1,10-Phenanthroline) iron (Ferroin)

Its oxidized form $(Phen)_3Fe^{3+}$ is pale blue, while the reduced form $(phen)_3Fe^{2+}$ is red. The transition potential is about 1.11 V. Among the substituted derivatives of phenanthrolines, 5-methyl- and 5-nitro-1,10-phenanthroline complexes of iron have found wide applications in redox titrations.

Sulfonic acid derivatives of diphenylamine and diphenylbenzidine are suitable indicators in many redox titrations. The structures of these indicators are shown below:

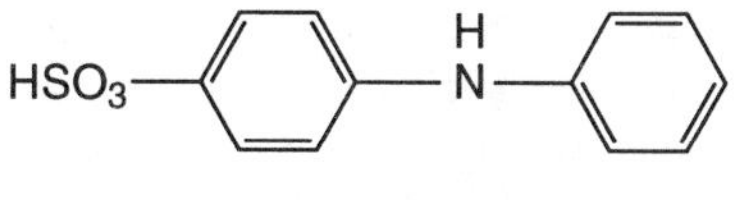

HSO_3 H N

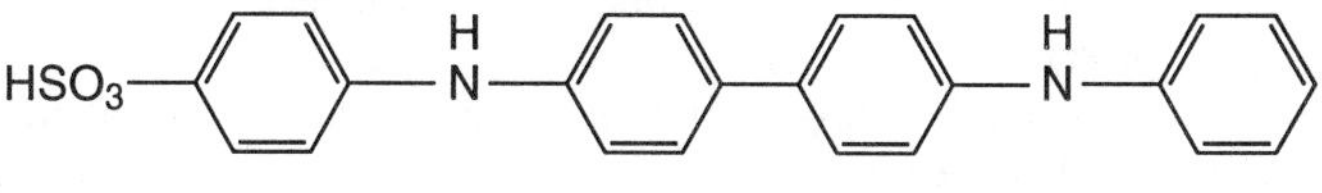

HSO_3 H N H N

The sodium salt of these acids may be used to prepare aqueous solutions of indicators. Other examples of redox indicators include starch, potassium thiocyanate, methylene blue, and phenosafranine. Some selected general indicators in redox titrations are listed in Table 1.6.3. The properties of starch as an indicator in iodometric titration are discussed in the following section.

IODOMETRIC TITRATION

Iodometric titration involves the reaction of iodine with a known amount of reducing agent, usually sodium thiosulfate ($Na_2S_2O_3$) or phenylarsine oxide (PAO). Starch solution is used as an indicator to detect the end point of the titration. Thus, the exact amount of iodine that would react with a measured volume of sodium thiosulfate of known strength is determined. From this, the concentration of the analyte in the sample, which is proportional to the amount of iodine reacted with thiosulfate or PAO, is then calculated.

Iodometric titration is performed out in one of the following ways:

1. Potassium iodide solution in excess is added to the sample. Analyte such as chlorine liberates iodine from KI under acidic condition. The liberated iodine is directly titrated against standard $Na_2S_2O_3$ or PAO. The concentration of the iodine liberated is proportional to the concentration of the analyte in the sample. The reaction of chlorine with potassium iodide is as follows:

$$Cl_2 + 2\ KI \Rightarrow 2\ KCl + I_2 \uparrow$$

$$2\ S_2O_3^{2-} + I_2 \rightarrow S_4O_6^{2-} + 2\ I^-$$

   Thus, the concentration of the analyte in the sample should be equal to:

$$\text{mg analyte/L} = \frac{\text{mL titrant} \times \text{its normality} \times \text{milliequivalent wt. of analyte}}{\text{mL sample}}$$

2. The iodine produced may be consumed by other substances present in the wastewater before it reacts with the standard reducing agent. Thus, there is a good chance that some iodine may be lost; the whole amount of released iodine might not fully react with the standard reducing agent, $Na_2S_2O_3$ or PAO, which can produce erroneous results. This may be avoided by performing a back titration in which an accurately measured amount of the reducing agent is added to a measured volume of the sample. If the analyte is an oxidizing agent like chlorine, it would react with thiosulfate or PAO. The excess unreacted reducing agent is then measured by titrating with standard iodine or iodate titrant. At the end point, the solution turns from colorless to blue. The amount of reducing agent that reacts with the analyte is thus measured by subtracting the unreacted amount from the starting amount added to the sample. From this, the concentration of the analyte in the sample is calculated.
3. A measured volume of sample is added to a known quantity of standard iodine solution estimated to be in excess over the amount of analyte (e.g., sulfide) in the sample. The standard iodine solution should contain an excess of potassium iodide. The analyte reacts with iodine. This would cause a lowering of strength of iodine solution after the reaction. The normality of iodine is then determined from titration against a standard solution of sodium thiosulfate. Concentration of analyte in the sample, which is proportional to the amount of iodine consumed, is calculated as follows:

$$\text{mg analyte/L} = \frac{[(A \times B) - (C \times D)]}{\text{mL sample}} \times \text{ mg equivalent wt. of analyte}$$

where A = mL iodine solution,
B = normality of iodine solution,
C = mL $Na_2S_2O_3$ solution, and
D = normality of $Na_2S_2O_3$ solution.

The above equation is derived as follows:

The amount of iodine consumed by the analyte =
Starting amount of iodine – Leftover amount of iodine after reaction

The starting amount of iodine is equivalent to $A \times B$. The leftover amount of iodine is equivalent to $A \times B_1$, where B_1 is the normality of iodine solution after reaction (the volume of iodine is constant).

Volume of iodine × its normality = volume of thiosulfate × its normality; that is,

$$A \times B_1 = C \times D$$

or,

$$B_1 = \frac{C \times D}{A}$$

Thus, the amount of iodine consumed = $(A \times B) - (A \times B_1)$
or,

$$\left[(A \times B) - \left(A \times \frac{C \times D}{A}\right)\right]$$

or,

$$\left[(A \times B_1) - (C \times D)\right]$$

If one gram equivalent weight of iodine reacts with one gram equivalent weight (or 1000 mg equivalent weight) of the analyte present in a measured volume of sample, the concentration of analyte, as mg/L analyte

$$= \frac{\left[(A \times B_1) - (C \times D)\right] \times \text{milliequivalent wt. of analyte}}{\text{mL sample taken}}$$

Example

In the titrimetric analysis of sulfide (S^{2-}), 100 mL of wastewater was added to 20 mL of 0.025 N iodine solution which contained potassium iodide and was acidified. The solution was titrated against 0.025 N $Na_2S_2O_3$ solution using starch indicator. The end point in the titration was obtained after the addition of 18.7 mL of titrant. Determine the concentration of sulfide in the sample.

$$\text{mg } S^{2-}/L = \frac{[(20 \times 0.025) - (18.7 \times 0.025)]}{100} \times 16{,}000$$

$$= 5.2$$

The equivalent weight of sulfide as per the reaction $S + 2\ I_2 \rightarrow S^{2-} + I^-$, is 32/2 or 16, because 1 mol S^{2-} (or 32 g S^{2-}) loses 2 mol electrons. Therefore, the milliequivalent weight is 16,000.

Role of Potassium Iodide

Potassium iodide, which is used in unmeasured but excess amounts in iodometric titration, is the source of iodine for many types of reactions. It dissociates to iodide anion, which then reacts with the analyte to produce iodine. Hypochlorite reaction is shown below as an example. By measuring the amount of iodine released, the concentration of the analyte in the sample can be determined.

$$2\ KI \rightarrow 2\ K^+ + 2\ I^-$$

$$OCl^- + 2\ I^- + 2\ H^+ \rightarrow Cl^- + I_2 + H_2O$$

$$I_2 + 2\ S_2O_3^{\ 2-} \rightarrow 2\ I^- + S_4O_6^{\ 2-}$$

Potassium iodide also enhances the solubility of iodine in water.

Reducing Agent

Sodium thiodulfate, $Na_2S_2O_3$ is the most commonly used reductant in iodometric titration. It is a moderately strong reducing agent. It reduces iodine to iodide ion. The thiosulfate ion is quantitatively oxidized to tetrathionate ion. The half-reaction is:

$$2\ S_2O_3^{\ 2-} \rightarrow S_4O_6^{\ 2-} + 2\ e$$

The equivalent weight of $Na_2S_2O_3$ in the above reaction is the same as its molar mass, i.e., 158 (as 2 mol of thiosulfate generates 2 mol electrons). Thus, to prepare 1 *N* $Na_2S_2O_3$, 158 g salt is dissolved in reagent grade water and the solution is made up to 1 L. The solution of appropriate strength must be standardized before titration. One advantage of $Na_2S_2O_3$ over other reducing agents is that its solution is resistant to air oxidation. However, it may be noted that low pH, presence of Cu(II) ions, presence of microorganisms, solution concentration, and exposure to sunlight may cause thiosulfate to decompose to sulfur and hydrogen sulfite ion, as follows:

$$2\ S_2O_3^{2-} + H^+ \rightarrow HSO_3^- + S_{(s)}$$

For this reason, the thiosulfate solutions need to be restandardized frequently.

Phenylarsine oxide, $C_6H_5As = O$, is as effective as sodium thiosulfate in reducing iodine. It is more stable than thiosulfate. An advantage is that it is stable even in dilute solution. This substance is, however, highly toxic and is a suspected carcinogen. Because of its toxicity, its application is limited. One such application is in the amperometric titration of residual chlorine. The oxidation-reduction reaction of PAO is similar to thiosulfate. Its equivalent weight in iodine reaction is 168.

Standardization of $Na_2S_2O_3$

Potassium iodate (KIO_3), potassium hydrogen iodate ($KHIO_3$), potassium dichromate ($K_2Cr_2O_7$), and potassium ferricyanide ($K_3Fe(CN)_6$) are some of the primary standards commonly used to standardize sodium thiosulfate titrant.

An accurately weighted amount of primary standard is dissolved in water containing an excess of potassium iodide. Upon acidification, stoichiometric amounts of iodine are liberated instantly, which are titrated with thiosulfate titrant of unknown strength, decolorizing the blue starch-iodine complex at the end point. With potassium iodate, the ionic reaction is as follows:

$$IO_3^- + 5\ I^- + 6\ H^+ \rightarrow 3\ I_2 + 3\ H_2O$$

Therefore, 1 mol $IO_3^- \equiv 3$ mol I_2.

Since 1 mol iodine reacts with 2 mol thiosulfate as per the reaction

$$I_2 + 2\ S_2O_3^{2-} \rightarrow S_4O_6^{2-} + 2\ I^-$$

therefore, 1 mol $IO_3^- \equiv 3$ mol $I_2 \equiv 6$ mol $S_2O_3^{2-}$.

As mentioned, other oxidizing primary standards are equally efficient. Thus, the molecular reaction of potassium dichromate with potassium iodide in the presence of a strong acid produces a stoichiometric amount of iodine as follows:

$$K_2Cr_2O_7 + 6\ KI + 14\ HCl \rightarrow 8\ KCl + 2\ CrCl_3 + 3\ I_2 + 7\ H_2O$$

$$1\ \text{mol}\ Cr_2O_7^{2-} \equiv 3\ \text{mol}\ I_2 \equiv 6\ \text{mol}\ S_2O_3^{2-}$$

Example

0.1750 g KIO_3 was dissolved in water. A large excess of KI was added into the solution, which was acidified with HCl. The liberated iodine was titrated with $Na_2S_2O_3$ of unknown strength using starch indicator. 57.8 mL of thiosulfate solution was added to obtain the end point. Determine the molarity of $Na_2S_2O_3$.

$$1 \text{ mol } KIO_3 \equiv 3 \text{ mol } I_2 \equiv 6 \text{ mol } Na_2S_2O_3$$

$$0.1750 \text{ g } KIO_3 \times \frac{1 \text{ mol } KIO_3}{214.0 \text{ g } KIO_3} \times \frac{6 \text{ mol } Na_2S_2O_3}{1 \text{ mol } KIO_3}$$

$$= 0.004907 \text{ mol } Na_2S_2O_3 = 4.907 \text{ mmol } Na_2S_2O_3$$

(the formula weight of KIO_3 is 214.0.)

$$\text{The molarity of } Na_2S_2O_3 = \frac{4.907 \text{ mmol}}{57.8 \text{ mL}} = 0.0849\ M$$

Starch Indicator

Starch solution is the preferred indicator in iodometric titration. Although iodine itself has a perceptible color at a concentration of 5×10^{-6} *M* (which is the equivalent of adding one drop of 0.05 *M* iodine solution in 100 mL), it cannot serve as its own indicator because the environmental samples are often dirty and not so clear.

Starch is composed of macromolecular components, α-amylose and β-amylose. The former reacts irreversibly with iodine to form a red adduct. β-Amylose, on the other hand, reacts with iodine forming a deep blue complex. Because this reaction is reversible, β-amylose is an excellent choice for the indicator. The undesired alpha fraction should be removed from the starch. The soluble starch that is commercially available, principally consists of β-amylose. β-Amylose is a polymer of thousands of glucose molecules. It has a helical structure into which iodine is incorporated as I_5^-.

When thiosulfate or PAO titrant is slowly added to the deep blue solution of starch-iodine complex, the reducing agent takes away iodine from the helix. At the end point, when all the iodine is lost, the solution becomes colorless.

A large excess of iodine can cause irreversible breakdown of starch. Therefore, in indirect iodometric analysis, addition of starch is delayed. It is added before the end point when there is not enough iodine left in the solution. This is indicated when the deep brown color of the solution (due to iodine) becomes pale yellow on gradual addition of the reducing agent titrant. When the reducing agent is titrated directly with iodine, starch may be added at the beginning because the iodine liberated is not enough to decompose the starch.

ARGENTOMETRIC TITRATION

Argentometric titration involves the titrimetric determination of an analyte using silver nitrate solution as titrant. Its application in environmental analysis is limited to the determination of chloride and cyanide in aqueous samples. The principle of the method is described below.

Ag^+ preferentially reacts with the analyte to form a soluble salt or complex. During this addition, Ag^+ reacts with the analyte only, and not the indicator. But when all the analyte is completely consumed by Ag^+ and no more of it is left in the solution, addition of an excess drop of silver nitrate titrant produces an instant change in color because of its reaction with the silver-sensitive indicator. Some of the indicators used in the argentometric titrations are potassium chromate or dichlorofluorescein in chloride analysis and *p*-dimethylaminobenzalrhodanine in cyanide analysis. Silver nitrate reacts with potassium chromate to form red silver chromate at the end point. This is an example of precipitation indicator, where the first excess of silver ion combines with the indicator chromate ion to form a bright red solid. This is also known as Mohr method.

Another class of indicators, known as adsorption indicators, adsorb to (or desorb from) the precipitate or colloidal particles of the silver salt of the analyte at the equivalence point. The indicator anions are attracted into the counterion layer surrounding each colloidal particle of silver salt. Thus, there is a transfer of color from the solution to the solid or from the solid to the solution at the end point. The concentration of the indicator, which is an organic compound, is not large enough to cause its precipitation as a silver salt. Thus, the color change is an adsorption and not a precipitation process. Fluorescein is a typical example of an adsorption indicator in argentometric titration.

Often, greater accuracy may be obtained, as in Volhard type titration, by performing a back titration of the excess silver ions. In such a case, a measured amount of standard silver nitrate solution is added in excess to a measured amount of sample. The excess Ag^+ that remains after it reacts with the analyte is then measured by back titration with standard potassium thiocyanate (KSCN). If the silver salt of the analyte ion is more soluble than silver thiocyanate (AgSCN), the former should be filtered off from the solution. Otherwise, a low value error can occur due to overconsumption of thiocyanate ion. Thus, for the determination of ions (such as cyanide, carbonate, chromate, chloride, oxalate, phosphate, and sulfide, the silver salts of which are all more soluble than AgSCN), remove the silver salts before the back titration of excess $Ag.^+$ On the other hand, such removal of silver salt is not necesary in the Volhard titration for ions such as bromide, iodide, cyanate, thiocyanate, and arsenate, because the silver salts of these ions are less soluble than AgSCN, and will not cause any error. In the determination of chloride by Volhard titration, the solution should be made strongly acidic to prevent interference from carbonate, oxalate, and arsenate, while for bromide and iodide analysis titration is carried out in neutral media.

Example

A standard solution was prepared by dissolving 6.7035 g $AgNO_3$ in water to a volume of 1 L. What volume of potassium thiocyanate of strength 0.0957 *M* would be needed to titrate 50 mL of this solution?

$$AgNO_3 + KSCN \rightarrow AgSCN + KNO_3$$

(1 mol) (1 mol) (1 mol) (1 mol)

The molar mass of $AgNO_3 = 169.87$. Thus, 6.7035 g $AgNO_3$

$$= 6.7035 \text{ g } AgNO_3 \times \frac{1 \text{ mol } AgNO_3}{169.87 \text{ g } AgNO_3}$$

$$= 0.03946 \text{ mol } AgNO_3$$

$$\text{The molarity of this solution} = \frac{0.03946 \text{ mol}}{1 \text{ L}} = 0.03946\ M$$

The amount of $AgNO_3$ contained in 50 mL standard solution $= 0.05 \text{ L} \times 0.03946 \text{ mol} = 0.001973 \text{ mol } AgNO_3$, which would require 0.001973 mol $AgNO_3 \times \dfrac{1 \text{ mol KSCN}}{1 \text{ mol } AgNO_3} = 0.001973$ mol KSCN. Therefore, the volume of 0.0957 *M* KSCN solution that would contain 0.001973 mol KSCN =

$$0.001973 \text{ mol KSCN} \times \frac{1 \text{ L}}{0.0957 \text{ mol KSCN}} \times \frac{1000 \text{ mL}}{1 \text{ L}} = 20.6 \text{ mL.}$$

Example

To analyze sulfide (S^{2-}) in an aqueous sample, 150 mL of sample was made ammoniacal before argentometric titration. The sample required 10.45 mL of 0.0125 *M* $AgNO_3$ solution in the titration. Determine the concentration of the sulfide in the sample.

The reaction can be written as follows:

$$2\ Ag^+ + S^{2-} \rightarrow Ag_2S\ (s)$$

The number of moles of solute = molarity × volume (L). Therefore, the mole amount of $AgNO_3$ contained in 10.45 mL solution of 0.0125 *M* strength

$$0.0125 \times 0.01045 \text{ L } AgNO_3 = 0.0001306 \text{ mol } AgNO_3$$

This is also equal to 0.0001306 mol Ag^+.

The mass of S^{2-} that would react with 0.0001306 mol Ag^+

$$= 0.0001306 \text{ mol } Ag^+ \times \frac{1 \text{ mol } S^{2-}}{2 \text{ mol } Ag^+} \times \frac{32.07 \text{ g } S^{2-}}{1 \text{ mol } S^{2-}}$$

$$= 0.0020935 \text{ g } S^{2-} = 2.0935 \text{ mg } S^{2-}$$

Thus, 2.0935 mg S^{2-} was present in 150 mL sample. Therefore, the concentration of sulfide in the sample

$$\frac{2.0935 \text{ mg } S^{2-}}{150 \text{ mL}} \times \frac{1000 \text{ mL}}{1 \text{ L}} = 13.96 \text{ mg } S^{2-}/\text{L}$$

COMPLEXOMETRIC TITRATIONS

A complexometric titration is a rapid, accurate, and inexpensive method to analyze metal ions. However, its application in metal analysis in environmental samples is very much limited, because the more common atomic absorption/emission spectrometry method gives a lower detection limit.

Complexometric methods involve titrations based on complex formations. A ligand may have a single donor group, that is a lone pair of electrons, or two or more donor groups, to form coordinate bonding with metal ions. Such ligands are termed unidentate, bidentate, tridentate, and so on, depending on the number of binding sites in them. Thus, ammonia NH_3, having a lone pair of electrons, is an example of unidentate ligand, while glycin, an amino acid having two binding sites (one on the nitrogen atom and the other on the oxygen atom of the -OH group, after its H^+ is removed), is a bidentate ligand. A particular class of coordination compounds is known as a chelate. A chelate is produced when a ligand having two or more donor groups binds to the metal ion, forming a five- or six-membered heterocyclic ring. Ethylenediaminetetraacetic acid (EDTA) is probably the most widely used chelate in complexometric titrations. It is an aminopolycarboxylic acid having the following structure:

```
CH3COŌ                        ŌOCCH3
      \                      /
       N—CH2—CH2—N
      /                      \
CH3COŌ                        ŌOCCH3
```

The molecule is a hexadentate ligand because it has six donor sites: the two nitrogen atoms and the four carboxyl groups. EDTA forms chelates with all metal ions combining in a 1:1 ratio, irrespective of the charge on the metal ion. Such chelates have high stability too, because of their cage-like structures in which the ligand surrounds the metal ion, protecting the latter from solvation with solvent molecules. Another common chelate similar to EDTA is nitrilotriacetic acid (NTA), which has the following structure:

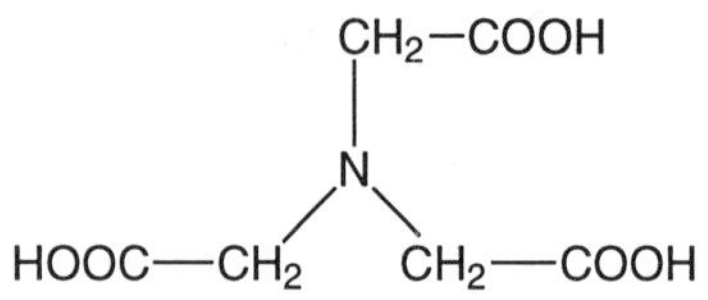

$$
\begin{array}{c}
CH_2-COOH \\
| \\
N \\
\diagup \quad \diagdown \\
HOOC-CH_2 \qquad CH_2-COOH
\end{array}
$$

The cations may be analyzed either by direct titration or back titration. In the former method, a small amount of a metal that forms a less stable complex with EDTA than the analyte metal ion is added to the EDTA titrant. For example, magnesium or zinc forms a chelate with EDTA that is less stable than calcium chelate. Thus, in the determination of calcium in an aqueous sample, add a small quantity of magnesium into the standard EDTA titrant. Add a few drops of the indicator Eriochrome Black T into the sample. When the titrant is added onto the sample, calcium in the sample displaces magnesium from the EDTA-magnesium chelate, forming a more stable chelate with EDTA. The liberated magnesium ions combine with the indicator producing a red color. The titration is continued. When all the calcium ion is complexed with EDTA, any addition of excess EDTA titrant would result in the formation of magnesium-EDTA complex, causing a breakdown of the Mg-indicator complex. The latter breaks down because Mg-EDTA complex is more stable than the Mg-Eriochrome-black complex. Thus, at the end point of the titration, the red color of the solution decolorizes. Often, the titration is performed at pH 10.0, which is maintained by the addition of a buffer. At this pH, the color at the end point is blue. When the pH is greater than 10.0, metal hydroxides may precipitate out.

In the back-titration method, a measured amount of an excess standard EDTA solution is added to the sample. The analyte ion combines with EDTA. After the reaction is complete, the excess EDTA is back-titrated against a standard solution of magnesium or zinc ion. Eriochrome Black T or Calmagite is commonly used as an indicator. After all the remaining EDTA chelates with Mg^{2+} or Zn^{2+}, any extra drop of the titrant solution imparts color to the indicator signifying the end point. The cations that form stable complexes or react slowly with EDTA can also be measured by the back-titration method.

Example

For the determination of Ca^{2+} in an aqueous sample, 50 mL of 0.01 *M* EDTA standard solution was added to a 100 mL sample. The excess EDTA was back-titrated against 0.01 *M* EDTA standard solution to Eriochrome Black T end point. The volume of titrant required was 35.7 mL. Determine the concentration of Ca^{2+} in the sample.

$$50 \text{ mL } 0.01\, M \text{ EDTA} \equiv \frac{0.01 \text{ mol}}{1 \text{ L}} \times 0.05 \text{ L}$$

$$= 0.005 \text{ mol or } 0.500 \text{ mmol EDTA}$$

$$35.7 \text{ mL } 0.01\ M\ Zn^{2+} \equiv \frac{0.01 \text{ mol}}{1 \text{ L}} \times 0.0375 \text{ L}$$

$$= 0.00375 \text{ mol or } 0.375 \text{ mmol } Zn^{2+}$$

Because EDTA combines with Zn^{2+} at 1:1 molar ratio, 0.375 mmol titrant (Zn^{2+}) would consume 0.375 mmol EDTA. Thus, the amount of EDTA that combined with Ca^{2+} in the sample is 0.500 mmol – 0.375 mmol = 0.125 mmol, which also reacts in 1:1 molar ration with Ca.$^{2+}$ Therefore, 100 mL sample contains 0.125 mmol. Thus, the concentration of Ca^{2+} is

$$\frac{0.125 \text{ mmol}}{100 \text{ mL}} = 0.00125\ M$$

We can express the above concentration of Ca^{2+} in mg/L.

$$\frac{0.00125 \text{ mol } Ca^{2+}}{1 \text{ L}} \times \frac{40 \text{ g } Ca^{2+}}{1 \text{ mol } Ca^{2+}} \times \frac{1000 \text{ mg } Ca^{2+}}{1 \text{ g } Ca^{2+}} = 50 \text{ mg } Ca^{2+}/\text{L}$$

The mono- or disodium salt of EDTA may also be used in complexometric titrations. The formula weights of these salts in their dehydrated form are 372.3 and 349.3, respectively. EDTA, or its sodium salt, is standardized against a standard calcium solution. The indicators that are commonly used for such titrations are organic dyes. The structures of two common indicators are as follows:

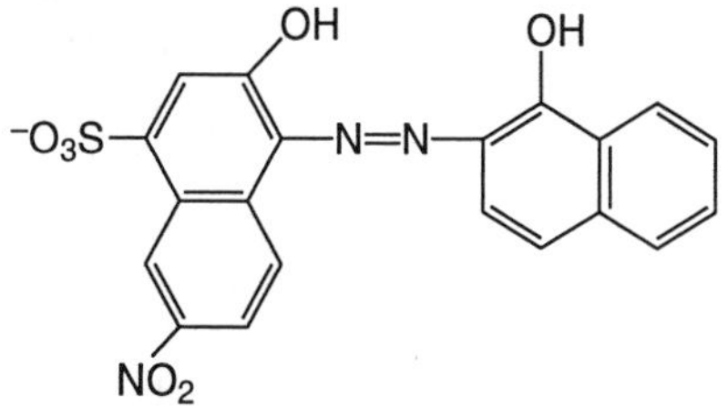

OH OH
$^{-}O_3S$ —N=N—
NO_2

(Eriochrome Black T)

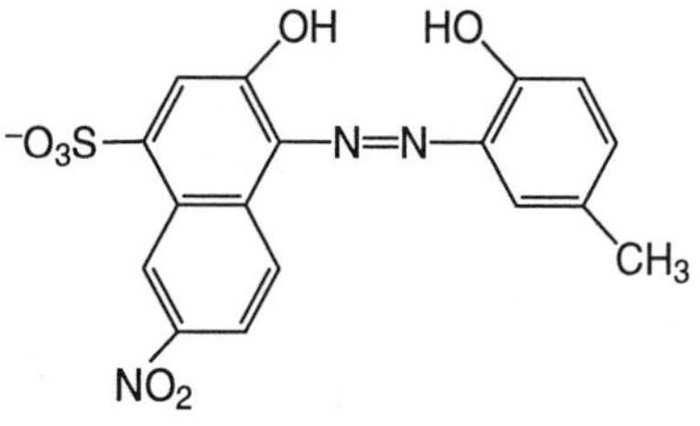

OH HO
$^{-}O_3S$ —N=N—
CH_3
NO_2

(Calmagite)

POTENTIOMETRIC TITRATION

Potentiometric titration can determine the end point more accurately than the color indicators. Thus, the quantitative consumption of a titrant in an acid-base neutralization, oxidation-reduction reaction, or complex formation reaction can be determined precisely and very accurately by potentiometric titration. The titration involves the addition of large increments of the titrant to a measured volume of the sample at the initial phase and, thereafter, adding smaller and smaller increments as the end point approaches. The cell potential is recorded after each addition. The potential is measured in units of millivolt or pH. As the end point approaches, the change in electrode potential becomes larger and larger. The apparatus for potentiometric titration consists of an indicator electrode, a saturated calomel electrode, and a millivoltmeter or a pH meter with millivolt scale. The selection of electrodes depends on the analytes and the nature of titration. For oxidation-reduction titrations, the indicator electrode should be made of an inert metal, such as platinum, palladium, or gold. Occasionally, silver and mercury are also used. Glass electrodes are commonly used in acid-base titrations. Mercury indicator electrode may be used in complexometric titrations involving EDTA. The stability of the EDTA complex with the analyte metal ion should be lower than the mercury-EDTA complex. Both the metallic, as well as the membrane, electrodes have been used in general complex formation titrations. In argentometric titrations involving the precipitation reactions of silver nitrate, a silver indicator electrode is used.

End Point Detection

The end point in a potentiometric titration can be determined by one of the following three methods: Direct plot, first-derivative curve, and second-derivative curve.

The direct plot method is most commonly used to determine the end point. A titration curve is drawn by plotting the electrode potential in the Y-axis against mL titrant added in the X-axis. Near the end point, there is a sharp increase in potential. The mid-point of the steeply rising portion of the curve is taken as the end point of the titration (Figure 1.6.4).

A first-derivative curve may alternatively be constructed by plotting change in potential per unit volume, i.e., $\Delta E/mL$ of titrant, added against the mL of the titrant. The curve would appear like a sharp peak, as shown in Figure 1.6.5. The end point is the volume corresponding to the highest $\Delta E/mL$, i.e., the tip of the peak.

The second-derivative curve method is similar to the first-derivative method. The only change is that the difference of ΔE, or the change in potential per mL (or Δ^2E/mL) of consecutive additions, are plotted against the volume of titrant added. Figure 1.6.6 shows the shape of the curve. It may be noted from the figure that very near after the end point, the data changes sign. The volume of the titrant in the X-axis that corresponds to the tip of the peak is the end point. Alternatively, the inflection point on the X-axis, at which Δ^2E/mL changes its sign, may be considered as the end point.

As an example, a set of data from a potentiometric titration is presented in Table 1.6.4. The method by which the calculations are performed is illustrated along with the data.

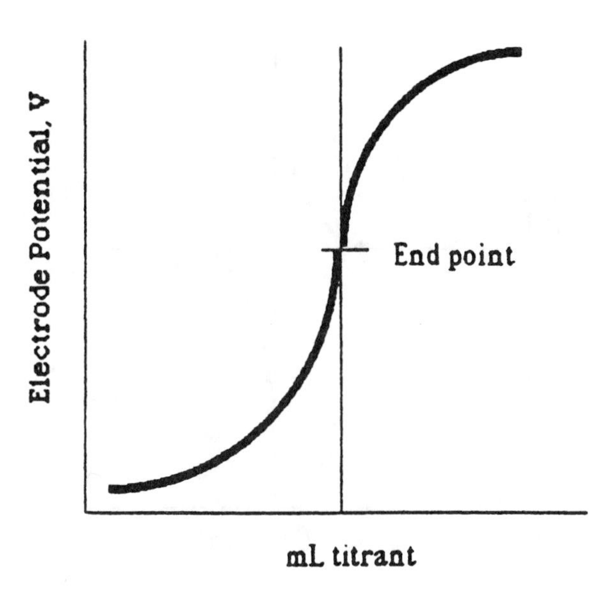

Figure 1.6.4 Direct plot.

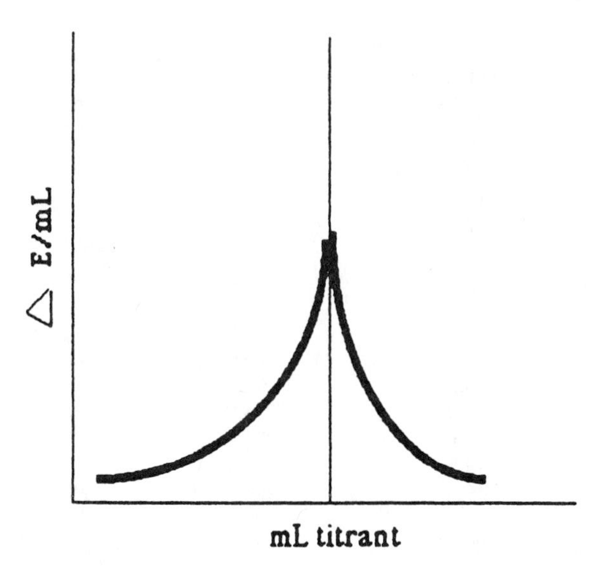

Figure 1.6.5 First-derivative curve.

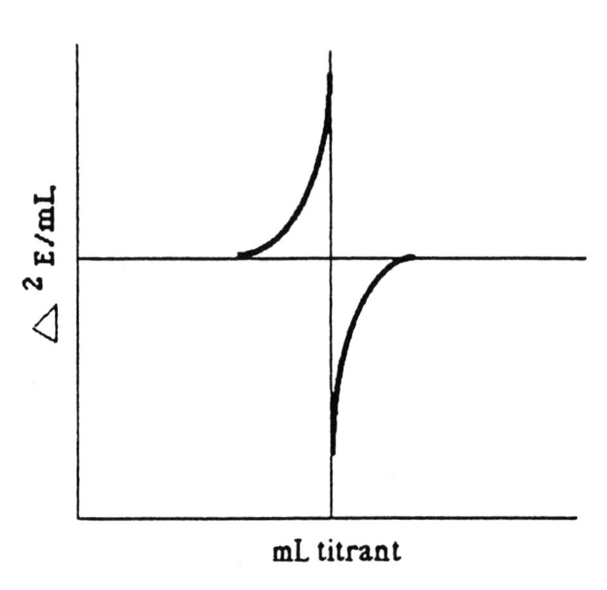

Figure 1.6.6 Second-derivative curve.

Table 1.6.4 A Set of Potentiometric Titration Data as an Example

Titrant volume (mL)	Electrode potential (mV)	ΔE/mL
5.0	58.2	3
10.0	72.3	4
15.0	91.4	8
17.0	107.0	14
18.0	121.5	14
18.5	132.5	22
18.8	144.8	34
19.00	156.0	60
19.10	170.0	140
19.20	202.0	320
19.30	284.2	820
19.40	251.1	330
19.50	235.4	160
19.60	228.3	70
19.80	220.3	40
20.10	212.4	24
20.60	208.0	16
21.60	200.0	8
23.60	190.2	5

1.7 COLORIMETRIC ANALYSIS

Colorimetric methods are most common and widely employed in environmental wet analysis. Most anions, all metals, and many physical and aggregate properties can be determined by colorimetric technique, which is fast and cost-effective. The method may, however, be unreliable for dirty and colored samples. Often, the presence of certain substances in samples can interfere with the test. In addition, if the color formation involves a weak color such as yellow, additional confirmatory tests should be performed. Despite these drawbacks, colorimetry is often the method of choice for a number of wet analyses.

Absorbance and *transmittance* are two important terms used in absorption measurement. If a beam of radiation of intensity P_0 passes through a layer of solution, a part of the light is absorbed by the particles of the solution; thus, the power of the beam weakens. The transmittance T of the solution is measured as the fraction of incident light transmitted by the solution. If the intensity of the transmitted light is P, then transmittance is measured as,

$$T = \frac{P}{P_0}$$

Absorbance is measured in logarithmic scale and is defined by the following equation:

$$A = -\log_{10} T = \log \frac{P_0}{P}$$

where A and T are the absorbance and transmittance of the solution, respectively. The absorbance of a solution is related to its concentration c by the following equation:

$$A = \log \frac{P_0}{P} = abc$$

where *b* is the thickness of the solution (or the pathlength of radiation) and *a* is a proportionality constant called absorptivity. This is known as Beers law. When the concentration is expressed in moles/L and *b* in cm, then *a* is termed as molar absorptivity, ε. Beer's law is modified to

$$A = \varepsilon bc$$

Thus, Beer's law exhibits a linear relationship between the absorbance of a solution to its pathlength and to its concentration. The relationship is very linear for pathlength, but not so for concentration. Beer's law works well for dilute solutions only. Solutions having concentrations greater than 0.01 *M* show significant deviation from Beer's law. Even in dilute solutions, the presence of an electrolyte, such as metal ions or chloride, sulfate or other anions, in large amounts, or the presence of large organic molecules can cause significant deviation from linearity. If the analytes dissociate or react with solvent, or form adducts then the molar absorptivity can change, causing departures from Beer's law. However, despite these limitations, Beer's law is extensively applied in environmental analysis, where the analyte concentrations encountered in samples are generally well below 0.01 *M*. If the concentration is above linearity range, the sample then must be diluted for measuring absorbance.

Beer's law applies only to monochromatic radiation, i.e., light of one wavelength and not polychromatic radiation. The absorbance is measured by a spectrophotometer. Any polychromatic radiation source may be employed from which the desired single band wavelengths can be filtered by the grating or prism for the measurement of the absorbance. A tungsten lamp or tungsten/halogen lamp is used as the light source. If transmittance is to be measured, then a filter photometer should be used. In environmental wet analysis, either a spectrophotometer or a filter photometer providing a light path of 1 cm or longer is used. It may be noted that absorbance and transmittance are inversely related. The greater the absorbance of a solution, the lower is its transmittance and vice-versa.

Beer's law still applies even if a solution has more than one kind of absorbing substance. The total absorbance of the system would then be the sum of the absorbance of individual components, i.e., $A_{total} = \varepsilon_1 bc_1 + \varepsilon_2 bc_2 + \ldots + \varepsilon_n bc_n$ where ε_1, ε_2, ε_n and c_1, c_2, c_n are the molar absorptivity and concentrations, respectively, of the absorbing components.

In colorimetric analysis, a reagent is selected that would form a colored complex or derivative with the analyte. Often, the analyte is extracted from the aqueous sample into an organic solvent before adding color-forming reagent. Such extractions become necessary, especially for organic analytes such as phenols, lignin, and tannin.

Table 1.7.1 lists some common pollutants and their characteristic wavelengths (at which their absorbance or transmittance should be measured).

The first step in any colorimetric analysis is to prepare a standard calibration curve (i.e., a series of standard solutions of the analyte is made at a specific concentration range), which are then treated with the color-forming reagent, the absorbance or transmittance of which is then measured. The lowest calibration

Table 1.7.1 Common Pollutants

Analyte	Colored complex/derivative	Wavelength nanometer
Ammonia	Yellow mercuric salt, Hg_2OINH_2	415
	Indophenol	630
Bromide	Bromoderivative of phenol red	590
Chlorine residual	An oxidized form of *N,N*-diethyl-*p*-phenylenediamine;	515
	An oxidized form of syringaldizine	530
Cyanide	Glutaconaldehyde derivative of barbituric acid	578
	Glutaconaldehyde derivative of pyrazolone	620
Fluoride	Zirconium dye	570
Iodide/iodine	Iododerivative of methylidynetris(*N,N*-dimethylaniline)	592
Nitrate	Azo derivative of *N*-(naphthyl) ethylenediamine dihydrochloride	540
Ozone	Oxidized indigo	600
Phenols	4-Aminoantipyrine phenol complex	460
Phosphorus	Vanadomolybdophosphoric acid and its aminonaphthol sulfonic acid derivative	400-470 815[a] 650-815[b]
Silica	Molybdosilicic acid	410
Sulfate	Barium sulfate	420
Sulfide	Methylene blue derivative	664[a] 600[b]
Sulfite	*Tris*(1,10-phenanthroline)iron(II)	510
Surfactants (anionic)	Methylene blue derivatives of sulfate esters and alkylbenzene sulfonates	652
Tannin and lignin	Phenol derivative of tungsto- and molybdophosphoric acid	700

[a] Absorbance only.
[b] Transmittance only.

standard should be the minimum analyte concentration that would produce the least measurable absorbance or transmittance. That should be the detection limit for the analyte. The highest calibration standard should be 10 to 15 times this concentration, and it must be within the linearity range. A plot is made between the absorbance or transmittance vs. the concentration (or the microgram mass of the analyte). The concentration of the analyte in the sample is then read from the graph.

The presence of other substances in the sample that may react with color-forming reagent can interfere with the test. These substances must be removed by precipitation or their interference effect must be suppressed by other means (such as pH control). These are discussed in detail throughout Part 2 of this text under colorimetric procedures for respective analytes.

1.8 ANALYSIS OF METALS BY ATOMIC ABSORPTION AND EMISSION SPECTROSCOPY

Of the elements in the Periodic Table more than two thirds are metals. Although many of these metals are toxic, only some metals are major environmental pollutants, because of their widespread use. U.S. EPA has classified 13 metals as priority pollutants: aluminum, antimony, arsenic, beryllium, cadmium, chromium, copper, lead, mercury, nickel, selenium, silver, and zinc. The Resource Conservation and Recovery Act has listed eight metals whose mobility in the soil is measured to determine the characteristic of toxic wastes. These metals include arsenic, barium, cadmium, chromium, lead, mercury, selenium, and silver—all but one from the above list of priority pollutant metals.

Metals in general can be analyzed by the following techniques:

1. Colorimetry
2. Atomic absorption or atomic emission spectrophotometry

In addition, some metals may be determined by other methods, including ion-selective electrode, ion chromatography, electrophoresis, neutron activation analysis, redox titration, and gravimetry. Atomic absorption or emission spectrophotometry is the method of choice, because it is rapid, convenient, and gives the low detection levels as required in the environmental analysis. Although colorimetry methods can give accurate results, they are time consuming and a detection limit below 10 μg/L is difficult to achieve for most metals.

SAMPLE DIGESTION FOR METALS

Aqueous and nonaqueous samples must be digested with an acid before their analysis by atomic absorption or atomic emission spectrophotometry. The metals and their salts present in the sample are converted into their nitrates due to the fact that the nitrates of all metals are soluble in water. Therefore, concentrated nitric acid by itself or in conjunction with hydrochloric acid, sulfuric acid, perchloric acid, or hydrofluoric acid is used in sample digestion for the determination of total metals. Nitric acid alone is, however, adequate for digestion of most metals. Table 1.8.1 lists the combinations of acids that may be helpful in sample digestion.

Table 1.8.1 Acid Combination Suggested for Sample Preparation

Acids combination	Suggested use
HNO_3-HCl	Sb, Sn, Ru and readily oxidizable organic matter
HNO_3-H_2SO_4	Ti and readily oxidizable organic matter
HNO_3-$HClO_3$	Difficult to oxidize organic materials
HNO_3-HF	Siliceous materials

Acid digestion is performed using a small volume (5 to 10 mL) of nitric acid alone or in conjunction with one of the previously mentioned acids on a hot plate. Alternatively, a laboratory-grade microwave unit, specifically designed for hot acid digestion, can be used. When the sample is boiled with acid, the latter should not be allowed to dry. The acid extract after boiling and cooling is diluted with water to a measured final volume for analysis.

ATOMIC ABSORPTION SPECTROMETRY

An atomic absorption spectrophotometer consists primarily of a light source to emit the line spectrum of an element (i.e., the element to be analyzed), a heat source to vaporize the sample and dissociate the metal salts into atoms, a monochromator or a filter to isolate the characteristic absorption wavelength, and a photoelectric detector associated with a microprocessor and a digital readout device for measuring the absorbance due to the metal at its corresponding concentration. The light source usually is a hollow cathode lamp or an electrodeless discharge lamp composed of the element to be measured. The heat source is an air-acetylene or air-nitrous oxide flame or a graphite furnace. In flame atomic absorption spectrometry, the heat source is a flame. The sample is aspirated into the flame and atomized. The light beam is directed through the flame. The metal atoms absorb energy at their own characteristic wavelength. The energy absorbed is proportional to the concentration of the element in the sample.

An atomic absorption spectrometer equipped with a graphite furnace or an electrically heated atomizer instead of the standard burner head gives better sensitivity and much lower detection limit than what is obtained with the flame technique (Table 1.8.2). The principle of this technique is the same as for the flame method. A small volume of sample is aspirated into a graphite tube which is heated in several stages. First, the sample is dried by low current heating. Then, it is charred at an intermediate temperature to destroy the organic matters and volatilize the compounds. Finally, it is heated to incandescence by a high current in an inert atmosphere to atomize the element. Atoms in their ground state absorb monochromatic radiation from the source. The intensity of the transmitted light is measured by a photoelectric detector. The intensity is inversely proportional to the number of ground state atoms in the optical path, i.e., the greater the quantity of ground state atoms in the optical path, the greater the absorbance (in other words, the less the amount of light transmitted through).

Table 1.8.2 Recommended Wavelength, Flame Type, and Technique for Flame Atomic Absorption Analysis

Element	Wavelength (nm)	Flame	Technique
Aluminum	309.3	N_2O-acetylene	DA, CE
Antimony	217.6	Air-acetylene	DA
Arsenic	193.7	Air-hydrogen	H
Barium	553.6	N_2O-acetylene	DA
Beryllium	234.9	N_2O-acetylene	DA, CE
Bismuth	223.1	Air-acetylene	DA
Cadmium	228.8	Air-acetylene	DA, CE
Cesium	852.1	Air-acetylene	DA
Chromium	357.9	Air-acetylene	DA, CE
Cobalt	240.7	Air-acetylene	DA, CE
Copper	324.7	Air-acetylene	DA, CE
Iridium	264.0	Air-acetylene	DA
Iron	248.3	Air-acetylene	DA, CE
Lithium	670.8	Air-acetylene	DA
Lead	283.3, 217.0	Air-acetylene	DA, CE
Magnesium	285.2	Air-acetylene	DA
Manganese	279.5	Air-acetylene	DA, CE
Molybdenum	313.3	N_2O-acetylene	DA
Nickel	232.0	Air-acetylene	DA, CE
Osmium	290.9	N_2O-acetylene	DA
Platinum	265.9	Air-acetylene	DA
Rhodium	343.5	Air-acetylene	DA
Ruthenium	349.9	Air-acetylene	DA
Silver	328.1	Air-acetylene	DA, CE
Selenium	196.0	Air-hydrogen	H
Silicon	251.6	N_2O-acetylene	DA
Strontium	460.7	Air-acetylene	DA
Tin	224.6	Air-acetylene	DA
Titanium	365.3	N_2O-acetylene	DA
Vanadium	318.4	N_2O-acetylene	DA
Zinc	213.9	Air-acetylene	DA, CE

Note: DA, direct aspiration; CE, chelation extraction; H, hydride generation.

The primary advantage of the graphite furnace technique over the conventional flame method is that the former requires a smaller volume of sample and the detection limit is much lower. Many metals can be determined at a concentration of 1 μg/L. A disadvantage of the graphite furnace technique, however, is that because of high sensitivity, interference due to other substances present in the sample can cause a problem. Such interference may arise from molecular absorption or from chemical or matrix effects. This can be reduced or eliminated by correcting for background absorbance and by adding a matrix modifier into the sample. Some common matrix modifiers are listed in Table 1.8.3. Certain metals such as molybdenum, vanadium, nickel, barium, and silicon react with graphite at high temperatures, thus forming carbides. Such chemical interaction may be prevented by using pyrolytically coated tubes. Graphite tubes with L'rov platforms should be used. To prevent the formation of metallic oxides and minimize oxidation of furnace tubes, argon should be used as a purge gas.

Table 1.8.3 Substances Added to the Sample for the Removal of Interference in Graphite Furnace Atomic Absorption Method

Element	Matrix modifiers
Aluminum	$Mg(NO_3)_2$
Antimony	$Mg(NO_3)_2$ and $Ni(NO_3)_2$
Arsenic	$Mg(NO_3)_2$, $Ni(NO_3)_2$
Beryllium	$Mg(NO_3)_2$, $Al(NO_3)_3$
Cadmium	$Mg(NO_3)_2$, $NH_4H_2PO_4$, $(NH_4)_2SO_4$, $(NH_4)_2S_2O_8$
Chromium	$Mg(NO_3)_2$
Cobalt	$Mg(NO_3)_2$, $NH_4H_2PO_4$, ascorbic acid
Copper	NH_4NO_3, ascorbic acid
Iron	NH_4NO_3
Lead	$Mg(NO_3)_2$, NH_4NO_3, $NH_4H_2PO_4$, $LaCl_3$, HNO_3, H_3PO_4, ascorbic acid, oxalic acid
Manganese	$Mg(NO_3)_2$, NH_4NO_3, ascorbic acid
Nickel	$Mg(NO_3)_2$, $NH_4H_2PO_4$
Selenium	$Ni(NO_3)_2$, $AgNO_3$, $Fe(NO_3)_3$, $(NH_4)_6Mo_7O_{24}$
Silver	$(NH_4)_2HPO_4$, $NH_4H_2PO_4$
Tin	$Ni(NO_3)_2$, NH_4NO_3, $(NH_4)_2HPO_4$, $Mg(NO_3)_2$, ascorbic acid

Chelation-Extraction Method

Many metals at low concentrations can be determined by chelation-extraction technique. These metals include cadmium, chromium, cobalt, copper, iron, lead, manganese, nickel, silver, and zinc. A chelating agent such as ammonium pyrrolidine dithiocarbamate (APDC) reacts with the metal, forming the metal chelate which is then extracted with methyl isobutyl ketone (MIBK). A 100-mL aliquot of aqueous sample is acidified to pH 2 to 3 and mixed with 1 mL APDC solution (4% strength). The chelate is extracted with MIBK by shaking the solution vigorously with the solvent for 1 min. The extract is aspirated directly into the air-acetylene flame. Calibration standards of metal are similarly chelated and extracted in the same manner and the absorbances are plotted against concentrations.

APDC chelates of certain metals such as manganese are not very stable at room temperature. Therefore, the analysis should be commenced immediately after the extraction. If an emulsion formation occurs at the interface of water and MIBK, use anhydrous Na_2SO_4.

The chelation-extraction method determines chromium metal in hexavalent state. In order to determine total chromium, the metal must be oxidized with $KMnO_4$ under boiling and the excess $KMnO_4$ is destroyed by hydroxylamine hydrochloride prior to chelation and extraction.

Low concentrations of aluminum and beryllium can be determined by chelating with 8-hydroxyquinoline and extracting the chelates into MIBK and aspirating into a N_2O-acetylene flame.

Hydride Generation Method

Arsenic and selenium can be determined using the hydride generation method. These metals in HCl medium can be converted to their hydrides by treatment with sodium borohydride. The hydrides formed are purged by nitrogen or argon into the atomizer for conversion into the gas-phase atoms.

The reaction with $NaBH_4$ is rapid when the metals are in their lower oxidation states as As(III) and Se(IV), respectively. Sample digestion with nitric acid, however, oxidizes these metals to their higher oxidation states, producing As(V) and Se(VI). These metals are reduced to As(III) and Se(IV) by boiling with 6 *N* HCl for 15 min. The digested sample is then further acidified with conc. HCl, treated with sodium iodide (for As determination only), and heated. Add to this solution 0.5 mL $NaBH_4$ solution (5% in 0.1 *N* NaOH, prepared fresh daily) and stir. The hydride generated (arsine or selenium hydride) is purged with the carrier gas such as argon and transported into the atomizer. Standard solutions of these metals are treated with $NaBH_4$ in the same manner for the preparation of the standard calibration curve. The presence of other substances in the samples causes little interference because hydrides are selectively formed and are removed from the solution. Commercially available continuous hydride generator units make the operation simpler than the manual method outlined above.

Cold Vapor Method for Mercury Determination

Cold vapor atomic absorption spectrophotometric method is applicable only for the mercury analysis. The principle of this method is described below.

After acid digestion with nitric acid, mercury and its salts are converted into mercury nitrate. Treatment with stannous chloride reduces mercury into its elemental form which volatilizes to vapors. Under aeration the vapors of mercury are carried by air into the absorption cell. The absorbance is measured at the wavelength 253.7 nm. Prior to reduction, any interference from sulfide and chloride are removed by oxidizing the extract with potassium permanganate. Free chlorine produced from chloride is removed by treatment with hydroxylamine sulfate reagent and by sweeping the sample gently with air. After the sample is acid digested with conc. H_2SO_4 and HNO_3, the extract is treated with two strong oxidizing agents: $KMnO_4$ and potassium persulfate ($K_2S_2O_8$). After adding 15 mL of $KMnO_4$ solution (5%) to the acid extract, let the solution stand for 15 min. To this solution, add about 10 mL of 5% $K_2S_2O_8$ solution. Heat the mixture to boiling for 2 h in a water bath. The excess of $KMnO_4$ is destroyed by adding NaCl-hydroxylamine sulfate solution [12% concentration of each, or a 10% hydroxylamine hydrochloride solution instead of $(NH_2OH)_2{\cdot}H_2SO_4$].

Calibration standards are made from a soluble mercury salt, such as, mercuric chloride. The standard solutions are analyzed first prior to the sample, following acid digestion, oxidation, and reduction, as described above. A standard calibration curve is constructed by plotting absorbance vs. concentrations of Hg (or mg Hg). The concentration of Hg in the sample is then determined by comparing the absorbance with that in the calibration curve.

ACCURACY IN QUANTITATION

Gross error and inaccuracy can often result at a very low level of detection, especially near the instrument detection limit (IDL). This is illustrated in the following data:

Concentration, µg/L	10	20	50	100
Absorbance	0.01	0.02	0.06	0.12

A calculation based on an absorbance reading of 0.01 at the IDL for an unknown sample can be inaccurate because there could be a large degree of fluctuation in absorbance readings at that level. For example, if 5 aliquots of a sample extract are aspirated into the flame, and if the absorbance is at the IDL, 0.01, there is a possibility of getting an erroneous reading. Let us suppose that the relative standard deviation of the measurement is 50%, which is not uncommon at the level of IDL. Then the true absorbance reading may oscillate anywhere from 0.005 to 0.015. A concentration corresponding to a digital display of absorbance ranging from 0.00 to 0.02, as determined from the above calibration data, would correspondingly range between <10 and 20 µg/L. Thus, if 10 g of a sludge sample having a total solid content of 2% is acid digested and diluted to 1 L, then we may end up with a final result ranging between <50 and 200 mg/kg (dry weight corrected). Thus, we see how a minor deviation in the absorbance reading near the IDL can cause such a noticeable large deviation in the final result. Such errors occur frequently in routine analysis for metals in sludge samples. This can be prevented by increasing the sample concentration, i.e., either by reducing the final volume of the extract, as in the above example, from 1 to 0.1 L or by increasing the amount of sample aliquot, say from 10 to 100 g for the digestion. A ten-time increase in the concentration of metal in the extract solution would produce an absorbance far above the IDL range, causing little deviation. In the above example, an absorbance reading in the range from 0.08 to 0.12 would correspond to a final, dry weight corrected result in the range of 33.5 to 60 mg/kg; and a reading between 0.09 and 0.12 would be equivalent to from 37.5 to 46 mg/kg.

Standard Addition

The method of standard addition should be performed to achieve accurate results. The method involves spiking an equal volume of standard solutions, at least three different concentrations to equal volumes of reagent grade water and sample aliquots, respectively. The absorbance is recorded and plotted in y-axis against the concentration in x-axis. The linear curve is extended through the y-axis. The distance from the point of intersection on the x-axis to the origin is equal to the concentration of the metal in the sample. An example is illustrated below in Figure 1.8.1.

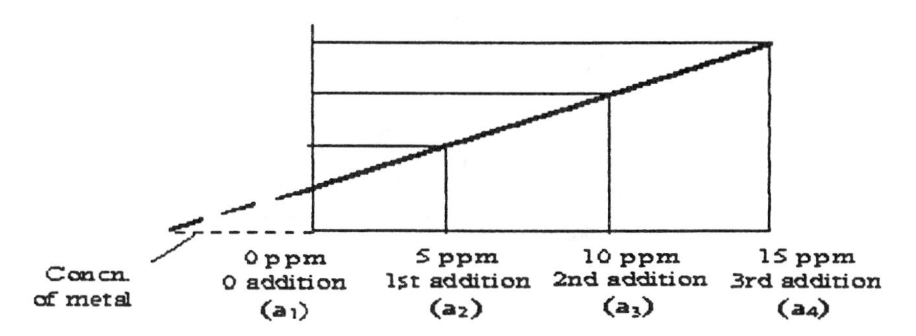

Figure 1.8.1 Linear curve showing standard addition method.

Note: a_1, 100 mL sample + 100 mL reagent grade water; a_2, 100 mL sample + 100 mL 5 ppm standard; a_3, 100 mL sample + 100 mL 10 ppm standard; a_4, 100 mL sample + 100 mL 15 ppm standard.

ATOMIC EMISSION PLASMA SPECTROSCOPY

Metals can be conveniently determined by emission spectroscopy using inductively coupled plasma (ICP). A great advantage of ICP emission spectroscopy as applied to environmental analysis is that several metals can be determined simultaneously by this method. Thus, multielement analysis of unknown samples can be performed rapidly by this technique. Another advantage is that, unlike atomic absorption spectroscopy, the chemical interference in this method is very low. Chemical interferences are generally attributed to the formation of molecular compounds (from the atoms) as well as to ionization and thermochemical effects. The principle of the ICP method is described below.

The apparatus consists primarily of an ICP source and a spectrometer. The spectrometer may be of simultaneous type (polychromator) or a sequential type (monochromator). The ICP source consists of a radio frequency generator that can produce at least 1.1 kW of power. It also has other components, which include torch, coil, nebulizer, spray chamber, and drain.

A flowing stream of argon gas is ionized by an applied oscillating radio frequency field, which is inductively coupled to the ionized gas by a water-cooled coil. The coil surrounds a quartz torch that confines the plasma. A sample aerosol generated in the nebulizer and spray chamber is injected into the ICP, at a high temperature of 6000 to 8000 K of plasma. These high temperatures ionize atoms producing emission spectra. Also, at this temperature, molecules of compounds formed completely dissociate, thus reducing any chemical interferences. The light emitted is focused onto a monochromator or a polychromator. The latter uses several exit slits to simultaneously monitor all configured wavelengths for multielement detection.

Spectral interferences from ion–atom recombination, spectral line overlaps, molecular band emission, or stray light can occur that may alter the net signal intensity. These can be avoided by selecting alternate analytical wavelengths and making background corrections.

The detection limit varies from element to element and is mostly in the low ppb range. Table 1.8.4 presents recommended wavelengths for metals analysis and their approximate detection limits.

Table 1.8.4 Recommended Wavelength and the Instrument Detection Level in Inductively Coupled Plasma Emission Spectrometry

Element	Wavelength recommended (nm)	Alternate wavelength (nm)	Approximate detection limit (μg/L)
Aluminum	308.22	237.32	50
Antimony	206.83	217.58	30
Arsenic	193.70	189.04	50
Barium	455.40	493.41	2
Beryllium	313.40	234.86	0.5
Boron	249.77	249.68	5
Cadmium	226.50	214.44	5
Calcium	317.93	315.89	10
Chromium	267.72	206.15	10
Cobalt	228.62	230.79	10
Copper	324.75	219.96	5
Iron	259.94	238.20	10
Lead	220.35	217.00	50
Lithium	670.78	—	5
Magnesium	279.08	279.55	30
Manganese	257.61	294.92	2
Molybdenum	202.03	203.84	10
Nickel	231.60	221.65	15
Potassium	766.49	769.90	100
Selenium	196.03	203.99	75
Silica (SiO_2)	212.41	251.61	20
Silver	328.07	338.29	10
Sodium	589.00	589.59	25
Strontium	407.77	421.55	0.5
Thallium	190.86	377.57	50
Vanadium	292.40	—	10
Zinc	213.86	206.20	2

1.9 ION-SELECTIVE ELECTRODES ANALYSIS

Many ions, which include both metals and anions, may be analyzed rapidly and with a good degree of accuracy by ion-selective electrodes. In addition, dissolved gases, such as oxygen, carbon dioxide, ammonia, and oxides of nitrogen, could be analyzed by this technique using a gas sensor electrode.

A selective-ion electrode system constitutes the two half-cells which are a sensing electrode and a reference electrode, a readout meter, and a solution containing the specific ion to be analyzed. The sensing electrode could be a solid state, a liquid membrane, or a gas-sensing electrode, or the most familiar type, a glass electrode. The reference electrode should be either a single junction or a double junction type electrode containing a freely flowing filling solution and should produce a stable potential. A filling solution completes the electrical circuit between the sample and the internal cell. The filling solutions commonly used in the reference electrodes are KCl, AgCl, and KNO_3. A filling solution, however, must not contain the ion to be analyzed.

When a sensing electrode is immersed in a solution containing the same ion to which it is selective, a potential develops across the surface of its membrane. This potential is measured as voltage and is proportional to the concentration of the ion in the solution. The voltage caused by the sensing electrode is compared against a stable potential produced by a reference electrode.

The principle of electrode analysis is based on the Nernst equation which can be written as:

$$E = E_o + S \log C$$

where E = the measured voltage
E_o = reference potential
S = the slope of the electrode
C = the concentration of the ion in the solution

In other words, the voltage difference between the sensing and the reference electrodes can be a measure of the concentration of the analyte ion, selective to the sensor electrode. The above equation can be written as follows:

$$C_x + C_s \times 10^{\Delta E/S}$$

where C_x and C_s = the concentrations of the unknown and the standardizing solutions, respectively
ΔE = the difference of potentials between that of the standardizing solution and the sample solution
S = the slope of the electrode that measures the change in electrode potential per tenfold change in concentration

ENVIRONMENTAL SAMPLE ANALYSIS

Selective-ion electrode technology has been successfully use to analyze many common pollutant anions, metal ions and dissolved gas molecules in wastewaters, drinking waters, surface- and groundwaters, sludges, soils, and solid wastes. The general analytical methods for solid matrix are the same as for water, except that the solid sample is first extracted in water or into an aqueous phase with proper pH adjustment, prior to analysis.

Selective-ion electrodes are commercially available from Orion Research and many other manufacturers. Follow the manufacturer's instruction manual for the use and operation of the electrodes. The analysis may be performed by one of the following methods:

- Standard calibration method
- Standard addition method
- Sample addition method

The electrode should be equipped with a readout device, such as a millivoltmeter or a microprocessor.

Standard Calibration Method

In this method, a few milliliters of ionic strength adjuster (ISA) and/or a pH adjuster is added to both standards and samples before measurement. This sets the ionic strength of the samples and the standards to a constant level.

Prepare at least two standards with one at concentrations below and one at concentrations above the expected range of sample concentrations. The concentration of higher standard should be ten times greater than that of the lower standard. More standards may be prepared to plot calibration curves outside the linearity range.

Add ISA solution (or pH adjustor, as recommended in the instruction manual) to the most diluted standard. Set the instrument mode to read millivolts. Place the electrodes in the solution and record the stable millivolt reading. Repeat these steps and record the millivolt values for the higher concentration standard(s). Plot a calibration curve on a semilogarithmic paper, taking the millivolt values on the linear axis (*Y*-axis) and the concentrations of the standards on the logarithmic axis (*X*-axis).

Rinse the electrodes and place in the sample after adding ISA. Record the stable millivolt reading in the instrument mode. Determine the concentration of the sample from the calibration curve. Dilute the sample if the concentration of the analyte ion in the sample is high and falls outside the calibration plot. Alternatively, prepare a new calibration plot adding more standards to bracket the concentration(s) of sample(s).

Standard Addition Method

In this method no calibration plot is required. The electrode potentials of the sample and then the sample plus a known amount of standard are recorded. The electrode slope is determined and the concentration of the analyte ion is calculated.

Determine the electrode slope following the instruction manual. Place electrodes in 100 mL or any suitable volume of the sample. Add a few milliliters of ISA and stir throughout. Set the instrument mode to read millivolts and record the stable millivolt reading, E_1.

Add a small, but accurately measured volume of standard to the sample which would produce a reading that is about twice the sample millivolt reading. The volume of standard should not exceed 15% of the volume of sample taken. The concentration of the standard is calculated as follows:

$$\text{Conc. of the standard to be spiked} = \frac{\text{Sample volume} \times \text{Estimated sample conc.}}{\text{Volume of standard to be spiked}}$$

For example, if the sample volume is 100 mL and the analyte concentration is estimated to be in the range of 2 ppm, add 5 mL of 40 ppm or 10 mL of 20 ppm standard.

Record the millivolt reading, E_2, after adding the standard. Record the value when the reading is stable.

The concentration, C_x, of the analyte ion in the sample may be calculated from the following equation:

$$C_x = \frac{\rho C_y}{\left[(1+\rho)\times 10^{\Delta E/S} - 1\right]}$$

where C_y = the concentration of the standard

ρ = the ratio of the volume of spiked standard to the volume of sample taken

S = the electrode slope

ΔE = the millivolt difference, $E_2 - E_1$

Example

Five mL of 75 ppm chloride standard was added to 100 mL of a wastewater sample for chloride analysis. The initial millivolt reading for the sample was 17.5 mV. After adding the standard, the reading was 35.3 mV. The slope was measured previously as 58.1 mV. Calculate the concentration of chloride in the sample.

$$C_y = 75 \text{ ppm}$$

$$\rho = \frac{5 \text{ mL}}{100 \text{ mL}} = 0.05$$

$$\Delta E = 35.3 \text{ mV} - 17.5 \text{ mV} = 17.8 \text{ mV}$$

$$S = 58.1 \text{ mV}$$

$$C_x = \frac{0.05 \times 75 \text{ ppm}}{\left[(1 + 0.05) \times 10^{17.8 \text{ mV}/58.1 \text{ mV}} - 1\right]}$$

$$= \frac{3.75 \text{ ppm}}{\left[\left(1.05 \times 10^{0.306}\right) - 1\right]}$$

$$= \frac{3.75 \text{ ppm}}{\left[(1.05 \times 2.02) - 1\right]}$$

$$= \frac{3.75 \text{ ppm}}{1.124}$$

$$= 3.34 \text{ ppm}$$

The concentration of chloride in the sample was 3.34 ppm.

Sample Addition Method

This method is very similar to the standard addition method described above. The only difference is that instead of adding the standard to the sample, the sample is added to the standard.

Determine the electrode slope before analysis as per the instruction manual Place electrodes in 100 mL of the standard and add the ISA, stirring throughout. Set the instrument mode to read millivolts. When the reading is stable, record the reading E_1.

Add a small but accurately measured volume of sample to the standard. The volume of sample to be spiked should not exceed 15% of the volume of the standard taken. Select the concentration of the standard as follows:

$$\text{Conc. of the standard} = \frac{\text{Volume of sample to be spiked} \times \text{Estimated sample conc.}}{\text{Volume of standard taken}}$$

Read the millivolt value E_2 after adding the sample. Record the value when the reading is stable. Determine the concentration C_x of the analyte in the sample from the following equation:

$$C_x = \left[(1+\rho)\times 10^{\Delta E/S} - \rho\right]\times C_y$$

where C_y = the concentration of the standard
ρ = the ratio of the volume of the standard taken to the volume of sample spiked
ΔE = the millivolt difference $E_2 - E_1$
S = electrode slope

Table 1.9.1 Application of Ion-Selective Electrodes in Environmental Analysis

Analyte	Electrode type	Ionic strength adjuster
Ammonia	Gas sensing	pH adjusting ISA
Bromide	Solid state	5 *M* $NaNO_3$
Carbon dioxide	Gas sensing	CO_2 buffer [a]
Chloride	Solid state	5 *M* $NaNO_3$
	Liquid membrane	—
	Combination	—
Chlorine residual	Solid state	Iodide reagent
Cyanide	Solid state	5 *M* $NaNO_3$
Fluoride	Combination	TISAB [b]
	Solid state	—
Fluoroborate	Liquid membrane	2 *M* $(NH_4)_2SO_4$
Iodide	Solid state	5 *M* $NaNO_3$
Nitrate	Liquid membrane	2 *M* $(NH_4)_2SO_4$
Nitrogen oxide/nitrite	Gas sensing	Acid buffer
Oxygen, dissolved	Liquid membrane	—
Perchlorate	Liquid membrane	2 *M* $(NH_4)_2SO_4$
Sulfide	Solid state	SAOB [c]
Surfactant	Liquid membrane	—
Thiocyanate	Solid state	5 *M* $NaNO_3$

[a] Citrate buffer (sodium citrate/citric acid).
[b] Total ionic strength adjustment buffer. It contains a buffer along with 1,2-cyclohexylenedinitrilotetraacetic acid and citrate or tartrate.
[c] Sulfide antioxidant buffer solution contains 0.2 *M* ethylenediamine-tetraacetic acid disodium salt (Disodium EDTA), 2 *M* NaOH solution, and ascorbic acid.

1.10 APPLICATION OF HIGH PERFORMANCE LIQUID CHROMATOGRAPHY IN ENVIRONMENTAL ANALYSIS

High performance liquid chromatography (HPLC) is a common analytical technique used to determine a wide range of organic compounds. Its application has been widespread in industries such as dyes, paints, and pharmaceuticals. More than two thirds of all organic compounds can be analyzed using HPLC methods. Its application in environmental analyses, however, has been relatively recent. Only a limited number of U.S. EPA methods are based on HPLC techniques.

The basic components of an HPLC system are (1) a pump with a constant flow control; (2) a high-pressure injection valve; (3) a chromatographic column; (4) a detector; and (5) a strip-chart recorder or a data system for measuring peak areas and retention times. Calibration standards are prepared at various concentrations and the retention times and peak areas of the analytes are compared against the standard solutions of analytes for their identifications and quantitations.

The analytes are separated by adsorption to a polar or nonpolar support surface or by partition into a stationary liquid phase. Silica is the most common polar adsorbent. HPLC basically involves the separation of compounds by partition on a stationary liquid phase, bonded to a support. The support, such as silica, is derivatized with a functional group that is covalently attached to the surface and is more stable than any coated phase. Such bonded phases can be used with most solvents and buffers. A mobile liquid phase transports the sample into the column where individual compounds are selectively retained on the stationary liquid phase and thus separated.

In normal phase liquid chromatography, a bonded polar surface, such as cyano-, diol-, or amino group bound to silica is employed, while the mobile phase is nonpolar. Such technique is commonly used to separate steroids, aflatoxins, saccharides, and thalidomide. On the other hand, reversed phase liquid chromatography, a highly versatile technique, commonly applied in environmental analysis, uses a nonpolar surface and a polar mobile phase. Such nonpolar surfaces include octadecyl-, octyl-, methyl-, and diphenyl groups bound to silica or polymers and packed, usually as 5 μm particles. The pore sizes of these packings and the total surface area available determine the access of compounds in a certain range of molecular weights and their distribution or adsorption onto the surface.

As a result, they are some of the factors that govern the selectivity. Many resin-based polymeric supports have been used as alternatives to silica-based reversed phase columns for the analysis of basic compounds at high pH (Supelco, 1995; Altech, 1995). While the silica-based bonded phase packings are unstable above pH 7, new resin-based C-18 reversed phase columns are operational at a wide pH range of 2 to 13. Water-acetonitrile and water-methanol mixtures are commonly used as polar mobile phase in reversed phase liquid chromatography. A guard column is often used along with the primary column to protect the system and extend the life of the primary column. Various reverse phase columns under different trade names are now commercially available. Cation exchange resins have been employed for many analyses.

Postcolumn derivatization is usually performed to determine the compounds separated on the column. Such derivatization reaction can produce derivatives of the analytes that can be determined at a lower detection level at a greater sensitivity and free from interference.

Analytes are often derivatized to enhance the sensitivity of UV, fluorescence, conductivity, or electrochemical detector. Such derivatizing reagents include *p*-bromophenacyl reagent for UV detection, and *o*-phthalaldehyde and dansyl chloride for fluorescence enhancement. In a postcolumn reaction, the effluent from the column is mixed with a reagent to form a derivative before it enters the detector. Absorbance detectors such as UV rays, or a photodiode array detector (PDAD), or a fluorescence detector are used in HPLC determination. A conductivity detector is used to detect inorganic substances. When using PDAD, interference can occur from many organic compounds at the shorter wavelengths (200 to 230 nm), at which they absorb light.

Fluorescence detectors are commonly used in many HPLC analyses. Compounds absorb light from a monochromatic light source and release it as fluorescence emission. The detector equipped with filters responds only to the fluorescent energy. Presence of trace impurities that fluoresce can cause interference in the test.

The performance of the HPLC system should be evaluated from the column efficiency and the symmetry of the peak. The column efficiency is determined as the number of theoretical plates, N, which should be greater than 5000. It is calculated as follows:

$$N = 5.54\left(\frac{RT}{W_{1/2}}\right)$$

where N = column efficiency (number of theoretical plates)
RT = retention time of components (in seconds)
$W_{1/2}$ = width of component peak at half height (seconds)

The greater the value of N, the greater is the column efficiency. The peak asymmetry factor should be between 0.8 and 1.8, and is determined as shown in Figure 1.10.1.

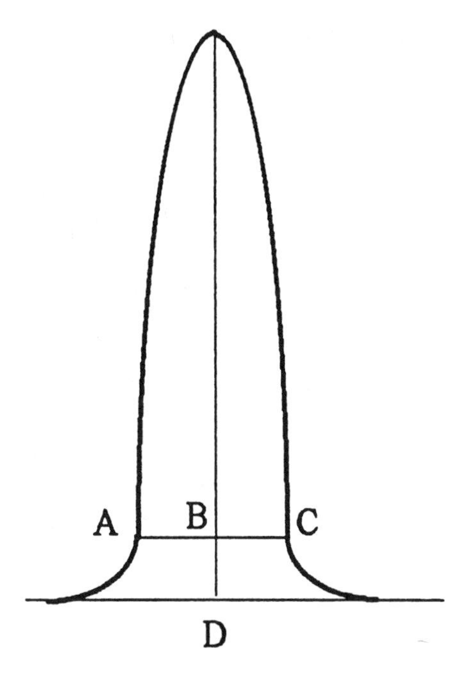

Figure 1.10.1 Asymmetry factor calculation. Peak height DE = 80 mm; 10% peak height = DB = 8 mm; peak width at 10% peak height = AC = 24 mm.

The asymmetry factor, therefore is

$$= \frac{AB}{BC} = \frac{13 \text{ mm}}{11 \text{ mm}} = 1.2$$

The assymetry factor should be between 0.8 and 1.8.

The use of HPLC technique for analyzing various types of organic pollutants or its potential application is highlighted in Table 1.10.1.

Table 1.10.1 Determination of Organic Analytes by High Performance Liquid Chromatography

Analytes	Example(s)	Method
Aldehydes and ketones	Acrolein, methyl ethyl ketone	Sample extract buffered to pH 3 and derivatized with 2,4-dinitrophenylhydrazine; derivative extracted and determined by UV at 360 nm
Polynuclear aromatic hydrocarbons	Anthracene, benzo(*a*)pyrene	Separated on a C-18 reverse phase column and detected by UV or fluorescence detector
Carbamate pesticides	Aldicarb, carbaryl	Separated on a C-18 reverse phase column, postcolumn hydrolysis with NaOH and derivatization with *o*-phthalaldehyde and 2-mercaptoethanol and fluorescence detection

Table 1.10.1 Determination of Organic Analytes by High Performance Liquid Chromatography (Continued)

Analytes	Example(s)	Method
Triazine pesticides	Atrazine, simazine	Separated on C-18 reverse phase column and detected by UV at 254 nm; mobile phase; acetonitrile: 0.01 *M* KH_2PO_4 (65:35)
Glyphosphate herbicides	*N*-(phosphono-methyl)glycine	Postcolumn oxidation with CaOCl to glycine and derivatized with *o*-phthalaldehyde; fluorescence detection at 340 nm
Chlorophenoxy acid herbicides (also other chlorinated acids and chlorophenols)	2,4-D, silvex (dichlorobenzoic acid, picloram)	Sample acidified with H_3PO_4, analytes separated on an HPLC cartridge containing C-18 silica and measured by photodiode array UV detector
Nitroaromatics and nitramine explosives	TNT, tetryl, RDX	Separated on C-18 or CN reverse phase column and UV detection at 254 nm
Tetrazine explosive	Tetrazine	Separated on ion-pairing reverse phase column and UV detection at 280 nm
Phenols	Phenol, *o*-cresol, pentachlorophenol	Phenol and cresols converted to phenolates with NaOH; reverse phase HPLC, using an UV (at 274 nm), fluorescence, or electrochemical detector
Nitriles	Propionitrile, acrylonitrile	Separated on a C-18 reverse phase column and determined by UV detector
Amides	Acrylamide	Separated on a C-18 reverse phase column and UV detection
Phthalate esters	Diethyl phthalate, di-*n*-octyl phthalate	Separated on a C-18, or C-8, reverse phase high temperature bonded silica column and UV detection at 254 nm; also determined by gel permeation chromatography
Organic bases	Pyridine	Separated on a base-deactivated C-18 or C-8 reverse phase column and detected by UV at 254 nm
Azo dyes	Metanil yellow, Congo red	Separated on a C-18 reverse phase column and detected by an MS or UV detector
Chlorinated pesticides	Aldrin, endrin	Separated on a C-18, C-8 or CN- high-temperature bonded silica; mobile phase isooctane—ethyl acetate (97:3) and detected by UV at 254 nm
Benzidines	Benzidine	Separated on a C-18 reverse phase column and detected by an electrochemical detector
Barbiturates	Barbital, phenobarbital	Separated on a C-18 high carbon loaded silica column; mobile phase: methanol-water; detected by UV at 254 nm
Alkaloids	Cocaine, morphine	Separated on a C-18 high carbon loaded (20%) silica column; mobile phase: water (buffered with 0.02 *M* KH_2PO_4, pH 3)—acetonitrile (75:25); detected by UV at 254 nm

1.11 ION CHROMATOGRAPHY

Ion chromatography is a single instrumental technique widely used for the sequential determination of many common anions in environmental matrices. Anions, such as, NO_3^-, NO_2^-, PO_4^{3-}, SO_4^{2-}, SO_3^{2-}, F^-, Cl^-, Br^-, I^-, oxyhalides, and many carboxylate ions may be determined rapidly in a single sequential step. The main advantages of the ion chromatography technique are:

1. Several anions can be determined in a single analysis.
2. It distinguishes halides (Br^-, Cl^-, etc.) and anions in different oxidation states (e.g., SO_4^{2-} and SO_3^{2-}, or NO_3^- and NO_2^-); such anions often interfere with each other in the wet analysis.
3. It is simple and rapid.

The method involves chromatographic separation of water soluble analytes and detection of separated ions by a conductivity detector. It can also be used to analyze oxyhalides, such as perchlorate (ClO_4^-) or hypochlorite (ClO^-); weak organic acids, metal ions, and alkyl amine. The analytes that can be determined by ion chromatography are listed in Table 1.11.1.

The aqueous sample is injected into a stream of carbonate/bicarbonate eluant. The eluant is pumped through an ion exchanger (a resin-packed column). Sample ions have different affinities for the resin; therefore, they move at different rates through the column, thus resulting in their separation. A strongly basic low capacity anion exchanger is used for the analysis of anions in the sample. The eluant and the separated anions then flow through a suppressor column which is a strong acidic cation exchanger that suppresses the conductivity of the eluant to enhance the detection of the sample ions. The separated ions are now converted into their highly conductive acid form (i.e., NO_3^- converted to HNO_3) and measured by their conductivities. They are identified by the retention times and quantified by comparing their peak areas/peak heights against calibration standards. Conductivity detection offers a highly sensitive and specific detection mode for both inorganic and organic ions.

Sensitivity and detection levels of analyte ions can be dramatically improved by chemical suppression and auto-suppression techniques (Dionex, 1995). In

Table 1.11.1 Application of Ion Chromatography in Environmental Analysis

Analytes	Formula/examples
Common inorganic anions	F^-, Cl^-, Br^-, I^-, NO_3^-, NO_2^-, PO_4^{3-}, SO_4^{2-}, SO_3^{2-}, CO_3^{2-}, PO_3^{3-}, HPO_4^{2-}, CNO^-
Oxyhalides	ClO_4^-, ClO_3^-, ClO_2^-, ClO^-, BrO_4^-, BrO_3^-, IO_4^-, IO_3^-
Pyrophosphate, polyphosphates, and metaphosphates	$P_2O_7^{4-}$, $P_3O_{10}^{5-}$, $P_4O_{13}^{6-}$, $P_2O_6^{2-}$
Thiosulfate and thiocyanate	$S_2O_3^{2-}$, SCN^-
Miscellaneous inorganic anions	CrO_4^{2-}, BO_3^{3-}, AsO_4^{3-}, SeO_4^{2-}, SeO_3^{2-}, MnO_4^{2-}, WO_4^{2-}, etc.
Metal ions (alkali and alkaline earth metals)	Li^+, Na^+, K^+, Mg^{2+}, Ca^{2+}
Common transition metal ions[a]	Fe^{2+}, Zn^{2+}, Cu^{2+}, Mn^{2+}, Ni^{2+}
Ammonium ion	NH_4^+
Chromium hexavalent	Cr^{6+} [b]
Cyanide	CN^- [c]
Organic anions of carboxylic acid type	Acetate, formate, oxalate, maleate, phthalate, tartrate
Sulfonated organic acids	Benzenesulfonate, *p*-toluenesulfonate
Chlorophenoxy acid herbicides	2.4-D, silvex, 2,4,5-T
Amines	Ethylamine, trimethyl amine, isobutylamine, morpholine, cyclohexylamine

[a] Postcolumn reaction with UV detection at 530 nm; eluant: pyridine-2,6-dicarboxylic acid.
[b] Postcolumn colorimetric reaction and photometric detection, also by chemically suppressed ion chromatography with conductivity detection.
[c] Acid digestion followed by chemically suppressed ion chromatography.

chemical suppression, no electrical potential is applied across by the electrodes, while in auto suppression, such a potential is applied to enable the transport of ions. This technique maximizes the sensitivity by increasing the analyte conductivity and reducing background conductivity and sample counter ions. This process is depicted in Figure 1.11.1 in two typical ion chromatograms. Furthermore, the capacity of the ion exchanger column increases, i.e., the column can take a higher load of analyte ions without any peak distortion.

Eluants other than carbonate/bicarbonate have also found wide application in many environmental and nonenvironmental analyses. Some common eluants are listed in Table 1.11.2. Sodium hydroxide solution has now become an eluant of choice for many ion chromatography analyses using suppressed conductivity detection. The schematic representation of the method is outlined in Figure 1.11.2.

The Na ions from the eluant are removed at the suppressor column, and also the hydronium ions (from water electrolyzed) combine with the OH ions from the eluant to form water that is less conductive than the eluant NaOH. Thus, the net effects are, (1) conductivity of background NaOH eluant decreases, and (2) the conductivity of the analyte anions increases, because they are now in their acid form.

Both of these effects significantly enhance the signal-to-noise ratio, thus lowering the detection levels and increasing the sensitivity.

Ion chromatography is the most convenient analytical approach for the determination of most inorganic anions, including oxyhalides in potable and wastewaters and solid waste extracts. The method has been approved by

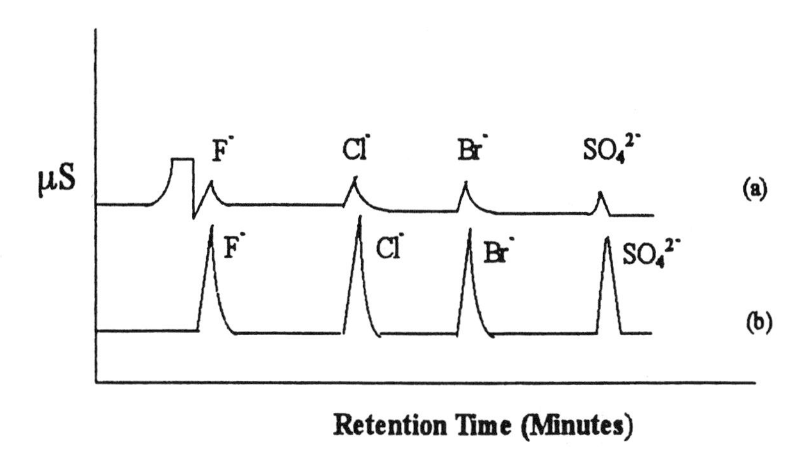

Figure 1.11.1 Ion chromatograms, (a) without suppression; (b) with suppression.

U.S. EPA and the Standard Methods (U.S. EPA, 1990; AWWA, 1993). Several columns are now commercially available, such as Ion Pac (Dionex 1995) or equivalent, that provide isocratic separation of common inorganic anions in less than 10 min using a carbonate/bicarbonate eluant and their detection at low ppm level employing a suppressed conductivity detector. In addition, by using gradient conditions and varying ionic strength by over two orders of magnitude, many mono-, di-, and trivalent ions can be separated in a single run. Such gradient ion chromatography with chemical suppression can prevent coelution and provide separation of a wide range of inorganic ions and organic ions. Many sulfonated organic acids, which are often found in leachates from hazardous waste dump sites, can be determined by ion chromatography by varying both solvent and ionic strength in a gradient elution profile.

Ion chromatography has also been used in air pollution studies to measure anions and cations in an air impinger solution and air particulate extracts.

Ion exchange columns are composed of microporous polymeric resins (or stationery phases) containing styrene or ethyl vinylbenzene crosslinked with divinylbenzene. This is a porous and chemically stable core. A layer of ion exchange coating is bonded onto this resin core, producing a reactive surface.

Table 1.11.2 Some Common Eluants Used in Ion Chromatography[a]

Eluants	General applications
Sodium carbonate/ sodium bicarbonate	Common inorganic anions: F^-, Cl^-, NO_3^-, PO_4^{3-}, and SO_4^{2-}
Sodium hydroxide	Common inorganic anions: F^-, Cl^-, SO_4^{2-}; organic anions: acetate, citrate, and fumarate
Sodium tetraborate/ boric acid	Common inorganic anions: F^-, NO_3^-, SO_4^{2-}; oxyhalides and carboxylates, such as acetate and formate
Methanesulfonic acid	Alkali and alkaline earth metal ions: Na^+, K^+, Ca^{2+}; ammonium
Methanesulfonic acid/ acetonitrile gradient	Alkali and alkaline earth metal ions; Li^+, Na^+, K^+, Mg^{2+}, ammonium ion; alkyl amines: methyl amine, triethyl amine, and morpholine

[a] Detector: Suppressed conductivity.

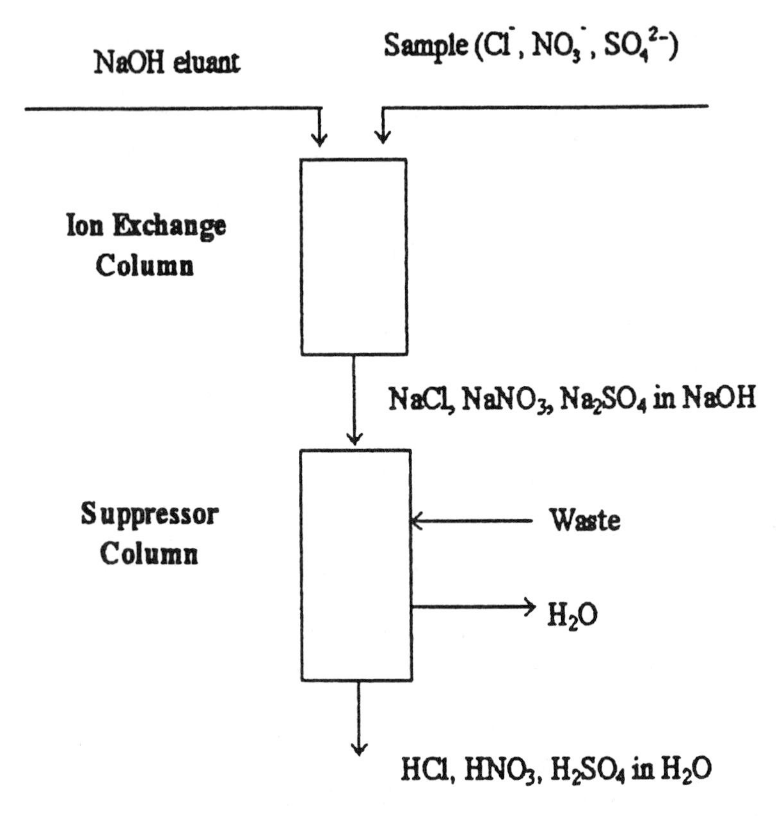

Figure 1.11.2 Schematic overview of sodium hydroxide elution method.

For an anion exchange column, the coating particles are functionalized with alkyl quaternary amine or alkanol quaternary ammonium groups. The particle size of the coating substances, the pellicular type structures, and the degree of crosslinking in the microporous resin all determine the degree of resolution of anions on the column (Dionex, 1995). Such columns are commercially available from many suppliers. A guard column is used to protect the separator column from organics and particulates.

1.12 AIR ANALYSIS

Analysis of air, whether indoor or outdoor air, primarily consists of three steps: (1) sampling of the air, (2) chemical analysis of the sampled air, and (3) quality assurance to ensure the precision and accuracy of measurement. Before performing the sampling, it is important to know the physicochemical properties of target contaminants. In addition, sampling for outdoor air requires a proper sampling plan which includes selection of proper sites for sampling based on weather condition, topography, and other factors.

SAMPLING PLAN

Concentrations of contaminants in the atmosphere may vary significantly from time to time due to seasonal climatic variation, atmospheric turbulence, and velocity and direction of wind. The most important meteorological factors are (1) wind conditions and the gustiness of wind, (2) the humidity and precipitation, (3) the temperature, which varies with latitude and altitude, (4) barometric pressure (varying with the height above the ground), and (5) solar radiation and the hours of sunshine, which vary with the season.

The concentrations of contaminants in the atmosphere and their dispersion in the air can be influenced by topography, i.e., whether the location of the polluting source is in a valley, plateau or mountain, or near a lake or the sea. For example, mountains can act as barriers to air flow while valleys will cause persistence in wind direction.

In addition to the climatic conditions and topography, planning for sampling the ambient atmosphere should include a study on any noticeable effect caused by the pollutants on health and vegetation, and evaluation of any background information available. Based on these studies, which are part of sampling planning, decision should be made on the sites selection, the number of samples to be collected, and the time and frequency of sampling. Such planning can help to compare the ambient air quality at different locations, geometric mean level of measured concentrations, and any trend or pattern observed.

For indoor air, a sampling plan should be prepared before performing the sampling. This is, however, somewhat different from the outdoor air sampling planning discussed above. The sampling scheme for indoor air should be based on the following information:

1. Toxic symptoms observed in the occupants of the building and their case histories
2. Location of the building
3. Wind direction
4. Ventilation in the room
5. Humidity
6. Background information, if available, on earlier studies (such as any previous air analysis report and the contaminants found)
7. The age of the house

The indoor air in very old houses may contain organics at trace levels due to what is known as the "sick building syndrome." The presence of low molecular weight organic compounds is attributed to degradation of wood releasing hydrocarbons, aldehydes, and ketones.

Thus, for both the atmospheric air and the indoor air, it is essential to prepare a proper sampling plan prior to sampling of the air.

Air Sampling

Air sampling involves collecting a measured volume of air for chemical analysis. This is done in two ways: (1) collecting the air from the site in a container, or (2) trapping the pollutants by passing a measured volume of air through a filter, or an adsorbent or absorbing solution. The principles of these methods are outlined below.

Direct Collection of Air

Air can be directly collected either in a Tedlar bag or a canister (e.g., SUMMA passivated canister) or a glass bulb. Canisters are best suited for air sampling. These are metal containers that can be chilled down to very low temperatures in a liquid nitrogen, helium, or argon bath. Air to be sampled can thus be condensed into the cryogenically cooled trap. Air may be collected either under high pressure using an additional pump or at a subatmospheric pressure by initially evacuating the canister. The canister can then be connected directly to the GC line. It may then be warmed or the sample air may be sucked in by a condensation mechanism into the GC column for analysis. Tedlar or other air collection bags or glass bulbs are equipped with inlet and outlet valves for passage of air. The container to collect air is connected to a sampling pump (or a vacuum pump) and the air is then pulled in for collection. No cooling or condensation is required in such samplings.

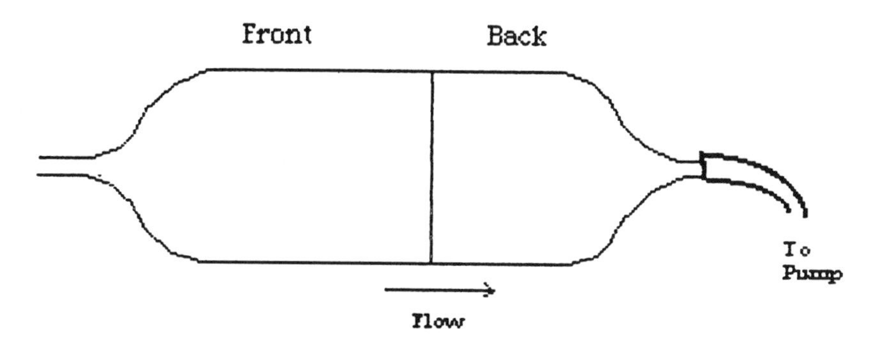

Figure 1.12.1 Adsorbent tube used for sampling of air for organic analysis.

Use of Adsorbent Tubes

Adsorbent tubes are commonly used for sampling air for organic analysis. Activated charcoal is one of the most widely used substances for trapping organics. Tenax and many other porous polymeric materials may be used for the same purpose. Silica gel is a common adsorbent for adsorption of polar compounds such as alcohol. Adsorbent tubes for air sampling are commercially available, or may be prepared. Air is passed through the adsorbent in the direction as shown in Figure 1.12.1 by connecting the tube to the sampling pump. The tube has two sections—front and back—separated by a thin pad. Both sections are packed with adsorbent, the larger quantity being filled in the front portion. If the amount of the analyte in the air is too high (more than what the front portion can hold to), or if the flow rate is too high, some of the analytes can pass through the front portion, but would be trapped over the adsorbent in the back. Thus, the total amount of analyte is the sum of the amounts that are trapped onto the front and the back sections, respectively.

For accurate analysis, we must know the total volume of air passed through the adsorbent tube, the mass of the analyte trapped, and the desorption efficiency of the solvent. Before sampling, the pump must be calibrated using a bubblemeter, a rotameter, or a gasometer to determine the flow rate. Using the flow rate and the time sampled, the total volume of air sampled can be determined.

The mass of the analyte can be measured from the chromatographic analysis. The analyte can be desorbed from the adsorbent surface either by a solvent (solvent desorption) or by heating (thermal desorption). In the former case, selection of a solvent should be based on its desorption efficiency (DE), miscibility with the analyte, and chromatographic response. For example, carbon disulfide is a solvent of choice for many organics because of its high desorption efficiency and poor chromatographic response to GC-flame ionization detector (GC-FID). Similarly, isooctane and hexane are good solvents for chlorinated pesticides, because of their high DE and poor response to GC-electron capture detector (GC-ECD). On the other hand, a chlorinated solvent such as methylene chloride or chloroform should not be used if the analysis is to be performed by

GC-ECD. This is because these solvents exhibit excellent response toward ECD, producing large peaks that can bury the small analyte peaks.

The DE of an appropriate solvent should be more than 85% and should be accounted for in the final calculation. If the DE for any specific compound is not known or not found in the literature, it can be determined by the following simple experiment:

> Inject 5 μL of the analyte onto the adsorbent taken in a vial. Allow it to stand for 20 to 30 min. Add 10 mL of solvent and allow it to stand for an hour. Inject the eluant solution into the GC and record the area response, A_1. Now inject 5 μL of the analyte into another 10 mL portion of the solvent. Inject this solutin onto the GC and record the area response, A_2. Dilute the ample and standard if the peak-overloading occurs. The DE is calculated as:

$$DE = \frac{A_1}{A_2} \times 100$$

Air Sampling for Particulates and Inorganic Substances

Dusts, silica, metal powder, carbon particles, and particulate matter are collected over membrane filters of appropriate pore size. Filter cassettes are used for this process. A membrane filter having the same diameter as the cassette is placed inside the cassette, one end of which is connected to the sampling pump. A measured volume of air is then sampled. While air can pass through the pores of the filter, the suspended particles get deposited on the filter.

Many water soluble analytes can be sampled by bubbling the air through water in an impinger. Acid vapors, alkali vapors, or their dusts can be collected in water and their aqueous solutions analyzed by wet methods. Often, water is made basic or acidic to trap acidic or basic analytes, respectively. Other solvents can be used in the impinger, depending on the solubility of the analytes and the vapor pressure of the solvents. Certain organics in the air can also be trapped in impingers if one uses the proper solvents.

Flow Rate

An important step in air sampling is to determine the total volume of air to be sampled, to select the proper flow rate, and to measure the flow rate accurately. If the concentration of the analyte in the air is expected to be low, a large volume of air needs to be sampled. A high flow of air can reduce the contact time of analyte molecules over the adsorbent surface, thus reducing their adsorption. Similarly, dissolution of the analyte in the impinger solvent is reduced under high flow rate. On the other hand, a low flow rate would require a longer sampling time. Thus, the flow rate and the total volume of air to be sampled depends on the nature of the analyte and its expected concentration. If the analysis is required for unknown contaminants of organic nature and if adsorbent tubes are used for sampling, a volume of 100 to 200 L air at a flow rate of 0.5 to 2 L/min should be suitable. If suspended particles are to be collected over membrane filters, a larger volume or air should be collected for accurate weight determination. In such cases, air-flow rate may be increased to reduce the sampling time.

As mentioned earlier, the personal sampling pump should be calibrated before and after sampling to determine accurately the flow rate and the volume of air sampled. A soap bubblemeter or a gasometer should be used for such calibration.

CHEMICAL ANALYSIS

Most target organic compounds in air can be determined by gas chromatography using a flame ionization detector. The analytes are desorbed into carbon disulfide, an aliquot of which is then injected into the GC-FID. Halogenated organics can be determined by GC using an electron capture detector or any halogen specific detector. Likewise, phosphorus and sulfur compounds can be determined by flame photometric detector. For GC analysis of nitrogen compounds, a nitrogen-phosphorus detector in nitrogen mode is most effective. Many aromatics and olefins can be determined by photoionization detector. Low molecular weight gaseous hydrocarbons such as methane, ethane, and ethylene in air can be analyzed by GC using a thermal conductivity detector. A detailed discussion on GC analysis including the use of columns and detectors is presented in Chapter 3.

Unknown organic compounds can be identified and quantitatively measured by gas chromatography/mass spectrometry (GC/MS). Analytes can be thermally desorbed from the adsorbent surface and analyzed by GC/MS. Alternatively, the bulk air from the site collected in a Tedlar or a canister can be injected or introduced by heating or suction into the system for separation and analysis.

High performance liquid chromatography, infrared spectroscopy, UV and visible spectrophotometry, and polarography are some of the other major analytical techniques used to determine many diverse classes of compounds.

Dusts, silica, and other suspended particles in the air are measured by gravimetry. The filter in the cassette should be weighted before and after sampling for accurate mass determination of deposited particles. For metal analysis, the metal dusts deposited on the filter must be acid-digested and analyzed by atomic absorption spectrophotometry or inductively coupled plasma spectrometry. The inorganic anions in the impinger solution in their varying oxidation states can simultaneously be measured by ion chromatography. Colorimetry and ion-specific electrode methods can also be applied to determine individual ions. Among other analytical techniques, titrimetry methods involving acid-base titrations and redox tritrations are often used to determine acidic, basic, and other inorganic analytes in the aqueous solutions in the impinger.

Analysis of specific groups of substances and individual compounds are discussed in Sections 2 and 3 of this text, respectively.

CALCULATION

The concentration of pollutants in air is generally expressed as mg/m^3 air. This unit is used to express the concentrations for all kinds of analytes including organic compounds, metal ions, inorganic anions, and particulate matter. Another unit is parts per million (ppm), which is often used to express concentrations of a specific compound. Conversion of ppm to mg/m^3 is as follows:

$$1 \text{ ppm} = \left(\frac{\text{Molecular weight of the compound}}{22.4}\right) \text{mg/m}^3 \text{ at STP}$$

where STP is the standard temperature and pressure which is 273°K and 1 atm (760 torr), respectively. The molar volume at STP is 22.4 L.

The above relationship is derived as follows:

$$\text{Molecular density} = \frac{\text{One mole mass}}{\text{Molar volume of analyte in air at STP}}$$

$$= \frac{1 \text{ g MW}}{22.4 \text{ L}} \times \frac{1000 \text{ mg}}{1 \text{ g}} \times \frac{1000 \text{ L}}{\text{m}^3}$$

$$= \frac{\text{MW}}{22.4} \times 10^6 \times \frac{\text{mg}}{\text{m}^3}$$

or $$\frac{1}{10^6} \times \frac{\text{mass}}{\text{volume}} = \frac{\text{MW}}{22.4} \text{mg/m}^3 \quad \left(\text{substituting } \frac{\text{mass}}{\text{volume}} \text{ for density}\right)$$

where MW = molecular weight.

That is,

$$1 \text{ part per } 10^6 \text{ part (mass/volume)} = \frac{\text{MW}}{22.4} \text{mg/m}^3$$

or,

$$1 \text{ ppm (weight/volume) at STP} = \frac{\text{MW}}{22.4} \text{mg/m}^3$$

or

$$1 \text{ mg/m}^3 = \frac{22.4}{\text{MW}} \text{ppm at STP}$$

For example, 1 ppm cyclohexane at STP = $\frac{84}{22.4}$ or 3.75 mg/m^3, or 2.93 mg/m^3 benzene at STP = $2.93 \text{ mg/m}^3 \times \frac{22.4}{78} \times \frac{\text{ppm}}{1 \text{ mg/m}^3}$ = 0.85 ppm.

Conversion of ppm to mg/m^3 or vice versa at any other temperature and pressure can be performed using one of the following two ways as shown below.

For example, air sampling of a compound is performed at an altitude where the temperature is 7°C and the pressure is 725 torr. Conversion of ppm (W/V) to mg/m^3 can be calculated using either the combined gas equation or by the ideal gas equation as illustrated below.

From Boyles', Charles' and GayLussacs' laws:

$$\frac{P_1V_1}{T_1} = \frac{P_2V_2}{T_2}$$

where P_1 = initial pressure
V_1 = initial volume
T_1 = initial temperature
P_2 = final pressure
V_2 = final volume
T_2 = final temperature

Temperature is always expressed in Kelvin scale.

At STP P_1 = 1 atm, V_1 = 22.4 L (molar volume) and T_1 = 0°C or 273 K. In the given problem,

$$P_2 = 725 \text{ torr} = 725 \text{ torr} \times \frac{1 \text{ atm}}{760 \text{ torr}} = 0.954 \text{ atm}$$

$$V_2 = ?$$

$$T_2 = 7°\text{C or } (7 + 273) = 280 \text{ K}$$

Therefore,

$$V_2 = \frac{1 \text{ atm} \times 22.4 \text{ L} \times 280 \text{ K}}{0.954 \text{ atm} \times 273 \text{ K}} = 24.08 \text{ L}$$

Thus, the molar volume at 7°C and 725 torr = 24.08 L.
Therefore, 1 ppm analyte in air at 7°C and 725 torr =

$$\frac{\text{MW of analyte}}{24.08} \text{ mg/m}^3$$

Thus, we need to determine the molar volume first at the given temperature and pressure conditions. This can also be determined from the ideal gas equation as follows:

$$PV = nRT$$

where P = pressure
V = volume
T = temperature
n = number of moles of analyte
R = ideal gas constant which is 0.082 L atm/mol K

In the above problem, $P = 0.954$ atm, $T= 280$ K and for 1 mol, $n = 1$. Therefore,

$$V = \frac{nRT}{P}$$

$$= 1 \text{ mol} \times \frac{0.082 \text{ L atm}}{\text{mol K}} \times \frac{280 \text{ K}}{0.954 \text{ atm}}$$

$$= 24.07 \text{ L}$$

Example 1

The concentration of sulfur trioxide in air at 15°C and 740 torr was found to be 1.57 mg/m^3. Express this concentration in ppm (weight/volume).

At first we have to determine the molar volume at 15°C and 740 torr either using combined gas laws or ideal gas equation. From the ideal gas equation:

$$V = \frac{nRT}{P}$$

$$= 1 \text{ mol} \times \frac{0.082 \text{ L atm}}{\text{mol K}} \times \frac{288 \text{ K}}{0.974 \text{ atm}}$$

$$= 24.25 \text{ L}$$

where $P = 740 \text{ torr} \times \dfrac{1 \text{ atm}}{760 \text{ torr}}$ or 0.974 atm
$T = 273 + 15 = 288$ K

Therefore, ppm SO_3 at the pressure and temperature conditions of measurement

$$= \frac{\text{conc. measured as mg/m}^3 \times \text{molar volume}}{\text{MW}}$$

$$= \left(\frac{1.57 \times 24.25}{80}\right) \text{ppm} = 0.476 \text{ ppm}$$

$$\left[\text{i.e., } \frac{1.57 \text{ mg}}{\text{m}^3} \times \frac{24.25 \text{ L}}{80 \text{ g}} \times \frac{1 \text{ g}}{1000 \text{ mg}} \times \frac{1 \text{ m}^3}{1000 \text{ L}} \times \frac{1{,}000{,}000 \text{ part}}{1 \text{ part}}\right]$$

Example 2

A total volume of 300 L air was sampled for toluene using activated charcoal adsorbent. The analyte was desorbed with 2 mL carbon disulfide. An aliquot of the eluant was analyzed by GC. The concentration of toluene in the eluant was found to be 13.7 mg/L. Determine its concentration in the air as mg toluene/m^3 air.

Volume of air sampled = 300 L
Volume of CS_2 extract = 2 mL = 0.002 L
Toluene found in extract = 13.7 mg/L

Mass of toluene in the extract = $\dfrac{13.7\text{ mg toluene}}{1\text{ L extract soln.}} \times 0.002$ L extract soln. = 0.0274 mg

which came from 300 L air. Therefore, the concentration of toluene in the air

$$= \frac{0.0274\text{ mg}}{300\text{ L air}} \times \frac{1000\text{ L air}}{\text{m}^3\text{ air}}$$

$$= 0.091\text{ mg/m}^3\text{ air}$$

1.13 APPLICATION OF IMMUNOASSAY TECHNIQUES IN ENVIRONMENTAL ANALYSIS

Enzyme immunoassay kits are now available for qualitative field testing or for laboratory screening and semiquantitative analysis of pesticides, herbicides, polychlorinated biphenyls (PCBs), mononuclear and polynuclear aromatic hydrocarbons, pentachlorophenol, nitroorganics, and many other compounds in aqueous and soil samples. Certain analytes may be quantitatively determined as well, with a degree of accuracy comparable to gas chromatography or high performance liquid chromatography determination. The method is rapid and inexpensive.

The analytical procedure consists of three steps: (1) sample extraction, (2) assay, and (3) color formation for visual or spectrophotometric determination of the analyte in the sample. If a semiquantitative or quantitative analysis is desired, the concentration of the analyte can be determined from a calibration curve prepared by plotting absorbance or optical density against concentrations of a series of standards. The principle of immunoassay testing is described below.

Polyclonal antibodies of different types are known to show affinity for specific compounds. Thus, antibodies that can bind to a specific substance to be analyzed are immobilized to the walls of the test tubes, plates, or microwells. Such test tubes and plates are commercially available and supplied in the test kit. A measured amount (between 10 and 50 μL) of sample or sample extract is added to one such test tube containing an assay diluent (a phosphate buffer). An equal volume of analyte-enzyme conjugate (commercially available and supplied in the kit) is then added to the test tube. The enzyme conjugate is a solution containing the same analytes covalently bound to an enzyme. The solution mixture is incubated or allowed to stand for a specific amount of time. During this period, the enzyme conjugate competes with the analyte molecules for a limited number of antibody binding sites in the test tube.

After the incubation, the unbound molecules are washed away, leaving behind the bound ones, anchored onto the antibody sites. Now a clear solution of a color-forming reagent (chromogenic substrate) such as, 3,3′,5,5′-tetramethylbenzidine is added to the mixture. The enzyme conjugate, bound to antibody sites on the wall, reacts with the chromogenic substrate forming a blue color. The enzyme catalyzes the transformation of the substrate into a product which reacts with the chromogen, causing a blue color. Enzyme acts as a catalyst. Each enzyme molecule can rapidly catalyze the conversion of thousands of substrate molecules into product molecules that react with chromogen. In other words, the darker the color, the greater the amount of bound enzyme conjugate or, conversely, the lesser the amount of the analyte in the sample. Thus, the color intensity is directly proportional to enzyme conjugate concentration and therefore, inversely proportional to the analyte concentration in the sample.

For qualitative screening, a visual comparison of color with standards can be made. For semiquantitative determination, however, a spectrophotometer should be used to read absorbance to plot a calibration standard curve. The color should be read as soon as possible because it becomes unstable after 30 min. The required period for incubation varies from substance to substance but can range from 5 to 10 min to 1 or 2 hours. In certain analysis, the immunochemical reaction may require quenching after a specific amount of time. The reaction can be stopped by adding an acid, such as 1 *N* HCl, which turns the blue to yellow. The intensity of yellow too can be measured to determine the analyte concentration in the sample.

An alternative assay procedure involves the use of antibody-coupled magnetic particles. Antibodies specific to the analyte of interest are covalently bound to paramagnetic particles rather than test tubes. The particles are suspended in buffered saline with preservative and stabilizers. Enzyme conjugate and the antibody-coupled magnetic particles suspension are then combined with the sample extract. The mixture is incubated. A magnetic field is applied to separate the magnetic particles which are then washed and treated with a color reagent. Blue develops on incubation. The reaction is then stopped by adding HCl, at which time the color turns yellow and is read by a spectrophotometer.

Other assay procedures that can be employed involve the use of a coated particulate system or double antibody separation technique. These assay methods, including those discussed above, have both advantages and disadvantages when compared with each other. For example, in double antibody methods, an additional incubation period is required for the second antibody, which increases the analysis time. Particulate systems require centrifugation to separate the bound particles. The more common coated-tube method is simple and rapid and does not involve centrifugation. However, its disadvantage is that the surface of the tube or plate limits the number of antibody for the reaction. Also, there may be loss of antibody because of absorption into the solid surface. An advantage of the magnetic particle method is that the antibody is covalently bound to the solid particles of uniform size (1 μm) giving even distribution throughout the reaction mixture. The method, however, requires the use of a magnetic field for separation.

Detection limits in the range of low parts per billion (ppb) can be achieved by immunoassay testing for certain parameters in aqueous samples. For soil samples, detection limits of <10 ppm can be achieved for many contaminants.

The presence of substances having the same functional groups can interfere in the test, giving a false positive value. For example, 2,3,6-trichlorophenol may interfere in the test for pentachlorophenol. Such interference effect may, however, be reduced by using an antibody that is most selective for the target analyte.

EXTRACTION

Aqueous samples can be tested directly without any sample extraction steps. An aliquot of the sample (10 to 250 μL) should be suitable for the test. For high analyte concentrations, dilution of samples may be necessary. Follow the assaying procedure given in the kit using appropriate amounts of sample (or extract), enzyme conjugate, and coloring reagent.

Soil samples are extracted with methanol. A weighed amount of soil (e.g., 10 g) is shaken vigorously for 1 min with methanol and filtered. The filtrate is diluted (1:1000) with a diluent buffer solution. The extract is then immunoassayed for analysis.

CALCULATION

For soil, sediment, or solid waste samples, the concentration is calculated as follows:

$$\text{Conc. analyte, } \mu\text{g/g} =$$

$$\text{Assay result } (\mu\text{g/L or ppb}) \times \frac{\text{Volume of extract (L)}}{\text{Mass of sample (g)}} \times \text{Dilution factor}$$

Example

Soil weighing 10 g was extracted with methanol to a volume of 50 mL. A 100-μL aliquot of the extract was diluted to 25 mL. The assay result was 3.5 ppb. Determine the concentration of the analyte in the sample.

Assay result = 3.5 μg/L

Volume of extract = 0.05 L

Sample weight = 10 g

$$\text{Dilution factor} = \frac{25\text{ mL}}{100\ \mu\text{L}} = \frac{25000\ \mu\text{L}}{100\ \mu\text{L}} = 250$$

$$\text{Conc. analyte, } \mu\text{g/g (ppm)} = \frac{3.5\ \mu\text{g}}{\text{L}} \times \frac{0.05\text{ L}}{10\text{ g}} \times 250 = 4.375\ \mu\text{g/g (ppm)}$$

Analyte concentrations in aqueous samples can be directly read from the standard curve. If the sample is diluted, the results must be multiplied by the dilution factor.

2 SPECIFIC CLASSES OF SUBSTANCES AND AGGREGATE PROPERTIES

2.1 ALDEHYDES AND KETONES

Aldehydes and ketones constitute an important class of organic compounds containing carbonyl group. These compounds have the following structural features:

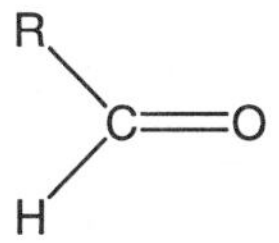

$$R(H)C{=}O \qquad R'(R)C{=}O$$

(Aldehyde) 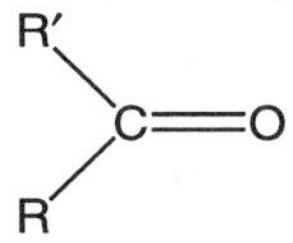 (Ketone)

where R and R′ are alkyl, alkenyl, alkynyl, or aryl group. In aldehyde, one hydrogen atom is attached to the carbonyl group, while in ketone no hydrogen is bound to the carbonyl group. Formaldehyde ($H_2C{=}O$) is the only aldehyde where the $>CO$ group is bound to the two H atoms. Thus, aldehydes and ketones are compounds of closely related classes exhibiting many similar chemical properties.

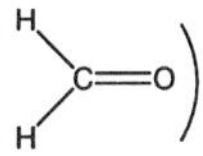

Low molecular weight aldehydes and ketones are used in the manufacture of resins, dyes, esters, and other organic chemicals. Many ketones including acetone and methyl ethyl ketone are industrial solvents. Some of the compounds that are used commercially are presented in Table 2.1.1. Many aldehydes and ketones are emitted into the atmosphere from chemical and combustion processes. Photochemical degradation of many organic substances also generates lower aldehydes and ketones.

Aldehydes and ketones of low molecular weights are volatile compounds and can be extracted out from the sample matrix by purge and trap or thermal desorption technique and determined by GC-FID or GC/MS. Compounds of low carbon numbers ($R < 4$) are soluble in water causing poor purging efficiency, and, therefore, producing elevated detection levels. Heating of the purging vessel may, therefore, become necessary to enhance the purging efficiency of such com-

Table 2.1 Some Commercially Used Aldehydes and Ketones

CAS no.	Compounds	Synonyms
	Aldehydes	
[75-07-0]	Acetaldehyde	Ethanal
[107-02-8]	Acrolein	2-Propenal
[100-52-7]	Benzaldehyde	Benzoic aldehyde
[123-72-8]	Butyraldehyde	Butanal
[104-55-2]	Cinnamaldehyde	3-Phenyl-2-propenal
[4170-30-3]	Crotonaldehyde	2-Butenal
[112-31-2]	Decyl aldehyde	Decanal
[5779-94-2]	2,5-Dimethylbenzaldehyde	—
[50-00-0]	Formaldehyde	Methanal
[107-22-2]	Glyoxal	Ethanedial
[111-30-8]	Glutaraldehyde	1,5-Pentanedial
[111-71-7]	Heptaldehyde	Heptanal
[66-25-1]	Hexaldehyde	Hexanal
[78-84-2]	Isobutyraldehyde	Isobutanal
[590-86-3]	Isovaleraldehyde	3-Methylbutanal
[78-85-3]	Methacrolein	Isobutenal
[18829-56-6]	Nonenaldehyde	2-Nonenal
[124-13-0]	Octaldehyde	Octanal
[123-38-6]	Propionaldehyde	Propanal
[620-23-5]	*m*-Tolualdehyde	3-Methylbenzaldehyde
[529-20-4]	*o*-Tolualdehyde	2-Methylbenzaldehyde
[104-87-0]	*p*-Tolualdehyde	4-Methylbenzaldehyde
[110-62-3]	Valeraldehyde	Pentanal
	Ketones	
[67-64-1]	Acetone	2-Propanone
[123-54-6]	Acetyl acetone	2,4-Pentanedione
[108-94-1]	Cyclohexanone	Cyclohexylketone
[120-92-3]	Cyclopentanone	Ketocyclopentane
[96-22-0]	Diethyl ketone	3-Pentanone
[108-83-8]	Diisobutyl ketone	Isovalerone
[123-19-3]	Dipropyl ketone	4-Heptanone
[541-85-5]	Ethyl amyl ketone	5-Methyl-3-heptanone
[106-35-4]	Ethyl butyl ketone	3-Heptanone
[591-78-6]	Methyl butyl ketone	2-Hexanone
[78-93-3]	Methyl ethyl ketone	2-Butanone
[110-12-3]	Methyl isoamyl ketone	5-Methyl-2-hexanone
[141-79-7]	Methyl isobutenyl ketone	Mesityl oxide
[108-10-1]	Methyl isobutyl ketone	4-Methyl-2-pentanone
[563-80-4]	Methyl isopropyl ketone	3-Methyl-2-butanone
[107-87-9]	Methyl propyl ketone	2-Pentanone

pounds. Alternatively, aqueous samples may be directly injected onto the GC column for separation and detection by an FID. This also gives a higher detection level. Two alternative methods may be used to analyze these compounds more accurately with a lower detection level. These methods are based on derivatization with 2,4-dinitrophenylhydrazine (DNPH) and determination of the derivatives by

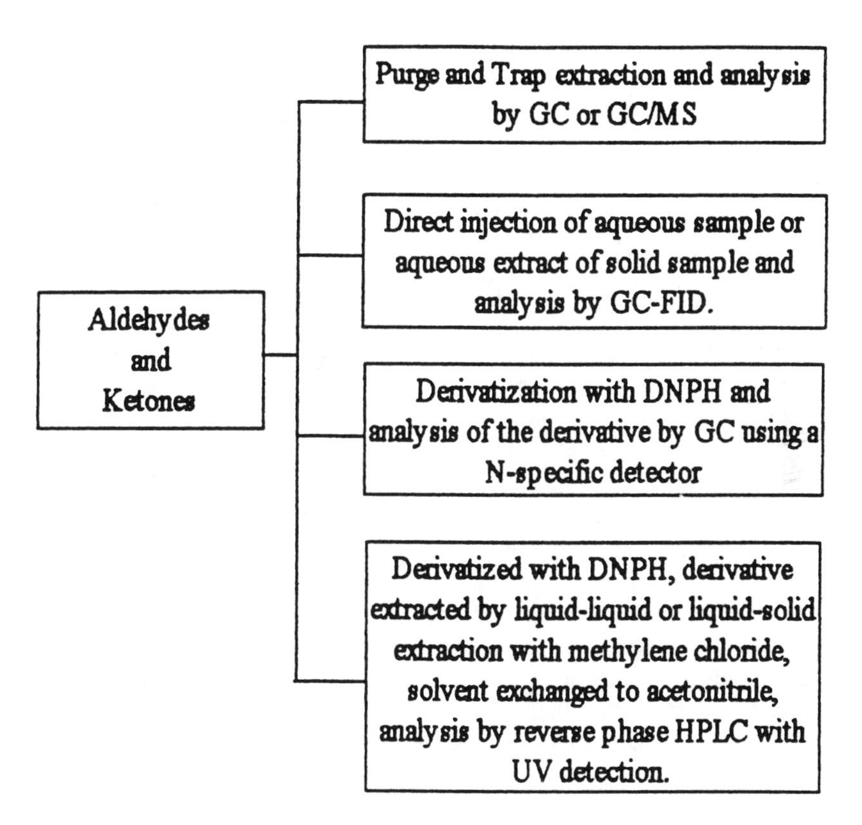

Figure 2.1.1 Schematic representation of the basic techniques for determining aldehydes and ketones.

(1) GC using a nitrogen-specific detector (NPD) (i.e., NPD in N-specific mode), and (2) reverse phase HPLC using an UV detector.

The basic techniques for the determination of aldehydes and ketones are summarized in Figure 2.1.1.

DERIVATIZATION WITH 2,4-DINITROPHENYLHYDRAZINE

The carbonyl group of an aldehyde or a ketone reacts with 2,4-dinitrophenylhydrazine (DNPH) to form a stable derivative according to the following equation:

$$\underset{\text{(Ketone)}}{RR'C{=}O} + \underset{\text{(DNPH)}}{H_2N{-}NH{-}C_6H_3(NO_2)_2} \xrightarrow{H^+}$$

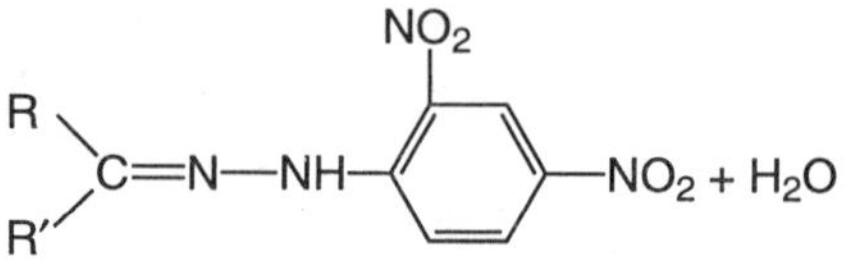

$$R(R')C{=}N{-}NH{-}C_6H_3(NO_2)_2 + H_2O$$

(DNPH-Derivative)

DNPH is often susceptible to formaldehyde or acetone contamination. It should, therefore, be crystallized with acetonitrile to remove any impurities. Repeated crystallization may further be performed to achieve the desired level of purity for DNPH. A 100-mL aliquot of aqueous sample is buffered with a citrate buffer and pH adjusted to 3 ± 0.1 with HCl or NaOH. The acidified sample is then treated with DNPH reagent and heated at 40°C for an hour under gentle swirling. The DNPH derivatives of aldehydes and ketones formed according to the above reaction are extracted with methylene chloride using liquid-liquid extraction. The extract is then solvent exchanged to acetonitrile for HPLC determination.

The DNPH derivatives may also be extracted by solid-phase extraction (liquid-solid extraction). The sorbent cartridge is conditioned with 10 mL of dilute 1 *M* citrate buffer solution (1:25 dilution) and 10 mL of saturated NaCl solution. The cartridge is then loaded with the extract and the derivatives eluted with acetonitrile.

If formaldehyde is the only compound to be analyzed, an acetate buffer may be used instead of a citrate buffer, and the sample may be pH adjusted to 5 ± 0.1.

Soils, sediments, and solid wastes may be extracted with the extraction fluid (sample to fluid mass ratio 1:20) in a rotating bottle at approximately 30 rpm for 18 hours. The extraction fluid is made from citrate buffer, pH adjusted to 3 ± 0.1 for DNPH derivatization. The derivatives are extracted with methylene chloride.

If the compounds are derivatized with DNPH and the derivatives are determined by GC or HPLC, the equivalent concentrations of the parent carbonyl compounds in the sample may be stoichiometrically determined as shown below in the following example.

Example 1

A 200-mL aliquot of an aqueous sample was pH buffered and derivatized with DNPH. The DNPH derivatives were extracted with methylene chloride and the extract was solvent exchanged to 50 mL acetonitrile. Analysis by HPLC-UV showed the presence of methyl ethyl ketone (MEK) derivative which was quantitated as 2.7 mg/L in the extract, using the calibration standards prepared from the solid derivative. Determine the concentration of the MEK in the sample.

The derivatization reaction is as follows:

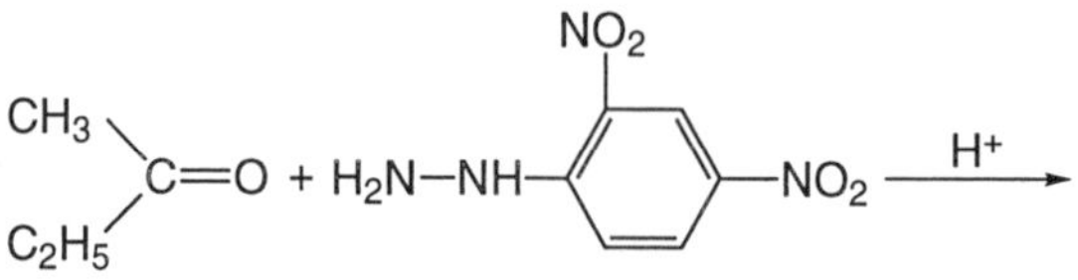

$$CH_3(C_2H_5)C{=}O + H_2N{-}NH{-}C_6H_3(NO_2)_2 \xrightarrow{H^+}$$

(MEK)
MW=72

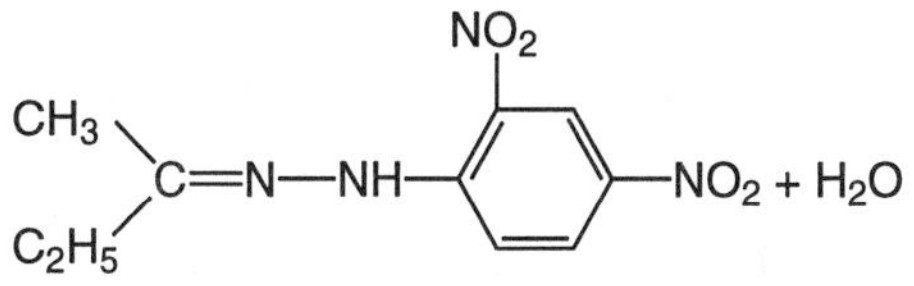

$$CH_3(C_2H_5)C{=}N{-}NH{-}C_6H_3(NO_2)_2 + H_2O$$

(MEK Derivative)
MW=252

The concentration of MEK derivative in 50 mL acetonitrile extract = 2.7 mg/L which is equivalent to:

$$\frac{2.7 \text{ mg MEK derivative}}{1 \text{ L extract}} \times \frac{72 \text{ g}}{252 \text{ g MEK derivative}}$$

or 0.77 mg MEK/L extract solution.

Therefore, the concentration of MEK in the sample

$$= \frac{0.77 \text{ mg MEK}}{1000 \text{ mL extract}} \times \frac{50 \text{ mL extract}}{200 \text{ mL sample}} \times \frac{1000 \text{ mL sample}}{1 \text{ L sample}}$$

$$= 0.19 \text{ mg MEK/L}$$

Chromatographic Columns and Conditions

HPLC column:	C-18 reverse phase column, such as Zorbax ODS or equivalent.
Mobile phase:	Acetonitrile/water (70:30) to 100% acetonitrile in 15 min.
Flow rate:	~ 1 mL/min.
Detector:	Ultraviolet, set at 360 nm.
Sample injection:	25 μL.
GC Column:	Aldehydes and ketones are polar compounds that can be separated on a polar or an intermediate polar column. Polyethylene glycol (PEG)-type phase, such as Carbowax 20 M, Supelcowax 10, VOCOL, DBWax, or equivalent are suitable for the purpose. Compounds may also be separated according to their boiling points on a nonpolar column. A 60 m long, 0.53 mm ID and 1 μm film or other appropriate dimension methyl silicone capillary columns, such as SPB-1, DB-1, or DB-5.

Sample Collection and Holding Time

Aldehydes and ketones are readily oxidized and must, therefore, be derivatized and extracted within 48 h of sample collection. The derivatized sample extracts should be analyzed within 3 days after preparation. Holding times exceeding more than 3 days have shown significant losses of compounds having seven or more C atoms. Samples should be collected without headspace and should be refrigerated and analyzed as soon as possible. Low molecular weight compounds may be stored for more than 3 days, but the analysis must be done within 7 days of sample collection.

AIR ANALYSIS

Aldehydes and ketones in ambient air may be determined by the U.S. EPA Method TO-5 or TO-11 or different NIOSH methods as listed in Appendices J and K at the end of this book. The EPA methods are based on derivatization of these carbonyl compounds in air with DNPH reagent and measuring the stable DNPH derivatives (2,4-dinitrophenylhydrazones) by reverse phase HPLC with UV detection. Air is drawn through a midget impinger containing 2 *N* HCl solution of 0.05% DNPH reagent and isooctane. The derivatives formed are soluble in iso-octane and dissolve into the organic layer. The aqueous layer is separated and extracted with hexane/methylene chloride (70:30). The extract is combined with iso-octane and the three solvent mixture is evaporated to dryness under a stream of nitrogen. The residue is dissolved in methanol and analyzed by HPLC.

The flow rate of air should be between 100 and 1000 mL/min and not greater than 1000 mL/min. We recommend a sample volume of 40 L and an air flow of 400 mL/min. A volume of 10 mL each of DNPH reagent and isooctane should be used in the impinger. However, a larger volume, 20 to 25 mL isooctane would be suitable for a higher flow rate, ensuring the solvent does not evaporate out during the process of sampling.

An alternative method suitable for formaldehyde and other aldehydes and ketones as well involves the use of DNPH coated silica gel adsorbent (U.S. EPA Method TO5) instead of the reagent solution taken in an impinger. A known volume of air is drawn through a prepacked silica gel cartridge coated with acidified DNPH. Florisil (magnesium silicate) 60 to 80 mesh may be used instead of silica gel. The flow rate may be between 500 and 1000 mL/min. Sample volume and sampling time should be selected based on the expected range of concentration of the pollutants in the air or their time-weighted average values. The DNPH derivatives of the analytes are eluted from the adsorbent with acetonitrile for HPLC determination.

The column and conditions for HPLC analysis are outlined earlier in this chapter. If isomeric aldehydes or ketones coelute, alternate HPLC columns or mobile phase composition should be used. The DNPH reagent solutions should be always freshly prepared. Calibration standards should be made in methanol from solid DNPH derivatives. Intermediate standards should be prepared according to the anticipated levels for each component.

NIOSH Methods 1300 and 1301 (NIOSH 1984) describe the determination of common ketones in air. The compounds evaluated in the study include acetone, methyl propyl ketone, methyl isobutyl ketone, methyl *n*-butyl ketone, diisobutyl ketone, cyclohexanone, ethyl butyl ketone, methyl amyl ketone, ethyl amyl ketone, mesityl oxide, and camphor [76-22-2]. The analysis involves adsorption of compounds over coconut shell charcoal, desorption with CS_2, and determination by GC-FID. The addition of a small amount of methanol (1%) into CS_2 gives better desorption for some of the ketones. A volume of 10 L air at a flow rate of 100 mL/min may be sampled for the analysis. Many other ketones (not listed above) may be analyzed in a similar manner. Ambersorb XE-347 is an effective adsorbent for methyl ethyl ketone (Method 2500, NIOSH 1984).

Simple aldehydes, such as formaldehyde or acrolein may be analyzed by derivatizing into a suitable derivative for GC-FID or -NPD determination (Methods 2502, 2501, NIOSH 1984). The derivatizing agents for these compounds are 2-(benzylamino)ethanol and 2-(hydroxymethyl)piperidine, respectively, coated on a support. Formaldehyde may also be determined using colorimetry and polarography (Methods 3500 and 3501, NIOSH 1984). See Part 3 of this book.

2.2 ALKALINITY

Alkalinity of water is a measure of its acid-neutralizing ability. The titrable bases that contribute to the total alkalinity of a sample are generally the hydroxides, carbonates, and bicarbonates. However, other bases such as phosphates, borates, and silicates can also contribute to the total alkalinity. The alkalinity value depends on the pH end point designated in the titration. The two end points commonly fixed in the determination of alkalinity are the pH 8.3 and pH 4.5 (or between 4.3 and 4.9, depending on the test conditions). When the alkalinity is determined to pH 8.3, it is termed *phenolphthalein alkalinity.* In such alkalinity titration, phenolphthalein or metacresol purple may be used as an indicator. On the other hand, the *total alkalinity* is measured by titrating the sample to pH 4.5 using bromocresol green as the indicator. Alkalinity may also be determined by potentiometric titration to the preselected pH. An acid standard solution, usually 0.02 N H_2SO_4 or HCl, is used in all titrations.

The procedure for potentiometric titration is presented in Chapter 1.6. In this titration, a standard acid titrant is added to a measured volume of sample aliquot in small increments of 0.5 mL or less, that would cause a change in pH of 0.2 unit or less per increment. The solution is stirred after each addition and the pH is recorded when a constant reading is obtained. A titration curve is constructed, plotting pH vs. cumulative volume titrant added. The volume of titrant required to produce the specific pH is read from the titration curve.

CALCULATION

$$\text{Alkalinity, mgCaCO}_3\text{/L} = \frac{V \times N \times 50,000}{\text{mL sample}}$$

where V is mL standard acid titrant used and N is normality of the standard acid.

Since the equivalent weight of $CaCO_3$ is 50, the milligram equivalent is 50,000. The result is, therefore, multiplied by the factor of 50,000 to express the alkalinity as mg $CaCO_3$/L.

Alkalinity of a sample is essentially caused by the presence of three principal ions: hydroxide (OH^-), carbonate (CO_3^{2-}), and bicarbonate (HCO_3^-). The contribution of each of these ions may be determined from the following relationship:

Hydroxide alkalinity. When $P = T$, all alkalinity is due to OH^-, i.e., hydroxide alkalinity is T. Also, when $P > 1/2\ T$ (but $<T$), hydroxide alkalinity is $2P - T$.

Carbonate alkalinity. When P is $1/2\ T$ or less (but greater than 0), carbonate alkalinity is equal to $2P$; on the other hand, when $P > 1/2T$, the alkalinity due to $CO_3^{2-} = 2(T - P)$.

Bicarbonate alkalinity. When $P = 0$, there is no OH^- or CO_3^{2-} in the solution. The total alkalinity measured is all due to HCO_3^-. Thus, bicarbonate alkalinity = T.

When P is not 0, but $< 1/2\ T$, alkalinity due to $HCO_3^- = (T - 2P)$.

The alkalinity relationship may be directly read from a set of chart diagrams called nomographs, which are available from The American Water Works Association. Nomographical computations require that the following be known: the temperature, the pH, total dissolved solids, and the total alkalinity of the samples.

2.3 BROMIDE

Bromide (Br^-) is the anion of the halogen bromine, containing an extra electron. It is produced from the dissociation of bromide salts in water. It may occur in ground and surface waters as a result of industrial discharges or seawater intrusion.

Bromide in water may be analyzed by one of the following three methods:

1. Phenol red colorimetric method
2. Titrimetric method
3. Ion chromatography

While the first method is used for low level detection of bromide in the range 0.1 to 1 mg/L, the concentration range for the titrimetric method is between 2 and 20 mg/L. The samples may be diluted appropriately to determine bromide concentrations at higher range. Ion chromatography is used to analyze many anions including bromide and is discussed in Section 1.8.

PHENOL RED COLORIMETRIC METHOD

Bromide ion reacts with a dilute solution of sodium *p*-toluenesulfonchloramide (chloramine-T) and is oxidized to bromine which readily reacts with phenol red at pH 4.5 to 4.7. The bromination reaction with phenol red produces a color that ranges from red to violet, depending on the concentration of bromide ion. An acetate buffer solution is used to maintain the pH between 4.5 and 4.7. The presence of high concentration of chloride ions in the sample may seriously interfere in the test. In such cases, the addition of chloride to the pH buffer solution or the dilution of the sample may reduce such interference effect. Remove free chlorine in the sample by adding $Na_2S_2O_3$ solution. In addition, the presence of oxidizing and reducing agents in the sample may interfere in the test.

Procedure

A 50-mL sample aliquot is treated with 2 mL buffer solution followed by 2 mL phenol red indicator and 0.5 mL chloramine-T solution. Shake well after each addition. Allow it to stand for 20 min. Add immediately 0.5 mL of sodium thiosulfate solution. Determine the concentration of bromide as mg/L from a calibration curve made by plotting bromide standards against absorbance at 590 nm. The color developed may also be compared visually in Nessler tubes against bromide standards.

If the bromide concentration is too high, dilute the sample to a final concentration range of 0.1 to 1.0 mg Br^-/L. Multiply the results with dilution factor if any.

Reagents

- pH buffer soln. (pH 4.6 to 4.7): dissolve 6.8 g sodium acetate trihydrate, $CH_3COONa \cdot 3H_2O$, and 9.0 g NaCl in distilled water. Add 3.0 mL glacial acetic acid and make the volume to 100 mL.
- Chloramine-T soln.: 500 mg in 100 mL distilled water.
- Phenol red indicator soln.: 20 mg in 100 mL distilled water.
- Sodium thiosulfate soln.: a 2 *M* solution is suitable. Dissolve 31.6 g anhydrous $Na_2S_2O_3$ or 49.6 g $Na_2S_2O_3 \cdot 5\ H_2O$ in distilled water and dilute to 100 mL.
- Bromide standards: anhydrous KBr, 0.7446 g is dissolved in distilled water and the final volume is made to 1 L. 1.0 mL ≡ 500 μg bromide.
- Dilute 10 mL of stock soln. to 1 L to prepare secondary standard (1.0 mL ≡ 5 μg Br^-) from which working standards are made. Dilute 0, 2, 5, 10, 15, and 20 mL of the above secondary standard to 100 mL with distilled water to obtain working standard solutions of 0, 0.1, 0.25, 0.5, 0.75, and 1.0 mg Br^-/L.

TITRIMETRIC METHOD

This titrimetric procedure (EPA Method 320.1) is similar to that of iodide. The sample is divided into two aliquots: One aliquot is analyzed for bromide plus iodide while the other aliquot is analyzed for iodide only. The difference of these two gives the concentration of bromide in the sample.

The sample is stored at 4°C and analyzed as soon as possible. Before analysis the sample is pretreated with calcium oxide (CaO) to remove the interference effects of iron, manganese, and organic matter.

Determination of Bromide Plus Iodide

Bromide and iodide are oxidized to bromate (BrO_3^-) and iodate (IO_3^-) by treatment with calcium oxychloride [$Ca(OCl)_2$]. The reactions are as follows:

$$Br^- + 3\ Ca(OCl)_2 \Rightarrow BrO_3^- + 3\ CaCl_2$$

$$I^- + 3\ Ca(OCl)_2 \Rightarrow IO_3^- + 3\ CaCl_2$$

The bromate and iodate so formed react with I^- (obtained from the dissociation of KI) in acid medium, liberating iodine, as shown below in the following equation.

$$BrO_3^- + 6\,I^- + 6\,H^+ \Rightarrow 3\,I_2 + Br^- + 3\,H_2O$$

$$IO_3^- + 5\,I^- + 6\,H^+ \Rightarrow 3\,I_2 + 3\,H_2O$$

The liberated iodine is titrated against standard sodium thiosulfate or phenylarsine oxide using starch indicator (see Section 1.6).

Procedure

Add 5 to 10 g CaO to 500 mL of sample. Shake vigorously and filter. Discard the first 100 mL of the filtrate.

In 100-mL aliquot of the above filtrate dissolve 5 g of NaCl with stirring. Add HCl solution dropwise to adjust the pH to approximately 7 or slightly less. A pH meter should be used to measure the pH. Transfer the sample to a 250-mL iodine flask or a wide mouth conical flask.

Add 20 mL of 3.5% (m/v) calcium hypochlorite solution followed by 1 mL of 1:4 HCl solution. Add 0.2 to 0.3 g powdered $CaCO_3$. Heat the solution to boiling. Add 4 mL sodium formate solution (50% w/w) slowly. Heat to boiling for 10 min, occasionally washing down the sides with distilled water.

Allow the solution to cool. If precipitate forms due to iron, add potassium fluoride dihydrate. Add a few drops of sodium molybdate ($Na_2MoO_4 \cdot 2H_2O$) solution (1%).

Dissolve 1 g of KI in the above solution. Add 10 mL of 1:4 H_2SO_4. Allow the solution to stand in the dark for 5 min. Titrate this solution with standardized sodium thiosulfate or phenylarsine oxide solution, using starch indicator. Add the indicator when the solution becomes pale straw following the addition of $Na_2S_2O_3$ or PAO. At the end point, the blue color disappears. Disregard any reappearance of blue color. Run a distilled water blank.

Determination of Iodide

The titrimetric analysis of iodide is discussed in detail in Section 2.16.

Calculation

$$Br^-(mg/L) = (\text{conc. of bromide } + \text{ iodide}) - (\text{conc. of iodide})$$

As shown in the equation above, the overall stoichiometry of the bromide reaction is

$$1 \text{ mol } Br^- \equiv 1 \text{ mol } BrO_3^- \equiv 3 \text{ mol } I_2 \equiv 6 \text{ mol } S_2O_3^{2-} \text{ or PAO}$$

$$\text{Or, } 1 \text{ mol titrant } S_2O_3^{2-} \text{ or PAO}$$

$$\equiv \frac{1}{6} \text{ mol } Br^- \equiv \frac{79.9\,g}{6} \text{ or } 13.32 \text{ g or } 13.320 \text{ mg } Br^-$$

Similarly, for idodide reaction:

$$1 \text{ mol titrant } S_2O_3^{2-} \text{ or PAO}$$

$$\equiv \frac{1}{6} \text{ mol I}^- \equiv \frac{126.9\text{ g}}{6} \text{ or } 21.15 \text{ g}$$

or 21,150 mg I^- (See also Chapter 2.16.)

$$\text{Conc. of Br}^-(\text{mg/L}) = (\text{conc. of bromide + iodide}) - (\text{conc. of iodide})$$

$$\equiv 13{,}220\left(\frac{A \times B}{C}\right) - 21{,}150\left(\frac{D \times E}{F}\right)$$

where A = Volume of $Na_2S_2O_3$ or PAO in milliliters required to titrate the sample for bromide plus iodide (blank corrected),

B = Normality of $Na_2S_2O_3$ or PAO needed to titrate the sample for bromide plus iodide,

C = Sample volume in milliliters used in the titration to determine bromide plus iodide in the sample,

D = Volume of $Na_2S_2O_3$ or PAO in milliliters required to titrate the sample for iodide (blank corrected),

E = Normality of the titrant $Na_2S_2O_3$ or PAO used to titrate the sample for iodide, and

F = Sample volume in milliliters used in the titration for iodide determination.

For blank corrections, subtract the milliliters of the titrant used in the blank anaylsis from the titrant readings *A* and *D*.

2.4 CHLORIDE

Chloride (Cl^-) is one of the most commonly occurring anions in the environment. It can be analyzed using several different methods, some of which are listed below:

1. Mercuric nitrate titrimetric method
2. Argentometric titrimetric method
3. Automated ferricyanide colorimetric method
4. Gravimetric determination
5. Ion-selective electrode method
6. Ion chromatography

Methods 1 and 3 are EPA approved (Methods 325.3 and 325.1-2, respectively) for chloride determination in wastewater. For multiple ion determination, ion chromatography technique should be followed (see Section 1.8).

MERCURIC NITRATE TITRIMETRIC METHOD

Chloride reacts with mercuric nitrate to form soluble mercuric chloride. The reaction is shown below for calcium chloride ($CaCl_2$) as a typical example.

$$CaCl_2 + Hg(NO_3)_2 \Rightarrow HgCl_2 + Ca(NO_3)_2$$

The analysis may be performed by titrimetry using a suitable indicator. Diphenyl carbazone is a choice indicator that forms a purple complex with excess mercuric ions in the pH range of 2.3 to 2.8. Therefore, the pH control is essential in this analysis. Xylene cyanol FF is added to diphenyl carbazone to enhance the sharpness of the end point in the titration. Nitric acid is used to acidify the indicator to the required low pH range.

Other halide ions, especially bromide and iodide, are interference in this analysis. Acidify alkaline samples before analysis. Fe^{3+}, CrO_4^{2-}, and SO_3^{2-} at concentrations above 10 mg/L are often used to interfere with the analysis.

Procedure

To 100-mL sample, add 1 mL acidified indicator reagent. The solution should become greenish-blue. If the color is pure blue, the pH is above 3.8, while a light green color indicates a pH less than 2. The pH adjustment is very crucial in this analysis. Titrate the sample with 0.0141 N $Hg(NO_3)_2$. Prior to the end point, the solution turns blue; at the end point, the color becomes purple. Perform a blank titration using distilled water.

$$\text{mg Cl}^- / \text{L} = \frac{(A - B) \times N \times 35,450}{\text{Volume of sample (mL)}}$$

where A = Volume of the titrant, $Hg(NO_3)_2$ required to titrant the sample,
B = Volume of the titrant required for the blank titration, and
N = Normality of $Hg(NO_3)_2$.

The number 35,450 is used in the above equation because the equivalent weight of chlorine is 35.45 and, therefore, the milligram equivalent weight is 35,450.

Dilute the sample if it is highly concentrated in chloride (>100 mg/L). Alternately, use a smaller sample portion or $Hg(NO_3)_2$ titrant of greater strength.

Reagents

- Mercuric nitrate titrant (0.0141 N): Dissolve 2.288 g $Hg(NO_3)_2$ or 2.425 g $Hg(NO_3)_2 \cdot H_2O$ in distilled water containing 0.25 mL conc. HNO_3. Dilute to 1 L. 1 mL of this titrant ≡ 500 μg Cl^-. (The exact normality of the above prepared solution may be determined by titrating this solution against NaCl standard using acidified diphenyl carbazone-xylene cyanol FF to purple end point.) Dissolve 824 mg NaCl (dried at 140°C) in 1 L distilled water to produce 0.0141 N NaCl. 1 mL of this solution ≡ 500 μg Cl^-. Use 100-mL NaCl solution in this titration against $Hg(NO_3)_2$ titrant. Run a blank using distilled water.

$$\text{Normality of } Hg(NO_3)_2 = \frac{1.41}{\text{mL titrant } Hg(NO_3)_2 \text{ (blank subtracted)}}$$

- Indicator solution. To 100 mL 95% ethanol or isopropyl alcohol, add 250 mg *s*-diphenyl carbazone, 4 mL conc. HNO_3, and 30 mg xylene cyanol FF. Store in a refrigerator in an amber bottle.

ARGENTOMETRIC TITRIMETRIC METHOD

Silver nitrate reacts with chloride in neutral or slightly alkaline solution, quantitatively precipitating silver chloride as shown below:

$$AgNO_3 + NaCl \Rightarrow AgCl + NaNO_3$$

Potassium chromate also reacts with $AgNO_3$ to form red silver chromate.

$$K_2CrO_4 + 2\ AgNO_3 \Rightarrow Ag_2CrO_4 + 2\ KNO_3$$

The above reaction, however, is less favorable than the former reaction. Therefore, K_2CrO_4 can indicate the end point of $AgNO_3$ chloride titration. At the end point, when no free chloride is left in the solution, addition of a drop of $AgNO_3$ titrant results in the formation of Ag_2CrO_4 producing a pink end point.

Several ions interfere in this analysis. These include bromide, iodide, cyanide, sulfide, sulfite, and thiosulfate. The latter three ions may be removed by treatment with H_2O_2. The sample should be diluted when iron and orthophosphate are present at concentrations above 10 mg/L.

Procedure

Adjust the pH of the sample in the range 7 to 10. To a 100-mL sample, add 1 mL 30% H_2O_2 and stir. Add 1 mL of K_2CrO_4 indicator. Titrate the sample with standard $AgNO_3$ (0.0141 *N*) titrant to pink end point. Run a blank using distilled water.

$$\text{mg Cl}^- / \text{L} = \frac{(A - B) \times N \times 35,450}{\text{Volume of sample (mL)}}$$

where A = Volume of $AgNO_3$ titrant (mL),
B = Volume of $AgNO_3$ titrant (mL) required for blank titration, and
N = Normality of the titrant, $AgNO_3$.

Reagents

- Standard $AgNO_3$ (0.0141 *N*): Dissolve 2.395 g $AgNO_3$ in 1 L distilled water. This titrant solution is standardized against standard NaCl solution using K_2CrO_4 indicator (See above; by substituting the known concentration for Cl^- in the above equation, the exact normality, *N*, of $AgNO_3$ can be determined. The concentration of chloride, mg Cl^-/L, in the prepared NaCl standard = mg Cl^-/L × 0.6066).
- Potassium chromate indicator solution: Dissolve 10 g K_2CrO_4 in approximately 20 to 30 mL distilled water. Add $AgNO_3$ solution until a red precipitate forms. Allow it to stand for a day. Filter and dilute the volume to 200 mL.

AUTOMATED FERRICYANIDE COLORIMETRIC METHOD

It is a rapid colorimetric determination in which about 15 to 30 samples may be analyzed per hour using an automated analytical equipment. Chloride at concentration range 1 to 250 mg/L may be analyzed using this method.

Chloride ion reacts with mercuric thiocyanate forming unionized mercuric chloride, liberating thiocyanate ion (SCN^-). The liberated thiocyanate ion reacts with Fe^{3+} to form a highly colored ferric thiocyanate. These reactions are shown below:

$$2\ Cl^- + Hg(SCN)_2 \Rightarrow HgCl_2 + 2\ SCN^-$$

$$3\ SCN^- + Fe^3 \Rightarrow Fe\ (SCN)_3$$

The amount of $Fe(SCN)_3$ formed is proportional to the original concentration of the chloride in the sample. Thus, the intensity of the color due to $Fe(SCN)_3$ formed is proportional to the chloride content in the sample.

Procedure

The apparatus used for the analysis is a continuous flow automated analytical instrument such as Technicon Autoanalyzer. Follow the manufacturer's instructions to set up the manifold and for general operation.

Chloride standards are prepared in the concentration range 1 to 250 mg/L from the stock standard solution (1.6482 g NaCl dried at 140°C dissolved in 1 L distilled water; 1 mL ≡ 1 mg Cl^-.) Prepare a standard calibration curve by plotting peak heights against the chloride concentrations of the standards from which the concentrations of Cl^- in the unknown sample is determined.

Reagents

- $Hg(SCN)_2$ soln: 4.17 g in 1 L methanol.
- Ferric nitrate soln: 202 g $Fe(NO_3)_3 . 9H_2O$ in 1 L distilled water containing 21 mL HNO_3. Mix, filter, and store the solution in an amber bottle.
- Color reagent: mix 150 mL $Hg(SCN)_2$ with 150 mL $Fe(NO_3)_3$ soln. prepared above. Dilute to 1 L. Add 0.5 mL Brij 35 (polyoxyethylene 23 lauryl ether).

GRAVIMETRIC DETERMINATION OF CHLORIDE

Gravimetric analysis for chloride in wastewaters does not give accurate results at concentrations below 100 mg/L. This method is rarely applied in routine environmental analysis because it is lengthy and rigorous.

This method is based on the fact that $AgNO_3$ reacts with chloride anions in the solution, thus precipitating out AgCl. Actually the latter is produced first as a colloid which then coagulates on heating. Such coagulation is favored by HNO_3 and a small excess of $AgNO_3$. Avoid adding an excess of $AgNO_3$ because it may coprecipitate with AgCl. Also avoid exposure of AgCl to sunlight which can cause photodecomposition.

Procedure

To a 100-mL sample acidified with HNO_3 and taken in a beaker, add slowly with stirring 0.2 *N* $AgNO_3$ solution until AgCl is observed to coagulate. Then add an additional 5 mL of $AgNO_3$ solution. Heat the sample to boiling for 5 min. Add 1 mL of $AgNO_3$ solution to confirm that precipitation is complete. If the supernatant solution turns white or additional precipitation is observed, keep on adding $AgNO_3$ slowly to the sample mixture till the precipitation is complete. Place the beaker in the dark and allow it to stand overnight or at least for 3 h.

Filter the precipitate using a clean, dry, and accurately weighed sintered glass or porcelain filtering crucible following the decantation of the supernatant liquids first through the filtering crucible. The AgCl should be quantitatively transferred from the beaker to the filtering crucible. Wash the precipitate 3 to 4 times with distilled water acidified with a few drops og HNO_3. The washing should be free of Ag^+ ion. If the final washing still contains Ag^+ ion, the precipitate should be subjected to more washings. The presence of Ag^+ in washing may be tested by adding a few drops of HCl to a small volume of washing in a test tube. The solution would turn turbid in the presence of Ag^+ ion.

Calculation

Dry the precipitate at 105°C for an hour. Place the crucible in a desiccator to cool. Weigh it with its contents. Repeat the steps of heating, cooling, and weighing until two consecutive weights are constant.

$$\text{mg/L Cl}^- = \frac{(B-A)}{V} \times \frac{1\text{ mol Cl}^-}{1\text{ mol AgCl}} \times 1000$$

$$= \frac{(B-A)}{V} \times 0.247 \times 1000$$

where A = Weight of empty dry filtering crucible,
B = Weight of the empty dry filtering crucible plus AgCl precipitate, and
V = mL of sample.

2.5 CYANATE

The formula of cyanate is CNO^-. It is a univalent anion formed by partial oxidation of cyanide (CN^-).

$$CN^- \xrightarrow{O} CNO^-$$

Under neutral or acidic conditions, it may further oxidize to CO_2 and N_2.

The analysis of cyanate is based on its total conversion to an ammonium salt. This is achieved by heating the acidified sample. The reaction is shown below:

$$2\ KCNO + H_2SO_4 + 4\ H_2O \Rightarrow (NH_4)_2SO_4 + 2\ KHCO_3$$

The concentration of ammonia (or ammonium) minus nitrogen before and after the acid hydrolysis is measured and the cyanate amount is calculated as equivalent to this difference.

Calculation

Thus, the amount of NH_3–N produced from mg CNO^-/L = $A - B$

where A = mg NH_3–N/L in the sample portion that was acidified and heated, and
B = mg NH_3–N/L in the original sample portion.

Therefore, the concentration of cyanate as mg CNO^-/L = $3.0 \times (A - B)$.

[In the above calculation for cyanate, the concentration of ammonia–nitrogen was multiplied by 3 because the formula weight of CNO^- is (12 + 14 + 16) or 42, which is three times 14 (the atomic weight of N, as ammonia–N).]

Procedure

Add 0.5 to 1 mL of 1:1 H_2SO_4 to a 100-mL portion of sample acidifying to pH 2 to 2.5. Heat it to boiling for 30 min. Cool to room temperature and bring up to the original volume by adding NH_3-free distilled water.

The analysis for NH_3–N after this may be performed by titrimetric, colorimetric, or selective-ion electrode method (see Chapter 2.14). Analyze for NH_3–N in an equal volume portion of the untreated original sample.

Treat the sample with $Na_2S_2O_3$ to destroy oxidizing substances (such as Cl_2 residual chlorine) that may react with cyanate. Add EDTA to complex metal ions which may be present in the sample and which may interfere by forming colored complexes with Nessler reagent.

Preserve the sample with caustic soda to pH >12 immediately after sampling to stabilize CNO^-.

2.6 CYANIDE, TOTAL

Cyanides are metal salts or complexes that contain the cyanide ion (CN^-). These cyanides could be subdivided into two categories: (1) simple cyanides such as NaCN, NH_4CN, or $Ca(CN)_2$ containing one metal ion (usually an alkai or alkaline-earth metal or ammonium ion) in its formula unit, and (2) complex cyanides such as $K_4Ce(CN)_6$ or $NaAg(CN)_2$ containing two different metals in their formula unit, usually one is an alkali metal and the other a heavy metal. The complex cyanide dissociates to metal and polycyanide ions. The latter may further dissociate to CN^- which forms HCN. The degree and rate of dissociation of complex cyanides depend on several factors, including the nature of the metal, pH of the solution, and dilution. Cyanide ion and HCN are highly toxic to human beings, animals, and aquatic life.

Cyanide in water may be determined by the following methods:

1. Silver nitrate titrimetric method
2. Colorimetric method
3. Ion-selective electrode method
4. Ion chromatography

SILVER NITRATE TITRIMETRIC METHOD

Cyanide reacts with silver nitrate as shown below forming the soluble cyanide complex, $Ag(CN)^{2-}$.

$$2\ CN^- + AgNO_3 \Rightarrow Ag(CN)_2^- + NO_3^-$$

When all the CN^- ions in the sample are complexed by Ag^+ ions, any further addition of a few drops of titrant, $AgNO_3$, can produce a distinct color with an indicator that can determine the end point of the titration. Thus, in the presence of a silver-sensitive indicator, *p*-dimethylaminobenzalrhodamine, Ag^+ ions at first combine preferentially with CN^-. When no more free CN^- is left, little excess of Ag^+

added reacts with the indicator, turning the color of the solution from yellow to salmon. Cyanide concentrations greater than 1 mg/L can be determined by titrimetry.

Alternately I^- may be used as an indicator. Titrate to first appearance of turbidity (due to the formation of AgI).

Procedure

The alkaline distillate of the sample (see below in Colorimetric Method, Procedure, for sample distillation), containing 1 mL of indicator solution is titrated against standard $AgNO_3$ titrant from yellow to a salmon color. Perform a blank titration using the same amount of water and caustic soda. Dilute the sample if necessary with dilute caustic soda solution.

$$\text{mg CN}^-/\text{L} = \frac{(A-B)\times 1000}{\text{Volume of original sample (mL)}} \times \frac{\text{Final volume of alkali distillate}}{\text{mL portion of distillate used}}$$

where A = mL of standard $AgNO_3$ soltion required to titrate the sample and
B = mL of standard $AgNO_3$ required in the blank titration.

Reagents

- Standard $AgNO_3$ solution: See Chapter 2.4 for standardization of $AgNO_3$.
- Indicator solution: 10 mg *p*-dimethylaminobenzalrhodamine in 50 mL acetone.
- Caustic soda dilution solution. 2 g NaOH in 1 L distilled water (0.05 *N*).

COLORIMETRIC METHOD

Cyanide is converted to cyanogen chloride. CNCl, by treatment with chloramine-T at pH below 8 without hydrolyzing to cyanate, CNO^-. On addition of pyridine-barbituric acid or pyridine-pyrazolone reagent, CNCl, reacts with pyridine to form an intermediate nitrile which hydrolyzes to glutaconaldehyde. The latter reacts with barbituric acid or pyralozone to give a blue color — the intensity of which is proportional to the concentration of cyanide in the sample. The reaction steps are shown below.

$$2\,CN^- + Cl_2 \Rightarrow 2\,CNCl$$

$$CNCl + C_5H_5N \Rightarrow C_5H_5N^+{-}CN + Cl^-$$

(Pyridine) (Intermediate Nitrile)

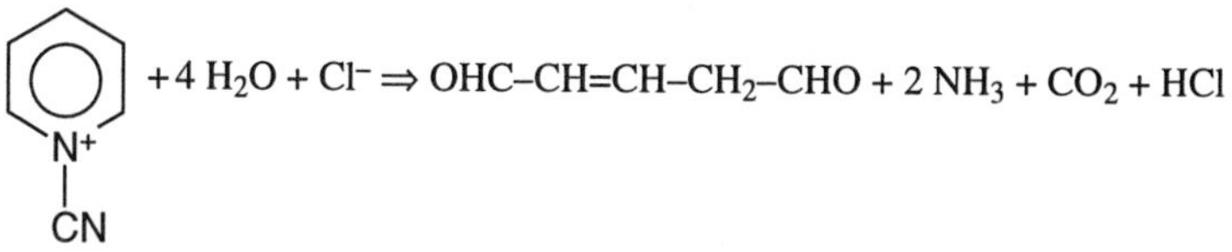

$$C_5H_5N^+\text{–}CN + 4\,H_2O + Cl^- \Rightarrow OHC\text{–}CH{=}CH\text{–}CH_2\text{–}CHO + 2\,NH_3 + CO_2 + HCl$$

(Glutaconaldehyde)

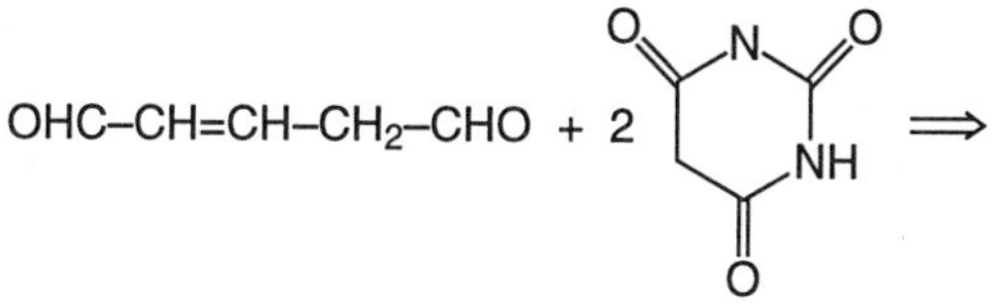

$$OHC\text{–}CH{=}CH\text{–}CH_2\text{–}CHO + 2\ (\text{barbituric acid}) \Longrightarrow$$

(Glutaconaldehyde) (Barbituric Acid)

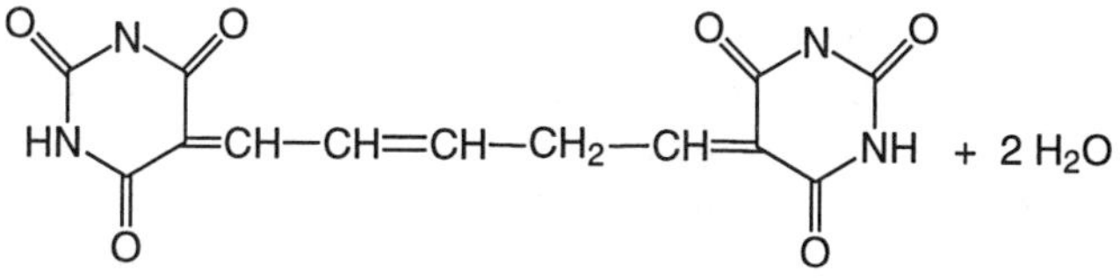

$$(\text{barbituric acid residue}){=}CH\text{—}CH{=}CH\text{—}CH_2\text{—}CH{=}(\text{barbituric acid residue}) + 2\,H_2O$$

(Blue Derivative)

$$OHC\text{–}CH{=}CH\text{–}CH_2\text{–}CHO + 2\ (\text{pyrazolone}) \Longrightarrow$$

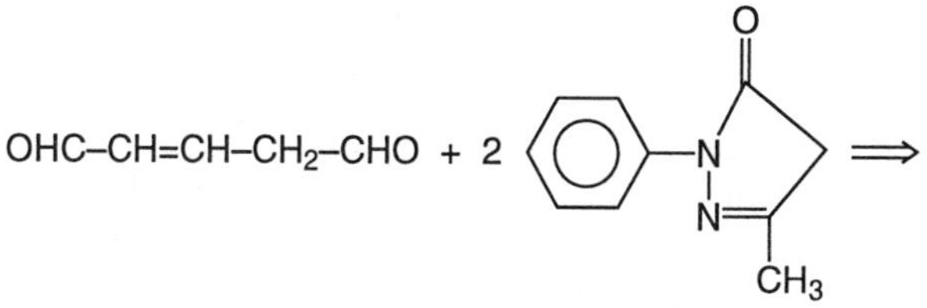

(Glutaconaldehyde) (Pyrazolone)

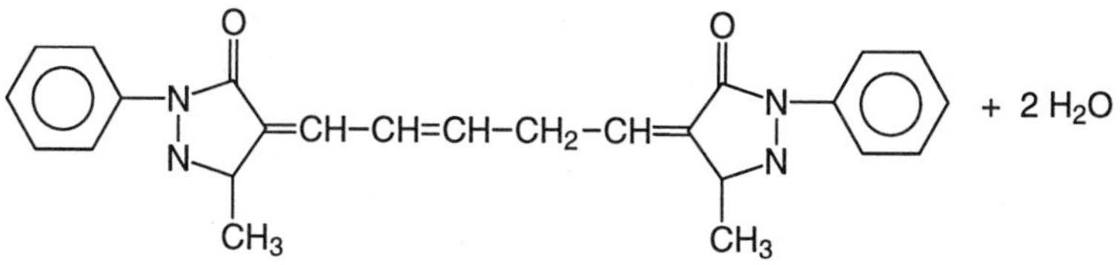

$$(\text{pyrazolone residue}){=}CH\text{–}CH{=}CH\text{–}CH_2\text{–}CH{=}(\text{pyrazolone residue}) + 2\,H_2O$$

(Blue Derivative)

Procedure

Aqueous samples must be distilled ro remove interference before analyzing for cyanide. A measured volume of a sample portion is acidified with dilute H_2SO_4 and distilled in a typical cyanide distillation unit. In an acid medium all cyanides are converted into HCN, which is collected over a diluted NaOH solution in a receiving flask. (HCN is absorbed in caustic soda solution.) A portion of distillate is then analyzed for cyanide. HCN is extremely toxic, and therefore, all distillations must be carried out in a hood.

To 80 mL alkali distillate or a portion of the sample in a 100-mL volumetric flask, add 1 mL of acetate buffer and 2 ml of chloramine-T solution and mix by inversion. Some samples may contain conpounds in addition to cyanide that may consume chlorine too. Therefore, perform a test for residual chlorine using KI-starch paper, 1 min after adding chloramine-T reagent. (If the test is negative, add 0.5 mL chloramine-T and recheck for residual chlorine 1 min later.) Allow the solution to stand for 2 min. Add 5 mL of pyridine-barbituric acid or pyridine-pyrazolone reagent and mix. Dilute to the mark with distilled water and let stand for 8 min. Read absorbance against distilled water at 578 nm using a 1-cm cell. Measure absorbance after 40 min at 620 nm when using pyridine-pyrazolone reagent.

Perform a blank analysis using the same volume of caustic soda dilution solution. Prepare a series of cyanide standards and plot a calibration curve from μg CN^- vs. absorbance. The efficiency of the sample distillation step may be checked by distilling two of the standards and then performing the above colorimetric test. Distilled standards must agree within ±10% of undistilled standards.

$$\mu g\ CN^-/L = \frac{A \times 1000 \times V_1}{V_2 \times V_3}$$

where A = μg CN^- read from standard curve,
V_1 = Final volume of alkali distilled solution (in mL),
V_2 = Volume of original sample taken for distillation (in mL), and
V_3 = Volume of alkali distillate portion taken for colorimetric analysis (in mL).

Alternately, the calibration curve may be plotted as ppb concentration of CN^- of working standards vs. their corresponding absorbance. In such case, retain the volume of standard solutions same as that of the sample distillate. For example, if the volume of the alkali distillate of the sample is 80 mL before adding reagents, use the same volume of working standard solutions (i.e., 80 mL), for color development and measuring the absorbance for plotting the standard calibration curve. Dilute the sample distillate if the CN^- concentration exceeds the range of the calibration curve.

$$\text{ppb } CN^- = \frac{\text{ppb } CN^- \text{ read from standard curve} \times V_1}{V_2} \times \text{dilution factor (if any)}$$

PREPARATION OF CYANIDE STANDARDS AND CALIBRATION CURVE

Stock cyanide solution. Dissolve 2.510 g KCN and 2.0 g KOH (or 1.6 g NaOH) in 1 L water. Standardize this solution against standard $AgNO_3$ solution (See the first section in this chapter, Silver Nitrate Tetrimetric Method).

$$1 \text{ mL} = 1 \text{ mg } CN^-$$

Secondary cyanide solution. Dilute 1 mL of the above stock solution to 1 L with caustic soda dilution solution.

$$1 \text{ mL} = 1 \ \mu\text{g } CN^-$$

Working CN^- standard solutions. Pipette 0, 1, 2, 5, 10, 20, and 40 mL of secondardy standard into 100-mL volumetric flasks and dilute first to 90 mL with caustic soda dilution solution and then to a final volume of 100 mL with distilled water after adding buffer and reagents.

Secondary std soln., mL 1 mL = 1 μg CN^-	Microgram CN^- in 100 mL of working std soln.	Microgram CN^- in 80 mL of working std soln.	Concentration of CN^- in working std soln. ppb
0	0	0	0
1.0	1.0	0.8	10.0
2.0	2.0	1.6	20.0
5.0	5.0	4.0	50.0
10.0	10.0	8.0	100.0
20.0	20.0	16.0	200.0
40.0	40.0	32.0	400.0

Reagents

- Acetate buffer: Dissolve 82 g sodium acetate trihydrate in 100 mL water. Acidify to pH 4.5 with glacial acetic acid.
- Chloramine-T solution: 1 g powder in 100 mL distilled water.
- Pyridine-barbituring acid: Take 15 g barbituric acid in a 250-mL volumetric flask. Add about 15 to 20 mL water to wash the side of the flask and wet barbituric acid. Add 75 mL pyridine and shake to mix. This is followed by addition of 15 mL conc. HCl. Shake well. When the flask is cool, add distilled water diluting to 250 mL. Swirl the flask to dissolve all barbituric acid. Store the solution in an amber bottle in a refrigerator. The solution should be stable for 4 to 5 months.
- Pyridine-pyrazolone solution: Prepared by mixing saturated solutions *a* and *b*. Solution *a* is made by dissolving 0.25 g of 3-methyl-1-phenyl-2-pyrazolin-5-

one in 50 mL distilled water by heating at 60°C and stirring. The solution is filtered. Solution *b* is made by dissolving 0.01 g 3,3-dimethyl-1,1-diphenyl-[4,4-*bi*-2-pyrazoline]-5,5-dione in 10 mL pyridine and then filtering. A pink color develops when both the filtrates are added.

- Caustic soda dilution solution: Dissolve 1.6 g NaOH in 1 L distilled water.

DETERMINATION OF CYANIDE IN SOIL, SEDIMENT, AND SOLID WASTE

Total cyanide content (which includes water soluble cyanides, iron cyanides, and other insoluble cyanides) in solid matrices may be determined as follows:

To 1 g sample, add 100 mL of 10% caustic soda solution and stir for 12 h. (This treatment is required only if iron cyanides are suspected to be present in the sample.) After this, adjust the pH to less than 8.0 with 1:1 H_2SO_4. Add about 0.2 g sulfamic acid to avoid nitrate/nitrite interference. This is followed by addition of 25 mg lead carbonate (to prevent interference from sulfur compounds). The mixture is distilled and collected over NaOH solution. This distillate is analyzed for CN^- by colorimetric, titrimetric, or ion-selective electrode method.

$$\text{CN}^- \text{ in the solid matrix, mg/kg} = \frac{\text{mg/L CN}^- \text{ in the leachate} \times \text{volume of leachate (L)} \times 1000}{\text{weight of sample (g)}}$$

CYANIDES IN AEROSOL AND GAS

Hydrogen cyanide and cyanide salts in aerosol and gas may be analyzed by NIOSH Method 7904 (NIOSH, 1984).

Between 10 and 180 L of air at a flow rate of 0.5 to 1 L/min is passed through a filter-bubbler assembly of a 0.8-µm cellulose ester membrane and 10 mL of 0.1 *N* KOH solution. While cyanide particulates retain over the filter membrane, HCN is trapped over the KOH solution in the bubbler. The membrane filter is then placed in 25 mL of 0.1 *N* KOH solution for 30 min to extract the cyanide particulates deposited on it. The KOH extract and the bubbler KOH solution are analyzed for cyanide by selective-ion electrode technique (see Chapter 1.9 for a detailed analytical procedure) using KCN standards. Calculate the concentration of particulate CN^- in the air sampled as follows:

$$\text{particulate CN}^- = \frac{M_f - B_f}{V} \text{ mg/m}^3$$

where Mf = Mass of CN^- present in the sample filter (µg),
Bf = Mass of CN^- in the average media blank filter (µg), and
V = Volume of air sampled (L).

Similarly, the concentration of HCN in the air sampled is determined as follows:

$$\text{mg HCN/m}^3 \text{ air} = \frac{(M_b - B_b) \times 1.04}{V} \text{ mg/m}^3$$

where M_b = Mass of CN^- in the bubbler (µg),
B_b = Mass of CN^- in the blank bubbler, and
V = Volume of air sampled (L).

The stoichiometric conversion factor from CN^- to HCN is the ratio of the formula weight of HCN to that of CN^-, which is $\frac{27}{26}$ or 1.04.

To express mg/m^3 HCN in ppm concentration at 25°C and 1 atm pressure, calculate as follows:

$$\text{ppm HCN} = \frac{\text{mg/m}^3 \text{ HCN}}{1.1}$$

2.7 CYANIDE AMENABLE TO CHLORINATION

This test is performed to determine the amount of cyanide in the sample that would react with chlorine. Not all cyanides in a sample are amenable to chlorination. While HCN, alkali metal cyanides, and CN^- of some complex cyanides react with chlorine, cyanide in certain complexes that are tightly bound to the metal ions are not decomposed by chlorine. Calcium hypochlorite, sodium hypochlorite, and chloramine are some of the common chlorinating agents that may be used as a source of chlorine. The chlorination reaction is performed at a pH between 11 and 12. Under such an alkaline condition, cyanide reacts with chlorine to form cyanogen chloride, a gas at room temperature, which escapes out. Cyanide amenable to chlorination is therefore calculated as the total cyanide content initially in the sample minus the total cyanide left in the sample after chlorine treatment.

Cyanide amenable to chlorination = Total CN^- before chlorination – Total CN^- left after chlorination.

Procedure

Two 500-mL aliquots or the volume diluted to 500 mL are needed for this analysis. Perform the test for total CN^- in one aliquot of the sample following distillation.

To the other aliquot, add calcium hypochlorite solution (5 g/100 mL) dropwise while maintaining the pH between 11 and 12 with caustic soda solution. Perform a test for residual chlorine using KI-starch paper. The presence of excess chlorine is indicated from the iodide-starch paper turning to a distinct blue color when a drop of the solution is poured on the paper. If required, add additional hypochlorite solution.

Agitate the solution for 1 h. After this period, remove any unreacted chlorine by adding one of the following: (a) 1 g of ascorbic acid, (b) a few drops of 2% sodium arsenite solution, or (c) 10 drops of 3% H_2O_2, followed by 5 drops of 50% sodium thiosulfate solution. Insure that there is no residual chlorine as indicated from no color change with KI-starch paper.

Distill this chlorine treated sample aliquot for cyanide analysis using colorimetric, titrimetric, or ion-selective electrode method outlined in the preceding sections.

Alternative Procedure

This is an easy and short method that does not require distillation of the sample. It is applicable when the CN^- concentration is less that 0.3 mg/L. For higher concentration, dilute the sample. This method determines the HCN and cyanide complexes that are amenable to chlorination. Thiocyanate is positive interference in this test. High concentration of total dissolved solids (greater than 3000 mg/L) may affect the test result. To compensate for this, add an equivalent amount of NaCl in NaOH solution.

Adjust the pH of the sample to between 11.5 and 12. To 20 mL of sample in a 50 mL volumetric flask, add 5 mL of phosphate buffer and 2 drops of EDTA solution. Add 2 mL chloramine-T solution and stir well. Test for residual chlorine using KI-starch paper. If required, add more chloamine-T so that there is enough residual chlorine in the solution. Allow the solution to stand for exactly 3 min. After this, add 5 mL pyridine-barbituric acid and mix well. Dilute to 50 mL mark. Let the solution stand for 8 min. Measure the absorbance at 578 nm in a 1 cm cell against distilled water. Determine the CN^- concentration from the calibration curve.

$$\text{Cyanide amenable to chlorination, } \mu g/L \text{ (ppb)}$$

$$= \frac{\text{ppb } CN^- \text{ read from calibration curve} \times 50}{20 \text{ mL}}$$

The above procedure is similar to the analysis of total cyanide in the alkaline distillate described in the preceding section, except that the sample pH are different. In the above method, chlorination is done on the alkaline sample at a pH between 11.5 and 12. On the other hand, in the determination of cyanide (total), the pH of the distillate is maintained below 8 with acetate buffer before adding chloramine-T solution.

Thiocyanate interferes in the above test. If thiocyanate is found to be present in the sample (see Thiocyanate, spot test), cyanide is masked by adding formaldehyde solution. The sample is then analyzed to determine the amount of thiocyanate that would react with chlorine. Thus the cyanide amenable to chlorination is equal to the difference between CN^- concentrations obtained in the untreated and formaldehyde treated sample aliquots.

To a 20-mL aliquot of sample adjusted for pH between 11.5 and 12, add 2 to 3 drops of 10% formaldehyde solution. Stir and then allow the solution to stand for 10 min. After this, add the reagents and follow the procedure given above to determine the cyanide amenable to chlorination by colorimetric method.

Reagents

See also preceding section.

- Calcium hypochlorite solution (5%): 5 g in 100 mL distilled water; store in dark in an amber glass bottle.
- EDTA solution 0.05 *M*: 1.85 g ethylenediaminetetraacetic acid disodium salt in water to 100 mL.
- Phosphate buffer: 13.8 g sodium dihydrogen monohydrate, $NaH_2PO_4 \cdot H_2O$ dissolved in water to 100 mL.
- Formaldehyde solution (10%): Dilute 27 mL of 37% commercial grade solution to 100 mL with water.

2.8 FLUORIDES

Fluoride (F^-) is a halogen ion that occurs in many potable and wastewaters. It may also occur in soils, sediments, hazardous waste, aerosol, and gas. While a low concentration of fluoride (below 1 ppm at controlled level in drinking water) is beneficial for reducing dental caries, a higher content is harmful. Fluoride in water may be determined by one of the following methods:

1. Colorimetric SPADNS method
2. Colorimetric automated complexone method
3. Ion-selective electrode method
4. Ion chromatography

Methods 1 and 2 are colorimetric techniques based on the reaction between fluoride and a dye. Methods 3 and 4 are discussed in Chapters 1.9 and 1.11, respectively.

COLORIMETRIC SPADNS METHOD

In acid medium, fluoride reacts instantaneously with zirconyl-dye lake which is composed of zirconyl chloride octahydrate, $ZrOCl_2 \cdot 8\ H_2O$, and sodium 2-(parasulfophenylazo)-1,8-dihydroxy-3,6-naphthalene disulfonate (SPADNS), displacing the Zr^{2+} from the dye lake to form a colorless complex anion, ZrF_6^{2-} and the dye. As a result, the color of the solution lightens as the concentration of F^- increases.

Chlorine, carbonates, bicarbonates, and hydroxides are common interferences. The former is removed by adding a drop of 1% solution of sodium arsenite ($NaAsO_2$). If the sample is basic, neutralize it with HNO_3. Sample may be diluted to reduce the interference effect.

Procedure

Prepare a standard calibration curve using fluoride standards from 0 to 1.5 mg F^-/L. A 50-mL volume of standard solutions is treated with 10 mL of zirconyl-acid-SPADN reagent mixture. The intensity of the color is measured either by a spectrophotometer at 570 nm or a filter photometer equipped with a greenish yellow filter having maximum transmittance from 550 to 580 nm. The absorbance reading of the standards and the sample are noted after setting the photometer to zero absorbance with the reference solution (zirconyl-acid-SPADN mixture). The concentration of fluoride is determined from the calibration standard curve, taking dilution of the sample, if any, into consideration.

Zirconyl-acid-SPADN reagent is made by combining equal volumes of sodium 2-(parasulfophenylazo)-1,8-dihydroxy-3,6-naphthalene disulfonate (SPADN) with zirconyl-acid reagent. The former is prepared by dissolving 1 g of it in 500 mL distilled water. The latter is made by dissolving 0.14 g zirconyl chloride octahydrate in 50 mL of distilled water followed by addition of 350 mL conc. HCl and dilution to 500 mL.

AUTOMATED COLORIMETRIC METHOD

The analysis is performed using a continuous flow analytical instrument such as Technicon Autoanalyzer or equivalent. The sample is distilled and the distillate, free of interference, is reacted with alizarin fluorine blue $[C_{14}H_7O_4 \cdot CH_2N(CH_2 \cdot COOH)_2]$-lanthanum reagent to form a blue complex. The absorbance is measured at 620 nm. Standard fluoride solutions are prepared in the range from 0.1 to 2.0 mg F^-/L using the stock fluoride solution.

2.9 HALOGENATED HYDROCARBONS

Halogenated hydrocarbons or halocarbons are halogen-substituted hydrocarbons. These substances contain carbon, hydrogen, and halogen atoms in the molecules. They are widely used as solvents, dry cleaning and degreasing agents, refrigerants, fire extinguishers, surgical anesthetics, lubricants and intermediates in the manufacture of dyes, artificial resins, plasticizers, and pharmaceuticals. Because of their wide applications, these compounds are found in the environment in trace quantities and constitute an important class of regulated pollutants. Most halogenated hydrocarbons are liquids at ambient temperature and pressure. Some low-molecular weight compounds such as methyl chloride or vinyl chloride are gases; compounds of higher molecular weight, such as iodoform, are solids at ambient conditions. Several halogenated hydrocarbons have been listed as priority pollutants by U.S. EPA and their methods of analysis are documented (U.S. EPA 1984-1992; Methods 601, 612, 624, 625, 501, 502, 503, 524, 8240, and 8260). In this chapter, halogenated hydrocarbons are defined as halogen-substituted compounds of alkane, alkene, cycloalkane, and aromatic classes, but exclude polychlorinated biphenyls and chlorinated pesticides like BHC isomers. Their methods of analyses are based on their certain physical properties, such as volatility, boiling point, and water solubility. Halogenated hydrocarbons may be analyzed using any one of the following methodologies:

1. Purge and trap concentration (or thermal desorption) from the aqueous matrices (aqueous samples or aqueous extracts of nonaqueous samples or methanol/acetone extract of nonaqueous samples spiked into reagent-grade water), separation of the analytes on a suitable GC column and their determination using a halogen-specific detector or a mass spectrometer.
2. Liquid–liquid extraction or liquid–solid extraction for aqueous samples (and soxhlett extraction or sonication for nonaqueous samples), followed by sample concentration, cleanup, and determination by GC or GC/MS. Direct injection, waste dilution, or other extraction techniques, depending on the sample matrices, may be used.

In general, the purge and trap technique is applied to analyze substances that have boiling points below 200°C and are insoluble or slightly soluble in water.

Heating of the purging chamber may be required for more soluble compounds. Soils, sediments, and solid wastes should be extracted with methanol or acetone, and an aliquot of the extract is spiked into reagent-grade water in the purging vessel for purge and trap extraction. Alternatively, an aliquot of the aqueous extract of the solid matrix may be subjected to purge and trap concentration. In either case, highly volatile compounds are susceptible to mechanical loss during such two-step extraction processes. Such highly volatile analytes in solid matrices may be determined by thermal desorption of a weighed aliquot of the sample, as is, under an inert gas purge.

The purge and trap system consists of a purging device, a trap, and a desorber. Such systems are commercially available from several sources. The purging device is a glass vessel that can accept samples with a water column at least 5 cm deep. The headspace above the sample should be less than 15 L if the sample volume is 5 mL. The base of the sample chamber should have a glass frit so that the purge gas passes through the water column, producing bubbles with a diameter of 3 mm at the origin.

The trap should consist of 2,6-diphenylene oxide polymer (Tenax GC grade), silica gel, and coconut charcoal, each constituting a one third portion. A methyl silicone packing at the inlet can extend the life of the trap. Tenax alone may be used if only compounds boiling above 35°C are to be analyzed. A silica gel trap is required for highly volatile compounds, while charcoal effectively adsorbs dichlorodifluoromethane and related compounds.

The desorber is a heating device capable of rapidly heating the trap to 180°C. Trap failure may be noted from poor bromoform sensitivity or characterized from a pressure drop over 3 psi across the trap during purging. The Tenax section of the trap should not be heated over 200°C.

U.S. EPA's analytical procedures mention an 11-min purge with nitrogen or helium at a flow rate of 40 mL/min.; and 4-min desorption at 180°C, backflushing the trap with an inert gas at 20 to 60 mL/min. A 5-mL sample volume is recommended for purging. A larger volume of sample may be required to obtain a lower detection level. Other conditions may be used if precision and accuracy of analysis are met.

The analytes separated on GC column are determined by a halogen-specific detector, such as an electrolytic conductivity detector (ELCD) or a microcoulometric detector. An ECD, FID, quadrupole mass selective detector, or ion trap detector (ITD) may also be used. A photoionization detector (PID) may also be used to determine unsaturated halogenated hydrocarbons such as chlorobenzene or trichloroethylene. Among the detectors, ELCD, PID, and ECD give a lower level of detection than FID or MS. The detector operating conditions for ELCD are listed below:

- ELCD: Tracor Hall model 700-A detector or equivalent
- Reactor tube: Nickel 1/16 in. outer diameter
- Reactor temperature: 810°C
- Reactor base temperature: 250°C
- Electrolyte: 100% n-propyl alcohol

- Electrolyte flow rate: 0.8 mL/min
- Reaction gas: Hydrogen at 40 mL/min
- Carrier gas: Helium at 40 mL/min
- GC column: A capillary column of intermediate polarity can give adequate resolution of isomers as well as unsaturated compounds. Many such columns are commercially available. These include 105 m long × 0.53 mm ID, Rtx-502.2; 60 m long × 0.75 mm ID, VOCOL; 60 m long × 0.53 mm ID, DB-62 or equivalent. Other capillary columns include 95% dimethyl-5% diphenyl polysiloxane coated columns, such as DB-5, SPB-5, Rtx-5, AT-5, or equivalent. These columns are also suitable for separation of a number of nonhalogenated organic compounds of intermediate polarity.

Table 2.9.1 lists some commonly used volatile halogenated hydrocarbons. Most of these are U.S. EPA listed pollutants. The term "volatile" indicates that these substances may be extracted by purge and trap technique. This also includes a few compounds of relatively moderate boiling range, responding adequately to the purge and trap method. The characteristic masses for GC/MS determination are also presented in the table.

SOLVENT EXTRACTION

Low volatile, high molecular weight halogenated compounds can be extracted with hexane or iso-octane and determined by GC-ECD. Methylene chloride may be used for extraction if the analysis is done by GC/MS. Purge and trap efficiency will be poor for such compounds, especially those boiling over 200°C. Soils, sediments, and solid wastes may be extracted with methylene chloride by sonication or soxhlett extraction. Interferences from acidic compounds, such as chlorophenol, may be removed by acid-base partitioning cleanup. The extract is then concentrated and analyzed by GC/MS or exchanged to hexane and analyzed by GC-ECD.

Nonpurge and trap extraction, such as liquid-liquid microextraction, can be used for many volatile compounds with boiling points well below 200°C. Similarly, the purge and trap method can be used for compounds boiling well over 200°C, especially when the purging chamber is heated.

Table 2.9.2 presents some of the halogenated hydrocarbons that can be effectively extracted out from aqueous and nonaqueous matrices by liquid-liquid or other nonpurge and trap extraction. Such compounds include di-, tri-, tetra-, penta-, and hexahalo- substituents aromatics and cyclohexanes, and haloderivatives of alkanes with carbon number greater than 6. Characteristic mass ions for GC/MS identification are also presented for some of these listed compounds.

SAMPLE COLLECTION AND PRESERVATION

Samples should be collected in a glass container without headspace and refrigerated. If volatile compounds are to be analyzed by the purge and trap method, the analysis must be done within 14 days of sampling. If substances of

Table 2.9.1 Purgeable Volatile Halogenated Hydrocarbons and Their Characteristic Masses

		Characteristic masses for GC/MS identification	
CAS no.	Compounds determined by ELCD/PID	Primary ion (m/z)	Secondary ions (m/z)
[107-05-1]	Allyl chloride [b]	76	41, 39, 78
[100-44-7]	Benzyl chloride [b]	91	126, 65, 128
[108-86-1]	Bromobenzene [b]	156	77, 158
[74-97-5]	Bromochloromethane	128	49, 130
[75-27-4]	Bromodichloromethane	83	85, 127
[460-00-4]	4-Bromofluorobenzene	95	174, 176
[75-25-2]	Bromoform	173	175, 254
[56-23-5]	Carbon tetrachloride	117	119
[108-90-7]	Chlorobenzene [b]	112	77, 114
[126-99-8]	2-Chloro-1,3-butadiene [b]	56	49
[124-48-1]	Chlorodibromomethane	129	208, 206, 127
[67-66-3]	Chloroform	83	85
[544-10-5]	1-Chlorohexane	84	49, 51
[563-47-3]	3-Chloro-2-methyl propene [b]	55	39, 90
[126-99-8]	Chloroprene	53	88, 90, 51
[107-05-1]	3-Chloropropene [b]		—
[95-49-8]	2-Chlorotoluene	91	126
[106-43-4]	4-Chlorotoluene	91	126
[96-12-8]	1,2-Dibromo-3-chloropropane [c]	75	155, 157
[106-93-4]	1,2-Dibromoethane	107	109, 188
[74-95-3]	Dibromomethane	93	95, 174
[541-73-1]	1,2-Dichlorobenzene	146	111, 148
[106-47-7]	1,4-Dichlorobenzene	146	111, 148
[1476-11-5]	*cis*-1,4-Dichloro-2-butene	75	53, 77, 124, 89
[110-57-6]	*trans*-1,4-Dichloro-2-butene [c]	53	75, 88
[75-71-8]	Dichlorodifluoromethane	85	87
[75-34-3]	1,1-Dichloroethane	63	65, 83
[107-06-2]	1,2-Dichloroethane	62	98
[75-35-4]	1,1-Dichloroethene [b]	96	61, 63
[156-59-4]	*cis*-1,2-Dichloroethene [b]	96	61, 98
[156-60-5]	*trans*-1,2-Dichloroethene [b]	96	61, 98
[78-87-5]	1,2-Dichloropropane	63	112
[142-28-9]	1,3-Dichloropropane	76	78
[590-20-7]	2,2-Dichloropropane	77	97
[563-58-6]	1,1-Dichloropropene [b]	75	110, 77
[542-75-6]	1,3-Dichloropropene [b]	75	77, 110
[10061-01-5]	*cis*-1,3-Dichloropropene [b]	75	77, 39
[10061-02-6]	*trans*-1,3-Dichloropropene [b]	75	77, 39
[540-36-3]	1,4-Difluorobenzene [b]	114	—
[74-83-9]	Ethyl bromide	108	110
[75-00-3]	Ethyl chloride	64	66
[106-93-4]	Ethylene dibromide	107	109, 188
[462-06-6]	Fluorobenzene [b]	96	77
[87-68-3]	Hexachlorobutadiene [b]	225	223, 227
[67-72-1]	Hexachloroethane	201	166, 199, 203
[74-83-9]	Methyl bromide	94	96
[74-87-3]	Methyl chloride	50	52

Table 2.9.1 Purgeable Volatile Halogenated Hydrocarbons and Their Characteristic Masses (Continued)

CAS no.	Compounds determined by ELCD/PID	Characteristic masses for GC/MS identification	
		Primary ion (m/z)	Secondary ions (m/z)
[75-09-2]	Methylene chloride	84	86, 49
[74-88-4]	Methyl iodide	142	127, 141
[76-01-7]	Pentachloroethane	167	130, 132,
[363-72-4]	Pentafluorobenzene [b]	168	165, 169
[630-20-6]	1,1,1,2-Tetrachloroethane	131	133, 199
[79-34-5]	1,1,2,2-Tetrachloroethane	83	131, 85
[127-18-4]	Tetrachloroethylene [b]	164	129, 131, 166
[87-61-6]	1,2,3-Trichlorobenzene [b]	180	182, 145
[120-82-1]	1,2,4-Trichlorobenzene [b]	180	182, 145
[71-55-6]	1,1,1-Trichloroethane	97	99, 61
[79-00-5]	1,1,2-Trichloroethane	83	97, 85
[79-01-6]	Trichloroethylene [b]	95	97, 130, 132
[75-69-4]	Trichlorofluoromethane	101	103, 151, 153
[96-18-4]	1,2,3-Trichloropropane	75	77
[75-01-4]	Vinyl chloride [b]	62	64

[a] Electron-impact ionization mode.
[b] Compounds that also show PID response.
[c] Poor purging efficiency.

Table 2.9.2 Some Solvent Extractable Halogenated Hydrocarbon Pollutants and Their Characteristic Masses

CAS no.	Compound determined by GC-ECD and GC/MS	Characteristic ions for GC/MS identification	
		Primary (m/z)	Secondary (m/z)
[100-39-0]	Benzyl bromide	91	170, 172
[100-44-7]	Benzyl chloride	91	126, 65, 128
[108-86-1]	Bromobenzene	156	77, 158
[90-13-1]	1-Chloronaphthalene	162	127, 164
[91-58-7]	2-Chlorotoluene	91	126
[106-43-4]	4-Chlorotoluene	91	126
[96-12-8]	1,2-Dibromo-3-chloropropane	75	155 ,157
[95-50-1]	1,2-Dichlorobenzene	146	148, 111
[541-73-1]	1,3-Dichlorobenzene	146	148, 111
[106-46-7]	1,4-Dichlorobenzene	146	148, 111
[106-93-4]	Ethylene dibromide	107	109, 188
[118-74-1]	Hexachlorobenzene	284	142, 249
[87-68-3]	Hexachlorobutadiene	225	223, 227
[77-47-4]	Hexachlorocyclopentadiene	237	235, 272
[67-72-1]	Hexachloroethane	117	201, 199
[70-30-4]	Hexachlorophene	196	198, 209, 406
[1888-71-7]	Hexachloropropene	213	211, 215, 117
[608-93-5]	Pentachlorobenzene	250	252, 108, 248, 215
[95-94-3]	1,2,4,5-Tetrachlorobenzene	216	214, 179, 108, 143
[120-82-1]	1,2,4-Trichlorobenzene	180	182, 145

low volatility are to be determined following solvent extraction, the extraction should be performed within 14 days and analysis within 30 days from extraction. U.S. EPA methods, however, mention a 7-day holding time for solvent extraction (for semivolatile organic pollutants that include some halogenated hydrocarbons), and a 45-day holding time for the analysis after extraction.

If residual chlorine is present, add sodium thiosulfate (~100 mg/L) for its removal. Ascorbic acid may also be added to reduce residual chlorine. The latter is preferred if the analysis includes gaseous halocarbons.

The use of plastic containers should be avoided because phthalate impurities can interfere in ECD determination.

INTERNAL STANDARDS/SURROGATES AND TUNING COMPOUNDS

Several compounds, including many deuterated and fluoro derivatives, have been used in the published literature. These include fluorobenzene, pentafluorobenzene, 1,2-dichlorobenzene-d_4, 1-chloro-2-fluorobenzene, 1,4-difluorobenzene, 1,2-dichloroethane-d_4, 1,4-dichlorobutane, and 2-bromo-1-chloropropane. U.S. EPA has set the tuning criteria for bromofluorobenzene and decafluorotriphenylphosphine as tuning compounds for volatile and semivolatile organic compounds. (See Chapter 1.4)

AIR ANALYSIS

Common halogenated hydrocarbons in air can be determined by NIOSH, ASTM, and U.S. EPA methods.

The NIOSH methods, in general, are based on adsorption of compounds in the air over a suitable adsorbent, desorption of the adsorbed analytes into a desorbing solvent, and, subsequently, their determination by GC using a suitable detector. A known volume of air is drawn through a cartridge containing coconut shell charcoal. The adsorbed compounds are desorbed into carbon disulfide, propanol, benzene, toluene, hexane, or methylene chloride. An aliquot of the solvent extract is then injected onto the GC column. FID is the most commonly used detector. Other detectors, such as ECD, ELCD, or PID have been used, however, in the method development of certain compounds. NIOSH method numbers and the analytical techniques are presented in Table 2.9.3.

Sampling of halogenated hydrocarbons in ambient air as per U.S. EPA methods (TO-1, TO-2, TO-3, and TO-14) may be performed using one of the three general techniques: (1) adsorption, (2) cryogenic cooling, and (3) canister sampling. In the adsorption method of air sampling, a measured volume of air is drawn through a cartridge containing an adsorbent such as Tenax (GC grade) or carbon molecular sieve. The former is effective in adsorbing substances that have boiling points in the range of 80 to 200°C. Carbon molecular sieve, on the other hand, is suitable for highly volatile halocarbons having boiling points in the range of –15 to +120°C. Unlike the NIOSH methods that require

Table 2.9.3 NIOSH Methods for Air Analysis for Halogenated Hydrocarbons

Compounds	NIOSH method	Desorbing solvent	GC detector
Benzyl chloride, bromoform, carbon tetrachloride, chlorobenzene, chlorobromomethane, *o*-and *p*-dichlorobenzene, 1,1-dichloroethane, 1,2-dichloroethane, ethylene dichloride, hexachloroethane, tetrachloroethylene, 1,1,1-trichloroethane, 1,1,2-trichloroethane, 1,2,3-trichloropropane	1003	CS_2	FID
Allyl chloride	1000	Benzene	FID
Bromotrifluoromethane	1017	CH_2Cl_2	FID
Dibromodifluoromethane	1012	Propanol	FID
Dichlorodifluoromethane, 1,2-dichlorotetrafluoroethane	1018	CH_2Cl_2	FID
Dichlorofluoromethane	2516	CS_2	FID
1,2-Dichloropropane	1013	Acetone/cyclohexane (15:85)	ELCD
Ethyl bromide	1011	2-Propanol	FID
Ethyl chloride	2519	CS_2	FID
Ethylene dibromide	1008	Benzene/methanol (99:1)	ECD
Hexachlor-1,3-cyclopentadiene	2518		ECD
Methyl bromide	2520	CS_2	FID
Methyl chloride	1001	CH_2Cl_2	FID
Methyl iodide	1014	Toluene	FID
Pentachloroethane	2517	Hexane	ECD
1,1,2,2-Tetrabromoethane	2003	Tetrahydrofuran	FID
1,1,2,2-Tetrachloroethane	1019	CS_2	FID
1,1,2,2-Tetrachloro-2,2-difluoroethane, 1,1,2,2-tetrachloro-1,2-difluoroethane	1016	CS_2	FID
Trichloroethylene	1022/3701	CS_2	FID/PID [a]
Trichlorofluoromethane	1006	CS_2	FID
1,1,2-Trichloro-1,2,2-trifluoroethane	1020	CS_2	FID
Vinyl bromide	1009	Ethanol	FID
Vinyl chloride	1007	CS_2	FID
Vinylidene chloride	1015	CS_2	FID

Note: FID, flame ionization detector; ECD, electron capture detector; ELD, electrolytic conductivity detector; GC, gas chromatograph; PID, photoionization detector.

[a] Air directly injected into a portable GC, equipped with a PID.

desorption of analytes with a solvent, these methods (TO-1 and TO-2) are based on the thermal desorption technique. The cartridge is placed in a heated chamber and purged with an inert gas. The desorbed compounds are transferred onto the front of a GC column held at low temperature (–70°C) or to a specially designed cryogenic trap from which they are flash evaporated onto the precooled GC column.

Many volatile, nonpolar organics that have boiling points in the range of –10 to +200°C, which include several halogenated hydrocarbons, can be collected in a trap placed in liquid argon or oxygen. After the sampling (drawing a measured volume of air) through the trap, the liquid cryogen is removed. The contents of the trap are swept with a carrier gas under heating to a precooled GC column. Alternatively, air may be collected in a passivated canister (SUMMA Model) initially evacuated at subatmospheric pressure or under pressure using a pump. The collected air in the canister is transferred and concentrated in a cryogenically cooled trap from which it is transported onto a GC column by the techniques described above.

The compounds are separated on the GC column, temperature programmed, and determined by an ECD, FID, or MS. Although these methods were developed for a limited number of specific compounds, the same methods may be applied to related compounds. Selection of the methods should be based on the boiling points of the halogenated hydrocarbons. Table 2.9.4 presents U.S. EPA's Method numbers for air analysis of halogenated hydrocarbons. Analysis of compounds not listed in Table 2.9.4, however, may be performed by similar procedures, based on their boiling points.

Table 2.9.4 U.S. EPA Methods for the Air Analysis of Halogenated Hydrocarbons

Compounds	U.S. EPA methods
Allyl chloride	TO-2, TO-3
Benzyl chloride	TO-1, TO-3, TO-14
Carbon tetrachloride	TO-1, TO-2, TO-3, TO-14
Chlorobenzene	TO-1, TO-3, TO-14
Chloroprene	TO-1, TO-3
1,2-Dibromoethane	TO-14
1,2-Dichlorobenzene	TO-14
1,3-Dichlorobenzene	TO-14
1,4-Dichlorobenzene	TO-1, TO-14
1,1-Dichloroethane	TO-14
1,2-Dichloroethane	TO-14
1,2-Dichloroethylene	TO-14
1,2-Dichloropropane	TO-14
1,3-Dichloropropane	TO-14
Ethyl chloride	TO-14
Freon-11, -12, -113, and -114	TO-14
Hexachlorobutadiene	TO-14
Methyl chloride	TO-14
Methylene chloride	TO-2, TO-3, TO-14
Tetrachloroethylene	TO-1, TO-2, TO-3, TO-14
1,2,3-Trichlorobenzene	TO-10, TO-14
1,2,4-Trichlorobenzene	TO-14
1,1,1-Trichloroethane	TO-1, TO-2, TO-3, TO-14
1,1,2-Trichloroethane	TO-14
Vinyl chloride	TO-2, TO-3, TO-14
Vinyl trichloride	TO-14
Vinylidine chloride	TO-2, TO-3, TO-14

2.10 HARDNESS

Hardness of a water sample is a measure of its capacity to precipitate soap. The presence of calcium and magnesium ions in water essentially contributes to its hardness. Other polyvalent ions, such as aluminum, also cause hardness. Their effect, however, is minimal, because these polyvalent ions occur in water often in complex forms and not as free ions. As a result, they cannot precipitate soap. Although calcium is not the only cation causing hardness, for the sake of convenience, hardness is expressed as mg $CaCO_3$/L. Similarly, anions other than carbonate, such as bicarbonate, also cause hardness in water. To distinguish the contributions of such anions from carbonates, hardness is sometimes termed as "carbonate hardness" and "noncarbonate hardness." This can be determined from alkalinity. The relationship is as follows:

When the total hardness measured in the sample is numerically greater than the sum of both carbonate alkalinity and bicarbonate alkalinity, then

$$\text{Carbonate hardness} = \text{carbonate alkalinity} + \text{bicarbonate alkalinity}$$

and

$$\text{Noncarbonate hardness} = \text{total hardness} - \text{carbonate hardness}$$

or

$$\text{Total hardness} - (\text{carbonate alkalinity} + \text{bicarbonate alkalinity})$$

When total hardness is equal to or less than the sum of carbonate and bicarbonate alkalinity, all hardness is noncarbonate hardness only and there is no carbonate hardness.

Hardness can be measured by either (1) calculation from the concentration of calcium and magnesium ions in the ample, or (2) EDTA titration.

HARDNESS DETERMINATIONS

Calculation

Analyze the metals, calcium, and magnesium in the sample using atomic absorption spectrophotometry or any other suitable technique and determine their concentrations.

Compute hardness as mg equivalent $CaCO_3$/L, as follows:

$$\text{Hardness, mg equivalent } CaCO_3/L$$

$$= 2.497 \times (\text{conc. of Ca, mg/L}) + 4.118 \times (\text{conc. of Mg, mg/L})$$

The factors 2.497 and 4.118 are obtained by dividing the formula weight of $CaCO_3$ (100.09) by atomic weights of Ca (40.08) and Mg (24.30), respectively.

Titration

Hardness can be measured precisely by EDTA titration. The principle of this titrimetric method is discussed extensively under Titrimetric Analysis in Chapter 1.6.

Ethylenediaminetetraacetic acid (EDTA) and its sodium salt readily react with calcium, magnesium, and certain other metal cations to form soluble chelates. Certain dyes, such as Calmagite or Eriochrome-Black T used as color indicators, also react with these metal ions, especially Ca^{2+} and Mg^{2+} forming colored complexes. Thus, Ca and Mg ions combine with the indicator molecules, producing a wine red color at the pH 10. Thus, the sample is pH adjusted to 10 using a buffer and is made wine red before the titration by adding the indicator solution to it. The addition of titrant forms more stable Ca-EDTA and Mg-EDTA complexes, displacing Ca^{2+} and Mg^{2+} from their respective chelates with the indicator. Thus, the end point of titration signifies the completion of chelation of all Ca and Mg ions in the sample with the titrant EDTA. This results in the dissociation of all metal-indicator complex molecules. The wine red color turns blue. To enhance the sharpness of the end point, a small amount of magnesium salt of EDTA is added to the buffer.

2.11 HERBICIDES: CHLOROPHENOXY ACID

Chlorophenoxy acids are one of the most important classes of chlorinated herbicides. In these compounds, chlorosubstituted benzene rings are attached to lower carboxylic acids via an oxygen atom, as shown in the following structures:

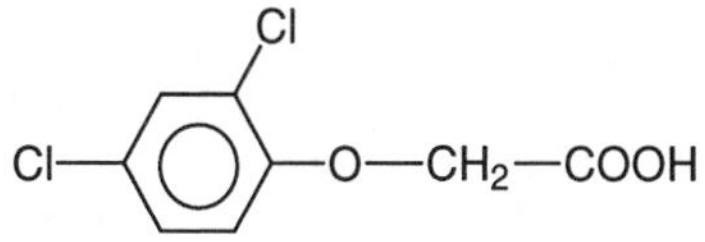

Cl

Cl— —O—CH_2—COOH

(2,4-Dichlorophenoxy)Acetic Acid or 2,4-D

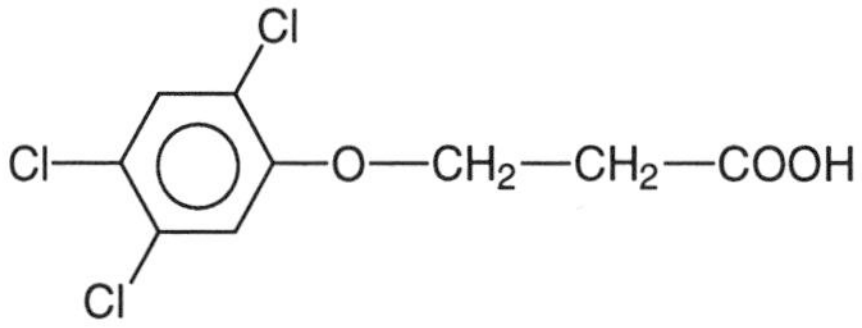

Cl

Cl— —O—CH_2—CH_2—COOH

Cl

(2,4,5-Trichlorophenoxy)-2-Propionic Acid or Silvex

Some of the common chlorophenoxy acid herbicides are listed in Table 2.11.1.

Table 2.11.1 Some Common Chlorophenoxy Acid Herbicides

CAS no.	Common name	Chemical name
[94-75-7]	2,4-D	(2,4-dichlorophenoxy) acetic acid
[93-76-5]	2,4,5-T	(2,4,5-trichlorophenoxy) acetic acid
[93-72-1]	Silvex	(2,4,5-trichlorophenoxy)propionic acid
[94-82-6]	2,4-DB	4-(2,4-dichlorophenoxy) butyric acid
[94-74-6]	MCPA	(4-chloro-2-methylphenoxy)acetic acid
[120-36-5]	Dichlorprop	2-(2,4-dichlorophenoxy)propionic acid
[1918-00-9]	Dicamba	3,6-dichloro-2-methoxybenzoic acid

The method of analysis primarily involves four basic steps: (1) extraction of herbicides from the sample into an organic solvent, (2) hydrolysis of the extract, (3) esterification, and (4) gas chromatographic (GC) determination of the herbicide esters formed.

ANALYSIS

Sample Extraction

Aqueous samples are extracted with diethyl ether while soils, sediment, and solid wastes are extracted with acetone and diethyl ether. Prior to extraction, the sample is acidified with HCl to a pH below 2. Such acidification is necessary due to the fact that in nature or in the environmental matrix, herbicides may occur as acids, salts, or esters. Acidification converts all these forms into chlorophenoxy acids.

Aqueous sample after acidification is extracted thrice with ether. The herbicide acids are now in the ether phase (top layer). The aqueous phase is discarded.

In the case of solids, the sample (~50g) is moistened with water and acidified with HCl under stirring. The content is then mixed with 25 mL acetone and shaken for several minutes. To this, add about 100 mL diethyl ether and shake the mixture further for several minutes. The extract is decanted and collected. The above extraction steps are repeated two more times. The extracts are combined together and the pH is checked. HCl is added if required to maintain the pH below 2. Allow the layers to separate. The aqueous layer is discarded.

2,4-Dichlorophenylacetic acid is recommended as a surrogate standard in the U.S. EPA Method 8151.

Hydrolysis

Herbicide acids extracted into the ether are now hydrolyzed with KOH and water. Such a hydrolysis step is important for removing most extraneous organic materials from the sample. Add 20 to 30 mL organic free reagent grade water and a few milliliters of 37% KOH solution into the ether extract and evaporate the ether on a water bath. Use boiling chips in all heatings. Hydrolysis converts the chlorophenoxy acids into their potassium salts which are soluble in water. The aqueous solution of the potassium salts of herbicides are shaken repeatedly with ether in the separatory funnel. Ether wash removes most extraneous organic matter. The top ether layer is discarded. The whole purpose of alkaline hydrolysis as mentioned is to remove the organic interference and cleanup the sample.

The potassium salts of the herbicides are then converted back to their acids by treatment with H_2SO_4. The aqueous solution is acidified with cold 1:3 H_2SO_4 to pH below 2. The chlorophenoxy acids regenerated are then extracted into ether in a separatory funnel by repeat extractions. The aqueous phase is discarded in order to achieve complete esterification of herbicide acids. The ether extract containing herbicides must be completely free from moisture even at trace level. Therefore, add acidified anhydrous Na_2SO_4 to the extract in excess amount (~10 g). The mixture is shaken well and allowed to stand for at least a few hours.

After drying (removal of water), the extract is quantitatively transferred into a Kuderna-Danish flask equipped with a concentrator tube and a Snyder column for sample concentration. The apparatus is placed in a water bath and ether is evaporated out. Use boiling chips in all heating operations. The volume of the extract is concentrated down to 1 to 2 mL.

Esterification

Chlorophenoxy acids after being extracted out from the sample matrix, separated from organic interferences and concentrated down into a small volume of ether, are now converted into their methyl esters. Such esterification of herbicides is essential for their determination by GC. While chlorophenoxy acids themselves show poor response, their ester derivatives produce sharp peaks with good resolution.

Esterification may be performed by using either diazomethane (CH_2N_2) or borontrifluoride-methanol (BF_3-CH_3OH). Other esterifying reagents include BCl_3-methanol, BCl_3-butanol, and pentafluorobenzyl bromide. The latter two produce butyl and pentafluorobenzyl derivatives, respectively. All U.S. EPA methods mention the use of diazomethane for esterification. An advantage of diazomethane is that the reaction goes to completion and also that many chlorinated herbicides (containing carboxylic groups) other than the chlorophenoxy types, esterify efficiently. The presence of water mars the reaction. A major disadvantage of using diazomethane, however, is that the compound is a carcinogen and can explode under the following conditions: heating over 90°C, grinding or stirring its solution, or contact with alkali metals.

Diazomethane reacts with chlorophenoxy acid herbicides at room temperature to form the methyl esters. It is generated by combining 2 mL ether, 1 mL carbitol, 1 to 2 mL 37% KOH, and 0.2 g Diazald. Diazomethane formed is purged with nitrogen at a flow of 10 mL/min, and bubbled through the ether extract of the herbicides for about 10 min. On the other hand, a Diazald Kit may be used to produce diazomethane. About 2 mL of the diazomethane solution obtained from the generator kit is allowed to stand with the herbicide extract for 10 min with occasional swirling. The solvent is evaporated at room temperature. Residue formed is dissolved in hexane for GC analysis. Any unreacted diazomethane is destroyed by adding 0.2 g silicic acid and allowing it to stand until there is no more evolution of nitrogen gas.

Methanol can be used instead of diazomethane for esterification of herbicides. The reaction is catalyzed by BF_3. To 1 mL of herbicide extract, add an equal amount of toluene or benzene. This is followed by 1 mL of BF_3-methanol. The solution is heated in a water bath for a few minutes. Ether evaporates out. Addition of a few milliliters of water partitions unreacted methanol and BF_3 into the aqueous phase, while the methyl esters of herbicides remain in the upper layer of benzene or toluene. The extraction steps discussed above are summarized in the following schematic diagram.

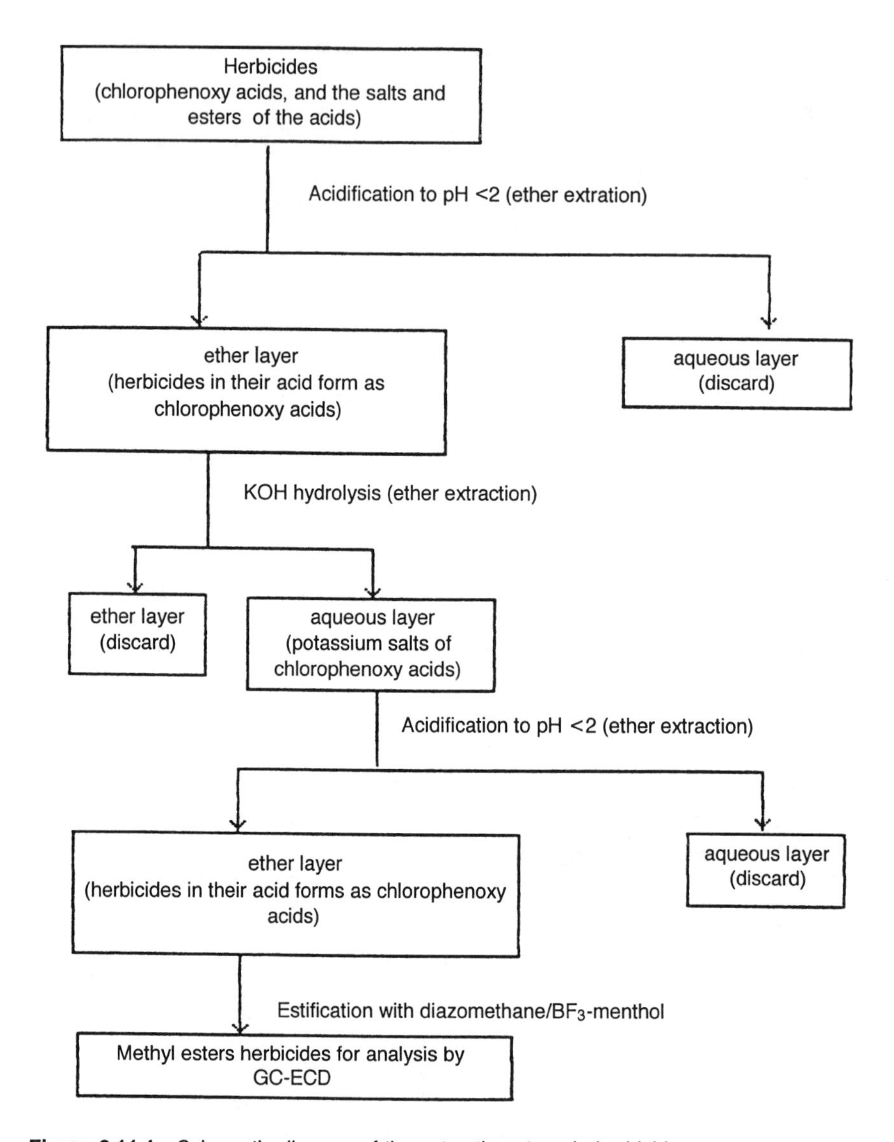

Figure 2.11.1 Schematic diagram of the extraction steps in herbicides.

Gas Chromatographic Analysis

The methyl esters of herbicides are analyzed by GC using an electron capture detector. Microcoulometric detector or electrolytic conductivity detector may alternatively be used. GC/MS, if available, should be employed to confirm the presence of analytes. Other instrumental techniques include GC-FID and HPLC. The latter can measure the acids for which esterification is not required.

Several packed and capillary columns have been reported in the U.S. EPA methods, research literature, and manufacturers' product catalogs. Some common columns are described below.

Packed column: 1.8 m × 4 mm ID glass packed with either (1) 1.5% SP 2250/1.95% SP-2401 on Supelcoport (100/120 mesh); (2) 5% OV-210 on Gas Chrom Q (100/120 mesh); or (3) 1.5% OV-17/1.95% QF-1. Efficient separation can also be achieved on a 2 mm ID column. Use 5% argon-95% methane carrier gas at flow rate of 60 to 70 mL/min when using a 4 mm ID column; while for a 2 mm ID column the carrier gas flow rate should be 20 to 30 mL/min. An oven temperature isothermal between 160° and 190°C should be suitable. Under the above conditions of temperature and carrier gas flow rate, most chlorophenoxy acid methyl esters elute within 10 min on a 4 mm ID packed column.

Fused silica capillary columns give better separation than packed columns. Columns having inside diameters of 0.25, 0.32, and 0.53 mm and film thickness between 0.25 and 1 μm have found use in herbicides analysis. The stationary phase is generally made out of phenyl silicone, methyl silicone, and cyanopropyl phenyl silicone in varying compositions. Some common columns are DB-5, DB-1701, DB-608, SPB-5, SPB-608, SPB-1701, Rtx-5, AT-1701, HP-608, BP-608, or equivalent. Use helium as carrier gas; flow rate 30 cm/s on narrowbore columns with 0.25 or 0.32 mm ID and 7 mL/min for megabore 0.53 ID columns.

The methyl esters can be also determined by GC-FID. Using a 30 m × 0.32 mm ID × 0.25 μm (film thickness) capillary column, such as DB-1701 or equivalent, the compounds can be adequately separated and detected by FID. The recommended carrier gas (helium) flow rate is 35 cm/s, while that of the makeup gas (nitrogen) is 30 cm/min. All of the listed herbicides may be analyzed within 25 min. The oven temperature is programmed between 50 and 260°C, while the detector and injector temperatures should be 300 and 250°C, respectively. The herbicides may alternatively converted into their trimethylsilyl esters and analyzed by GC-FID under the same conditions. FID, however, gives a lower response as compared with ECD. The detection level ranges from 50 to 100 ng. For quantitation, either the external standard or the internal standard method may be applied. Any chlorinated compound stable under the above analytical conditions, which produces a sharp peak in the same RT range without coeluting with any analyte, may be used as an internal standard for GC-ECD analysis. U.S. EPA Method 8151 refers the use of 4,4′-dibromooctafluorobiphenyl and 1,4-dichlorobenzene as internal standards. The quantitation results are expressed as acid equivalent of esters. If pure chlorophenoxy acid neat compounds are esterified and used for calibration, the results would determine the actual concentrations of herbicides in the sample. Alternatively, if required, the herbicide acids can be stoichiometrically calculated as follows from the concentration of their methyl esters determined in the analysis:

$$\text{conc. of herbicide acid} = \text{conc. of methyl esters} \times \frac{\text{molecular wt. of acid}}{\text{molecular wt. of ester}}$$

The molecular weights of herbicide acids are presented under individual compounds listed alphabetically in Part 3. To determine the molecular weights of the respective methyl esters, add 14 to the molecular weights of the corresponding chlorophenoxy acids.

AIR ANALYSIS

Analysis for 2,4-D and 2,4,5-T in the air may be performed using NIOSH Method 5001. Other chlorophenoxy acid herbicides such as 2,4,5-TP, 2,4-DB, and MCPA can be analyzed in a same general way. The method involves HPLC determination of herbicides in the form of acids or salts but not their esters.

Between 20 and 200 L air at a flow rate of 1 to 3 L/min is passed through a glass fiber filter. The herbicides and their salts, deposited on the filter, are desorbed with methanol, and the chlorophenoxy acid anions are determined by HPLC using UV detector at 284 to 289 nm. The LC eluent is a mixture of $NaClO_4$-$Na_2B_4O_7$ at 0.001 *M* concentration. Other eluent composition and UV detector wavelength may be used. A stainless steel column 50 cm x 2 mm ID, packed with Zipax Sax™ or equivalent may be used at ambient temperature and 1000 psi.

As mentioned, the HPLC-UV method determines the herbicides as acids and salts only. If the esters are also to be measured, the analytes should be desorbed with acetone-water mixture, acidified, extracted with ether, and analyzed by GC-ECD following hydrolysis and esterification as described earlier in this chapter.

2.12 HYDROCARBONS

Hydrocarbons are organic compounds containing only C and H atoms in the molecules. The broad class of organic compounds includes alkanes, alkenes, alkynes, cycloalkanes (naphthenes), and aromatics. Hydrocarbons occurring in nature are natural gases (mainly methane); crude oil (which is a complex mixture of alkanes, aromatics, and naphthenes of wide-ranging carbon numbers); and coals and coal-tars, many of which consist of fused polyaromatic rings. Cracking or distillation of crude oil produces various petroleum fractions for their commercial applications. These fractions, as shown below, are distinguished by their carbon numbers, boiling range, and commercial uses and include gasoline, kerosene, jet fuel, diesel oil, lubricating oil, and wax.

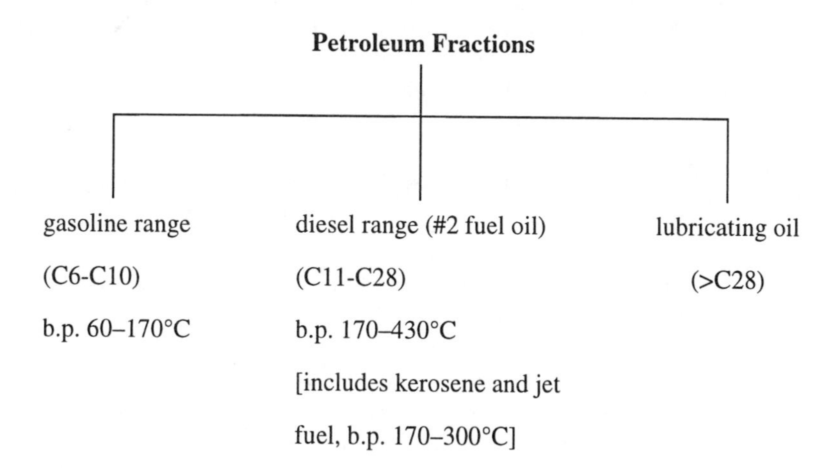

These petroleum products may be found in the environment from trace to significant amounts in groundwaters, industrial wastewaters and sludges, soils, sediments, and solid wastes.

ANALYSIS

Total petroleum hydrocarbons (TPH) and the various petroleum and coal-tar fractions in aqueous and nonaqueous samples can be determined by methods based on infrared (IR) spectrometry and gas chromatography (GC) techniques as summarized below.

- Total petroleum hydrocarbons — Freon extraction followed by IR spectroscopy; or extraction with methanol or methylene chloride and determination by GC-FID.
- Gasoline — Purge and trap extraction and determination by GC using an FID, or a PID and FID in series, or a mass spectrometer (MS).
- Diesel range organics (including kerosene and jet fuel) — Extraction with methylene chloride or an appropriate solvent and analysis by GC-FID.
- Polynuclear aromatic hydrocarbons (PAH) (a coal-tar distillate fraction) — Solvent extraction and determination by HPLC, GC-FID or GC/MS; enzyme immunoassay testing may be applied for semiquantitive determination.

Total Petroleum Hydrocarbons

Analysis for TPHs in aqueous and nonaqueous samples involves extraction of sample aliquots with a fluorocarbon solvent, such as freon, followed by determination of the hydrocarbons in the extract by IR spectroscopy measuring absorbance at the wavelength 2930 cm^{-1}. Aqueous samples are repeatedly extracted with freon in a separatory funnel. The freon extract (bottom layer) is separated and dried with anhydrous Na_2SO_4 before the IR analysis. Nonaqueous samples are mixed with anhydrous Na_2SO_4 and extracted with freon in a Soxhlett apparatus for 3 to 4 h. Dark, dirty, or colored extract is treated with silica gel to remove the color before IR measurement. Supercritical fluid extraction or sonication may be performed instead of Soxhlett extraction. The efficiency of extraction, however, must be established by performing duplicate analysis and determining the percent spike recovery. A calibration curve should be prepared for quantitation (plotting absorbance against concentration of calibration standards), which may be made from chlorobenzene, iso-octane, and tetradecane or from any petroleum fraction.

GC analysis involves two separate analyses: (1) for gasoline range organics by purge and trap extraction followed by FID or PID-FID determination (see Chapter 2.9 for details of purge and trap method), and (2) for diesel range organics by methylene chloride extraction and FID detection. The results of both analyses are summed to calculate TPH. If gasoline range organics are not determined separately by the purge and trap method, methylene chloride at room temperature can extract most components of gasoline except the highly volatile ones, and the result for TPHs can be fairly determined by GC-FID.

Gasoline Range Organics

Gasoline is a mixture of C-5 to C-10 hydrocarbons containing alkanes, aromatics, and their alkyl substituents, boiling in the range of 60 to 170°C. While the major alkane components of gasoline include pentane, hexane, methylhexane, isooctane, and their isomers, the aromatic components primarily constitute benzene, toluene, ethylbenzene, xylenes, and other alkyl benzenes. The GC chromatograms of gasoline range organics should generally constitute the peaks eluting between 2-methylpentane and 1,2,4-trimethylbenzene.

Aqueous samples can be analyzed by purge and trap extraction, followed by gas chromatography determination using FID or PID-FID connected in series. The aromatic fractions can be determined by PID or FID.

An inert gas is bubbled through the sample. The volatile hydrocarbons are transferred into the vapor phase and trapped over a sorbent bed containing 2,6-diphenylene oxide polymer (Tenax GC). A methyl silicone (3% OV-1 on Chromosorb-W, 60/80 mesh) packing protects the trapping material from contamination. Other adsorbents such as Carbopack B and Carbosieve S III may also be used. If pentane and other low boiling hydrocarbons need to be detected, the sorbent trap should be filled with activated charcoal, silica gel, and Tenax, respectively, in equal amounts.

The gasoline components are desorbed from the trap by heating the trap at 180°C. If Carbopack B and Carbosieve S III are used as adsorbents, the desorption and baking temperatures of the trap should be increased to 250 and 300°C, respectively. The desorbed components are separated on a capillary column and determined by FID, PID-FID in series, or MSD.

The capillary column should be able to resolve 2-methylpentane from the methanol solvent front (when gasoline in soil is to be determined) and ethylbenzene from m/p-xylene. A 30 m × 0.53 mm ID x 1.0 μm film fused silica capillary column, such as DB-5, SPB-5, AT-5, or equivalent should give adequate separation. Columns with other dimensions can also be used. A longer column, e.g., Rtx 502.2 (105 m × 0.53 mm × 0.3 μm) gives excellent resolution, but with a longer analysis time.

Diesel Range Organics

Diesel or fuel oil is a mixture of alkanes in the range C-10 to C-28 boiling between approximately 170 and 430°C.

The method of analysis involves extraction of 1 L of aqueous sample (liquid–liquid extraction) or 25 g of soil (sonication or Soxhlett extraction or supercritical fluid extraction) or an appropriate amount of the sample with methylene chloride. The extract is dried, concentrated to a volume of 1 mL, and injected into a capillary GC column for separation and detection by FID. For quantitation, the area or height response of all peaks eluting between C-10 and C-28 are summed and compared against the chromatographic response of the same peaks in a #2 Fuel or Diesel Oil standard. A 10-component *n*-alkanes mixture containing even numbered alkanes ranging between 10 and 28 C atoms has been recommended as an alternative calibration standard. These alkanes occur in all types of diesel oils, and each compound constitutes approximately a 1% total mass of diesel fuel, i.e., 1 g of diesel fuel contains about 10 mg each of any of the above alkanes. Therefore, when using the latter as a calibration standard, the result must be multiplied appropriately by 100.

Example

A 25-g soil sample was extracted with methylene chloride and the extract was concentrated to a final volume of 2 mL. A 10-component alkane mixture at a concentration of each component as 50 μg/mL was used in quantitation performed by external standard method. 1 μL of the extract and the standard were injected onto the column for analysis. Determine the concentration of diesel range organics in the sample from the following data:

Sum of the area response for the 10-component standard = 370,000
Sum of the area response for the same ten components in the sample extract = 290,000

Thus, the concentration of these alkanes in the extract

$$= \frac{290,000}{370,000} \times 50\ \mu\text{g/mL} = 39.2\ \mu\text{g/mL of each component}$$

This is approximately equivalent to 39.2 μg/mL × 100 ≡ 3920 μg/mL of total diesel fuel in the sample extract. Therefore, the concentration of diesel fuel in the sample is:

$$= \frac{3920\ \mu\text{g}}{\text{mL}} \times \frac{2\ \text{mL}}{25\ \text{g}} \times \frac{1000\ \text{g}}{1\ \text{kg}} = 313,600\ \mu\text{g/kg}\ \text{ or }\ 313.6\ \text{mg/kg}$$

Unlike gasoline, the chromatographic response for all alkanes in diesel, especially the ones greater than C-22, are not similar. However, a fairly accurate result may be obtained by comparing the area or height response of alkanes below C-24 (between C-10 and C-22) in both the standard and the sample extract (Cavanaugh, 1995).

An internal standard may be added onto the extract and the quantitation may be performed by IS method. The surrogates recommended are *o*-terphenyl and 5A-Androstane.

For quantitation using an IS, determine the response factor (RF) as follows:

$$\text{RF} = \frac{\text{Total area (or height) or 10 diesel components in the sample extract}}{\text{Area (or height) of IS}}$$

$$\times \frac{\text{Concentration of IS (mg/L)}}{\text{Concentration of 10 component standard (mg/L)}}$$

The RF should be determined using standards at various concentrations, and an average RF value should be used in the calculation. Both the external standard and internal standard calibration methods for GC analysis are fully discussed in Chapter 1.3.

Thus the concentration of 10 components in the sample is

$$\frac{A_s \times C_{is} \times V_t \times D}{A_{is} \times \mathrm{RF} \times V_s}$$

where A_s and A_{is} = the area (or height) response for the analyte and IS, respectively,
V_t = mL of final extract,
V_s = volume of sample extracted in L (or kg), and
D = dilution factor ($D = 1$, if the extract is not diluted).

As mentioned earlier, if a 10-component alkane mixture is used as standard, the result obtained from IS or extraction standard calibration must be multiplied by 100 because each alkane constitutes about 1% mass of diesel fuel. If a #2 fuel oil or a diesel oil standard is used instead of a 10-component alkane mixture, and when all the major peaks in the sample extract and standards are taken into consideration, then the result must not be multiplied by 100.

2.13 HYDROCARBONS, POLYNUCLEAR AROMATIC

Polynuclear aromatic hydrocarbons (PAH) are aromatic compounds that contain two or more benzene rings fused together. These substances may be analyzed by HPLC, GC, GC/MS and enzyme immunoassay techniques. The latter is a rapid screening method that may be applied for a qualitative or semiquantitative determination Test kits are commercially available for such screening. U.S. EPA (1995) has specified a method (Draft Method 4035) that detects a range of PAHs to different degrees and measures the composite of individual responses to determine the total PAHs in the sample.

U.S. EPA has listed 16 PAH as priority pollutants in wastewaters and 24 PAH in the category of soils, sediments, hazardous wastes, and groundwaters. Some common PAH compounds including the ones listed by U.S. EPA as priority pollutants are presented in Table 2.13.1. All these analytes, as well as any other compound that has a polyaromatic ring, may be analyzed by similar methods. The analytical steps include extraction of the sample with methylene chloride or an appropriate solvent, concentration of the solvent extract into a small volume, cleanup of the extract using silica gel (for dirty samples), and determination of PAH by HPLC, GC, or GC/MS. The HPLC method is superior to packed column GC analysis which suffers from a coelution problem.

The latter does not adequately resolve the following four pairs of compounds:

- Anthracene and phenathrene
- Chrysene and benzo(*a*)anthracene
- Benzo(*b*)fluoranthene and benzo(*k*)fluoranthene
- Dibenzo(*a,h*)anthracene and indeno(1,2,3-*cd*)pyrene

These and some other coeluting substances may, however, be separated on a 30 or 60 m fused silica capillary column with 0.32 or 0.25 mm ID. On the other hand, HPLC analysis is faster than the capillary GC determination. The presence of a compound detected must be confirmed on an alternate column or

Table 2.13.1 Common Polynuclear Aromatic Hydrocarbons

CAS no.	Compounds
[83-32-9]	Acenaphthene
[208-96-8]	Acenaphthylene
[120-12-7]	Anthracene
[53-96-3]	2-Acetylaminofluorene
[56-55-3]	Benz(*a*)anthracene
[205-99-2]	Benzo(*b*)fluoranthene
[207-08-9]	Benzo(*k*)fluoranthene
[191-24-2]	Benzo(*g,h,i*)perylene
[50-32-8]	Benzo(*a*)pyrene
[192-97-2]	Benzo(*e*)pyrene
[90-13-1]	1-Chloronaphthalene
[91-58-7]	2-Chloronaphthalene
[218-01-9]	Chrysene
[191-07-1]	Coronene [a]
[224-42-0]	Dibenz(*a,j*)acridine
[53-70-3]	Dibenz(*a,h*)anthracene
[132-64-9]	Dibenzofuran
[192-65-4]	Dibenzo(*a,e*)pyrene
[57-97-6]	7,12-Dimethylbenz(*a*)anthracene
[206-44-0]	Fluoranthene
[86-73-7]	Fluorene
[193-39-5]	Indeno(1,2,3-*cd*)pyrene
[56-49-5]	3-Methylcholanthrene
[91-57-6]	2-Methylnaphthalene
[91-20-3]	Naphthalene
[602-87-9]	5-Nitroacenaphthene
[198-55-0]	Perylene [a]
[85-01-8]	Phenanthrene
[213-46-7]	Picene [a]
[129-00-0]	Pyrene

[a] Not listed under U.S. EPA priority pollutants.

by GC/MS. A major problem with GC/MS identification is, however, that some of the compounds produce the same characteristic masses, as shown below:

1. Phenanthrene and anthracene
2. Chrysene and benzo(*a*)anthracene
3. Benzo(*b*)fluoranthene, benzo(*k*)fluoranthene, and benzo(*a*)pyrene
4. Fluoranthene and pyrene
5. Benzo(*g,h,i*)perylene and indeno(1,2,3-*cd*)pyrene

Benzo(*a*)pyrene and the compounds in the pairs 4 and 5 may, however, be distinguished from their retention times.

HPLC COLUMNS AND CONDITIONS

- Reverse phase column: HC-ODS Sil-X 25 cm x 2.6 mm ID and 5 μm particle diameter or Vydac 201 TP with similar dimensions.

- Detector: UV (at 254 nm) or fluorescence.
- Mobile phase: 40% to 100% acetonitrile in water.

GC COLUMNS AND CONDITIONS

- GC packed column: 1.8 m × 2 mm ID packed with 3% OV-17 on Chromosorb W-AW-DCMS (100/120 mesh) or equivalent.
- Detector-FID: 100°C for 4 min, 8°C/min to 280°C Carrier gas N_2 flow rate 40 mL/min. Other column and conditions may be used.
- GC capillary column: 30 m × 0.25 mm ID × 0.25 µm film fused silica capillary column such as PTE-5, DB-5, or equivalent.

MS CONDITIONS

- Scanning range: 35 to 500 amu.
- Scanning time: 1 s/scan using 70 V (nominal) electron energy in the electron impact ionization mode.

The characteristic mass ions for PAH are presented in Table 2.13.2.

SAMPLE EXTRACTION AND CLEANUP

Aqueous samples are extracted with methylene chloride by liquid-liquid extraction in a separatory funnel or a liquid-liquid extractor. The extract is concentrated to 1 mL for GC analysis. If HPLC analysis were to be performed, methylene chloride should be exchanged to acetonitrile by evaporating the solvent extract with a few mL of acetonitrile and adjusting the final volume to 1 mL.

If interferences are suspected to be present or if the sample is dirty, a silica gel cleanup should be performed.

Aqueous samples may alternatively be extracted by solid phase extraction using reversed phase C-18 stationary phase column, such as Supelclean ENVI-18. The column must be conditioned with toluene-methanol (10:1), methanol, and deionized water, respectively, prior to sample addition. PAH analytes are eluted with toluene-methanol (10:1) mixture.

Soils, sediments, and solid wastes are mixed with anhydrous Na_2SO_4 and extracted with methylene chloride by sonication or Soxhlett extraction. Supercritical fluid extraction (SFE) may be performed using CO_2 under the following conditions (Supelco, 1995):

- Solvent: CO_2
- Temperature and pressure: 100°C at 4000 psi
- Flow rate: 600 mL/min (for 15 min)
- Restrictor: Fused silica, 30 cm × 50 µm ID
- Collection: 4 mL, methylene chloride

Table 2.13.2 Characteristic Ions for Polynuclear Aromatic Hydrocarbons

Compounds	Primary ions	Secondary ions
Acenaphthene	154	152, 153
Acenaphthylene	152	151, 153
Anthracene	178	176, 179
2-Acetylaminofluorene	181	152, 180, 223
Benz[*a*]anthracene	228	226, 229
Benzo[*b*]fluoranthene	252	125, 253
Benzo[*k*]fluoranthene	252	125, 253
Benzo[*g,h,i*]perylene	276	138, 277
Benzo[*a*]pyrene	252	125, 253
1-Chloronaphthalene	162	127, 164
2-Chloronaphthalene	162	127, 164
Chrysene	228	226, 229
Dibenz[*a,j*]acridine	279	250, 277, 280
Dibenz[*a,h*]anthracene	278	139, 279
Dibenzofuran	168	139
1,2:4,5-Dibenzopyrene	302	150, 151
7,12-Dimethylbenz[*a*]anthracene	256	120, 239, 241
Fluoranthene	202	101, 203
Fluorene	166	165, 167
Indeno[1,2,3-*cd*]pyrene	276	138, 227
3-Methylcholanthrene	268	126, 134, 252
2-Methylnapthalene	142	141
Naphthalene	128	127, 129
5-Nitroacenaphthene	199	141, 152, 169
Phenanthrene	178	176, 179
Pyrene	202	200, 203
	Surrogates/IS	
Acenaphthene-d_{10}	164	162, 160
Chrysene-d_{12}	240	120, 236
Naphthalene-d_8	136	68
Perylene-d_{12}	264	260, 265
Phenanthrene-d_{10}	188	94, 80

AIR ANALYSIS

PAH have very low vapor pressure at room temperature. These substances, however, may deposit on the dusts in the air. PAH may be produced during pyrolysis of organic materials. These compounds can contaminate the air near coke ovens, as well as during loading and unloading of pencil pitch.

The analysis of particle bound PAH involves collection of PAH bound to dust particles on 0.8 μm glass fiber or silver membrane filters, desorption of PAH from the particles into a suitable organic solvent, and analysis of the extract by a capillary GC using an FID. Between 500 and 1000 L air at a flow rate of 120 L/h is recommended for sampling, which can give a detection limit of 0.15 to 0.50 $\mu g/m^3$ for each compound (Riepe and Liphard, 1987). The method suggests the installation of an absorber resin, such as XAD-2 or Tenax, after the filter if

PAH vapors were to be trapped (at temperature over 50°C). Cyclohexane or toluene is recommended as desorption solvent.

Air analysis may be performed by U.S. EPA Method TO13 (U.S. EPA 1988), which is quite similar to the above method. PAH-bound particles and vapors (many compounds may partially volatilize after collection) may be trapped on a filter and adsorbent (XAD-2, Tenax, or polyurethane foam), and then desorbed with a solvent. The solvent extract is then concentrated and analyzed by HPLC (UV/Fluorescence detection), GC-FID, or GC/MS (preferably in SIM mode). Because of very low level of detection required for many carcinogenic PAHs, including benzo(*a*)pyrene, the method suggests the sampling of a very high volume of air (more than 300,000 L).

Analysis of PAH in air may be done by NIOSH Methods 5506 and 5515 (NIOSH, 1985). Air is passed over a sorbent filter consisting of PTFE and washed XAD-2. The sample volume and flow rate should be 200 to 1000 L and 2 L/min, respectively. The PAH compounds are extracted with a solvent and analyzed by HPLC using a UV or Fluorescence detector. UV absorbance is set at 254 nm. A fluorescence detector is set at excitation 340 nm or emission 425 nm. The solvent extract may be alternatively analyzed by GC-FID using a capillary column.

2.14 NITROGEN (AMMONIA)

Ammonia (NH_3) occurs in varying concentrations in groundwater, surface water, and waste water. Its occurrence in waters and sludge is primarily attributed to its formation resulting from the reduction of nitrogen-containing organics, deamination of amines and hydrolysis of urea, and also to its use in water treatment plants for dechlorination. Its concentration in groundwaters is relatively low because of its adsorption to soil.

Ammonia-nitrogen (NH_3-N) may be analyzed by the following methods:

1. Colorimetric nesslerization method
2. Colorimetric phenate method
3. Titrimetric method
4. Ammonia-selective electrode method

Sample distillation is often required before analysis, especially for waste waters and sludges where interference effect is significant. Distillation, however, may not be necessary for potable waters or clean and purified samples where the concentration of ammonia is expected to be low. When the titrimetric method is followed, the sample must be distilled.

SAMPLE DISTILLATION

Distillation of sample is often necessary for the removal of interfering contaminants. The sample is buffered at pH 9.5 with borate buffer prior to distillation. This decreases hydrolysis of cyanates (CNO^-) and organic nitrogen compounds.

Distillation is performed in a 2 L borosilicate glass apparatus or one with aluminum or tin tubes condensing units. Before the sample is distilled, clean the apparatus until it is free from trace ammonia. This is done by distilling 500 mL NH_3-free distilled water containing 20 mL borate buffer and adjusted to pH 9.5 with NaOH.

To 500 mL sample, add 25 mL borate buffer and adjust the pH to 9.5 with 6 *N* NaOH. If there is residual chlorine in the sample, dechlorinate by treating with $Na_2S_2O_3$ before adding borate buffer.

Distill the sample at a rate of 5 to 10 mL/min. Collect between 200 and 350 mL distillate over 50 mL of one of the following solutions, depending on the method of analysis:

- Boric acid — colorimetric nesslerization method
- Boric acid-indicator solution — titrimetric analysis
- Sulfuric acid (0.05 *N*) — colorimetric phenate method or ammonia electrode method

Reagents

- Ammonia-free distilled water: pass distilled water through a strongly acidic cation-exchange resin. Alternatively, add a few drops of conc. H_2SO_4 and redistill.
- Borate buffer solution: to 9.5 g sodium tetraborate ($Na_2B_4O_7 \cdot 10\ H_2O$) in 500 mL water (0.025 *M*), add 88 mL 0.1 *N* NaOH solution and dilute to 1 L.
- Sodium thiosulfate: dissolve 3.5 g $Na_2S_2O_3 \cdot 5\ H_2O$ in 1 L water.
- Boric acid solution: 2 g boric acid in 100 mL water.

TITRIMETRIC METHOD

This method is based on the titration of basic ammonia with standard sulfuric acid using methyl red-methylene blue indicator to pale lavender end point. Distill 100 mL sample into 50 mL boric acid mixed indicator solution. Titrate ammonia in this distillate solution with standard H_2SO_4 (0.02 *N*) until the color turns to pale lavender. Perform a blank titration using distillate obtained from reagent grade water under similar conditions. Calculate the concentration of NH_3-N in the sample as follows:

$$\text{mg NH}_3\text{-N/L} = \frac{(V_s - V_b) \times 280}{\text{mL sample}}$$

where V_s is mL H_2SO_4 required for the titration of sample distillate, and V_b is mL H_2SO_4 required in the blank titration.

Since normality of H_2SO_4 used is 0.02, which is equivalent to 0.02 *N* NH_3 solution or 0.02 × 17g/L or 340 mg/L ammonia (because the equivalent weight of NH_3 is 17.) This is equal to $\left(\frac{340\text{ mg}}{\text{L}} \times \frac{14}{17}\right)$ or 280 mg/L of nitrogen (because one mole of NH_3 contains one mole or 14 g N). If in the titration we use H_2SO_4 having different normality, then substitute 280 with $140x$ in the calculation where x = normality of H_2SO_4.

Reagents

- Methyl red-methylene blue indicator solution: Dissolve 200 mg methyl red in 100 mL 95% ethanol. Separately dissolve 100 mg methylene blue in 50 mL 95% ethanol. Mix both these solutions.
- Boric acid-indicator solution: Dissolve 20 g boric acid in distilled water. Add 10 mL methyl red-methylene blue indicator solution. Dilute to 1 L.
- Sulfuric acid standard (0.02 *N*): Dilute 3 mL of conc. H_2SO_4 to 1 L with CO_2-free distilled water. This solution would have a strength of approximately 0.1 *N*. Dilute 200 mL of this 0.1 *N* H_2SO_4 to 1 L with CO_2-free distilled water. Determine the normality of this solution by titrating with 0.02 *N* standard Na_2CO_3 solution using methyl red or methyl red-methylene blue indicator. Na_2CO_3 solution of 0.02 *N* is made by dissolving 1.060 g anhydrous Na_2CO_3 dried in an oven at 140°C, in distilled water, and diluted to 1 L.

Alternatively, standardize 0.1 *N* commercially available H_2SO_4 or the acid of same strength prepared above against 0.1 *N* Na_2CO_3 (5.300 g anhydrous Na_2CO_3 dissolved in distilled water and diluted to 1 L) using methyl red indicator (the color changes from pink to yellow at the end point) or methyl red-methylene blue indicator.

$$\text{Normality of } H_2SO_4 = \frac{\text{mL } Na_2CO_3 \text{ titrant (blank subtracted)} \times 0.1\,N}{\text{mL } H_2SO_4 \text{ taken}}$$

Dilute 200 mL of H_2SO_4 standardized above to 1 L, which should give a normality in the range 0.02. Record the exact strength of this solution.

Standardization of 0.1 *N* H_2SO_4 may alternatively be performed by potentiometric titration to pH of about 5 using Na_2CO_3 standard.

COLORIMETRIC NESSLERIZATION METHOD

Clean samples may be directly analyzed by this method without distillation. However, a distillate portion should be analyzed for samples containing colored matters, or in presence of interferences.

Under alkaline conditions, Nessler reagent reacts with ammonia to produce a yellow mercuric salt. The intensity of the color produced is proportional to the concentration of ammonia in the sample. The reaction is as follows:

$$2\,K_2HgI_4 + NH_3 + 3\,KOH \Rightarrow Hg_2OINH_2 + 7\,KI + 2\,H_2$$

Procedure

If undistilled sample is used, the addition of zinc sulfate in presence of NaOH would precipitate out iron, magnesium, calcium, and sulfide as a heavy flocculent, leaving a clear and colorless supernate. To 100 mL sample, add 1 mL $ZnSO_4$ solution and mix. Add 0.5 mL NaOH solution. Adjust the pH to 10.5 by further addition of NaOH solution, if needed. Filter out any heavy flocculent precipitate formed. Discard the first 20 mL filtrate. To 50 mL filtrate, add 1 drop of EDTA reagent. If Ca^{2+}, Mg^{2+}, or other metal ions are present in the sample, EDTA would form salt with these ions, thus preventing their reaction with Nessler reagent. After this, add 2 mL of Nessler reagent and mix. Allow the mixture to stand for 10 min. Measure the absorbance or transmittance of the color developed. Measure the color of the blank and standards.

If the sample is distilled, neutralize a 50 mL portion of boric acid distillate with NaOH and then add 1 mL Nessler reagent. Let the solution stand for 10 min, after which measure the absorbance or transmittance against a reagent blank. Prepare a calibration curve under same conditions as samples. The reagent blank and the standards should be distilled, neutralized, and Nesslerized before color measurements.

Calculation

For undistilled sample,

$$\text{mg } NH_3\text{-N/L} = \frac{\mu\text{g } NH_3\text{-N read from calibration curve}}{\text{mL of pretreated sample taken}}$$

When the sample is distilled,

$$\text{mg } NH_3\text{-N/L} = \frac{\mu\text{g } NH_3\text{-N read from the curve}}{\text{mL of original sample} \times \dfrac{V_1}{V_2}}$$

where V_1 is the total volume of distillate collected, mL (including boric acid adsorbent, and V_2 is the mL distillate portion taken for Nesslerization.

Reagents and Standards

- Use ammonia-free distilled water in the preparation of all the reagents and standards.
- Nessler reagent: Dissolve 100 g HgI_2 and 70 g KI in 250 mL distilled water. Add this mixture slowly with stirring to a solution of NaOH (160 g in 500 mL distilled water). Dilute to 1 L.
- EDTA reagent: Add 50 g disodium ethylenediamine tetraacetate dihydrate to 100 mL of 10% NaOH solution (10 g NaOH/100 mL distilled water). Heat if necessary to dissolve.
- Zinc sulfate solution: Dissolve 10 g $ZnSO_4 \cdot 7\,H_2O$ in distilled water and dilute to 100 mL.

- Ammonia-nitrogen standards: Dissolve 3.819 g anhydrous NH_4Cl, dried at 100°C, in distilled water and dilute to 1 L. The strength of this solution is 1000 mg NH_3-N/L or 1 mL = 1 mg NH_3-N = 1.22 mg NH_3.

Dilute 10 mL of stock solution prepared above to 1L. The strength of this secondary standard is 10 mg NH_3-N/L or 1 mL = 10 μg NH_3-N = 12.2 μg NH_3.
Prepare a series of Nessler tube standards as follows:

Secondary standard taken (mL)	NH_3–N/50 mL (μg)
0.0	0.0
1.0	10.0
2.0	20.0
4.0	40.0
6.0	60.0
8.0	80.0
10.0	100.0

COLORIMETRIC PHENATE METHOD

Ammonia reacts with hypochlorite to form monochloroamine. The latter reacts with phenol to form an intensely blue compound, indophenol. The reaction is catalyzed by $MnSO_4$. The reaction steps are as follows:

$$NH_3 + OCl^- \rightarrow NH_2Cl + OH^-$$

(Monochloroamine)

$$NH_2Cl \xrightarrow[\text{catalyst}]{\text{phenol}} O{=}C_6H_4{=}N{-}C_6H_4{-}OH$$

(Indophenol)

The intensity of color developed, which depends on the amount of indophenol produced, is proportional to the concentration of ammonia in the sample. The intensity of the color is measured by a spectrophotometer or a filter photometer (providing a light path of 1 cm) at 630 nm. The concentration of NH_3-N is determined from a calibration standard curve.

Procedure

Take 10 mL sample in a beaker and place it on a magnetic stirrer. Add 1 drop of $MnSO_4$ solution. Add 0.5 mL hypochlorous acid reagent, followed by immediate addition of 0.5 mL (10 drops) phenate reagent dropwise while stirring vigorously.

Allow the solution to stand for 10 min for maximum color development. Read the absorbance at 630 nm after zeroing the spectrophotometer to reagent blank. Run a distilled water blank following the same procedure. Prior to the sample analysis, prepare a calibration standard curve following exactly the above procedure. Run one of the standards through the procedure with each batch of sample.

Calculation

$$\mu\text{g NH}_3\text{-N/L} = \frac{\mu\text{g NH}_3\text{-N read from the standard curve}}{\text{mL sample used}} \times 1000$$

If the results are to be expressed in mg/L, the above equation simplifies to:

$$\text{mg NH}_3\text{-N/L} = \frac{\text{mg NH}_3\text{-N read from the standard curve}}{\text{mL sample used}}$$

In routine analysis, one standard may be run in a batch to check the calibration curve. Within a linear range, the concentration of NH_3–N in the sample may be determined from a single standard as follows:

$$\text{mg NH}_3\text{-N/L} = \frac{A \times B}{C \times S}$$

where A = Absorbance of sample,
B = µg NH_3–N in standard,
C = Absorbance of standard, and
S = mL sample used.

Subtract any blank value, if positive, from the result.

If the sample is turbid or colored or has a high alkalinity (>500 mg $CaCO_3$/L), distill the sample and use a portion distillate. When sample distillate is used in the analysis, multiply the result by $\frac{V_1}{V_2}$, where V_1 is mL of total distillate collected and V_2 is mL distillate used for color development. Use any one of the above equations in the calculation.

Reagents and Standards

- Phenate reagent: Dissolve 2.5 g NaOH and 10 g phenol in 100 mL distilled water.
- Hypochlorous acid: Add 5 mL of 5 to 6% NaOCl solution to 20 mL water. Adjust pH to 6.5 to 7.0 with HCl. The reagent is unstable after a week.
- Manganous sulfate solution (0.003 *M*): 50 mg $MnSO_4 \cdot H_2O$ in 100 mL water.

- Calibration standards: Prepare the stock NH_3-N standard solution by dissolving 0.382 g anhydrous NH_4Cl, dried at 100°C in distilled ammonia-free water, and diluting to 1 L. The strength of this solution is 122 μg NH_3/mL = 100 μg N/mL.
- Secondary standard is made by diluting 5 mL stock solution to 1000 mL. The strength of this solution is 0.500 μg N/mL or 0.605 μg NH_3/mL.
- Prepare the calibration curve using the following standards and record the corresponding absorbance Plot absorbance against μg NH_3-N in the standards.

Taken in beaker	Microgram mass of NH_3-N present
0.5 mL secondary std diluted to 10 mL	0.25
1 mL secondary std + 9 mL water	0.50
2 mL secondary std + 8 mL water	1.0
5 mL secondary std + 5 mL water	2.5
10 mL secondary std + no water	5.0

ION-SELECTIVE ELECTRODE

This method can measure the concentration of NH_3-N in water in the range 0.03 to 1400 mg/L. Color and turbidity do not affect the measurement. Distillation of sample, therefore, is not necessary. High concentration of dissolved solids in the sample, however, can cause error. Also, certain complex forming ions, such as mercury or silver, which form complex with ammonia, interfere with the test. Presence of amines in the sample can give high value.

Use an ammonia electrode (Orion Model 95-10, Beckman Model 39565 or equivalent) along with a readout device, such as a pH meter with expanded millivolt scale between –700 mV and +700 mV or a specific ion meter. The electrode assembly consists of a sensor glass electrode and a reference electrode mounted behind a hydrophobic gas-permeable membrane. The membrane separates the aqueous sample from an ammonium chloride internal solution. Before analysis, the sample is treated with caustic soda to convert any NH_4^+ ion present in the sample into NH_3. The dissolved NH_3 in the sample diffuses through the membrane until the partial pressure of NH_3 in the sample becomes equal to that in the internal solution. The partial pressure of ammonia is proportional to its concentration in the sample. The diffusion of NH_3 into the internal solution increases its pH, which is measured by a pH electrode. The chloride level in the internal standard solution remains constant. It is sensed by a chloride ion-selective electrode which serves as the reference electrode.

Procedure

Follow the manufacturer's instructions for the operation of the electrode.

Prepare a series of NH_3-N standards from the stock solution covering the expected range of its concentrations in the samples. Place 100 mL of each standard solution in 150 mL beakers. To calibrate the electrometer, immerse the electrode into the lowest standard first and add 1 mL of 10 *N* NaOH. This should raise the pH of the solution to above 11. Add NaOH solution only after immersing the electrode.

Stir the solution throughout the measurements at a low but constant rate using a magnetic stirrer. Maintain a constant temperature throughout. Record the millivolt reading when the solution is stable. Repeat this procedure with the remaining standard, measuring the millivolt reading for each standard, proceeding from lowest to highest concentration.

Prepare a calibration curve on a semilogarithmic paper, plotting concentrations of NH_3–N against their corresponding millivolts response. Plot concentrations on the log axis and millivolts on the linear axis.

Record the potentials in millivolts for samples (using 100-mL aliquots or portions diluted to 100 mL for highly concentrated samples) using the above procedure. Determine the concentration of NH_3–N in the samples from the calibration curve. If samples were diluted, multiply the results by the dilution factors.

Alternatively, the analysis may be performed by standard addition method (see Chapter 1.9) in which no calibration curve is required. Prepare an NH_3 standard solution that is about 10 times as concentrated as the estimated concentration of NH_3–N in the sample. Determine the electrode slope following the instruction manual. To 100 mL sample, add 1 mL 10 *N* NaOH and immerse the electrode and stir the solution. Record the millivolt value E_1 when the reading is stable. Add 10 mL of standard solution into the sample. Mix thoroughly and record the stable millivolt reading E_2. Calculate the millivolt difference ΔE as $E_2 - E_1$ and determine the concentration, C_x mg NH_3–N/L, from the following expression:

$$C_x = Q \times C_y$$

where C_y is the concentration of added standard and Q is a reading corresponding to ΔE value which may be found in the known addition table (given in the instruction manual for the electrode).

C_x alternatively may be calculated from the following equation:

$$C_x = \frac{\rho C_y}{\left[(1+\rho)\times 10^{\Delta E/S} - 1\right]}$$

where ρ is the ratio of the volume of spiked standard to the volume of sample taken and S is the electrode slope (see Chapter 1.9 for a typical calculation).

Reagents and Standards

- Use NH_3-free distilled water to prepare all reagents and standards.
- 10 *N* NaOH: Dissolve 40 g NaOH in distilled water and dilute to 100 mL.
- NH_3–N standard: Dissolve 3.819 g anhydrous NH_4Cl in distilled water and dilute to 1 L. Concentration of this solution is 1000 mg NH_3–N/L. Prepare 0.1, 1, 10, and 100 mg/L standards by dilutions. Prepare the calibration curve in the concentration range 0.1 to 1000 mg NH_3–N/L.

2.15 NITROGEN (NITRATE)

Nitrate, which is produced by oxidation of nitrogen, is a monovalent polyanion having the formula NO_3^-. Most metal nitrates are soluble in water and occur in trace amounts in surface- and groundwaters. Nitrate is toxic to human health and, chronic exposure to high concentrations of nitrate, may cause methemoglobinemia. Maximum contaminant limit in potable water imposed by U.S. EPA is 10 mg nitrate as nitrogen/L.

Nitrates in water may be analyzed by the following methods:

1. Ion chromatography
2. Nitrate selective electrode method
3. Cadmium reduction method
4. Miscellaneous reduction method
5. Brucine method

Methods 3 and 4 are colorimetric procedures based on reduction of nitrate to nitrite, followed by diazotization and then coupling to an azo dye. The analysis may be performed manually or by use of an automated analyzer. Method 2 is applicable in the range 10^{-5} to 10^{-1} M NO_3^-. The colorimetric method 5 has been found to give inconsistent results.

CADMIUM REDUCTION METHOD

In the presence of cadmium, nitrate (NO_3^-) is reduced to nitrite (NO_2^-). The nitrite produced is diazotized with sulfanilamide. This is followed by coupling with N-(1-naphthyl)-ethylenediamine to form a highly colored azo dye. The intensity of the color developed is measured by a spectrophotometer or a filter photometer at 540 nm. The concentration of oxidized N/L (NO_3^--N plus NO_2^--N) is read from a standard curve prepared by plotting absorbance (or transmittance) of standard against NO_3^--N concentrations.

The reactions are as follows:

$$NO_3^- + Cd + 2H^+ \longrightarrow NO_2^- + Cd^{2+} + H_2O$$

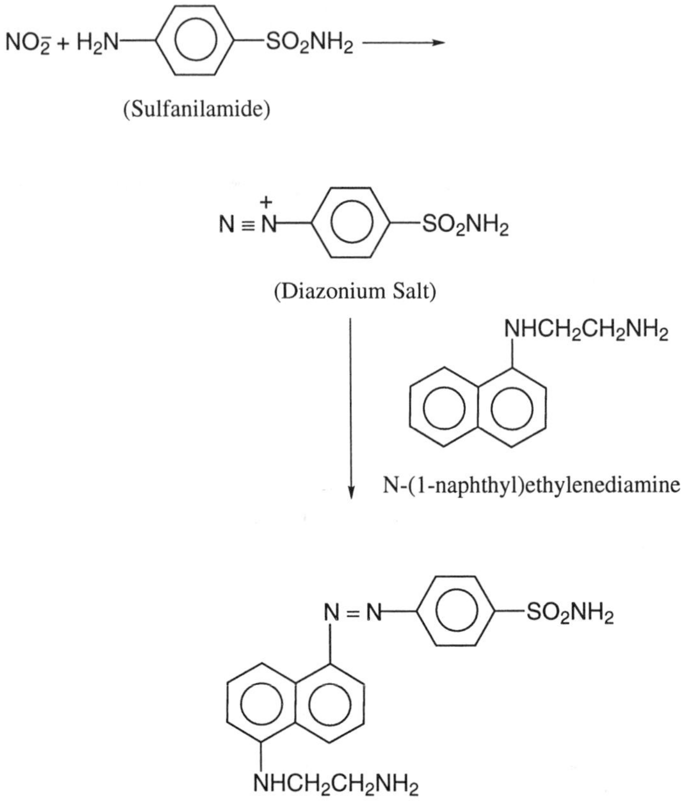

$$NO_2^- + H_2N\text{—}C_6H_4\text{—}SO_2NH_2 \longrightarrow$$

(Sulfanilamide)

$$N \equiv \overset{+}{N}\text{—}C_6H_4\text{—}SO_2NH_2$$

(Diazonium Salt)

NHCH2CH2NH2 (on naphthalene ring)

N-(1-naphthyl)ethylenediamine

$$N = N\text{—}C_6H_4\text{—}SO_2NH_2$$ (N=N attached to naphthalene ring bearing $NHCH_2CH_2NH_2$)

(Colored Azo Dye)

Nitrite is determined separately by the colorimetric method on another aliquot of the sample not subjected to Cd reduction. Subtract NO_2^--N value to determine the concentration of NO_3^--N in the sample.

Procedure

Mix 25 mL sample with 75 mL of NH_4Cl-EDTA solution. The sample mixture is passed through the cadmium column at a rate of 5 to 10 mL/min. This reduces nitrate to nitrite. Discard the first 25 mL and collect the remaining solution.

To 50 mL of solution passed through cadmium column, add 2 mL of color reagent and mix. Allow the solution to stand for 30 min. Measure the absorbance at 543 nm against a distilled water-reagent blank. Read the concentration of NO_3^--N from the prepared standard curve. Standards should be reduced exactly like samples.

Apparatus, Reagents, and Standards

- Cadmium column: Insert glass wool plug at the bottom of the column. Add Cd-Cu granules into the column producing a height of 18.5 cm. Fill the column with water and maintain water level above the granules to prevent any entrapment of air. Wash the column with a dilute solution of NH_4Cl–EDTA. Before beginning analysis, activate the column bypassing 100 mL solution containing 1 mg NO_3^--N standard in NH_4Cl-EDTA solution (25:75).
- Ammonium chloride-EDTA solution: Dissolve NH_4Cl (13 g) and disodium EDTA (1.7 g) in groundwater. Adjust the pH to 8.5 with conc. NH_4OH and dilute to 1 L.
- Color reagent: Dissolve 10 g sulfanilamide in dilute phosphoric acid solution (100 mL 85% H_3PO_4 in 800 mL water). Add 1 g *N*-(1-naphthyl) ethylenediamine dihydrochloride into this solution. Mix well and dilute to 1 L.
- Standards: Dissolve 7 g dried potassium nitrate (KNO_3) in distilled water and dilute to 1 L [1 mL ≡ 1 mg (1000 ppm)]. Dilute 10 mL of the above stock solution to 1 L to produce a secondary standard, 10 mg NO_3^--N/L.

  Dilute 0.5, 1.0, 2.0, 5.0, and 10.0 mL of the secondary standard to 100 mL, respectively. This would give a series of standards of strength 0.05, 0.1, 0.2, 0.5, and 1.0 mg NO_3^--N/L, respectively. Check the efficiency of the column by comparing one of the reduced nitrate standards to a nitrite standard at the same concentration. If the reduction efficiency falls below 75%, use freshly prepared Cu-Cd granules.
- Copper-cadmium granules: Mix 25 g 40 to 60 mesh Cd granules with 6 *N* HCl and shake well. Decant off HCl and rinse the granules with distilled water. Add 100 mL 2% $CuSO_4$ solution to the cadmium granules. Swirl for 5 to 7 min. Decant and repeat this step with fresh $CuSO_4$ solutions until a brown colloidal precipitate of Cu appears. Wash out the precipitate with water.

MISCELLANEOUS REDUCTION METHOD

Reducing agents other than Cd may be used to convert nitrate to nitrite. Some of those agents are hydrazine sulfate ($N_2H_4 \cdot H_2SO_4$), titanous chloride ($TiCl_3$), and stannous chloride ($SnCl_2$). The nitrite formed is diazotized with sulfanilamide and coupled with N-(1-naphthyl) ethylenediamine dihydrochloride to form a highly colored azo dye, the absorbance of which may be measured by a spectrophotometer. A calibration curve is plotted using NO_3^--N standards subjected to similar reduction. Nitrite is determined separately without reduction, and the value is subtracted from the total nitrate and nitrite as analyzed after reduction.

NITRATE ELECTRODE METHOD

Nitrate in water may be analyzed by a nitrate selective sensor. Chloride and bicarbonate ions at concentrations about ten times greater than nitrate interfere in this test. Sulfide, cyanide, and halide ions are eliminated by using a buffer solution containing $AgSO_4$. The buffer — boric acid at pH 3 — removes bicarbonate.

Procedure

Prepare three nitrate standards at concentrations 1, 10, and 50 mg/L. To 10 mL of each standard in 40 mL beakers, add 10 mL buffer solution, respectively. Immerse the tip of the electrode in the solution while stirring. Record the stable millivolt reading for each standard. Prepare a calibration curve plotting NO_3^--N concentrations in the abscissa and millivolts in the ordinate. The plot should be a straight line with a slope of 57 ± 3 mV/decade at 25°C. After this, rinse and dry the electrode; immerse in the sample-buffer mixture and record the stable potential reading. Determine the concentration of NO_3^--N in the sample from the calibration curve.

Reagents

- Buffer solution: Dissolve the following reagents in 700 to 800 mL distilled water: 17.3 g Al_2SO_4, 1.3 g boric acid, and 2.5 g sulfamic acid. Add 0.10 *N* NaOH slowly to adjust the pH to 3.0. Dilute to 1 L and store in a dark bottle.
- Reference electrode filling solution: 0.53 g $(NH_4)_2SO_4$ in 1 L distilled water.

2.16 NITROSAMINES

Nitrosamines or nitrosoamines are nitroso derivatives of amines in which a nitroso (NO) group is attached to the nitrogen atom of the amine. These compounds have the following general structure:

$$O{=}N{-}N\begin{matrix} R \\ R' \end{matrix}$$

where R and R′ are alkyl or aryl groups.

Nitrosamines are toxic compounds as well as potent animal and human carcinogens (Patnaik, 1992). These substances occur in trace quantities in tobacco smoke, meat products, and salted fish. Some of these compounds are classified by U.S. EPA as priority pollutants in industrial wastewaters, potable waters, and hazardous wastes. These nitrosamines are listed in Table 2.16.1. Such pollutants occurring in environmental samples can be determined by U.S. EPA's analytical procedures (U.S. EPA 1990, 1992).

Table 2.16.1 Nitrosamines Classified as Priority Pollutants by U.S. EPA under the Resource Conservation and Recovery Act

CAS no.	Compound
[62-75-9]	*N*-Nitrosodimethylamine [a]
[10595-95-6]	*N*-Nitrosomethylethylamine
[55-18-5]	*N*-Nitrosodiethylamine
[621-64-7]	*N*-Nitrosododi-*n*-propylamine [a]
[924-16-3]	*N*-Nitrosodibutylamine
[86-30-6]	*N*-Nitrosodiphenylamine [a]
[100-75-4]	*N*-Nitrosopiperidine
[930-55-2]	*N*-Nitrosopyrrolidine
[59-89-2]	*N*-Nitrosomorpholine

[a] Also classified as pollutants in the wastewater category.

The analysis involves extraction of these compounds into a suitable solvent followed by their detection on the GC using either a nitrogen-phosphorus detector (NPD), or a reductive Hall Electrolytic Conductivity Detector (HECD) or a Thermal Energy Analyzer (TEA). These compounds can be detected by a FID too, however, at a lower sensitivity. For example, the detector response to the lowest amount of *N*-nitrosodi-*n*-propylamine on FID is in the range 200 ng, in comparison to 5 ng on the NPD. When interferences are encountered, TEA or a reductive HECD should be used for the analysis. Both of these detectors offer high sensitivity and selectivity. The presence of nitrosamines in the sample extract must be confirmed either on a GC using a second column or by GC/MS preferably under high resolution MS conditions. *N*-Nitrosodiphenylamine breaks down to diphenylamine at a temperature above 200°C in the GC inlet and is measured as diphenylamine. The latter compound, if present in the sample, must, therefore, be removed by Florisil or Alumina cleanup before analysis.

EXTRACTION

A 1-L aliquot, or any appropriate volume of accurately measured aqueous sample, is extracted with methylene chloride by liquid-liquid extraction. The extract is concentrated to 1 mL or any small volume on a Kuderna-Danish setup. A florisil column cleanup may be necessary if the sample is dirty, or the presence of interferences is known or suspected or if *N*-nitrosodiphenylamine is to be determined.

Soils, sediments, and hazardous wastes should be extracted with methylene chloride by sonication or Soxhlett extraction. The extract is then concentrated and cleaned up for the removal of interferences as follows.

Prior to florisil column cleanup, the column should be preeluted with ether-pentane mixture (15:85 v/v). The extract is then transferred onto the column and the column is eluted again to remove any diphenylamine contaminant. The analytes *N*-nitrosodimethylamine, *N*-nitrosodi-*n*-propylamine, and *N*-nitrosodiphenylamine, and other similar aliphatic and aromatic nitrosamines are desorbed from the column with acetone-ether mixture (5:95 v/v). This fraction is then concentrated for analysis.

In alumina column cleanup, the column is first preeluted with ether-pentane mixture (30:70) before the sample extract is transferred onto the column. It is then successively eluted with ether-pentane mixture of 30:70 and 50:50% composition, respectively. This separates *N*-nitrosodiphenylamine. The latter elutes into the first fraction, from the interfering substance diphenylamine which goes into the second fraction along with the analytes *N*-nitrosodimethylamine and *N*-nitrosodi-*n*-propylamine. A small amount of the latter compound is also eluted into the first fraction. A cleanup procedure for other nitrosamines (not classified under U.S. EPA's priority pollutants) should generally be the same as described above. The composition of ether-pentane mixture and the elution pattern, however, must be established first before performing the cleanup.

Table 2.16.2 Characteristic Masses for Some Common Nitrosamine Pollutants

Compounds	Primary ion (m/z)	Secondary ions (m/z)
N-Nitrosodimethylamine	42	74, 44
N-Nitrosomethylethylamine	88	42, 43, 56
N-Nitrosodiethylamine	102	42, 57, 44, 56
N-Nitrosodi-*n*-propylamine	70	42, 101, 130
N-Nitrosodibutylamine	84	57, 41, 116, 158
N-Nitrosodiamylamine	98	57, 41, 186
N-Nitrosodiphenylamine [a]	169	168, 167
N-Nitrosopiperidine	114	42, 55, 56, 41
N-Nitrosopyrrolidine	100	41, 42, 68, 69

[a] Diphenylamine produces the same characteristic ions.

ANALYSIS

The sample extract is analyzed by GC-NPD and/or GC/MS. Other GC detectors, as mentioned earlier, may be used instead of NPD. If low detection level is desired, quantitation should be done from the GC analysis. The presence of any analyte found in the sample must be confirmed on an alternate GC column or preferably by GC/MS. If a thermoionic detector is used and if the sample extract is to be analyzed without any cleanup, it is necessary to exchange the solvent from methylene chloride to methanol.

Column and conditions are as follows:

- Packed column: 1.8 m × 4 mm ID glass packed with (1) 10% Carbowax 20 M/2% KOH on Chromosorb W-AW (80/100 mesh), or (2) 10% SP-2250 on Supelcoport (100/120 mesh) or equivalent at 110° to 220°C at 8°C/min, with the carrier gas at 40 mL/min flow rate.
- Capillary column: Fused silica capillary column, such as RTX-5, DB-5, SPB-5, or equivalent, 15 m × 0.53 mm ID × 1.5 μm film (direct injection) at 40° to 240°C at H_2 carrier gas linear velocity 80 cm/s (or flow rate 10 mL/min). A longer column 30 m with smaller ID and lesser film thickness may be used with split injection. Temperature and carrier gas flow conditions may be set accordingly.

The characteristic ions for GC/MS identification (under electron impact ionization at 70 eV nominal energy) for some common nitrosamine pollutants are tabulated in Table 2.16.2.

AIR ANALYSIS

Nitrosamines in air may be analyzed by NIOSH Method 2522 (NIOSH, 1989). Air at a flow rate between 0.2 and 2 L/min is passed through a solid sorbent tube containing Thermosorb/N as an adsorbent. The sample volume should be between 15 and 1000 L. If high concentrations of nitrosamines are expected, another backup Thermosorb/N tube should be used in sampling. The sorbent tube, however, can adsorb the analytes up to 1500 μg loading with no breakthrough. The analytes are desorbed with 2 mL methanol-methylene chloride mixture (1:3), allowed to stand for 30 min, and analyzed by GC using a TEA. The column and conditions employed in the method development are as follows:

- Column: Stainless steel, 10" × 1/8" packed with 10% Carbowax 20 *M* + 2% KOH on Chromosorb W-AW.
- Temperature: Injector 200°C, detector 550° to 600°C, column 110° to 200°C at 5°C/min.
- Gases: N_2 carrier, 25 mL/min; oxygen, 5 mL/min; ozone 0.2 mL/min.
- U.S. EPA Method TO7 describes the determination of N-nitrosodimethylamine in ambient air (U.S. EPA, 1986). The method is similar to the NIOSH method discussed above and uses Thermosorb/N as adsorbent. The air flow is 2 L/min and the sample volume recommended is 300 L air. The analyte is desorbed with methylene chloride and determined by GC/MS or an alternate selective GC system, such as TEA, HECD, or thermoionic nitrogen-selective detector. The latter detector and the TEA are more sensitive and selective than the other detectors. Therefore, the interference from other substances is minimal. Other nitrosamines in air may be determined in the same way.

2.17 OXYGEN DEMAND, BIOCHEMICAL

Biochemical oxygen demand (BOD) is an empirical test that measures the amount of oxygen required for microbial oxidation of organic compounds in aqueous samples. Such a test measures the amount of oxygen utilized during a specific incubation period (generally, 5 days) for biochemical oxidation of organic materials and oxidizable inorganic ions, such as Fe^{2+} and sulfide. The incubation is performed in the dark at 20 ± 1°C. The results of the BOD analyses are used to calculate waste loadings and to design wastewater treatment plants.

Different volumes of sample aliquots are placed in 300 mL incubation bottles and diluted with "seeded" dilution water. The bottles are filled to their full capacity without leaving any headspace, and tightly closed. The BOD bottles are then placed in a thermostatically controlled air incubator or a water bath at 20 ± 1°C in the dark to prevent any photochemical reaction.

The dilution water is prepared by adding 1 to 2 mL of phosphate buffer solution to an equal volume of $MgSO_4 \cdot 7\,H_2O$ (22.5 g/L), $FeCl_3 \cdot 6\,H_2O$ (0.25 g/L), and $CaCl_2$ (27.5 g/L) and diluting into desired volume of reagent grade water. The amounts of components per liter of phosphate buffer solution are KH_2PO_4 (8.5 g), K_2HPO_4(21.75 g), $Na_2HPO_4 \cdot 7\,H_2O$ (33.4 g), and NH_4Cl (1.7 g). The pH of this solution should be 7.2.

Because BOD measures the amount of oxygen needed by the microbes to oxidize the organics in the wastewater, this oxygen must, therefore, be supplied initially into the aqueous medium before incubation. The dilution water, therefore, must contain a sufficient quantity of dissolved oxygen (DO). At ambient conditions, oxygen is slightly soluble in water. Such an amount of oxygen, however, is often sufficient to oxidize trace organics found in relatively clean samples. To ensure availability of surplus oxygen in the medium, the dilution water should be aerated using an air compressor. This enhances the DO concentration.

The concentration of DO before and after incubation is measured. If the microbial population is not sufficiently large in the sample, microorganisms should be added into the dilution water from an outside source. Oxygen consumed by the organics is determined from the difference, and the BOD is calculated as follows:

$$\text{BOD, mg/L} = \frac{(A_1 - A_2)V_2}{V_1}$$

where A_1 = Initial conc. DO, mg/L in the dilution water prepared,
A_2 = Conc. DO, mg/L in the diluted sample after 5 days incubation,
V_1 = mL sample aliquot diluted, and
V_2 = Volume of the BOD bottle, mL.

The population of microorganisms should be adequately large to oxidize biodegradable organics. The dilution water must, therefore, be seeded, especially in the BOD determination of such organic-rich polluted water.

Domestic wastewaters, polluted surface waters, or the effluents from biological waste treatment plants are sources of high microbial populations. When such seeded dilution water is used, the BOD is calculated as follows:

$$\text{BOD, mg/L} = \frac{[(A_1 - A_2) - (B_1 - B_2)f] \times V_2}{V_1}$$

where B_1 = DO of seed control* before incubation, mg/L,
B_2 = DO of seed control after incubation, mg/L, and

$$f = \frac{\%\ \text{seed in diluted sample}}{\%\ \text{seed in seed control}}$$

GRAPHICAL CALCULATION

BOD may also be calculated by graphical method (Hach, 1982). These graphical calculations have the following advantages over the single point calculation outlined above:

1. DO of dilution water, even if >0.2 mg/L, would not affect the calculation.
2. The BOD value would be more accurate and reliable because it is based on the valid data points only, while the outlining bad data points are discarded.
3. The initial DO does not need to be measured.

Method

Prepare a series of five or six dilutions using different volumes of sample aliquots. No initial DO is required to be measured. These dilutions are incubated for 5 days. The DO remaining after incubation is measured. A graph is constructed, plotting mL samples diluted vs. mg/L DO remaining after incubation.

* The BOD of the seed is measured like any sample. The seeded dilution water should consume between 0.6 and 1.0 mg/L DO.

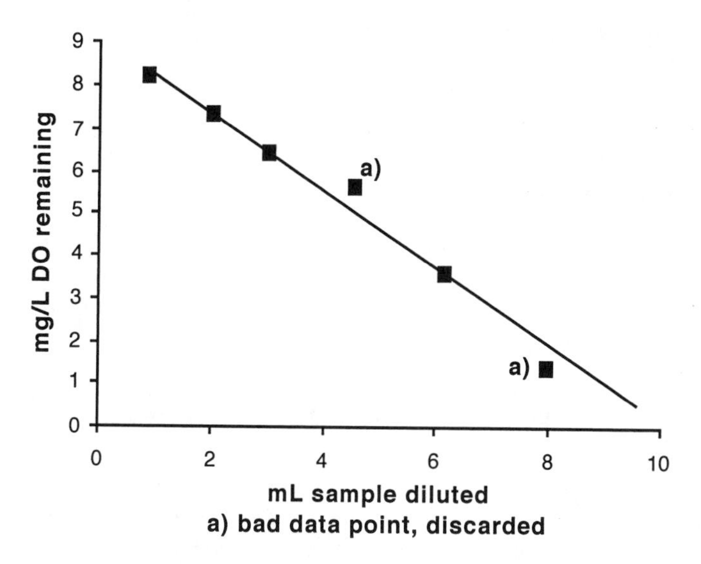

Figure 2.17.1 Example of a graph showing a series of results from an experiment.

mL Sample diluted	mg/L DO remaining
1	8.25
2	7.40
3	6.65
4	5.80
6	3.80
8	1.45

The graph is a straight line over the best fit points. The slope of the line is equal to the quantity of DO consumed (as mg/L) per mL of sample. The y-intercept should be equal to the DO in the dilution water after incubation. The BOD is calculated from the following equation:

$$\text{mg/L BOD} = (\text{slope} \times V_2) - y \text{ intercept} + \text{sample DO}$$

where V_2 is the volume of the BOD bottle, mL.

The graphical method (Figure 2.17.1) of BOD calculation is shown in the above table. Sample DO in this example was 3.0 mg/L, determined by potentiometric or Winkler titration. From the above graph (Figure 2.17.1):

$$\text{slope} = 9/10 = 0.90$$

$$y \text{ intercept} = 9.0 \text{ mg/L}$$

$$\text{If } V_2 = 300 \text{ mL (BOD bottle)}$$

$$\text{Therefore, BOD} = (0.90 \times 300) - 9 + 3$$

$$= 264 \text{ mg/L}$$

MEASUREMENT OF DISSOLVED OXYGEN (DO)

DO in water may be determined by the following two methods: iodometric titration (Winkler method) and electrode method.

Iodometric or Winkler Method

A measured volume of sample is taken in a glass-stoppered bottle. Add to this a divalent manganese solution ($MnSO_4$). This is followed by the addition of a strong base. This results in the precipitation of divalent $Mn(OH)_2$, as shown below in the following equation:

$$MnSO_4 + 2\ NaOH \rightarrow Mn(OH)_2 + Na_2SO_4$$

DO in the sample rapidly oxidizes an equivalent amount of dispersed Mn^{2+} precipitate to higher valent manganese(IV) oxide hydroxide floc, which has the formula $MnO(OH)_2$. This is shown below in the following equation:

$$2\ Mn^{2+} + O_2 + 4\ OH^- \rightarrow 2\ MnO(OH)_2$$

In the presence of iodide ion, and in acid medium, Mn^{4+} is reduced back to divalent Mn^{2+}, thus liberating I_2.

$$Mn^{4+} + 2\ I^- \xrightarrow{\text{acid}} Mn^{2+} + I_2$$

The iodine liberated is stoichiometrically equivalent to the DO in the sample. It is titrated against a standard solution of $Na_2S_2O_3$ or phenyl arsine oxide using starch indicator to a colorless end point.

$$\text{mg DO/L} = \frac{A \times N}{\text{mL sample}} \times 8000$$

where A is the mL titrant, $Na_2S_2O_3$ or PAO and N is the normality of the titrant.

In BOD measurement, the sample volume is usually 300 mL. The milli-equivalent weight for oxygen is 8000.

Various modifications of the original Winkler method have been developed to eliminate interferences. This includes adding azide to suppress any interference from NO_2^- in the sample.

In this method, sodium azide solution is added before acidification into the sample mixture containing Mn^{4+} flocculent. Azide prevents any possible reaction of nitrite with iodide. Interference from Fe^{3+} is overcome by adding a small amount of KF solution (1 mL of 40% soln.) before acidification.

If the normality of the $Na_2S_2O_3$ titrant is 0.0375 *N*, each mL titrant is equivalent to 1 mg DO, when the entire 300 mL content of the BOD bottle is titrated.

Electrode Method

Oxygen-sensitive membrane electrodes are commercially available. The electrode in such a system is covered with an oxygen-permeable plastic membrane, thus protecting it from impurities. The current is proportional to the activity of dissolved molecular oxygen, and at low concentration, to the amount of DO. Before the measurement of DO in the sample, calibrate the electrodes using standards of known DO concentrations (determined from iodometric titrations).

KINETICS OF BOD REACTION

The BOD reaction follows first-order kinetics. The reaction rate is proportional to the amount of oxidizable organic matter remaining at any time. Also, the reaction rate depends on the temperature. The population of microorganisms should be adequately large and stabilized. The BOD value for any incubation period may be theoretically determined from its ultimate BOD and the rate constant for the reaction from the following equation:

$$Bt = U\left(1 - 10^{-kt}\right)$$

where Bt = BOD-t for t days incubation period,
U = Ultimate BOD, and
k = Rate constant for the reaction.

In order to calculate the BOD at any time, the k value must be experimentally determined. For this, the DO concentrations should be measured for any two incubation periods and then calculated from the following equation for the first-order reaction:

$$k = \frac{ln\left(\frac{C_1}{C_2}\right)}{T_2 - T_1}$$

where C_1 = Conc. DO, mg/L after T_1 incubation time,
C_2 = Conc. DO, mg/L after T_2 incubation period, and
k = Rate constant for the reaction.

Because the ultimate BOD should be theoretically equal to the COD, the BOD at any time may also be approximately estimated from the COD value. This is shown below in the following example:

Example

Two aliquots of a sample were diluted equally and the DO contents were measured after 2 and 5 d incubation periods, respectively. The DO concentrations were 6.5 and 4.3 mg/L, respectively. Determine the rate constant *k* for the BOD reaction.

$$k = \frac{ln\left(\frac{6.5 \text{ mg/L}}{4.3 \text{ mg/L}}\right)}{(5-2) \text{ days}}$$

$$= 0.137$$

From the known *k* value, BOD may be calculated for any incubation period. However, to solve this problem, the ultimate BOD or the BOD for any other incubation period must be known. This is illustrated in the following problem.

Example

The ultimate BOD of a sample was measured as 540 mg/L after a 30-d incubation period. Determine the BOD-5, if the rate constant, *k*, is 0.05 for the reaction.

$$\text{BOD-5} = 540\left(1 - e^{-0.05 \times 5}\right) \text{ mg } O_2 / \text{L}$$

$$= 540(1 - 0.78) \text{ mg } O_2 / \text{L}$$

$$= 119 \text{ mg/L}$$

Alternatively, DO concentration after 5-d incubation may be theoretically calculated from the DO concentation measured after 30-d incubation. Such calculation requires the use of first-order rate equation. From the initial DO before incubation that was measured and the DO for a 5-d incubation that may be theoretically estimated, the BOD-5 may be determined. However, for any such calculations, the rate constant, *k*, must be known.

2.18 OXYGEN DEMAND, CHEMICAL

Chemical oxygen demand (COD) is a measure of the oxygen equivalent of organic matter in the sample that is susceptible to oxidation by a strong oxidizing agent. A boiling mixture of potassium dichromate ($K_2Cr_2O_7$)–H_2SO_4 can oxidize most types of organic matter and is generally used in the COD determination. Other strong oxidants, such as $KMnO_4$–H_2SO_4 are also effective.

Complete oxidation of organic compounds under such strong oxidizing conditions produces carbon dioxide and water. Other additional products such as HCl or NO_2 may, however, form if the organic compound contains a Cl or N atom, respectively, in its molecule. COD for any organic compound or any oxidizable inorganic ion (i.e., anions or metal ions in their lower oxidation states) can be theoretically calculated from writing a balanced equation. Some oxidation reactions are presented below and how COD may be calculated is shown in the following problem:

$$2\,C_5H_{10} + 15\,O_2 \xrightarrow[\text{acid}]{K_2Cr_2O_7} 10\,CO_2 + 10\,H_2O$$

(2 mol *n*-pentane reacts with 15 mol O_2)

$$C_5H_5OH + 3\,O_2 \xrightarrow[\text{acid}]{K_2Cr_2O_7} 2\,CO_2 + 3\,H_2O$$

(1 mol ethanol reacts with 3 mol O_2)

$$2\,Fe^{2+} + O_2 \xrightarrow[\text{acid}]{K_2Cr_2O_7} 2\,Fe^{3+} + 2\,O^{-}$$

(Metal ion is oxidized to higher oxidation state.)

Problem

A 250-mL aliquot of wastewater containing 72 mg/L of *n*-propanol (C_3H_7OH) was spiked with 50 mL of 100 mg/L acetone solution. Determine the COD as mg/L of the resulting solution.

To solve this problem, first we write the balanced oxidation reactions for propanol and acetone, which are as follows:

$$2\,C_3H_7OH + 9\,O_2 \rightarrow 6\,CO_2 + 8\,H_2O$$

$$(n\text{-Propanol})$$

$$CH_3COCH_3 + 4\,O_2 \rightarrow 3\,CO_2 + 3\,H_2O$$

$$(\text{Acetone})$$

$$250\text{ mL of }72\text{ mg/L }n\text{-propanol} = 0.25\text{ L} \times \frac{72\text{ mg}}{\text{L}} = 18\text{ mg }n\text{-propanol}$$

The oxygen requirement for this 18 mg *n*-propanol

$$= 0.018\text{ g }C_3H_7OH \times \frac{1\text{ mol }C_3H_7OH}{60\text{ g }C_3H_7OH} \times \frac{9\text{ mol }O_2}{2\text{ mol }C_3H_7OH} \times \frac{32\text{ g }O_2}{1\text{ mol }O_2}$$

$$= 0.0432\text{ g }O_2$$

or 43.2 mg O_2. Similarly,

$$50\text{ mL of }100\text{ mg/L acetone} = 0.05\text{ L} \times \frac{100\text{ mg}}{\text{L}}$$

or 5 mg acetone, the oxygen requirement for which

$$= 0.005\text{ g acetone} \times \frac{1\text{ mol acetone}}{58\text{ g acetone}} \times \frac{4\text{ mol }O_2}{1\text{ mol acetone}} \times \frac{32\text{ g }O_2}{1\text{ mol }O_2}$$

$$= 0.0011\text{ g }O_2\text{ or }11.0\text{ mg }O_2$$

Thus, the total amount of O_2 required to oxidize 18 mg *n*-propanol and 5 mg acetone = 43.2 mg O_2 + 11.0 mg O_2 = 54.2 mg O_2.

The total volume of sample + spiking solution = 250 mL + 50 mL = 300 mL, i.e., 54.2 mg O_2/300 mL solution, or

$$\frac{54.2\text{ mg }O_2}{300\text{ mL}} \times \frac{1000\text{ mL}}{1\text{ L}} = 180.7\text{ mg }O_2/\text{L}$$

Therefore, the COD of the resulting solution = 180.7 mg/L.

Certain inorganic substances also contribute to COD. These include oxidizable anions such as S^{2-}, SO_3^{2-}, NO_2^-, PO_3^{3-}, AsO_2^-, ClO^-, ClO_2^-, ClO_3^-, BrO_3^-, IO_3^-, and SeO_3^{2-} and such metal ions as Fe^{2+}, Cu^{1+}, Co^{2+}, Sn^{2+}, Mn^{2+}, Hg^{1+}, and Cr^{3+} (mostly transition metal ions in their lower oxidation states). If any of these ions is present in detectable concentration, its COD can be calculated from the balanced equation:

$$NO_2^- + \tfrac{1}{2}O_2 \rightarrow NO_3^-$$

Thus, 1 mol of nitrite ion (NO_2^-) would react with 0.5 mol of O_2 to form nitrate (NO_3^-), i.e., 46 g NO_2^- would require 16 g oxygen; or the COD due to 1 mg/L NO_2^- would be 0.35 mg/L. This is equivalent to a COD of 1.15 mg/L for every 1 mg/L of nitrite nitrogen (NO_2^-–N).

Ammonia present in the sample also contributes to COD. However, in presence of excess free Cl^-, it is converted into HH_4Cl and is not oxidized. Long chain aliphatic compounds are often difficult to oxidize. A catalyst, Ag_2SO_4, is therefore added to promote such oxidation. Halide ions present in the sample may, however, react with Ag_2SO_4 forming silver halide. Such halide interference may be partially overcome by adding $HgSO_4$.

An aliquot of sample is refluxed in a strong acid solution of a known excess of potassium dichromate ($K_2Cr_2O_7$) in the presence of Ag_2SO_4 and $HgSO_4$. Cr^{6+} (dichromate ion) is reduced to Cr^{3+} during oxidation. The net ionic reaction in the acid medium is as follows:

$$Cr_2O_7^{2-}(aq) + 14\ H^+(aq) + 6\ e^- \rightarrow 2\ Cr^{3+}(aq) + 7\ H_2O$$

After the digestion is completed, the amount of Cr^{6+} consumed is determined either by titration or by colorimetry. The initial amount of Cr^{6+}, and that which is left after the reaction, are determined by titrating against a standard solution of ferrous ammonium sulfate (FAS) using ferroin indicator. The initial amount of Cr^{6+} may also be calculated from the standard grade of $K_2Cr_2O_7$ used to prepare the solution.

The reaction of FAS with Cr^{6+} is as follows:

$$3\ Fe^{2+} + Cr^{6+} \rightarrow 3\ Fe^{3+} + Cr^{3+}$$

1,10-Phenanthroline (ferroin), which is used as an indicator in this titration, forms an intense red color with Fe^{2+}, but no color with Fe^{3+}. When all the Cr^{6+} is reduced to Cr^{3+}, Fe^{2+} reacts with the indicator forming ferroin complex. The color of the solution changes from greenish blue to reddish brown signaling the end point of titration.

The COD in the sample is calculated as follows:

$$\text{COD, mg/L} = \frac{(A - B) \times N \times 8000}{\text{mL sample}}$$

where A = mL FAS required for titration of blank,
B = mL FAS required for the titration of sample, and
N = Normality of FAS.

The multiplication factor 8000 is derived as follows:

> As we see from the reaction above, 3 mol of Fe^{2+} require 1 mol of Cr^{6+}, i.e., 0.333 mol $K_2Cr_2O_7$/each mol FAS titrant. Also, we used 0.0417 mol $K_2Cr_2O_7$. Multiplying by 1000 to convert grams to milligrams, we finally get a factor of $\frac{0.333 \times 1000}{0.0417}$ or 8000.

Standardization of FAS titrant:

98 g $Fe(NH_4)_2(SO_4)_2 \cdot 6\,H_2O$ in distilled water + 20 mL conc. H_2SO_4 and dilute to 1 L. This solution of approximately 0.25 *M* (or 0.25 *N*) strength is titrated against standard solution of $K_2Cr_2O_7$. The latter solution is made by dissolving 12.259 g $K_2Cr_2O_7$ (dried at 105°C for 2 h) in distilled water and diluting to 1 L. This solution has a strength of 0.0417 *M* or a normality of 0.25 *N*. (The equivalent weight of $K_2Cr_2O_7 = \frac{294.22 \text{ (Formula weight)}}{6}$ or 49.036.)

$$\text{Normality of FAS} = \frac{\text{mL } K_2Cr_2O_7 \times 0.25}{\text{mL FAS}}$$

Indicator solution

- Dissolve 1.485 g 1,10-phenanthroline monohydrate and 0.695 g $FeSO_4 \cdot 7\,H_2O$ in distilled water and dilute to 1 L.
- Ag_2SO_4: 5.5 g is dissolved in 1000 g of conc. H_2SO_4.

SAMPLE TREATMENT

To a 50-mL sample in a 500-mL refluxing flask, 1 g $HgSO_4$ is added. This is followed by the slow addition of 5 mL conc. H_2SO_4 (containing Ag_2SO_4), some glass beads, a 25 mL 0.0417 *M* $K_2Cr_2O_7$, and 70 mL conc. H_2SO_4 slowly. The mixture is refluxed for 2 h, cooled, diluted, and titrated against standard ferrous ammonium sulfate.

For samples containing low organics, $K_2Cr_2O_7$ solution of 0.00417 *M* strength should be used. This may be titrated against 0.025 *N* FAS. The sample digestion may be performed under open reflux conditions as described earlier, or in borosilicate ampules heated at 150°C on a heating block. When such culture tubes or ampules are used, sample volumes must be below 10 mL and the acid and $K_2Cr_2O_7$ solutions must be in small volumes accordingly. The total volume of sample plus reagents should not exceed two thirds volume of the container.

COLORIMETRIC ANALYSIS

Colorimetric measurement of COD is faster and easier to perform than the titrimetric analysis. In addition, additional reagents are not required. The sample is digested in an ampule, culture tube, or vial under closed reflux conditions. The concentration of dichromate is determined from its absorbance which is measured by a spectrophotometer set at 600 nm and compared against the standard calibration curve. Hach COD vials (Hach, 1994) are commercially available for COD measurements in low, medium, and high ranges. For low range COD analysis, the Hach method measures the amount of yellow Cr^{6+} left after the dichromate reduction. On the other hand, high range measurement determines the amount of green Cr^{3+} produced (Hach, 1989).

Potassium hydrogen phthalate (KHP) is used as a reference standard for COD analysis. The theoretical COD for 1 mg KHP is 1.175 mg, which is determined from either of the following equations:

$$2\ C_6H_4\begin{matrix}\text{—COOH}\\ \text{—COOK}\end{matrix} + 10\ K_2Cr_2O_7 + 41\ H_2SO_4 \rightarrow 16\ CC_2 + 46\ H_2O + 11\ K_2SO_4 + 10\ Cr_2(SO_4)_3$$

$$2\ C_6H_4\begin{matrix}\text{—COOH}\\ \text{—COOK}\end{matrix} + 15\ O_2 \rightarrow 16\ CO_2 + 4\ H_2O + 2\ KOH$$

As noted in the first equation, the oxidizing agent $K_2Cr_2O_7$ provides all the oxygen required for the oxidation of organic in a closed system.

2.19 PESTICIDES: CARBAMATE, UREA, AND TRIAZINE

A large number of nitrogen-containing pesticides are characterized by the carbamate or urea functional group or triazine ring in their structures. These substances constitute the three most common classes of nitrogen-containing pesticides. The structural features of these three distinct classes of substances are shown below:

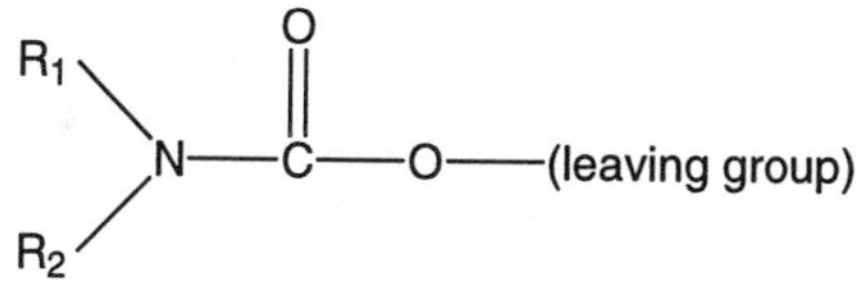

1. Carbamate

where R_1 and R_2 are alkyl or aryl groups or hydrogen atoms. Compounds containing a methyl group attached to the N atom are also known as N-methyl carbamates.

Substitution of one or both of the oxygen atoms with sulfur in the above structure gives thiocarbamate.

The structure of urea type pesticide is similar to carbamate, except the terminal oxygen atom is replaced by a nitrogen atom.

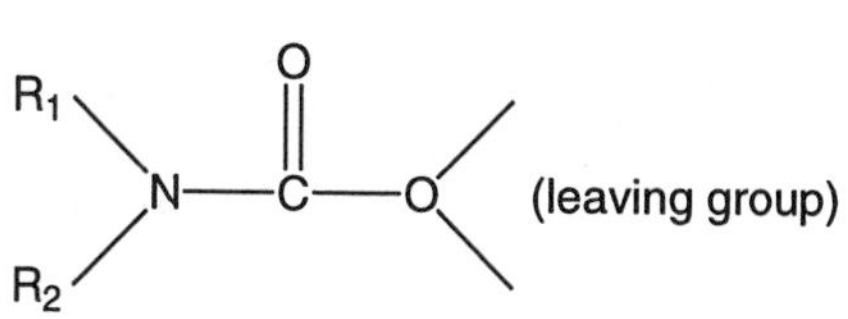

2. Urea Type

Triazine is a nitrogen heterocyclic ring containing three N atoms in the ring.

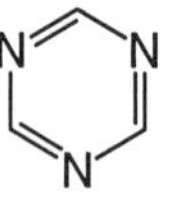

3. Triazine

Discussed below are some pesticides that fall under the above structural features. Compounds having the same type or closely related structures can be analyzed by the same methodologies. Certain instrumental techniques such as GC, GC/MS, or HPLC can be applied in general to analyze these substances irrespective of their structures. For example, using a capillary column of appropriate polarity and a nitrogen phosphorus detector (NPD) in N-specific mode, many types of nitrogen-containing pesticides can be analyzed. Only the pesticides of carbamate, urea, and triazine types are presented in this chapter. This chapter includes many pesticides that have not yet been added in the U.S. EPA priority pollutant list. Refer to Part 3 of this text for the analysis of individual substances including those that are not discussed here.

CARBAMATE PESTICIDES

Carbamate pesticides are best analyzed by HPLC using postcolumn derivatization technique. Some common carbamate pesticides are listed in Table 2.19.1. Compounds are separated on a C-18 analytical column and then hydrolyzed with 0.05 *N* sodium hydroxide. Hydrolysis converts the carbamates to their methyl amines which are then reacted with *o*-phthalaldehyde and 2-mercaptoethanol to form highly fluorescent derivatives. The derivatives are detected by a fluorescence detector. *o*-Phthaladehyde reaction solution is prepared by mixing a 10-mL aliquot of 1% *o*-phalaldehyde solution in methanol to 10 mL of acetonitrile containing 100 μL of 2-mercaptoethanol and then diluting to 1 L with 0.05 *N* sodium borate solution.

The following conditions are recommended for the liquid chromatograph (LC), fluorometric detection, and for the postcolumn hydrolysis and derivatization.

LC Columns

C-18 reverse phase HPLC column, e.g., 25 cm × 4.6 mm stainless steel packed with 5 μm Beckman Ultrasphere, or 15 cm × 2.9 mm stainless steel packed with 4 μm Nova Pac C18, or 25 cm × 4.6 mm stainless steel packed with 5 μm Supelco LC-1 or equivalent.

Chromatographic Conditions

- Solvent A — reagent grade water acidified with phosphoric acid (10 drops/1 L)
- Solvent B — methanol/acetonitrile (1:1)
- Flow rate — 1 mL/min

Table 2.19.1 Some Common Carbamate Pesticides

CAS no.	Pesticides	Alternate names(s)
[116-06-3]	Aldicarb	Temik
[1646-88-4]	Aldicarb sulfone	Aldoxycarb, Standak
[1646-87-3]	Aldicarb sulfoxide	—
[671-04-5]	Banol	Carbanolate
[101-21-3]	Chlorpropham	Chloro IPC
[63-25-2]	Carbaryl	Sevin, Arylam
[1563-66-2]	Carbofuran	Furadan
[17804-35-2]	Benomyl	Benlate
[2032-59-9]	Aminocarb	Matacil
[22781-23-3]	Bendiocarb	Ficam
[2032-65-7]	Methiocarb	Mercaptodimethur, Mesurol
[16752-77-5]	Methomyl	Lannate
[23135-22-0]	Oxamyl	—
[23135-22-0]	Pirimicarb	Pirimor
[2631-37-0]	Promecarb	Carbamult
[122-42-9]	Propham	IPC
[114-26-1]	Propoxur	Baygon, Aprocarb
[1918-18-9]	Swep	—
[3766-81-2]	Fenobucarb	Baycarb
[1918-11-2}	Terbucarb	Terbutol
[6988-21-2]	Dioxacarb	Elocron

Hydrolysis Conditions

- Solution: 0.05 *N* NaOH
- Flow rate: 0.7 mL/min
- Temperature: 95°C
- Residence time: 35 s

Postcolumn Derivatization Conditions

- Solution: *o*-phthalaldehyde/2-mercapthoethanol
- Flow rate: 0.7 mL/min
- Temperature: 40°C
- Residence time: 25 s

Fluorometer Conditions

- Excitation wavelength: 330 nm
- Emission wavelength: 418 nm cutoff filter

Extraction

Aqueous samples, especially those that are clean, may be directly injected into the reverse phase HPLC column. No extraction is required for this. A 200- to 400-μL aliquot of the sample may be injected. Wastewaters, aqueous industrial wastes, and leachates can be extracted with methylene chloride. Nonaqueous samples such as soils, sediment, solid wastes, sludges, oils, and nonaqueous liquids are extracted with acetonitrile. Solids and sludges are dried at 105°C and an accurately weighted dried sample is repeatedly extracted with acetonitrile. The mixture should be shaken at least for an hour using a platform shaker. Oils and oily wastes should be extracted with a mixture of hexane and acetonitrile for several hours. Transfer the solvents into a separatory funnel. Shake for a few minutes and then allow the solvent phases to separate. Collect the acetonitrile layer containing the carbamates in a volumetric flask. Discard the hexane phase.

For dirty and heavily contaminated samples, cleanup of the extracts may often become necessary. This is done by treating 20 mL of ethylene glycol and evaporating the solvent in a water bath at 50 to 60°C in a stream of nitrogen. The ethylene glycol residue is then combined with a few mL of methanol, and the mixture is then passed through a prewashed C-18 reverse phase cartridge. Analytes which are now adsorbed on the cartridge are eluted with methanol for HPLC analysis.

Solvent exchange is very crucial for HPLC analysis. The methylene chloride or acetonitrile extract is exchanged into methanol as described above, using ethylene glycol before passing through any C-18 reverse phase cartridge.

Certain substances, such as alkyl amines, produce fluorescence. Presence of such fluorescence compounds in the sample can produce positive interference. Base hydrolysis may result in the formation of such substances. Similarly, coeluting compounds that quench fluorescence may cause negative interference. Blanks should be run using organic-free reagent water without sodium hydroxide and *o*-phthalaldehyde to determine the presence of such interfering substances.

Air analysis for carbamate pesticides may be performed by sampling air over 1 μm PTFE membrane. The analytes collected over the membrane are extracted with methylene chloride, exchanged into methanol, and analyzed by HPLC using postcolumn derivatization technique as described above. Certain pesticides may be analyzed too by the colorimetric method (see Part 3 under individual compounds).

UREA PESTICIDES

Urea pesticides are structurally similar to carbamates. Some common pesticides of this class are listed in Table 2.19.2. These substances can be determined by reverse phase HPLC method. Aqueous samples can be analyzed by U.S. EPA Method 553 using a reverse phase HPLC column interfaced to a mass spectrometer with a particle beam interface. The outline of the method is described below.

Aqueous samples are extracted by liquid-liquid extraction with methylene chloride. The sample extract is concentrated by evaporating methylene chloride and exchanging the solvent to methanol. Alternatively, the sample may be extracted by solid-phase extraction, using a sorbent cartridge packed with C-18

Table 2.19.2 Some Common Pesticides Containing Urea Functional Group

CAS no.	Pesticides	Alternate name(s)
[13360-45-7]	Chlorbromuron [a]	Maloran
[1982-47-4]	Chloroxuron [a]	Tenoran
[330-54-1]	Diuron [a]	Karmex
[101-42-8]	Fenuron	Dybar
[4482-55-7]	Fenuron TCA [a]	Urab
[2164-17-2]	Fluometuron [a]	Cotoran
[34123-59-6]	Isoproturon	Arelon
[330-55-2]	Linuron [a]	Lorox
[3060-89-7]	Metobromuron [a]	Patoran
[150-68-5]	Monuron [a]	Telvar
[140-41-0]	Monuron TCA [a]	Urox
[555-37-3]	Neburon [a]	Kloben
[66063-05-6]	Pencycuron [a]	Monceren
[1982-49-6]	Siduron	Tupersan
[34014-18-1]	Tebuthiuron	EL-103

[a] Contains halogen atoms in the molecules.

impregnated on silica or a neutral polystyrene/divinylbenzene polymer. The sample is mixed with ammonium acetate (0.8 g/L) prior to extraction. The cartridge is flushed with methanol and then the sample is passed through the cartridge. The analytes are eluted from the cartridge with methanol and concentrated by evaporation. The samples should be extracted within 7 days. The extracts must be stored below 0°C and analyzed within 3 weeks.

The sample extract is injected into a HPLC containing a reverse phase HPLC column. The analytes are identified by comparing their retention times with that of the standard. A more reliable and confirmatory test is to use a mass spectrometer equipped with a particle beam and interfaced to the HPLC column. The compounds are identified from their mass spectra as well as retention times. The chromatographic column and conditions are described below.

Column

- C-18 Novapak (Waters) or equivalent, packed with silica, chemically bonded with octadecyldimethylsilyl groups.
- Condition the column by pumping acetonitrile through it for several hours at a rate of 1 to 2 drops/min.

Liquid Chromatograph

LC should be able to maintain flow rates between 0.20 and 0.40 mL/min and perform a gradient elution from 100% solvent A (75:25 v/v water: acetonitrile containing 0.01 ammonium acetate) to solvent B (acetonitrile).

Use an additional LC pump for postcolumn addition of acetonitrile at a constant rate (0.1 to 0.5 mL/min).

Mass Spectrometer:

Electron ionization must occur at a nominal electron energy of 70 eV. Scan range 45 to 500 amu with 1 to 2 s/scan. The MS should be tuned using a performance check solution of decafluorotriphenylphosphine oxide (DFTPPO). The latter is obtained by adding a slight excess of H_2O_2 to a solution of DFTPP. The product crystals are washed and dissolved in acetonitrile to prepare a 100 ppm solution. Inject 400 to 500 ng DFTPPO into the LC/MS to measure ion abundance criteria for system performance check, especially to check mass measuring accuracy and the resolution efficiency. The DFTPPO tuning criteria is given below in Table 2.19.3.

Table 2.19.3 DFTPPO Tuning Criteria for LC/MS System Performance Check

Mass m/z	Ion abundance criteria
77	Major ion, must be present
168	Major ion, must be present
169	4–10% of mass 168
271	Major ion, base peak
365	5–10% of base peak
438	Must be present
458	Molecular ion, must be present
459	15–24% of mass 458

The particle beam LC/MS interface must reduce the system pressure to about 1×10^{-6} to 1×10^{-4} torr at which electron ionization occurs.

Precision & Accuracy

The precision and accuracy data are not available for all the urea pesticides listed in the Table 2.19.3. However, a matrix spike recovery between 70 and 130% and a RSD below 30% should be achieved for aqueous samples. Samples should be spiked with one or more surrogates. Compounds recommended as surrogates are benzidine-d_8, 3,3-dichlorobenzidine-d_6, and caffeine-$^{15}N_2$. Surrogate concentrations in samples or blank should be 50 to 100 μg/L.

TRIAZINE PESTICIDES

Triazine pesticides (herbicides) can be determined primarily by all the major instrumental techniques: HPLC, GC, and GC/MS, following sample extraction. Detection limits in low ppb range can be achieved using GC-NPD in nitrogen specific mode. HPLC method using UV detection (254 nm) may be alternatively employed if such a low range of detection is not required. Analytes may alternatively be determined by GC-FID or GC/MS. A polyphenylmethylsiloxane capillary column (30 m length, 0.25 mm ID, and 0.25 μm film thickness) having intermediate polarity, such as DB-35, AT-35, SPB-35, Rtx-35, or equivalent, or

Table 2.19.4 Some Common Triazine Pesticides

CAS no.	Pesticides	Alternate names(s)
[834-12-8]	Ametryne	Gesapax
[101-05-3]	Anilazine	Kemate, Dyrene
[1610-17-9]	Atraton	Gestamin
[1912-24-9]	Atrazine	Atratol, Gesaprim
[64902-72-3]	Chlorsulfuron	Glean
[21725-46-2]	Cyanazone	Bladex
[4147-51-7]	Dipropetryne	Sancap
[21087-64-9]	Metribuzin	Sencor, Lexone
[32889-48-8]	Procyazine	Cycle
[1610-18-0]	Prometon	Primatol
[7287-19-6]	Prometryne	Caparol
[139-40-2]	Propazine	Gesamil, Milogard
[122-34-9]	Simazine	Amizine, Gesapun
[1014-70-6]	Simetryne	Gybon
[5915-41-3]	Terbuthylazine	Gardoprim
[886-50-0]	Terbutryne	Prebane
[64529-56-2]	Tycor	Ethiozin, Ebuzin
[51235-04-2]	Velpar	Hexazinone

a polycyanopropylphenylmethylsiloxane column such as OV-1701, DB-1701, AT-1701, SPB-7, or equivalent, can provide adequate resolution of triazines for GC or GC/MS analysis. Some common triazine pesticides are listed in Table 2.19.4.

Conditions for typical HPLC analysis are given below:

Column:	Accubond ODS (C-18) (15 cm × 4.6 mm ID × 5 μm film thickness)
Mobile phase:	Acetonitrile/0.01 *M* K_2HPO_4 (35:65)
Flow rate:	1.5 mL/min
Detector:	UV at 254 nm

Sample Extraction

Aqueous samples can be extracted by liquid-liquid extraction or by solid phase extraction. The latter offers certain advantages over the liquid-liquid extraction. The solid-phase extraction reduces the solvent consumption and eliminates the problem of emulsion. The procedure for this extraction is described below.

The cartridge is loaded first with 5 mL acetone. Vacuum is applied and the eluant is discarded. Repeat these steps first with 5 mL 0.1 *M* and then with 5 mL of 0.02 *M* potassium hydrogen phosphate (K_2HPO_4), respectively. The sorbent should not be allowed to go dry. Add the sample to the cartridge from a sample reservoir attached to the top of the cartridge. Remove the reservoir and rinse the cartridge with 3 mL of 40:60 methanol-water mixture while applying vacuum. Discard the eluant. Centrifuging the cartridge may be performed to remove additional water to concentrate the sample. Analytes are now eluted from the cartridge with 3 mL acetone. Use vacuum for this. Collect the eluant and concentrate the solution to dryness under a stream of nitrogen and mild heat. The residue is dissolved in 200 μL acetone for GC analysis or 200 μL acetonitrile for HPLC analysis.

2.20 PESTICIDES: ORGANOCHLORINE

Organochlorine pesticides refers to all chlorine-containing organics used for pest control. The term, however, is not confined to compounds of any single and specific type of chemical structures or organic functional group(s), but includes a broad range of substances. The grouping together of these compounds is more or less based on the similarity in their chemical analysis. Many chlorinated pesticides that were commonly used in the past are no longer being used now, because of their harmful toxic effect on human and contamination of the environment. Many such pesticides and their residues are still found in the environment in trace quantities in groundwaters, soils, sediments, and wastewaters. These substances are stable, bioaccumulative, and toxic, and some are also carcinogens (Patnaik, 1992). Table 2.20.1 presents some common chlorinated pesticides, most of which are listed as priority pollutants by U.S. EPA.

Chlorinated pesticides in aqueous and nonaqueous matrices may be determined by U.S. EPA Methods 608, 625, 505, 508, 8080, and 8270 (U.S. EPA 1984–1994). Analysis of these pesticides requires extraction of the aqueous or nonaqueous samples by a suitable organic solvent, concentration, and cleanup of the extracts, and determination of the analytes in the extracts, usually by GC-ECD or GC/MS. These steps are outlined below.

SAMPLE EXTRACTION

Aqueous samples are extracted with hexane or with methylene chloride by liquid-liquid extraction using a separatory funnel or a mechanical shaker, or by microextraction. Aqueous samples can also be extracted by solid phase extraction using a C-18 cartridge. Selection of sample volume should be based on the extent of sample concentration that may be needed to achieve the required detection level in the analysis, as well as the use of packed or capillary column. A larger sample concentration is required for packed column than that for capillary column analysis. U.S. EPA recommends the extraction of 1 L sample to a final volume of 1 mL for wastewater analysis performed on a packed column. For the analysis of potable water by GC-ECD on a capillary column, concentration of a 35-mL

Table 2.20.1 Some Common Chlorinated Pesticides and Their Degradation Products

CAS no.	Common name
[15972-60-8]	Alachlor
[309-00-2]	Aldrin
[1912-24-9]	Atrazine
[319-84-6]	α-BHC
[319-85-7]	β-BHC
[319-86-8]	δ-BHC
[58-89-9]	γ-BHC (Lindane)
[133-06-2]	Captan
[57-74-9]	Chlordane
[5103-71-9]	α-Chlordane
[5103-74-2]	γ-Chlordane
[2675-77-6]	Chlorneb
[501-15-6]	Chlorobenzilate
[2921-88-2]	Chlorothalonil
[8897-45-6]	DCPA (Dacthal)
[72-54-8]	4,4′-DDD
[72-55-9]	4,4′-DDE
[50-29-3]	4,4′-DDT
[60-57-1]	Dieldrin
[99-30-9]	Dichloran
[72-20-8]	Endrin
[7421-93-4]	Endrin aldehyde
[53494-70-5]	Endrin ketone
[959-98-8]	Endosulfan-I
[33213-65-9]	Endosulfan-II
[1031-07-8]	Endosulfan sulfate
[2593-15-9]	Etridiazole
[76-44-8]	Heptachlor
[1024-57-3]	Heptachlor epoxide
[118-74-1]	Hexachlorobenzene
[77-74-4]	Hexachlorocyclopentadiene
[72-43-5]	Methoxychlor
[2385-85-5]	Mirex
[39765-80-5]	*trans*-Nonachlor
[54774-45-7]	*cis*-Permethrin
[51877-74-8]	*trans*-Permethrin
[1918-16-7]	Propachlor
[122-34-9]	Simazine
[8001-35-2]	Toxaphene
[1582-09-8]	Trifluralin

sample aliquot to a final volume of 2 mL by microextraction is sufficient to produce the required detection range for the pesticides. If the sample extracts were to be analyzed by GC/MS, the extraction solvent can be methylene chloride or hexane. On the other hand, if a GC-ECD is used, a nonchlorinated solvent such as hexane or isooctane should be used for extraction. Alternatively, a chlorinated solvent, such as methylene chloride can be used, but the extract must be exchanged into hexane or isooctane if analyzed by GC-ECD. In solid phase extraction, the aqueous sample is passed through the cartridge after cleaning the

Table 2.20.2 Elution Patterns for Pesticides in Florisil Column Cleanup

Pesticides	Percent recovery by fraction		
	6% Ether in hexane	15% Ether in hexane	50% Ether in hexane
β-BHC	97	—	—
δ-BHC	98	—	—
γ-BHC(Lindane)	100	—	—
Chlordane	100	—	—
4,4′-DDD	99	—	—
4,4′-DDE	98	—	—
4,4′-DDT	100	—	—
Dieldrin	—	100	—
Endosulfan I	37	64	—
Endosulfan II	—	7	91
Endosulfan sulfate	—	—	100
Endrin	4	96	—
Endrin aldehyde	—	68	26
Heptachlor epoxide	100	—	—
Toxaphene	96	—	—

cartridge with acetone and methanol. Pesticides adsorbed on the sorbent are eluted with acetone. The eluant is concentrated to dryness. The residue is dissolved in acetone or any other suitable solvent for GC analysis.

Soils, sediments, and solid wastes samples are extracted by sonication or Soxhlett extraction using the solvents mentioned above. The sample should be mixed with anhydrous Na_2SO_4 before extraction.

The solvent extract should be subjected to one or more cleanup steps for the removal of interfering substances. The presence of phthalate esters, sulfur, or other chlorinated compounds can mask pesticide peaks. The extract should, therefore, be cleaned up from the interfering substances using a florisil column or by gel permeation chromatography (see Chapter 1.5). The distribution patterns for the pesticides in the florisil column fractions are presented in Table 2.20.2.

An additional cleanup step should be performed to remove sulfur using mercury or copper powder. Permanganate-sulfuric acid treatment is not recommended. Unlike PCBs, most pesticides are fully or partially oxidized by reaction with $KMnO_4$-H_2SO_4. Treatment with concentrated H_2SO_4 alone is not recommended, because pesticides such as dieldrin and endrin were found to be totally destroyed at 20 ng/mL and endrin aldehyde and endosulfan sulfate partially decomposed at 60 ng/mL concentrations, respectively, in the extract (Cavanaugh and Patnaik, 1995).

Accuracy of the extraction and analytical method may be evaluated from the surrogate spike recovery. Compounds recommended as surrogates are dibutylchlorendate, tetrachloro-*m*-xylene, 4,4′-dichlorobiphenyl, 2-fluorobiphenyl, 2,4,6-tribromophenol, hexabromobenzene, and *o,p*-DDE. The latter is found to be suitable in the packed column analysis without any coelution problem and with a retention time close to the pesticides in the midrange of the chromatogram of the pesticide mixture (Cavanaugh and Patnaik, 1995). Any other stable chlorinated organics that do not decompose or oxidize under the conditions of extraction and analysis and that exhibit good response to halogen-specific detectors may also be used as surrogates. The surrogate spike recovery should fall within 70 to 130%.

ANALYSIS

Organochlorine pesticides in the solvent extract are analyzed either by GC using a halogen-specific detector, most commonly ECD or by GC/MS. It may be also analyzed by GC-FID. All the U.S. EPA methods are based on GC-ECD and GC/MS determination. Compounds identified by GC on a specific column must be confirmed on an alternate column. Alternately, the presence of pesticides can be confirmed by analyzing on GC/MS. After qualitatively identifying and confirming the pesticide peaks, quantitation should be performed by GC-ECD using either internal standard or external standard method. When the internal standard method is followed, use more than one internal standard, all of which should have chromatographic response comparable to the analytes of interest and have retention times covering the entire range of chromatogram. Pentachloronitrobenzene, 4,4′-dichlorobiphenyl, and 4,4′-dioromooctafluorobiphenyl are some examples of internal standards.

A single point calibration may be used instead of a working calibration curve for quantitation by either external or internal standard method, if the response from the single point standards produces a response that deviates from the sample extract response by no more than 20%. The solvent for preparing calibration standards should preferably be the same one used to make the final sample extract. Hexane, isooctane, or methyl-*tert*-butylether is an appropriate solvent for the analysis of chlorinated pesticides by GC-ECD.

Chlorinated pesticides mixture can be separated on both packed and capillary columns. Some of the columns and conditions are as follows:

- Packed column: 1.8 m long × 4 mm ID glass column packed with (1) 1.5% SP-2250/1.95% SP-2401 on Supelcoport (100/120 mesh), or (2) 3% OV-1 on Supelcoport (100/120 mesh) or equivalent.
- Carrier gas: 5% methane/95% argon at 60 mL/min; Temperature: oven 200°C (isothermal), injector 250°C and ECD 320°C.
- Capillary column: Silicone-coated fused-silica 30 m (or other length) × 0.53 mm (or 0.32 or 0.25 mm) ID × 1.5 μm (or 0.83 μm) film thickness, such as DB-608, SPB-608, DB-5, Rtx-5, DB-17, or equivalent.
- Carrier gas: He 35 cm/s, makeup gas N_2 30 mL/min; Temperature: 140° to 260°C at 5° to 10°C/min, injector 250°C, and ECD 325°C. Other temperature and flow rate conditions suitable for the analysis may be used.

The presence of a pesticide determined on one column must be confirmed on an alternate column. Also, certain pesticides coeluting on one column can be separated by using another column. For example, 4,4′-DDD and Endosulfan-II coeluting on SP-2250/SP-2401 can be effectively separated on the OV-1 packed column. Conversely, 4,4-DDE and Dieldrin coeluting on OV-1 are better separated on SP-2250/SP-2401 column. Many chlorinated pesticides may coelute even on a 30 m long, 0.53 mm ID, and 1.5 μm film capillary column. This includes the following pairs on a DB-5 column (J&W Scientific, 1994):

1. Aldrin and DCPA
2. γ-Chlordane and *o,p*-DDE
3. α-Chlordane and endosulfan-I
4. Chlorobenzillate and Endrin
5. *p,p′*-DDT and endosulfan sulfate

These coeluting pairs, however, can be separated distinctly on the same column with a smaller film thickness, 0.83 μm.

Two common pesticides Endrin and DDT are susceptible to decomposition on dirty injection port. The former oxidizes to Endrin aldehyde and Endrin ketone, while the latter degrades to 4,4′-DDE and 4,4′-DDD. If degradation of either of these analytes exceeds 20%, corrective action should be taken before performing calibration. The percent breakdown is calculated as follows:

$$\%\text{ Breakdown for Endrin}$$
$$= 100 \times \frac{\text{Peak areas of Endrin aldehyde} + \text{Endrin ketone}}{\text{Peak areas of Endrin} + \text{Endrin aldehyde} + \text{Endrin keytone}}$$

$$\%\text{ Breakdown for DDT} = 100 \times \frac{\text{Peak areas of DDE} + \text{DDD}}{\text{Peak areas of DDT} + \text{DDE} + \text{DDD}}$$

A standard containing the pure compounds Endrin and DDT should be injected to determine the possible breakdown.

The pesticides in the solvent extract may also be analyzed by GC-FID. On a DB-5 capillary column (30 m × 0.25 mm ID × 0.25 μm), e.g., a pesticide mixture at 2 μL splitless injection may be analyzed at a Hydrogen (Carrier gas) flow of 43 cm/s, N_2 makeup gas flow of 30 mL/min and at an oven temperature programmed from 50 to 300°C. Other columns or conditions may be used. The instrument detection limit for FID is, however, higher than that for the ECD.

The confirmation of pesticides by GC/MS should be more reliable than that on the GC-ECD using an alternate column. Presence of stray interference peaks, even after sample cleanup, and the retention time shift and coelution problem, often necessitate the use of GC/MS in compounds identification. If a quantitative estimation is to be performed, select the primary ion or one of the major characteristic ions of the compounds and compare the area response of this ion to that in the calibration standard. Quantitation, however, is generally done from the GC-ECD analysis, because ECD exhibits a much greater sensitivity than the mass selective detector (MSD). For example, while ECD is sensitive to 0.01 ng dieldrin, the lowest MSD detection for the same compound is in the range of 1 ng. The primary and secondary characteristic ions for qualitative identification and quantitation are presented in Table 2.20.3. The data presented are obtained under MS conditions utilizing 70 V (nominal) electron energy under electron impact ionization mode.

Table 2.20.3 Characteristic Masses for Chlorinated Pesticides

Pesticides	Primary ion m/z	Secondary ions m/z
Aldrin	66	263, 220
Atrazine	200	215
α-BHC	183	181, 109
β-BHC	181	183, 109
δ-BHC	183	181, 109
γ-BHC (Lindane)	183	181, 109
Captan	79	149, 77, 119
α-Chlordane	375	377
γ-Chlordane	375	377
Chlorobenzilate	251	139, 253, 111, 141
4,4′-DDD	235	237, 165
4,4′-DDE	246	248, 176
4,4′-DDT	235	237, 165
Dieldrin	79	263, 279
Endosulfan I	195	339, 341
Endosulfan II	337	339, 341
Endosulfan sulfate	272	387, 422
Endrin	263	82, 81
Endrin aldehyde	67	345, 250
Endrin ketone	317	67, 319
Heptachlor	100	272, 274
Heptachlor epoxide	253	355, 351, 81
Hexachlorobenzene	284	142, 249
Hexachlorocyclopentadiene	237	235, 272
Methoxychlor	227	228, 152, 114, 274
Mirex	272	237, 274, 270, 239
trans-Nonachlor	409	
Simazine	201	
Toxaphene	159	231, 233
Trifluralin	306	43, 264, 41, 290
	Surrogates	
2-Fluorobiphenyl	172	171
2-Fluorophenol	112	64
Terphenyl-d_{14}	244	122, 212
2,4,6-Tribromophenol	330	332, 141

AIR ANALYSIS

Most of the organochlorine pesticides listed in this chapter may be analyzed by NIOSH and U.S. EPA Methods (U.S. EPA 1984–1988, NIOSH, 1984–1989). The method of analysis, in general, involves drawing a measured volume of air through a sorbent cartridge containing polyurethane foam or Chromosorb 102. The pesticides are extracted with an organic solvent such as toluene, hexane, or diethyl ether; the extract concentrated and analyzed by GC-ECD. The extract may be cleaned up by florisil to remove any interference (U.S. EPA Method 608).

A very low detection limit, >1 ng/m^3 may be achieved using a 24-h sampling period, sampling over 5000 L ambient air. Such a high volume sampler consists of a glass fiber filter with a polyurethane foam backup absorbent cartridge (U.S. EPA Method TO4). A low volume (1 to 5 L/min) sampler consisting of a sorbent cartridge containing a polyurethane foam may be used to collect pesticide vapors (U.S. EPA Method TO10). Pesticides are extracted with ether-hexane mixture and analyzed by GC-ECD. Pesticides in the extract may be determined by GC using other detectors as well, such as HECD or by a mass spectrometer. Certain pesticides may be analyzed by GC-FPD, GC-NPD, or by HPLC using an UV or an electrochemical detector.

2.21 PESTICIDES: ORGANOPHOSPHORUS

An important class of pesticides is organophosphorus compounds, which have the following general structure:

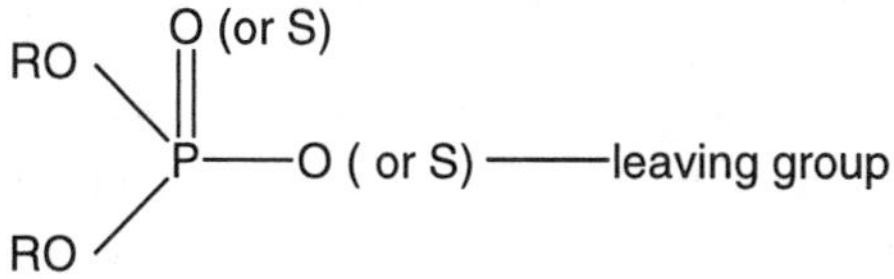

$$(RO)_2P(=O\ \text{(or S)})\text{—O (or S)—leaving group}$$

where R is an alkyl or aryl group. The phosphorus atom in such a compound is bound to one or two oxygen and/or sulfur atom(s). The leaving group may be any organic species that may cleave out from its oxygen or sulfur bond. Two typical examples are as follows:

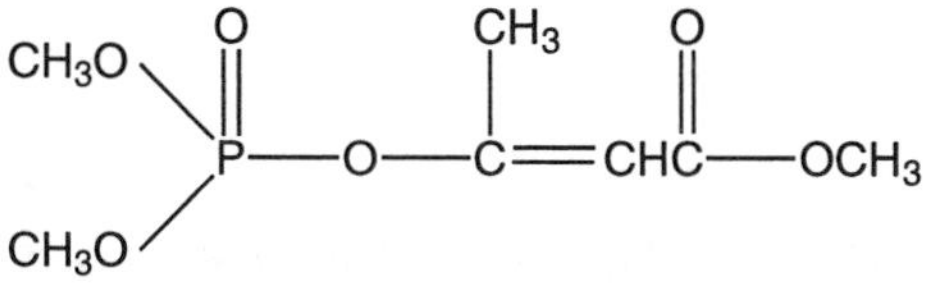

$$(CH_3O)_2P(=O)\text{—O—}C(CH_3)=CHC(=O)\text{—}OCH_3$$

(Mevinphos)

$$(C_2H_5O)_2P(=S)\text{—O—}C_6H_4\text{—}NO_2$$

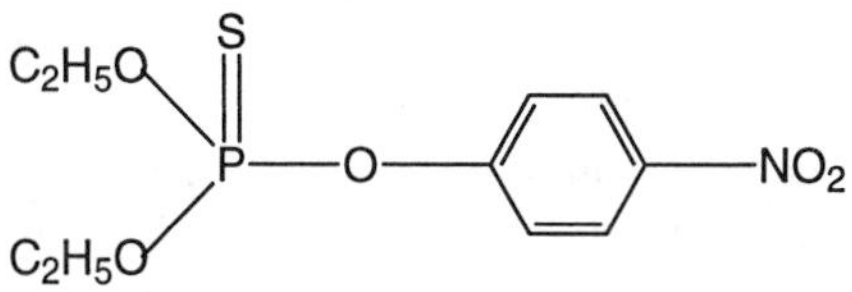

(Parathion)

Organic phosphates can cause moderate to severe acute poisoning. These substances inhibit the function of the enzyme, acetylcholinesterase by phosphorylating or binding the enzyme at its esteratic site. The symptoms of acute toxicity include tightening of the chest, increased salivation and lacrimation, nausea, abdominal cramps, diarrhea, pallor, elevation of blood pressure, headache, insomnia, and tremor. Ingestion of large quantities may cause convulsions, coma, and death. The toxicity, however, varies from substance to substance. Phorate, demeton, and disulfoton are among the most toxic organophosphorus insecticides, while malathion, ronnel, and tokuthion are much less toxic.

ANALYSIS OF AQUEOUS AND SOLID SAMPLES

GC and GC/MS techniques are the most common instrumental methods to analyze organophosphorus pesticides. These substances may also be analyzed by HPLC. However, there is no systematic precision and accuracy studies published on the HPLC determination of environmental samples. The detector required for GC analysis is either a nitrogen phosphorus detector (NPD) operated in the phosphorus specific mode, or a flame photometric detector (FPD) operated in the phosphorus specific mode.* Thus, any pesticide can be analyzed by GC-NPD or GC-FPD using a suitable capillary column as listed below. Some of the common pesticides are listed in Table 2.21.1. A halogen specific detector, such as electrolytic conductivity or micorcoulometric detector, may alternatively be used for GC analysis of only those pesticides that contain halogen atoms. Some of these halogen-containing organophosphorus pesticides are presented in Table 2.21.2. Analysis of organophosphorus pesticides by GC/MS should be the method of choice wherever possible. This is a confirmatory test to identify the compounds from their characteristic ions in addition to their retention times. Table 2.21.3 lists the primary and secondary characteristic ions of some common organophosphorus pesticides.

Sample Extraction

Aqueous samples are extracted with methylene chloride using a separatory funnel or a continuous liquid-liquid extractor. Solid samples are extracted with methylene chloride-acetone mixture (1:1) by either sonication or Soxhlett extraction. The methylene chloride extract should be finally exchanged to hexane or iso-octane or methyl tert-butyl ether. The latter solvents should be mixed with acetone during solvent exchange. The extracts should then be cleaned up by Florisil. Often Florisil cleanup reduces the percent recovery of analyte to less than 85%. A preliminary screening of the extract should, therefore, be done to determine the presence of interference and the necessity of florisil cleanup. Gel permeation cleanup also lowers the analyte recovery and thus is not recommended. If a FPD is used in the GC analysis, the presence of elemental sulfur can mask the analyte peaks. In such a case, sulfur cleanup should be performed. Sample extraction and cleanup procedures are described in Chapter 1.5.

* FPD measures the phosphorus or sulfur-containing substances. Because most organophosphorus pesiticides contain sulfur atoms, the FPD may be more sensitive and selective. Both these detectors may be used in a dual column dial detector system for analysis.

Table 2.21.1 Common Organophosphorus Pesticides

CAS no.	Pesticides	Alternate name(s)
[30560-19-1]	Acephate	Orthene
[21548-32-3]	Acconame	Geofos, Fosthietan
[1757-18-2]	Akton	—
[3244-90-4]	Aspon	—
[86-50-0]	Azinphos methyl	Guthion
[2642-71-9]	Azinphos ethyl	Guthion ethyl
[35400-43-2]	Bolstar	Sulprophos
[122-10-1]	Bomyl	Swat
[786-19-6]	Carbofenothion	Trithion
[470-90-6]	Chlorfenvinphos	Birlane
[56-72-4]	Coumaphos	Muscatox
[7700-17-6]	Crotoxyphos	Ciodrin
[8065-48-3]	Demeton-O,S	Systox
[333-41-5]	Diazinon	Neocidol
[2463-84-5]	Dicapthon	—
[97-17-6]	Dichlofenthion	Nemacide VC-13
[62-73-7]	Dichlorvos	Nerkol
[141-66-2]	Dicrotophos	Carbicron
[60-51-5]	Dimethoate	Cygon
[78-34-2]	Dioxathion	Navadel
[298-04-4]	Disulfoton	Thiodemeton
[2921-88-2]	Dursban	Trichlorpyrphos, Chlorpyrifos
[2104-64-5]	EPN	—
[563-12-2]	Ethion	Nialate
[16672-87-0]	Ethephon	Ethrel
[52-85-7]	Famphur	Famophos
[22224-92-6]	Fenamiphos	—
[122-14-5]	Fenitrothion	Folithion
[115-90-2]	Fensulfothion	Terracur P
[55-98-9]	Fenthion	Entex
[994-22-9]	Fonofos	Dyfonate
[42509-80-8]	Isazophos	Miral
[25311-71-1]	Isofenphos	Oftanol
[21609-90-5]	Leptophos	Lepton, Phosvel
[121-75-5]	Malathion	Cythion
[150-50-5]	Merphos	—
[10265-92-6]	Methamidophos	Tamaron
[950-37-8]	Methidathion	Supracide
[7786-34-7]	Mevinphos	Phosdrin
[6923-22-4]	Monocrotophos	Azodrin
[299-86-5]	Montrel	Crufomate, Ruelene
[300-76-5]	Naled	Dibrom
[301-12-2]	Oxydemeton-methyl	Meta Systox-R
[56-38-2]	Parathion-ethyl	Paraphos
[298-00-0]	Parathion-methyl	Nitran
[311-45-5]	Paraoxon	Phosphacol
[22224-92-6]	Phenamiphos	Nemacur
[298-02-2]	Phorate	Thimet
[2310-17-0]	Phosalone	Rubitox, Azofene
[947-02-4]	Phosfolan	Cylan
[13171-21-6]	Phosphamidon	Dimecron
[732-11-6]	Phosmet	Prolate, Imidan

Table 2.21.1 Common Organophosphorus Pesticides (Continued)

CAS no.	Pesticides	Alternate name(s)
[41198-08-7]	Profenofos	Curacron
[31218-83-4]	Propetamphos	Safrotin
[13194-48-4]	Prophos	Phosethoprop
[299-84-3]	Ronnel	Fenchlorphos, Etrolene
[3689-24-5]	Sulfotepp	—
[107-49-3]	TEPP	Tetron, Tetraethyl pyrophosphate
[13071-79-9]	Terbufos	Counter
[22248-79-9]	Tetrachlorvinphos	Stirophos
[3383-96-8]	Tetrafenphos	Abate, Temephos
[297-97-2]	Thionazin	Zinophos, Namafos
[34643-46-4]	Tokuthion	Protothiophos
[52-68-6]	Trichlorfon	Anthon, Chlorofos
[327-98-0]	Trichloronate	—

1. The pesticides listed above can be analyzed by GC-FPD (in P-mode) or by GC/MS following extractions. An open, tubular fused silica capillary column (35 m and 0.53 ID) gives better resolution and sensitivity than a packed column. Some compounds may coelute, which should be analyzed on an alternative column.
2. Precision and accuracy must be established for the analytes before their analysis.
3. Extraction should be performed immediately for substances such as TEPP, terbufos, and diazinon, which are unstable in water.
4. The recovery is poor for naled and trichlorfon, both of which may be converted to dichlorvos during extraction or on the GC column. The latter compound also shows poor recovery because its solubility is relatively high in water.
5. Demeton-O,S is a mixture of two isomers, giving two peaks. Merphos often gives two peaks due to its oxidation.
6. Many of the above substances are NOT U.S. EPA priority pollutants. No EPA methods are available for these unlisted compounds.

Table 2.21.2 Organophosphorus Pesticides Containing Halogen Atoms

CAS no.	Pesticides	Halogen atoms in the molecule
[1757-18-1]	Akton	3
[786-19-6]	Carbofenothion (Trithion)	1
[470-90-6]	Chlorfenvinphos	3
[56-72-4]	Coumaphos	1
[2463-84-5]	Dicapthen	1
[97-17-6]	Dichlofenthion	2
[62-73-7]	Dichlorvos	2
[2921-88-2]	Dursban (Chlorpyrifos)	3
[16672-87-0]	Ethephon (Ethrel)	1
[21609-90-5]	Leptophos	3
[5598-13-0]	Methyl dursban (Methyl chlorpyrifos)	3
[115-78-6]	Phosphan	3
[41198-08-7]	Profenofos	2
[299-84-3]	Ronnel (Fenchlorphos)	3
[22248-79-9]	Tetrachlorvinphos (Stirofos)	4
[52-68-6]	Trichlorfon (Anthon)	3

Table 2.21.3 Characteristic Ions for Identification of Some Common Organophosphorus Pesticides by GC/MS[a]

CAS no.	Pesticides	Primary ion, (m/z)	Secondary ions, (m/z)
[86-50-0]	Azinphos-methyl	160	132,93,104,105
[35400-43-2]	Bolstar	156	140,143,113,33
[786-19-6]	Carbofenothion	157	97,121,342,159,199
[470-90-6]	Chlorfenvinphos	267	269,323.325,295
[56-72-4]	Coumaphos	362	226,210,364,97,109
[7700-17-6]	Crotoxyphos	127	105,193,166
[298-03-3]	Demeton-O	88	89,60,61,115,171
[126-75-0]	Demeton-S	88	60,81,114,115,89
[333-41-5]	Diazinon	137	179,152,93,199, 304
[62-73-7]	Dichlorovos	109	185,79,145
[141-66-2]	Dicrotophos	127	67,72,109,193,237
[60-51-5]	Dimethoate	87	93,125,143,229
[298-04-4]	Disulfoton	88	97,89,142,186
[2921-88-2]	Dursban	197	97,199,125,314
[2104-64-5]	EPN	157	169,185,141,323
[563-12-2]	Ethion	231	97,153,125,121
[52-85-7]	Famphur	218	125,93, 109, 217
[115-90-2]	Fensulfothion	293	97,308,125,292
[55-38-9]	Fenthion	278	125,109,169,153
[21609-90-5]	Lephtophos	171	377,375,77,155
[121-75-5]	Malathion	173	125,127,93,158
[298-00-0]	Methyl parathion	109	125,263,79,93
[150-50-5]	Merphos	209	57,153,41,298
[7786-34-7]	Mevinphos	127	192,109,67,164
[6923-22-4]	Monocrotophos	127	192,67,97,109
[300-76-5]	Naled	109	145,147,301,79
[56-38-2]	Parathion	109	97,291,139,155
[298-02-2]	Phorate	75	121,97,93,260
[2310-17-0]	Phosalone	182	184,367,121,379
[732-11-6]	Phosmet	160	77,93,317,76
[13194-48-4]	Prophos	158	43,97,41, 126
[13171-21-6]	Phosphamidon	127	264,72,109,138
[3689-24-5]	Sulfotepp	322	97,65,93,121,202
[13071-79-9]	Terbufos	231	57,97,153,103
[22248-79-9]	Tetrachlorvinphos	109	329,331,79,333
[107-49-3]	Tetraethyl pyrophosphate	99	155,127,81,109
[297-97-2]	Thionazin	107	96,97,163,79,68
[34643-46-4]	Tokuthion	113	43,162,267,309
[512-56-1]	Trimethyl phosphate	110	79,95,109,140
[78-32-0]	Tri-*p*-tolyl phosphate	368	367,107,165,198
[126-72-7]	Tris (2,3-dibromopropyl) phosphate	201	137,119,217,219, 199

Note: A fused silica capillary column is recommended. Column type and conditions are presented earlier in this chapter.

[a] GC/MS conditions: Electron impact ionization; nominal energy 70 eV; mass range 35–500 amu; scan time 1 s/scan.

The extract should be then injected onto the GC column for GC or GC/MS analysis. If internal standard calibration is performed, spike three or more internal standards into the sample extract. Any organophosphorus pesticides whose analysis is not required or which is not found to be present in the screening test may be used as an internal standard. The chromatographic columns and conditions are presented below:

- Column: DB-5, DB-210, SPB-5, SPB-608, or equivalent (wide-bore capillary columns)
- Carrier gas: Helium, 5 mL/min
- Temperature: 50°C for 1 min; 5°C/min. to 140°C, held for 10 min; 10°C/min to 240°C, and held for 15 min.

If any pesticide is detected, its presence must be confirmed on a second GC column or by GC/MS. The detection limits, as well as the precision and accuracy of all analytes of interest must be determined before the analysis. Organophosphorus pesticides other than those listed under U.S. EPA's priority pollutants can also be analyzed by the above procedure if the precision and accuracy data of these substances are within the reasonable range of acceptance. Such acceptance criteria have not yet been published for a number of substances of this class. The spike recovery may vary from substance to substance, and is matrix dependent. However, in all cases, the precision, accuracy, and the detection limits of analytes should be established before the analysis.

AIR ANALYSIS

Air analysis for some of the individual pesticides of this class has been published by NIOSH. These pesticides include mevinphos, TEPP, ronnel, malathion, parathion, EPN, and demeton (NIOSH Methods 2503, 2504, 1450). In general, pesticides in air may be trapped over various filters, such as Chromosorb 102, cellulose ester, XAD-2, PTFE membrane (1 μm), or a glass fiber filter. The analyte(s) are extracted from the filter or the sorbent tube with toluene or any other suitable organic solvent. The extract is analyzed by GC (using a NPD or FPD) or by GC/MS. The column conditions and the characteristic ions for compound identifications are presented in the preceding section. Desorption efficiency of the solvent should be determined before the analysis by spiking a known amount of the analyte into the sorbent tube or filter and then measuring the spike recovery.

2.22 Ph AND Eh

pH is a measure of hydrogen ion $[H^+]$ concentration in an aqueous solution. It is defined as

$$pH = \log\frac{1}{[H^+]} = -\log[H^+]$$

Similarly, hydroxide ion $[OH^-]$ concentration may be expressed as pOH, which is

$$\log\frac{1}{[OH^-]} \quad \text{or} \quad -\log[OH^-]$$

The concentrations $[H^+]$ and $[OH^-]$ are expressed in molarity (*M* or mol/L). In a neutral solution, pH = pOH =7.00. The sum of pH and pOH is 14.0.

In an acidic solution, the hydrogen ion concentration is greater than 1.0×10^{-7} *M*, and thus the pH is less than 7.00. Similarly, in a basic solution, the $[H^+]$ is less than 1.0×10^{-7} *M* and, therefore, the pH is greater than 7.00.

The pH of a solution of a strong acid or a strong base can be calculated; conversely, $[H^+]$ and $[OH^-]$ can be determined from the measured pH of the solution. This is shown in the following examples.

Example 1

What is the pH of a solution of 0.005 *M* HCl?

Because HCl is a strong acid, it will completely dissociate into H^+ and Cl^- ions. Thus, 0.005 *M* HCl will dissociate into 0.005 *M* H^+ and 0.005 *M* Cl^- ions.

$$pH = -\log[H^+]$$

Substituting the value for $[H^+]$

$$pH = -\log[0.005] = 2.30$$

Example 2

0.655 g KOH was dissolved in reagent grade water and made up to a volume of 500 mL. What is the pH of this KOH solution prepared?

First, we calculate the molarity of this solution as follows:

$$\frac{0.655 \text{ g KOH}}{500 \text{ mL soln}} \times \frac{1 \text{ mol KOH}}{39 \text{ g KOH}} \times \frac{1000 \text{ mL soln}}{1 \text{ L soln}}$$

$$= 0.0336 \text{ mol KOH/L soln. or } 0.0336\ M \text{ KOH}$$

Because KOH is a strong base, it will completely dissociate into K^+ and OH^- ions; thus, 0.0336 *M* KOH will dissociate, producing 0.0336 *M* K^+ and 0.0336 *M* OH^- ions.

Thus, $$pOH = -\log[OH^-] \quad \text{or} \quad -\log[0.0336]$$

$$= 1.47$$

Therefore, $$pH = 14.00 - 1.47$$

$$= 12.53$$

This problem may also be solved as follows:

We know that $K_w = [H^+][OH^-]$. K_w is always $1.0 \times 10^{-14}\ M$; $[OH^-] = 0.0336\ M$; $[H^+] = ?$

$$[H^+] = \frac{1.0 \times 10^{-14}}{0.0336} \quad \text{or} \quad \frac{1.0 \times 10^{-14}}{3.36 \times 10^{-2}}$$

$$= 2.98 \times 10^{-13}\ M$$

$$pH = -\log(2.98 \times 10^{-13})$$

$$= -(0.47 - 13)$$

$$= 12.53$$

The pH of an aqueous sample may be measured by the electrometric or colorimetric method. The latter, which involves the use of pH indicator papers, could measure the pH between 2 and 12 within a fairly accurate degree of approximation. A pH below 10 may be determined to two significant figures and that of 10 and above may be measured to three significant figures. However, for environmental analysis, which requires a very high degree of accuracy and precision, pH is determined by the electrometric method. This electrometric technique determines the activity of the hydrogen ions by potentiometric measurement by means of a glass sensor electrode (of a hydrogen electrode) and a reference electrode (calomel or silver: silver chloride electrode). The electromotive force (emf) produced varies linearly with the pH.

The pH meter used for pH measurement consists of these glass and reference electrodes and a potentiometer for measuring electrode potential. The pH meter must be daily calibrated against the standard buffer solutions of pH 4, 7, and 10. pH measurements are affected by temperature and the presence of very high concentrations of suspended matter.

Soil pH is determined by mixing the soil with reagent grade water (1:1), stirring the solution, allowing the soil to settle down, and then recording the pH of the supernatent liquid. Experimental evidences, however, indicate a significant variation in pH values by changing the soil to water ratio (Milke and Patnaik, 1994). An accurate method of pH determination in soil involves ion exchange of H^+ ions present in the soil with Ca^{2+} ions added into the soil sample. A saturated solution of $CaCl_2$ in neutral water is added to an aliquot of soil. The mixture is stirred and allowed to stand. Ion exchange occurs between Ca^{2+} ions in the solution and H^+ ions in the soil. Hydrogen ions are released from the soil into the solution. The pH of the supernatant $CaCl_2$ solution containing the released H^+ ion is measured.

Eh

Eh is a measure of oxidation-reduction potential in the solution. The chemical reactions in the aqueous system depend on both the pH and the Eh. While pH measures the activity (or concentration) of hydrogen ions in the solution, Eh is a measure of the activity of all dissolved species. Aqueous solutions contain both oxidized and reduced species. For example, if iron is present in the solution, there is a thermodynamic equilibrium between its oxidized and reduced forms. Thus, at the redox equilibrium, the reaction is as follows:

$$Fe^{2+} \Leftrightarrow Fe^{3+} + e^-$$

The electron activity (or intensity) at redox equilibrium may be measured by a potentiometer. A pH meter or a millivolt meter may be used for measuring the potential difference between a reference electrode (such as a calomel electrode) and an oxidation-reduction indicator electrode (such as platinum, gold, or a wax-impregnated graphite electrode).

Samples must be analyzed immediately, preferably at the sampling site. Avoid exposure to air because it can oxidize any reduced species present in the sample.

Platinum electrode is most commonly used with Ag:AgCl reference electrode with KCl as the electrolyte. The electrode system should be first standardized against a standard redox solution before Eh of the sample is measured. The procedure for Eh determination is outlined below:

1. Prepare one of the following redox standard solutions as follows:
   Standard A: Dissolve 1.4080 g potassium ferrocyanide [$K_4Fe(CN)_6 \cdot 3H_2O$], 1.0975 g potassium ferricyanide [$K_3Fe(CN)_6$], and 7.4555 g KCl in reagent grade water and dilute the solution to 1 L
   Standard B: Dissolve 39.21 g ferrous ammonium sulfate [$Fe(NH_4)_2(SO_4)_2 \cdot 6\,H_2O$] and 48.22 g ferric ammonium sulfate [$Fe(NH_4)SO_4)_2 \cdot 12\,H_2O$] in reagent grade water; slowly add 56.2 mL conc. H_2SO_4; dilute the solution to 1 L.
   Redox standard solutions are also commercially available.
2. Measure the potential of this redox standard solution following the manufacturer's instructions for using pH meter or millivoltmeter.
3. Measure the potential of the sample in the same way.
4. Record the temperature of the sample.
   Eh can be determined from the following equation:

$$\text{Eh} = E_1 + E_2 - E_3$$

where E_1 = Sample potential relative to reference electrode,
E_2 = Theoretical Eh of reference electrode and the redox standard solution relative to the standard hydrogen electrode (E_2 may be calculated, see below), and
E_3 = Observed potential of the redox standard solution, relative to the reference electrode.

E_2, the theoretical Eh of the reference electrode, may be calculated from the stability constants. For each 1°C increase in temperature, the potential of $K_4Fe(CN)_6$-$K_3Fe(CN)_6$-KCl solution shows a decrease of about 2 mV. The Eh of this redox standard solution for platinum electrode vs. Ag:AgCl reference electrode may be calculated as follows:

$$\text{Eh(mV)} = 428 - 2.2(T - 25)$$

where T is the temperature of the solution.

The potentials of platinum electrode vs. various reference electrodes for redox standard solutions A and B are presented in Table 2.22.1.

Oxidation-reduction potential is sensitive to pH. Eh decreases with an increase in the pH and increases with a decrease in the pH, if H^+ ion or OH^- ion is involved in the redox half-cells.

Table 2.22.1 Potentials of Redox Standard Solutions for Selected Reference Electrodes

Electrode system (indicator-reference)	Potential in millivolts for redox standard solutions at 25°C	
	Standard A	Standard B
Platinum-Calomel (Hg: $HgCl_2$ saturated KCl)	+183	+430
Platinum-Ag:AgCl (1 *M* KCl)	+192	+439
Platinum-Ag:AgCl (saturated KCl)	+229	+476
Platinum-hydrogen	+428	+675

2.23 PHENOLS

Phenols are organic compounds containing an –OH group attached to an aromatic ring. The structure of phenol, the prototype compound of this class, is

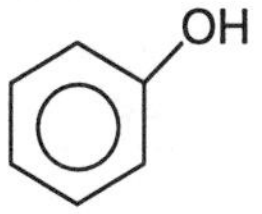

OH

Although the presence of other substituents in the ring can produce an array of diverse compounds of entirely different properties, the chemical analysis of most phenols, however, can be performed in the same way. This is attributed to (1) the acidic nature of the phenolic –OH group, and (2) that the –OH group can form derivatives.

Trace amounts of phenols may occur in many natural waters as well as in domestic and industrial wastewaters. Chlorination of such waters can produce chlorophenols.

Several phenolic compounds occurring in industrial wastewaters, soils, sediments, and hazardous wastes are classified as U.S. EPA priority pollutants. These are presented in Table 2.23.1.

The total phenolic compounds in an aqueous sample can be determined by a colorimetric method using 4-aminoantipyrine. This reagent reacts with phenolic compounds at pH 8 in the presence of potassium ferricyanide to form a colored antipyrine dye, the absorbance of which is measured at 500 nm. The antipyrine dye may also be extracted from the aqueous solution by chloroform. The absorbance of the chloroform extract is measured at 460 nm. The sample may be distilled before analysis for the removal of interfering nonvolatile compounds. The above colorimetric method determines only ortho- and meta-substituted phenols and not all phenols. When the pH is properly adjusted, certain para-substituted phenols, which include methoxyl-, halogen-, carboxyl-, and sulfonic acid substituents, may be analyzed too.

Table 2.23.1 Phenols Classified as U.S. EPA Priority Pollutants

CAS no.	Compounds
[108-95-2]	Phenol
[95-48-7]	2-Methylphenol
[108-39-4]	3-Methylphenol
[106-44-5]	4-Methylphenol
[105-67-9]	2,4-Dimethylphenol
[108-46-3]	Resorcinol
[95-57-8]	2-Chlorophenol
[120-83-2]	2,4-Dichlorophenol
[87-65-0]	2,6-Dichlorophenol
[59-50-7]	4-Chloro-3-methylphenol
[95-95-4]	2,4,5-Trichlorophenol
[88-06-2]	2,4,6-Trichlorophenol
[58-90-2]	2,3,4,6-Tetrachlorophenol
[87-86-5]	Pentachlorophenol
[88-75-5]	2-Nitrophenol
[100-02-7]	4-Nitrophenol
[51-28-5]	2,4-Dinitrophenol
[534-52-1]	2-Methyl-4,6-dinitrophenol

The individual phenolic compounds may be determined by GC-FID or GC/MS. Alternately, the phenol may be derivatized to a halogen derivative and measured by GC-ECD.

EXTRACTION

Aqueous samples are extracted with methylene chloride. If the sample is not clean or if the presence of organic interference is suspected, a solvent wash should be performed. For this, the pH of the sample is adjusted to 12 or greater with NaOH solution. The sample solution made basic is then shaken with methylene chloride. Organic contaminants of basic nature and most neutral substances partition into the methylene chloride phase, leaving phenols and other acidic compounds in the aqueous phase. The solvent layer is discarded. The pH of the aqueous phase is now adjusted to 2 or below with H_2SO_4, after which the acidic solution is repeatedly extracted with methylene chloride. Phenols and other organic compounds of acidic nature partition into the methylene chloride phase. The methylene chloride extract is then concentrated and exchanged into 2-propanol for GC analysis. For clean samples, a basic solvent wash is not necessary; however, the sample should be acidified before extraction. It may be noted that basic solvent wash may cause reduced recovery of phenol and 2,4-dimethylphenol.

Solvent exchange from methylene chloride to 2-propanol is necessary for derivatization of phenols for GC-ECD analysis. For GC-FID or GC/MS analysis, methylene chloride extract may be directly injected.

Soils, sediments, and solid wastes are extracted with methylene chloride by sonication or Soxhlett extraction. The extract may be subjected to an acid wash using reagent grade water acidified to a pH below 2. The basic organic interfering substances in the methylene chloride extract partition into the acidified water, leaving behind phenols and other acidic organics in the extract.

The surrogate and internal standards recommended for the analysis of phenols are 2-fluorophenol, 2,4,6-tribromophenol, 2-perfluoromethyl phenol, and pentafluorophenol.

ANALYSIS

Phenols are analyzed by either GC or GC/MS technique. For GC analysis, FID is the most suitable detector. Alternately, an ECD may be used following derivatization of the phenols to their bromoderivatives using pentafluorobenzyl bromide. Although the derivatization route is lengthy and time-consuming, an advantage of this method is that it eliminates interferences and, thus, any sample cleanup step may be avoided.

Derivatization

Derivatization reagent is prepared by mixing 1 mL of pentafluorobenzyl bromide and 1 g of 18-crown-6-ether and diluting to 50 mL with 2-propanol. 1 mL of this reagent is mixed with 1 mL of 2-propanol solution of the sample or the extract and 3 mg K_2CO_3, shaken gently and heated for 4 h at 80°C in a water bath. After cooling, the solution is mixed with 10 mL hexane and 3 mL of water and shaken. An accurately measured volume of the hexane extract of the derivative is now passed through silica gel (4 g) and anhydrous Na_2SO_4 (2 g) in a chromatographic column. Prior to this, the column is preeluted with hexane. After passing the extract, the phenol derivatives are eluted from the column using toluene-hexane or toluene-propanol mixture. The percent recovery of phenol derivatives for each fraction is presented in Table 2.23.2.

The column and conditions for GC analysis are listed below. Alternative columns and conditions may be employed to achieve the best separation:

- Packed column for underivatized phenol: 1.8 m long × 2 mm ID glass, packed with 1% SP-1240DA on Supelcoport (80/100 mesh) or equivalent; carrier gas N_2 at a flow rate 30 mL/min; oven temp: 80° to 150°C at 8°C/min.
- Packed column for derivatized phenols: 1.8 m long × 2 mm ID glass, packed with 5% OV-17 on Chromosorb W-AW-DMCS (80/100 mesh) or equivalent; carrier gas 5% methane/95% argon at 30 mL/min; column temp: 200°C, isothermal.
- Capillary column: Silicone-coated fused-silica 30 m (or 15 m) × 0.53 mm (or 0.32 or 0.25 mm) ID × 1 μm (or 1.5 μm) film such as DB-5, SPB-5, Rtx-5, or equivalent; carrier gas He 10 to 15 mL/min or H_2 5 to 10 mL/min; FID and injector temp between 250° and 310°C; column temp: 40° to 300°C at 8° to 10°C/min; 1 μL direct injection for 0.53 mm ID column or a split injection at a split ratio 50:1 for a column of lower ID.

Table 2.23.2 Elution Pattern of Phenol Derivatives Using Silica Gel

Parent phenol	Percent recovery			
	Fraction 1[a]	Fraction 2[b]	Fraction 3[c]	Fraction 4[d]
Phenol	—	90	10	—
2,4-Dimethylphenol	—	95	7	—
2-Chlorophenol	—	90	1	—
2,4-Dichlorophenol	—	95	1	—
2,4,6-Trichlorophenol	50	50	—	—
4-Chloro-3-methylphenol	—	84	14	—
Pentachlorophenol	75	20	—	—
2-Nitrophenol	—	—	9	90
4-Nitrophenol	—	—	1	90

[a] 15% toluene in hexane.
[b] 40% toluene in hexane.
[c] 75% toluene in hexane.
[d] 15% 2-propanol in toluene.

On FID-capillary column, a detection level at 5 ng for each phenolic compound can be achieved. For quantitation of derivatized phenol on the ECD, the standards should be derivatized as well, and the ratio of the final volume of hexane extract of the derivative to that loaded on the silica gel column should be taken into consideration in the final calculation.

The presence of phenols should be confirmed on a GC/MS, if available. The characteristic mass ions of some common phenols found in the environment are listed in Table 2.23.3.

AIR ANALYSIS

Most phenols have very low vapor pressures and are not likely to be present in gaseous state in ambient air. The particles or suspension in the air may, however, be determined by different sampling and analytical techniques. NIOSH and U.S. EPA methods discuss the analysis of only a few common phenols, which include phenol, cresols, and pentachlorophenol.

Air is drawn through a midget impinger or a bubbler containing 0.1 *N* NaOH solution. Phenol and cresols are trapped as phenolates. The pH of the solution is adjusted <4 by H_2SO_4. The compounds are determined by reverse-phase HPLC with UV detection at 274 nm. An electrochemical or fluorescence detector may also be used. The solution may be analyzed by colorimetric or GC-FID technique.

Pentachlorophenol is collected on a filter (cellulose ester membrane)-bubbler sampler. It is then extracted with methanol and analyzed by HPLC-UV. Use of alternate sampling train, Zelflour filter, and silica gel tube have also been reported (Vulcan, 1982).

Table 2.23.3 Characteristic Mass Ions for GC/MS[a] Determination of Some Common Phenol Pollutants

Phenols	Primary ion (m/z)	Secondary ions (m/z)
Phenol	94	65, 66
2-Methylphenol	107	108, 77, 79, 90
3-Methylphenol	107	108, 77, 79, 90
4-Methylphenol	107	108, 77, 79, 90
2,4-Dimethylphenol	122	107, 121
Resorcinol	110	81, 82, 53, 69
2-Chlorophenol	128	64, 130
2,4-Dichlorophenol	162	164, 98
4-Chloro-3-methylphenol	107	144, 142
2,4,6-Trichlorophenol	196	198, 200
2,4,5-Trichlorophenol	196	198, 97, 132, 99
2,3,4,6-Tetrachlorophenol	232	131, 230, 166, 234
Pentachlorophenol	266	264, 268
2-Nitrophenol	139	109, 65
2,4-Dinitrophenol	184	63, 154
2,6-Dinitrohphenol	162	164, 126, 98, 63
4-Nitrophenol	139	109 ,65
4,6-Dinitro-2-methyl phenol	198	51, 105
2-Cyclohexyl-4,6-dinitrophenol	231	185, 41, 193, 266
Surrogates		
2-Fluorophenol	112	64
Phenol-d_6	99	42, 71
2,4,6-Tribromophenol	330	332, 141

[a] Electron impact ionization mode using 70 V nominal electron energy.

2.24 PHOSPHORUS

Phosphorus occurs in natural waters, wastewaters, sediments, and sludges. The main sources of phosphorus released into the environment include fertilizers, many detergents and cleaning preparations, and boiler waters to which phosphates are added for treatment. From an analytical standpoint, phosphorus is classified into three main categories:

1. Orthophosphate, PO_4^{3-}, e.g., Na_3PO_4
2. Condensed phosphate including meta-, pyro-, and polyphosphates:
   $Na_2P_2O_6$, $Na_3P_3O_9$ (metaphosphate)
   $Na_4P_2O_7$ (pyrophosphate)
   $Na_5P_3O_{10}$ (tripolyphosphate)
3. Organically bound phosphorus

Orthophosphate and condensed phosphate are a measure of inorganic phosphorus. The latter is also termed as "acid-hydrolyzable phosphate." However, during mild acid hydrolysis, a small amount of phosphorus from organic phosphorus compounds may be released. To determine suspended and dissolved forms of phosphorus, the sample should be filtered through a 0.45 μm membrane filter, and the filtrate and the residue analyzed separately.

ANALYSIS

The analytical steps are outlined in Figure 2.24.1.

SAMPLE PREPARATION

For the determination of acid-hydrolyzable phosphorus content of the sample, which is the difference between the orthophosphate in the untreated sample and the phosphate found after mild acid hydrolysis, the sample is first acidified with H_2SO_4 and then hydrolyzed by boiling for 1.5 to 2 h. The sample may also be

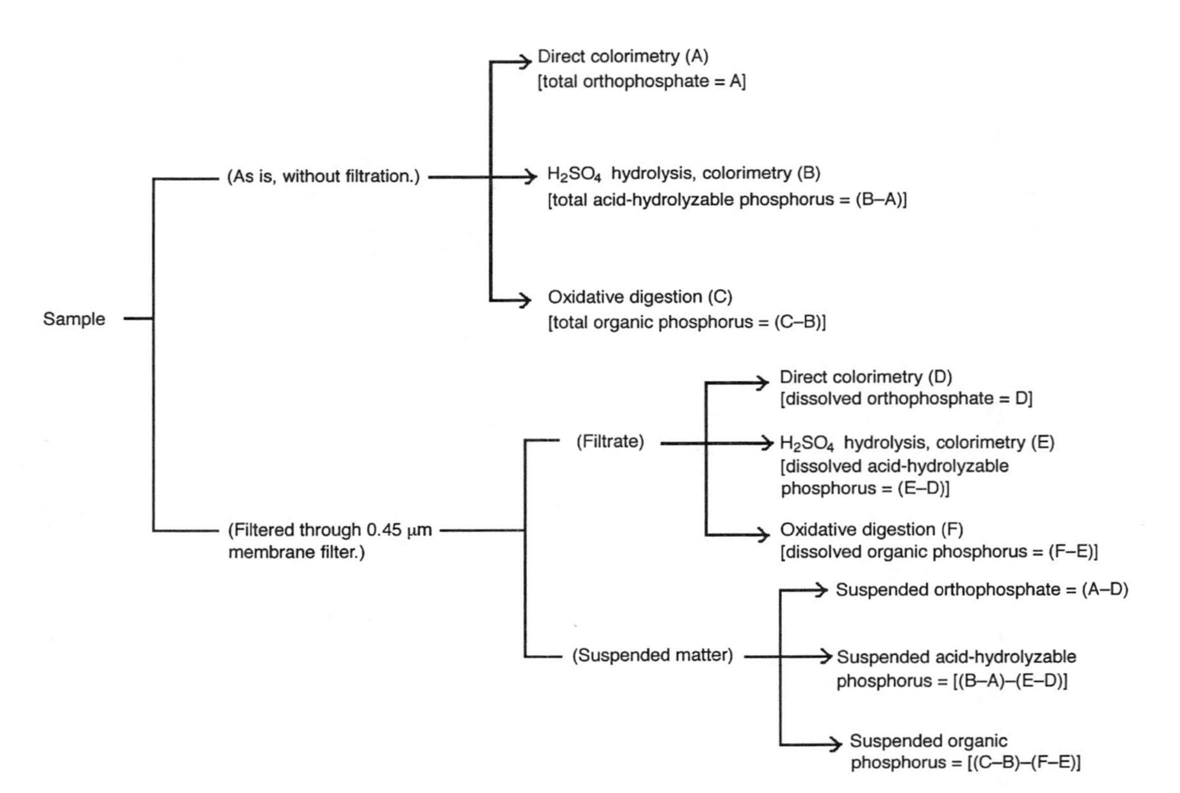

Figure 2.24.1 Schematic outline of analytical steps for phosphorus.

heated for 30 min in an autoclave at 1 to 1.4 atm. This converts condensed phosphates into orthophosphates as shown below in the reaction:

$$K_4P_2O_7 + 2\ H_2SO_4 + H_2O \rightarrow 2\ H_3PO_4 + 4\ K^+ + 2\ SO_4^{2-}$$

A 100-mL sample aliquot is acidified using phenolphthalein indicator and then treated with 1 to 2 mL of H_2SO_4(~10 *N*) prior to heating. After the hydrolysis, the sample is neutralized to a faint pink color with NaOH solution. The final volume is brought back to 100 mL with distilled water.

$$(RO)(R'O)P(=O){-}OR'' + (NH_4)_2S_2O_8 \xrightarrow{\text{acid}} H_3PO_4 + 2\ NH_4^+ + S_2O_8^{2-} + \text{organic fragments}$$

Oxidative digestion is performed to determine the organic phosphorus (which is the phosphorus measured after oxidative digestion minus phosphate determined after mild acid hydrolysis). It converts organically bound phosphorus into orthophosphate as shown in the example below:

$$R{-}O{-}P(=O)(OH){-}O{-}R' + H_2SO_4 + HNO_3 \rightarrow H_3PO_4 + S_4^{2-} + NO_3^- + \text{organic fragments}$$

Such oxidation is carried out using (1) nitric acid-perchloric acid, (2) sulfuric acid-nitric acid, or (3) persulfate. When nitric acid-perchloric acid is used, the sample is acidified with conc. HNO_3, followed by the addition of further acid and evaporation on a hot plate to a small volume. This is then treated with a 1:1 mixture of nitric and perchloric acid and evaporated gently until dense white fumes of perchloric acid just appear. When sulfuric acid-nitric acid is used, the sample is treated with a few mL of conc. H_2SO_4-HNO_3 mixture (1:5 volume ratio), evaporated to a small volume until HNO_3 is removed (the solution becomes colorless).

For persulfate digestion, the sample is acidified with conc. H_2SO_4, followed by the addition of 1 mL acid and then about 0.5 g of potassium persulfate or ammonium persulfate. The mixture is either boiled gently on a hot plate to a small volume or digested in an autoclave.

After digestion, the sample is neutralized with 6 *N* NaOH solution and brought back to the initial volume with distilled water for colorimetric determination.

COLORIMETRIC ANALYSIS

The principle of the colorimetric test is based on the reaction of orthophosphate with ammonium molybdate under acidic conditions to form a heteropoly

$$12\,MoO_3 + H_2PO_4^- \rightarrow (H_2PMo_{12}O_{40})^-$$

phosphomolybdate complex

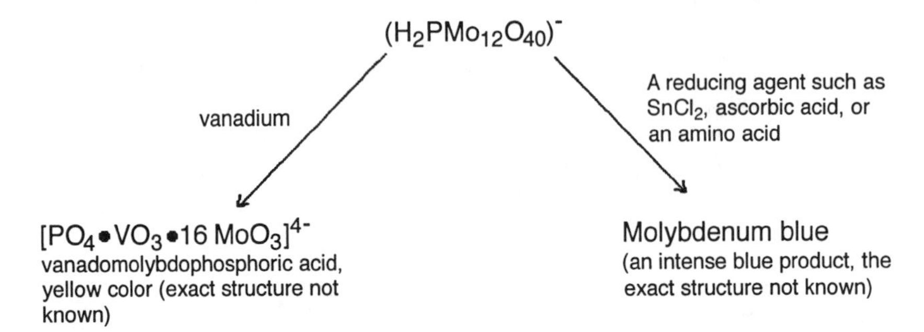

Figure 2.24.2 Schematic representation of the key reaction steps in the colorimetric analysis of phosphorus.

acid, molybdophosphoric acid, which, in the presence of vanadium, forms yellow vanadomolybdophosphoric acid. The intensity of the color is proportional to the concentration of phosphate. Often, yellow is not a strong color to measure, especially when the sample is dirty. Therefore, the molybdophosphoric acid formed may be reduced with stannous chloride to form molybdenum blue which has an intense color. The same dye may be formed by the reaction of orthophosphate with ammonium molybdate and potassium antimonyl tartrate in acid medium and reduction with ascorbic acid or an amino acid.

The key reaction steps are shown in Figure 2.24.2.

Interference

Silica, arsenate, sulfide, thiocyanate, thiosulfate, fluoride, chloride, and iron (II) interfere in the test. Most interferences are observed at high concentrations (>50 mg/L). Silica interference is removed by oxidation with Br water.

Calibration Standards

A series of calibration standards are made from anhydrous potassium dihydrogen phosphate, KH_2PO_4. Dissolve 0.2195 g salt in 1 L distilled water to produce PO_4^{3-} (as P) concentration of 50 mg/L. The strength of the calibration standards should be in the range of 0.05 to 1.0 mg PO_4^{3-} (as P) per liter. Prepare a calibration curve plotting absorbance (at 400 to 490 nm) against concentration. The standards should be subjected to the same treatment (i.e., acid hydrolysis or oxidative digestion) as sample. The concentration of P in the sample is read from the calibration curve.

The detection limit for measuring P as yellow vanadomolybdophosphoric acid is about ~200 μg/L in 1 cm spectrophotometer cell and ~10 μg/L when determined as molybdenum blue.

ION CHROMATOGRAPHY

Orthophosphate in an untreated or treated sample may be determined by ion chromatography (See Chapter 1.11). A detection limit of 0.1 mg/L may be achieved with a 100-μL sample loop and a 10 mmho full-scale setting on the conductivity detector. The column and conditions for a typical wastewater analysis are listed below. Equivalent column and alternate conditions may be used.

- Column: Ion Pac AS11 (Dionex) or equivalent; 4 × 250 mm for the column and 4 × 50 mm for the guard column (ion exchange group: alkanol quaternary ammonium)
- Eluent: 0.05 *M* NaOH in 40% methanol
- Flow rate: 1 mL/min
- Detection: Suppressed conductivity, chemical suppression mode

2.25 PHTHALATE ESTERS

Phthalates are the esters of phthalic acid having the following general structure:

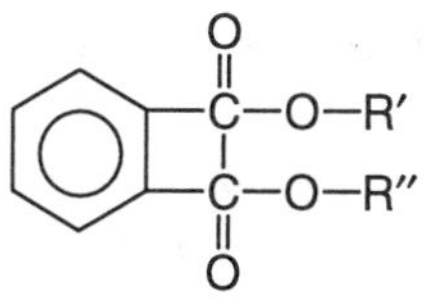

where R′ and F″ are alkyl, alkenyl, and aryl groups. These substances are used as plasticizers of synthetic polymers such as polyvinyl chloride and cellulose acetate. Lower aliphatic phthalates are used in the manufacture of varnishes and insecticides.

Many phthalates are found in trace quantities in wastewaters, soils, and hazardous wastes, often leaching out into the liquid stored in plastic containers or PVC bags.

The acute toxicity of phthalates is very low, exhibiting symptoms of somnolence and dyspnea in test animals only at high doses. Some of these substances are listed as U.S. EPA priority pollutants.

Phthalates are analyzed by GC, LC, and GC/MS techniques after the extraction and concentration of the samples. Aqueous samples may be directly analyzed by HPLC. Phthalates can be extracted from 1 L of aqueous sample by repeated extraction with methylene chloride using a separatory funnel. Phthalates having much greater solubility in methylene chloride, partition into this solvent and is separated. The extract is concentrated by boiling off methylene chloride and exchanged to hexane to a volume of 1 to 5 mL. Alternatively, 1 L aqueous sample is passed through a liquid-solid extraction cartridge containing octadecyl group bonded-silica. Phthalates adsorbed on the adsorbent surface are eluted with methylene chloride. Phthalates in soils and other solid matrices may be extracted by sonication or Soxhlett extraction using methylene chloride. The extract is exchanged to hexane during concentration. If the extract is to be analyzed by GC/MS, the solvent exchange to hexane is not necessary and the methylene chloride extract may be directly injected.

To eliminate the interference effect of other contaminants and for dirty sample extracts, cleanup may become necessary. The extract is either passed through a florisil column or an alumina column and the phthalate esters are eluted with ether-hexane mixture (20% ethyl ether in hexane, v/v).

The sample extract may be analyzed by GC or GC/MS. A 2- to 5-μL aliquot of the extract is injected into a GC and the phthalates are detected by an electron capture detector, a flame ionization detector (FID), or a photoionization detector. Some of the chromatographic packed or capillary columns that may be used for the phthalate analysis are listed below:

- Packed column: 1.5% SP-2250/1.96% SP-2401 on Supelcoport (100/120 mesh), 3% OV-1 on Supelcoport (100/120 mesh), and 3% SP-2100 on Supelcoport (100/120 mesh).
- Capillary column: Fused silica capillary column containing 95% dimethyl polysiloxane and 5% diphenyl polysiloxane (e.g., 30 m, 0.53 mm ID, 1.5 μm Rtx-5); 5% diphenyl polysiloxane, 94% dimethyl polysiloxane, and 1% vinyl polysiloxane (e.g., PTE-5, SPB-5, DB-5, or equivalent, 15 or 30 m × 0.53 mm ID).

The column temperatures in the analysis may be maintained between 150° and 220°C. Benzyl benzoate or *n*-butyl benzoate may be used as internal standard.

GC/MS analysis is a positive confirmatory test that identifies the compounds based on their characteristic ions. Table 2.25.1 lists the characteristic ions for some commonly occurring phthalates, which have been listed as U.S. EPA priority pollutants.

Aqueous samples containing phthalates at concentrations higher than 10 ppm may be directly injected into GC and measured by FID.

High performance liquid chromatography techniques may be successfully applied to analyze phthalate esters. A 15 or 25 cm column filled with 5 or 10 μm silica-based packings is suitable. Short columns (3.3 cm × 4.6 mm), commonly called 3 × 3 columns, offer sufficient efficiency and reduce analysis time and solvent consumption. Phthalate esters resolve rapidly on a 3 × 3 Supelcosil LC-8 column (3 μm packing) at 35°C and detected by a UV detector at 254 nm. Acetonitrile-water is used as mobile phase (flow rate: 2 ml/min; injection volume: 1 mL). Other equivalent columns under optimized conditions may be used.

AIR ANALYSIS

Analysis of phthalates in air may be performed by sampling 1 to 200 L air and collecting the esters over a 0.8 μm cellulose ester membrane. The phthalates are desorbed with carbon disulfide and the eluant is injected into a GC equipped with an FID. A stainless steel column 2 m × 3 mm OD, containing 5% OV-101 on 100/120 Chromosorb W-HP was originally used in the development of this method (NIOSH Method 5020) for dibutyl phthalate and *bis*(2-ethylhexyl)phthalate. Any other equivalent GC column listed above may also be used.

The EPA Methods in Table 2.25.2 list only a specific number of phthalates. Any other phthalates not listed under the methods, however, may be analyzed using the same procedures.

Table 2.25.1 Characteristic Ions for Phthalate Esters

CAS no.	Phthalate esters	Characteristic ions	
		Primary	Secondary
131-13-3	Dimethyl phthalate	163	164, 194
84-66-2	Diethyl phthalate	149	150, 177
84-74-2	Di-*n*-butyl phthalate	149	104, 150
84-69-5	Diisobutyl phthalate	149	150
131-18-0	Diamyl phthalate	149	150
84-75-3	Dihexyl phthalate	149	150
84-61-7	Dicyclohexyl phthalate	149	150
117-84-0	Di-*n*-octyl phthalate	149	43, 167
84-76-0	Dinonyl phthalate	149	150, 167
85-68-7	Butyl benzyl phthalate	149	91, 206
117-81-7	*bis*(2-ethylhexyl)phthalate	149	167, 279
146-50-9	*bis*(4-methyl-2-pentyl) phthalate	149	150, 167
75673-16-4	Hexyl 2-ethylhexyl phthalate	149	150, 167
117-82-8	*bis*(2-methoxyethyl) phthalate	149	150
605-54-9	*bis*(2-ethoxyethyl) phthalate	149	150
117-83-9	*bis*(2-*n*-butoxyethyl) phthalate	149	150

Table 2.25.2 EPA and NIOSH Methods for the Analysis of Phthalate Esters

Matrices	Methods	Extraction/concentration	Instrumental analysis
Drinking water	EPA Method 506	Liquid-liquid or liquid-solid extraction	GC-PID (capillary)
	EPA Method 525	Liquid-solid extraction	GC/MS (capillary)
Wastewater	EPA Method 606	Liquid-liquid extraction	GC-ECD
	EPA Method 625	Liquid-liquid extraction	GC/MS
Soil, sediment, sludges, hazardous waste	EPA Method 8060	Sonication or Soxhlett extraction	GC-ECD or GC-FID
	EPA Method 8061	Sonication or Soxhlett extraction	GC-ECD (capillary)
	EPA Method 8250	Sonication or Soxhlett extraction	GC/MS
	EPA Method 8270	Sonication or Soxhlett extraction	GC/MS (capillary)
Groundwater	EPA Methods 8060, 8061, 8250, 8270	Liquid-liquid extraction	GC or GC/MS techniques
Organic liquids	EPA Method 8060	Waste dilution or direct injection	GC-ECD or GC-FID
Air	NIOSH Method 5020[a]	Collected over cellulose ester membrane and desorped into CS_2	GC-FID

Note: GC, gas chromatography; GC/MS, gas chromatography/mass spectrometry; GC-PID, gas chromatography-photoionization detector; GC-ECD, gas chromatography-electron capture detector.

[a] NIOSH Method 5020 lists only dibutyl phthalate and *bis*(ethylhexyl) phthalate. Other phthalates in air may be analyzed by the same technique.

2.26 POLYCHLORINATED BIPHENYLS (PCBS)

Polychlorinated biphenyls (PCBs) are a class of chlorosubstituted biphenyl compounds that were once widely used as additives in transformer oils, lubricating oils, and hydraulic fluids. These substances have high boiling points, exhibiting high chemical and thermal stability, and flame resistance. However, because of their high toxicity and possible carcinogenic action in humans, these substances are no longer being used. In the U.S., PCBs were made under the trade name Aroclor. Table 2.26.1 presents the common Aroclors, their CAS numbers, and the chlorine contents. Each Aroclor is a mixture of several isomers. The general structure of the biphenyl ring is as follows:

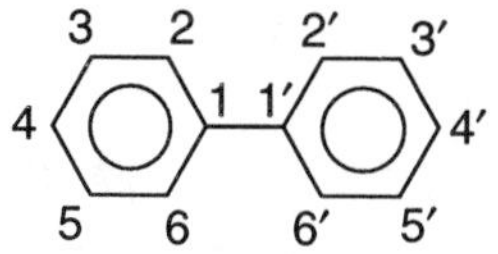

PCBs can be conveniently determined by most of the common analytical techniques which include GC-ECD, GC-HECD, GC-FID, GC/MS, HPLC, NMR, and enzyme immunoassay. Among these, GC-ECD and GC/MS are by far the most widely used techniques for the determination of PCBs in the environmental samples at a very low level of detection. While the former can detect the PCBs at subnanogram range, the mass selective detector (GC/MS) identifies the components relatively at a higher detection range, 10 to 50 times higher than the ECD detection level. GC/MS, however, is the best confirmatory method to positively confirm the presence of PCBs, especially in heavily contaminated samples. Aqueous and nonaqueous samples must be extracted into a suitable solvent prior to their analysis.

Being mixtures of several components, each Aroclor produces multiple peaks. The common GC columns that can readily separate the chlorobiphenyl components include 3% SP-2100, OV-1, DB-5, and SPB-5. Other equivalent columns or conditions can be used in the GC and GC/MS analyses. Table 2.26.2 presents some of the commonly used columns and conditions for analysis.

Table 2.26.1 PCBs and Their Chlorine Contents

Aroclor	CAS no.	Chlorine content (%wt)
1016	[12674-11-2]	41
1221	[11104-28-2]	21
1232	[11141-16-5]	32
1242	[53469-21-9]	42
1248	[12672-29-6]	48
1254	[11097-68-1]	54
1260	[11096-82-5]	60
1262	[37324-23-5]	62
1268	[11100-14-4]	68

Table 2.26.2 Common Analytical Columns and Conditions for GC Analysis of PCBs

Column	Dimensions	Conditions
	Packed	
1.5% SP-2250/1.95% SP-2401 on Supelcoport (100/120 mesh)	1.8 m × 4 mm	5% methane/95% argon carrier gas, 60 mL/min, ~200°C, isothermal
3% OV-1 on Supelcoport (100/120 mesh)	1.8 m × 4 mm	5% methane/95% argon carrier gas, 60 mL/min, ~200°C, isothermal
	Capillary	
DB-5, SPB-5, SE-54, Rtx-5, OV-73, OV-3, DC-200, DC-560, CP-Sil-8, DB-1701	30 m × 0.53 mm (1 or 1.5 μm film thickness)	

An oven temperature in the range of 200°C and detector and injector temperatures around 300°C and 250°C, respectively, should give good separation, sharpness of peaks, and fast analysis time. Electron capture detector (ECD) is the most commonly used detector for trace level analysis of PCBs by gas chromatography (GC), exhibiting a response to an amount below 0.1 ng PCBs. Thus, on a capillary column, an IDL in the range of 5 μg/L can be achieved. With proper sample concentration steps, a detection level several-fold lower to IDL may be obtained. Other halogen-specific detectors such as Hall electrolytic conductivity detector can also be used to analyze PCBs.

Because PCBs produce multiple peaks, extra care should be taken to identify the genuine peaks from any other contaminants, such as phthalate esters, sulfur or chlorinated pesticides, and herbicides, to avoid any false positive inference. The following steps should be taken for the qualitative determination:

1. All the major peaks of the reference Aroclor standard must be present in the GC chromatogram of the unknown sample extract.
2. The retention times of the sample peaks must closely match with that of the standard(s). One or more internal standards (such as dibutyl chlorendate, tetrachloro-*m*-xylene or *p*-chlorobiphenyl) should, therefore, be added into the sample extract, as well as Aroclor standards, to monitor any response time shifts.

3. When the chromatogram of the unknown sample has a number of peaks (some of which show matching with an Aroclor standard), then the ratios of the areas or heights of two or three major peaks of the unknown should be compared with the corresponding peak ratios in the standard at the same retention times. For example, if a sample is found to contain any specific Aroclor, measure the area or height ratio of largest to second largest peak in the sample extract and compare the same to that in the standard. The comparison of peak ratios, however, should not always be strictly followed, as it can lead to erroneous rejection. For example, because some of the same chlorobiphenyl components are common to most Aroclors, the presence of two or more Aroclors in the sample can alter the peak-ratio pattern. Similarly, the kinetics of oxidative or microbial degradation of chlorinated biphenyls can be different for each compound in the PCB mixture.
4. Finally, the presence of Aroclor found in the sample on the primary column must be redetermined and confirmed on an alternate column or by GC/MS using selective ion monitoring mode. The characteristic ions (under electron impact ionization) are 190, 222, 224, 256, 260, 292, 294, 330, 360, 362, and 394.

Quantitation

Although GC/MS is the most reliable technique for qualitative determination of PCBs, its quantitative estimation in environmental samples are more accurately done from the analysis by GC-ECD. The area counts or the heights of all major PCB peaks in the sample are first added up and then compared against the total area or the heights of the same number of peaks at the same retention times in the standard. A single-point calibration can be used if the peak areas of the PCBs are close to that of the standard. Often, many environmental samples show the presence of only some but not all characteristic PCB peaks. In such cases, if the peaks are confirmed to be those of an Aroclor, but their number is less than half the number of peaks in the corresponding standard, the PCBs in the sample should be considered and termed as WEATHERED. If some of the chlorinated biphenyl isomers found in the sample are in a measurable amount, their quantities should be determined by comparing the total area of those specific Aroclor peaks found in the sample to the total area of all the Aroclor peaks in the standard with the following qualifying statement: "sample contains weathered Aroclor; the concentration of all the chlorinated biphenyl components found in the sample is ______, determined as Aroclor ______."

SAMPLE EXTRACTION AND CLEANUP

Aqueous samples are extracted with methylene chloride by liquid-liquid extraction. The extract is concentrated and then exchanged to hexane. Soils, sediments, and solid wastes are extracted by sonication or Soxhlett extraction. Samples should be spiked with one or more surrogate standard solution to determine the accuracy of analysis. Some of the internal standards mentioned above may also be used as surrogates. If only the PCBs are to be analyzed, hexane instead of methylene chloride may be used throughout. Oil samples may be subjected to waste dilution, i.e., diluted with hexane or isooctane and injected onto the GC column for determination by ECD or HECD.

Sample cleanup often becomes necessary to remove interfering substances such as phthalate esters and many chlorinated compounds frequently found in wastewaters, sludges, and solid wastes. Most interfering contaminants can be removed from the solvent extract by gel permeation chromatography or by florisil cleanup (Chapter 1.5). In addition to these cleanup methods, the sample extract should be further cleaned up by shaking 1 mL extract with an equal volume of $KMnO_4$-H_2SO_4 ($KMnO_4$ in 1:1 H_2SO_4). Most organics at trace levels are oxidized under these conditions forming carbon dioxide, water, and other gaseous products, leaving behind PCBs in the hexane phase. The hexane phase is then washed with water, and the moisture in hexane is removed by anhydrous Na_2SO_4. If sulfur is known or suspected to be present in the sample, an aliquot of the cleaned extract after $KMnO_4$ treatment may be subjected to sulfur cleanup either by using mercury or copper powder (Chapter 1.5).

ALTERNATE ANALYTICAL METHODS

PCBs at high concentrations can be measured by GC-FID, NMR, and HPLC. Concentrations over 100 ppm can be determined by HPLC by UV detection at 254 nm. A normal phase HPLC technique with column switching can separate PCBs from chlorinated pesticides.

PCBs in soils and wastewaters can be rapidly screened on site or in the laboratory by immunoassay technique (Chapter 1.13). Immunoassay test kits are now commercially available from many suppliers. The samples can be tested at the calibration levels of 1 to 50 ppm. The kit primarily contains antibody-coated test tubes or magnetic particles, assay diluent, PCB-enzyme conjugate, a color-forming substance, and a solution to quench the reaction. The method does not distinguish accurately one Aroclor from another. PCBs can be measured semiquantitatively by comparing the optical density of the color formed in the sample against a set of calibration standards using a spectrophotometer.

AIR ANALYSIS

PCBs in air may be analyzed by NIOSH Method 5503. Using a personal sampling pump, 1 to 50 L air is passed though a 13 mm glass fiber filter and through a florisil column at a flow rate of 50 to 200 mL/min. A glass fiber filter is placed in a cassette that is connected to the florisil tube. The latter contains 100 and 550 mg florisil in the front and back sections of the tube, respectively.

PCBs collected on the filter or adsorbed onto the florisil are now desorbed with 5 mL hexane. The filter, the front section, and the back section of the florisil and the media blank are analyzed separately by GC-ECD under the chromatographic conditions as described above. The results are added up to determine the total PCBs. Also, a media blank is run to determine the background contamination. Correction is made for any PCBs found in the media blank.

Florisil used for adsorbing PCBs should have a particle size of 30/48 mesh and should be dried at 105°C. After cooling, add a small amount of distilled water to florisil (3 g water to 97 g florisil) prior to its use.

The column used in NIOSH study is a 2 mm ID packed glass column containing 1.5% OV-17 plus 1.95% QF-1 on 80/100 mesh Chromosorb WHP. Other equivalent column may be used. The flow rate of carrier gas nitrogen may be set at 40 to 50 mL/min (on a packed column).

2.27 POLYCHLORINATED DIOXINS AND DIBENZOFURANS

Polychlorinated dibenzo-*p*-dioxins and dibenzofurans are tricyclic aromatic compounds having the following structures:

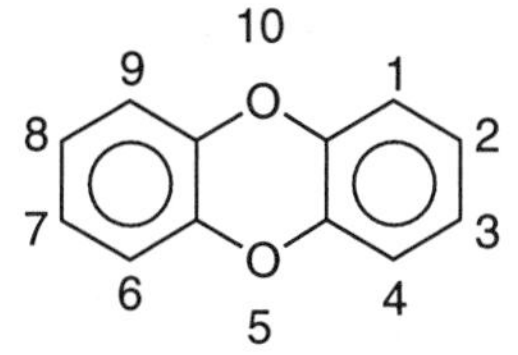

(Dibenzo-*p*-dioxin) 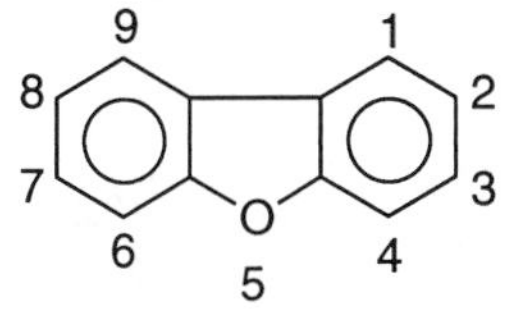 (Dibenzofuran)

Chlorosubstitutions in the aromatic rings can give rise to several isomers. For example, one to eight chlorine atoms can be attached onto different positions in the dibenzo-*p*-dioxin rings, thus producing a total of 75 isomers. Chlorosubstitution in positions 2, 3, 7, and 8 gives 2,3,7,8-tetrachlorodibenzo-*p*-dioxin (2,3,7,8-TCDD), CAS [1746-01-6], occurring in trace amount in chlorophenoxyacid herbicide, 2,4,5-T. 2,3,7,8-TCDD is an extremely toxic substance that can cause liver and kidney damage, ataxia, blurred vision, and acne-like skin eruption (Patnaik, 1992). This is also a known carcinogen and teratogen. It melts at 303°C and has a vapor pressure 1.5×10^{-9} torr. Its solubility in water is about 0.3 µg/L and in chloroform about 370 mg/L (Westing, 1984).

ANALYSIS

Chlorinated dioxins and dibenzofurans are best analyzed by GC/MS techniques, using both low- and high-resolution mass spectrometry. A measured amount of sample is extracted with a suitable solvent. The solvent extract containing the analytes is concentrated down to a small volume and then subjected to cleanup for the removal of interferences. The extract is injected onto the GC column for the separation of individual compounds. The chlorinated dioxins and dibenzofurans are then identified from their characteristic mass ions using a GC/MS in selective ion monitoring (SIM) mode.

SAMPLE EXTRACTION AND CLEANUP

Aqueous samples are extracted with methylene chloride. A 1-L volume of sample is repeatedly extracted in a separatory funnel. The methylene chloride extract is exchanged to hexane during concentration to a volume of 1 mL. Nonaqueous samples, such as soils, sediments, sludges, fly ash, and tissues may be extracted by Soxhlett extraction or sonication. Methylene chloride, toluene, hexane, or a combination of these solvents may be used for extraction. Sludges containing 1% or more solids should be filtered. The aqueous filtrates and the solid residues are extracted separately. They are then combined prior to cleanup and analysis.

Chlorinated compounds such as PCBs, haloethers, chloronaphthalenes, etc., which may be present in several orders of magnitude higher than the analytes of interest, can interfere in the analysis. These interfering substances may be removed from the solvent extracts as follows.

Acid-Base Partitioning

The hexane extract is shaken with 1:1 H_2SO_4 in a small separatory funnel for 1 min and the bottom H_2SO_4 layer is discarded. Such acid wash may be repeated two or three times. The extract is then repeatedly washed with 20% KOH solution. Contact time must be minimized because KOH could degrade certain chlorinated dioxins and dibenzofurans. If acid-base washing is performed, the sample extract should be washed with 5% NaCl solution each time after acid and base washes, respectively. Acid–base partitioning cleanup may, however, be omitted completely if the sample is expected to be clean.

Alumina Cleanup

This procedure is very important and should be performed for all samples. Activated alumina should be used. The alumina column containing anhydrous Na_2SO_4 on the top is first preeluted with 50 mL hexane and then 50 mL of 5% methylene chloride/95% hexane mixture. After this, the sample extract is completely transferred onto the column which is then eluted with 20% methylene chloride/80% hexane solution. The eluant is then concentrated for analysis.

Silica Gel Cleanup

This cleanup may be performed only if the alumina cleanup does not remove all interferences. A silica gel column containing anhydrous Na_2SO_4 on the top is preeluted with 50 mL 20% benzene/80% hexane solution. The extract is then loaded onto the column and the analytes are eluted with the above benzene-hexane mixture.

Carbon Column Cleanup

If significant amounts of interferences still remain in the sample extract after performing the above cleanup procedures, the extract may be subjected to a further cleanup step. Prepare a mixture of active carbon AX-21 and Celite 545, containing 8% and 92%, respectively. This is heated at 130°C for several hours and packed into a chromatographic column. This carbon column is preeluted with toluene and hexane, respectively. The sample extract is now loaded onto the column and the analytes are subsequently eluted with toluene. The eluant is concentrated for analysis.

GC/MS ANALYSIS

The polychlorinated dioxins and dibenzofurans are separated on a fused silica capillary column. A 60 m long and 0.25 μm ID DB-5, SP-2330, or equivalent column having 0.2 μm film thickness should adequately resolute most isomers.

The mass spectrometer must be operated in SIM mode. The ^{13}C-analogs of isomers may be used as internal standards. The analytes are identified from their relative retention times and characteristic masses. In low resolution MS, the characteristic masses for 2,3,7,8-TCDD are 320, 322, and 257. Use either $^{37}Cl_4$-2,3,7,8-TCDD or $^{13}C_{12}$-2,3,7,8-TCDD as an internal standard. The m/z for these two internal standards are 328 and 332, respectively.

When using high resolution mass spectrometry (HRMS), the characteristic m/z for 2,3,7,8-TCDD are 319.8965 and 321.8936. The m/z for the corresponding $^{37}C_{14}$- and $13C_{12}$-isomers in HRMS are 327.8847 and 331.9367, respectively.

The quantitation is performed by the internal standard method. The SIM response for the isomers at their primary characteristic m/z are compared against the internal standard(s). A detection limit in the range of 2 to 5 ppt (0.002 to 0.005 μg/L) can be achieved for aqueous samples concentrated as above and analyzed using low resolution mass spectrometry method. A lower detection limit in 0.01 to 1 ppt range may be achieved by HRMS technique.

SAMPLE PRESERVATION AND HOLDING TIME

Samples must be refrigerated and protected from light. If residual chlorine is present, add $Na_2S_2O_3$ (80 mg/L sample). Samples must be extracted within 7 days of collection and subsequently analyzed within 40 days.

2.28 SILICA

Silica (SiO_2) occurs in high abundance all over the earth. It occurs in the form of sand and quarts. It is also present in rocks and silicate minerals. It is found in natural waters at varying concentrations from 1 to 100 mg/L.

Silica in water may be analyzed by the following methods:

1. Gravimetric method
2. Ammonium molybdate colorimetric method
3. Atomic absorption spectrophotometric method

Methods 1 and 2 are presented below. Method 3 is discussed in brief under metal analysis.

GRAVIMETRIC METHOD

Silica and silicates (SiO_3^{2-} and SiO_4^{4-}) react with hydrochloric or perchloric acid to form silicic acid, H_2SiO_3. Evaporation of the solution precipitates dehydrated silica. Upon ignition, dehydration is completed. The reaction steps are summarized below:

$$SiO_2 \text{ or silicates} \xrightarrow{HCl/HClO_3} H_2SiO_3$$

$$\xrightarrow{\text{evaporation}} \text{partially dehydrated } SiO_2$$

$$\xrightarrow[\text{at } 1200°C]{HCl/HClO_3} SiO_2 + H_2O \uparrow$$

The residue after ignition is weighed along with its container. This residue consists of dehydrated silica plus any nonvolatile impurities in the sample.

Hydrofluoric acid is then added to this residue. This converts all silica in the residue to silicon tetrafluoride, SiF_4, as shown below:

$$SiO_4 + 4\,HF \Rightarrow SiF_4 + 2\,H_2O$$

Upon ignition, SiF_4 volatilizes leaving behind any nonvolatile impurities. The container is coded and weighed again. The difference in weight is equal to the amount of silica that volatilizes.

$$\text{mg } SiO_2\,/\,L = \frac{(W_1 - W_2) \times 1000}{\text{mL sample taken}}$$

where W_1 is the weight of crucible and contents before HF treatment and W_2 is the weight of crucible and contents after volatilization of SiF_4.

Procedure

Place 100 to 200 mL of sample in a platinum evaporating dish (or an acid-leached glazed porcelain dish without etching). Add 5 mL of 1:1 HCl. Evaporate to near dryness in a 110°C oven or on a hot plate. Add another 5 mL 1:1 HCl. Evaporate to dryness. Add a small volume of 1:50 hot HCl and rinse the residue. Filter the mixture through an ashless filter paper. Transfer all the residue quantitatively from the 110°C evaporating dish onto the filter paper using distilled water. Wash the residue several times with diluted water until the washings show no chlorine ion (test with $AgNO_3$; addition of a few drops of $AgNO_3$ reagent to the washing should not produce white precipitate or turbidity).

Transfer the filter paper and residue to a platinum crucible. Dry at 110°C. Heat the covered crucible (keeping a little opening) gradually to 1200°C. Cool in a desiccator and weigh. Repeat the ignition, cooling, and weighing until constant weight is attained.

Moisten the residue with distilled water. Add a few drops of 1:1 H_2SO_4 followed by 10 mL HF (48% strength). Evaporate the mixture slowly to dryness. Ignite the residue at 1200°C. Record the constant weight. Determine the concentration of dissolved silica in the sample from the above equation.

AMMONIUM MOLYBDATE COLORIMETRIC METHOD

Under acid condition (at pH ~1) silica reacts with ammonium molybdate to form a yellow colored heteropoly acid, silicomolybdic acid. The reactions are shown below:

$$SiO_2 + H_2O \Rightarrow \underset{\text{(silicic acid)}}{H_2SiO_3}$$

$$H_2SiO_3 + 3\ H_2O \Rightarrow \underset{\text{(silicic acid hydrate)}}{H_8SiO_6}$$

$$H_8SiO_6 + \underset{\text{(ammonium molybate)}}{12\ (NH_4)_2MoO_4} + 12\ H_2SO_4$$

$$\Rightarrow \underset{\text{(silicomolybdic acid)}}{H_8\left[Si\left(Mo_2O_7\right)_6\right]} + 12\ (NH_4)_2SO_4 + 12\ H_2O$$

Phosphate reacts with ammonium molybdate too, similarly forming yellow phosphomolybdic acid. The presence of phosphate, therefore, interferes in the test. The addition of oxalic acid or citric acid destroys phosphomolybdic acid complex but not silicomolybdic acid complex. The intensity of color developed is proportional to the concentration of silica in the sample.

An additional color development step may often be required to confirm the yellow color of silicomolybdic acid. The latter is reduced to a dark blue substance by treating with aminonaphthalo sulfonic acid. The color of heteropoly blue formed is more intense than the yellow color of silicomolybdic acid. The latter test is more sensitive and can give a detection limit of 50 μg silica/L when using a spectrophotometer.

Certain forms of silica and many polymeric silicates do not react with ammonium molybdate. These complex silicates may be decomposed to simple molybdate-reactive silica by high temperature digestion or fusion with sodium bicarbonate.

Procedure

Place 50 mL sample in a Nessler tube. Add 1 mL of 1:1 HCl and 2 mL ammonium molybdate reagent. Shake well. Let the solution stand for 5 min. Add 2 mL oxalic acid solution, shake thoroughly, and allow the solution to stand for another 5 min. Measure the absorbance of the yellow color developed at 410 nm. Run a blank using 50 mL distilled water following the above procedure. Compare the absorbance of the sample solution with the standards and determine the concentration of silica from the calibration curve.

As a more sensitive alternative test or an an additional confirmatory test, add 2 mL of aminosulfonic acid reducing reagent 5 min after adding oxalic acid solution. Let the solution stand for 5 min and then measure the absorbance at 815 or 650 nm (at the latter wavelength the sensitivity is reduced.) Plot a separate calibration curve. Read the concentration of analyte from the calibration curve.

Reagent

- Reducing agent: to 300 mL of $NaHSO_3$ solution (~20% strength), add 100 mL of solution containing 1 g 1-amino-2-naphthol-4-sulfonic acid and 2 g Na_2SO_3. Filter and store the reagent in a plastic bottle in the dark at 4°C. Discard this solution when it becomes dark.

2.29 SULFATE

Sulfate is a divalent polyanion having the formula SO_4^{2-}. It is a common contaminant, occurring widely in wastewaters, wastes, and potable waters. Sulfate in water may be analyzed by:

1. Ion chromatography
2. Gravimetric method
3. Turbidimetric method

Method 1 is suitable for measuring low concentrations of SO_4^{2-}, at less than 1 ppm level. The lower limit of detection is about 0.1 mg/L. Gravimetric method, on the other hand, is reliable only at a relatively high concentration range (>10 mg/L). Turbidimetric method is applicable in the concentration range of 1 to 40 mg/L. Samples with high SO_4^{2-} concentrations may be accordingly diluted prior to analysis.

GRAVIMETRIC METHOD

Sulfate is precipitated as barium sulfate when added to barium chloride solution. The reaction is shown below for sodium sulfate.

$$Na_2SO_4 + BaCl_2 \Rightarrow 2\ NaCl + BaSO_4$$

The reaction is performed at boiling temperature and HCl medium. An acid medium is required to prevent any precipitation of $BaCO_3$ and $Ba_3(PO_4)_2$.

The precipitate is filtered, washed free from Cl^-, dried, and then weighed. Alternately, the precipitate, along with the filter paper, is ignited in a platinum crucible at 800°C. $BaSO_4$ residue is weighed and SO_4^{2-} concentration is calculated.

Interference

Silica and suspended matter in the sample produce high results, as does NO_3^- which forms $Ba(NO_3)_2$ and is occluded with $BaSO_4$ precipitate. Heavy metals interfere in the test by precipitating as sulfates, thus giving low results. Sulfates and bisulfates of alkali metals may occlude with $BaSO_4$, yielding low results.

Procedure

A measured volume of sample is evaporated nearly to dryness in a platinum dish and then treated with 1 to 2 mL of HCl. Evaporation is continued. The sample residue is successively treated with HCl and evaporated. The aim is to extract the SO_4^{2-} out leaving behind the insoluble silica and suspended matter in the residue. The residue is then heated with distilled water and the washing is combined with the acid solution.

The pH of the clear filtrate solution free from silica and suspended matters is now adjusted to 4.5 to 5.0. The solution is heated. $BaCl_2$ solution is slowly added with stirring until precipitation is complete.

The precipitate, $BaSO_4$, is filtered through a fritted glass filter (pore size <5 μm) or a 0.45 μm membrane filter. It is washed a few times with small amounts of hot distilled water to remove any Cl^- ion that may be adhering to it. It is dried in an oven for several hours at 105°C, cooled in a desiccator, and weighed. Heating, cooling, and weighing is repeated until a constant weight is obtained.

Alternately, $BaSO_4$ precipitate along with the filter paper is placed in a weighed platinum crucible and ignited at 800°C for 1 to 2 h. It is then cooled in a dessiccator and weighed.

Calculation

$$\text{mg } SO_4^{2-}/L = \frac{\text{mg } BaSO_4 \times 0.4115}{\text{mL sample}}$$

The factor 0.4115 comes from the ratio of formula weight of SO_4^{2-} to that of $BaSO_4$, which is $\frac{96.04}{233.40}$.

TURBIDIMETRIC METHOD

Sulfate ion is precipitated as $BaSO_4$ in an acid medium, reacting with $BaCl_2$. The solution turns turbid due to white $BaSO_4$ precipitate. The turbidity is measured by a nephelometer. Alternatively, the light absorbance of the $BaSO_4$ suspension is measured at 420 nm by a spectrophotometer providing a light path of 2.5 to 10 cm. A filter photometer equipped with a violet filter may also be used to measure the light transmittance at 420 nm. Concentration of SO_4^{2-} in the sample is determined from a standard calibration curve.

Turbidimetric method is applicable when the SO_4^{2-} concentrations is between 5 and 50 mg/L. For concentrations above 50 mg/L, dilute the sample and analyze.

Presence of a large amount of suspended matter or color in the sample would interfere in the test. Filter the sample to remove suspended matter.

Procedure

To a 100-mL sample, add 5 mL conditioning reagent. Stir the solution and add a spoonful of $BaCl_2$ crystals. Stir it for a minute at a constant speed. Measure the turbidity of this solution.

Prepare a calibration curve by plotting turbidity (in NTU if a nephelometer is used) or absorbance or transmittance (if a spectrophotometer or filter photometer is used) of $BaSO_4$ formed against the corresponding concentrations of SO_4^{2-} standards. Determine the concentration of SO_4^{2-} in the sample from the standard calibration curve.

Calculation

$$\text{mg } SO_4^{2-} / \text{L} = \frac{\text{mg } SO_4^{2-} \text{ read from the calibration curve} \times 1000}{\text{mL sample}}$$

Reagents

- Conditioning agent: Mix 30 mL conc. HCl, 300 mL distilled water, 100 mL 95% ethanol, and 75 g NaCl in a container. Add 50 mL glycerol and mix.
- Sulfate standards: Dissolve 147.9 mg anhydrous Na_2SO_4 in 1 L distilled water. Concentration of this stock solution is 100 mg SO_4^{2-}/L. Prepare five calibration standards from this stock solution as follows:

Dilution	Concentration, mg/L
2 mL → 100 mL	2.0
5 mL → 100 mL	5.0
10 mL → 100 mL	10.0
20 mL → 100 mL	20.0
40 mL → 100 mL	40.0

2.30 SULFIDE

Sulfide (S^{2-}) is a bivalent monoanion produced from the decomposition of metal sulfide salts. It occurs in groundwaters, hot springs, and wastewaters. It is also formed from the bacterial reduction of sulfate. Sulfide salts in solid wastes in contact with an acid can produce hydrogen sulfide. H_2S, which is highly toxic. In an aqueous sample, sulfide may be present as dissolved H_2S and HS^-, dissolved metallic sulfide, and acid-soluble metallic sulfide contained in suspended particles. All these soluble and insoluble sulfides and dissolved H_2S and HS^- together are termed as total sulfide. The sulfide remaining after the removal of suspended solids is termed the dissolved sulfide. Copper and silver sulfides are insoluble even under acidic conditions. Therefore, these two sulfides are not determined in the following tests.

In a nonaqueous sample, such as soil, sediment, or hazardous waste, the sample is vigorously shaken with acidified water to leach out sulfide. Sulfide in aqueous samples or leachates may be analyzed by one of the following methods:

1. Iodometric method
2. Methylene blue colorimetric method
3. Silver-silver sulfide electrode method

Methods 1 and 2 are commonly used in environmental analysis. Method 3 uses a silver electrode to indicate the end point of potentiometric titration of dissolved sulfide with standard $AgNO_3$. The electrode response is slow.

ANALYSIS OF SULFIDE IN WATER (SEE FIGURE 2.30.1)

SCREENING TEST FOR SULFIDE

Before analysis, it is often useful to determine qualitatively the presence of sulfide in the sample and the concentration range at which it is present. The following tests may be performed:

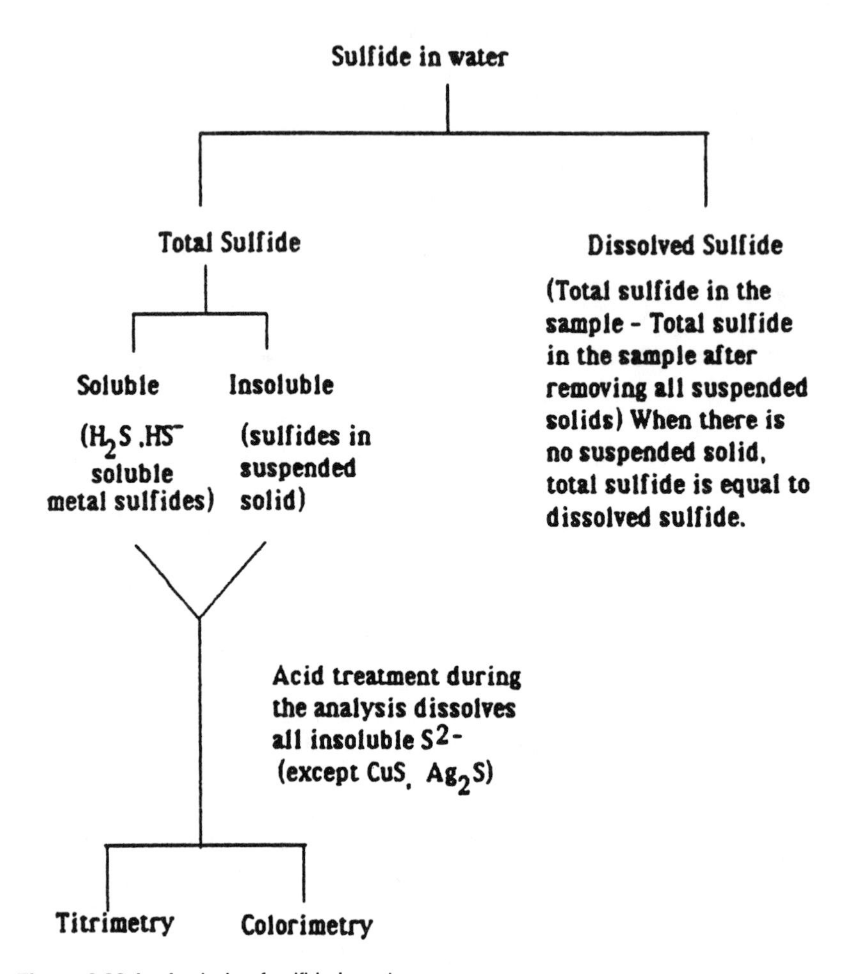

Figure 2.30.1 Analysis of sulfide in water.

1. *Antimony test.* To 100 mL of sample, add 5 drops of saturated solution of potassium antimony tartrate and 5 to 10 drops of 6 *N* HCl. A yellow antimony sulfide formation would confirm the presence of sulfide. Perform this test with known standards to match the color intensity to determine the concentrations of sulfide in the sample. The test is sensitive to concentration above 0.5 mg S^{2-}/L. Lead interferes in this test.
2. *Silver foil test.* H_2S evolved from slightly acidified sample can blacken a strip of silver foil due to the formation of silver sulfide. Silver is cleaned by dipping in NaCN solution prior to exposure.
3. *Lead acetate paper test.* A paper moistened with lead acetate solution turns black due to the formation of lead sulfide on exposure to H_2S.

SAMPLE PRETREATMENT

Sample pretreatment, although not always necessary, should be performed prior to analysis. This is required to remove the interfering substances and also to concentrate the sample. The presence of reducing agents such as sulfite, thiosulfate, iodide, and many organic substances interfere with both the iodometric and methylene blue methods. Addition of a zinc or cadmium salt precipitates out the sulfide that settles at the bottom. The supernatant liquid containing the interfering substances is removed and is replaced with smaller volume of distilled water, thus, to concentrate the sulfide. If sulfide concentration in the sample is found to be high from antimony screening test, do not concentrate. The supernatant liquid is removed and replaced by equal volume of distilled water or if dilution is required by a larger volume of water. Zinc sulfide precipitate, which is insoluble in water, dissolves after adding acid in the analysis.

Select a sample volume of 100 mL for colorimetric test or 400 mL for iodometric test. Add a few drops of zinc acetate solution (10% strength) into the flask and fill it with the sample. Add a few drops of 6 *N* NaOH solution to produce a pH above 9. Mix the solution vigorously using a shaker. If sulfide is present in the sample, it will form a precipitate. Allow the solution to stand for 15 to 20 min to let the precipitate settle. Decant and discard the supernatant liquid. Add a measured volume of distilled water. Note the final volume of solution containing the precipitate. Begin analysis immediately.

ANALYSIS OF DISSOLVED SULFIDE IN WATER

Analyze 1 aliquot of the sample for total sulfide by iodide titrimetric or the methylene blue colorimetric test (discussed in detail in the following sections).

Suspended solids are removed from the other aliquot of the sample as follows. Add 1 mL of 6 *N* NaOH to a 500-mL glass bottle. Fill the bottle with the sample. Add 1 mL of $AlCl_3$ solution (10 g $AlCl_3 \cdot 6\,H_2O$ in 15 mL water). Stopper the bottle. Rotate the bottle back and forth for 2 min to flocculate the contents. Allow the contents to settle for 10 to 15 min. Draw out the supernatant liquid to analyze sulfide. The difference in the sulfide values obtained from both the tests (i.e., before and after removing suspended solids from the sample), is equal to the concentration of dissolved sulfide.

IODOMETRIC METHOD

An excess amount of iodine is added to the sample which may or may not have been treated with zinc acetate. Iodine oxidizes sulfide to sulfur under acidic condition. This is shown in the following equations:

$$I_2 + S^{2-} \xrightarrow{\text{acid}} S + 2\,I^-$$

or,

$$I_2 + ZnS \xrightarrow{\text{acid}} S + ZnI_2$$

The excess iodine (unreacted surplus iodine) is back titrated with standard sodium thiosulfate solution using starch indicator. Phenylarsine oxide may be used as a titrant instead of sodium thiosulfate.

Procedure

A measured amount of standard iodine solution is placed in a 500-mL flask. The amount of iodine should be the excess over the expected quantity of sulfide in the sample. Add distilled water and bring the volume to 20 mL. Add 2 mL of 6 *N* HCl. Pipette 200 mL of sample into the flask. If the sulfide in the sample was precipitated as ZnS, transfer the precipitate with 100 mL distilled water into the flask. Add iodide solution and HCl. If iodine color disappears, add more iodine standard solution into the flask, until the color remains. Titrate with $Na_2S_2O_3$ standard. Before the end point, when the color changes to straw yellow or brown, add a few mL of starch solution and continue titration by dropwise addition of $Na_2S_2O_3$ standard until the blue color disappears. Record the volume of titrant added.

Calculation

$$\text{mg/L sulfide} = \frac{[(A \times B) - (C \times B)] \times 16{,}000}{\text{mL sample}}$$

where A = mL of standard iodine solution,
B = Normality of iodine solution,
C = mL of titrant, $Na_2S_2O_3$ solution, and
D = Normality of $Na_2S_2O_3$ solution.

(Since S^{2-} has two negative charges, its equivalent weight is 32 ÷ 2 or 16.) Equivalent weight expressed in mg, therefore, is 16,000, used in the above calculation.

If the normality of the standard iodine and standard $Na_2S_2O_3$ solutions are both 0.025, the above expression can be written as:

$$\text{mg/L sulfide} = \frac{(A - C) \times 16{,}000 \times 0.025}{\text{mL of sample}}$$

$$= \frac{(A - C) \times 400}{\text{mL of sample}}$$

In other words, 1 L of 0.025 *N* standard iodine solution would react with 400 mg of sulfide. This may also be calculated from the following equation:

$$\begin{array}{ccccccc} I_2 & + & S^{2-} & \Rightarrow & S & + & 2\,I^- \\ 254\text{ g} & & 32\text{ g} & & 32\text{ g} & & 254\text{ g} \end{array}$$

254 g I_2 reacts with 32 g S^{2-}. Therefore, 3.2 g I_2 weighed to make 0.025 *N* solution, would react with $\frac{32\text{ g} \times 3.2\text{ g}}{254\text{ g}}$ or 0.4 g or 400 mg of sulfide.

Phenylarsine oxide may also be used as the reducing agent in the titration, instead of sodium thiosulfate.

Reagents

- Standard iodine solution, 0.025 *N*: dissolve 20 to 25 g KI in a small amount of water. Add 3.2 g iodine and dissolve. Dilute with distilled water to 1 L. Standardize against 0.025 *N* $Na_2S_2O_3$ solution using starch indicator. In the equation, $I_2 + 2\ S_2O_3^{2-} \Rightarrow 2\ I^- + S_4O_6^{2-}$, 1 g equivalent weight iodine reacts with 1 g equivalent weight of $Na_2S_2O_3$. Therefore, the normality of standard iodine solution

$$= \frac{\text{mL of } Na_2S_2O_3 \text{ titrant} \times \text{its normality}}{\text{mL of iodine solution in titration}}$$

- Preparation and standardization of sodium thiosulfate/phenylarsine oxide solution; 0.025 *M*: dissolve 6.205 g sodium thiosulfate ($Na_2S_2O_3 \cdot 5\ H_2O$) or 4.200 g phenylarsine oxide ($C_6H_5\ As = 0$) [PAO] in distilled water. Add 0.5 g NaOH solid and dilute to 1 L. Standardize against potassium bi-iodate (potassium hydrogen iodate) solution as described below.

  To 2 g of KI in 100 mL distilled water, add a few drops of conc. H_2SO_4. Add 25 mL of standard bi-iodate solution. Dilute to 200 mL and titrate the liberated iodine against thiosulfate or PAO using starch indicator.

  The strength (molarity) of the titrant thiosulfate or PAO solution

$$= \frac{\text{molarity of bi-iodate soln.} \times 12 \times 25 \text{ mL}}{\text{mL of titrant}}$$

$$= \frac{0.0021\ M \times 12 \times 25 \text{ mL}}{\text{mL of titrant}}$$

  In the above calculation, the numerator is multiplied by 12, because 1 mol of bi-iodate = 12 mol of $S_2O_3^{2-}$ or PAO as determined from the following equation:

$$KH(IO_3)_2 + 10\ KI + 6\ H_2SO_4 \Rightarrow 6\ I_2 + 6\ H_2O + 5\ K_2SO_4 + KHSO_4$$

$$6\ I_2 + 12\ S_2O_3^{2-} \Rightarrow 12\ I^- + 6\ S_4O_6^{2-}$$

$$6\ I_2 + 12\ AsO^{2-} \Rightarrow 12\ I^- + 6\ As_2O_2^{2-}$$

- Standard potassium bi-iodate, $KH(IO_3)_2$, solution, 0.00208 *M*. The strength of potassium bi-iodate should be $\frac{0.025\ M}{12}$ or 0.00208 *M*. This is calculated from the thiosulfate strength 0.025 *M* and the following stoichiometry: 1 mol of thiosulfate ≡ 1/12 mol of bi-iodate. Dissolve 0.8125 g $KH(IO_3)_2$ in distilled water and dilute to 1 L.

METHYLENE BLUE COLORIMETRIC METHOD

Methylene blue is a dye intermediate that is also used as an oxidation-reduction indicator. Its molecular formula in the trihydrate form is $C_{16}H_{18}N_3SCl \cdot 3\ H_2O$. It is soluble in water, forming a deep blue solution. The structure of methylene blue is as follows:

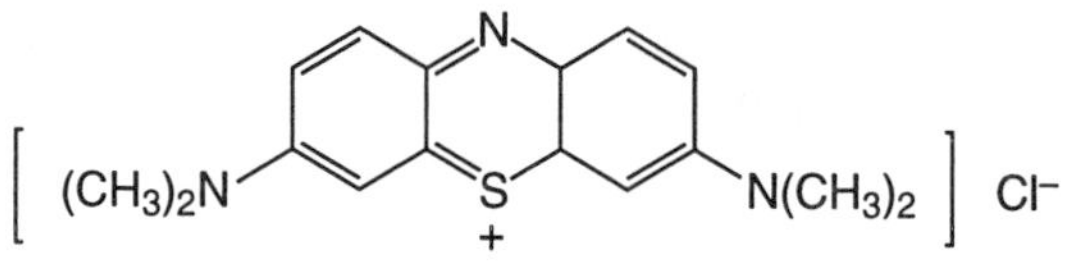

$[(CH_3)_2N$ — N — S^+ — $N(CH_3)_2]\ Cl^-$

Sulfide reacts with *N,N*-Dimethyl-*p*-phenylenediamine (*p*-aminodimethyl aniline) in the presence of ferric chloride to produce methylene blue:

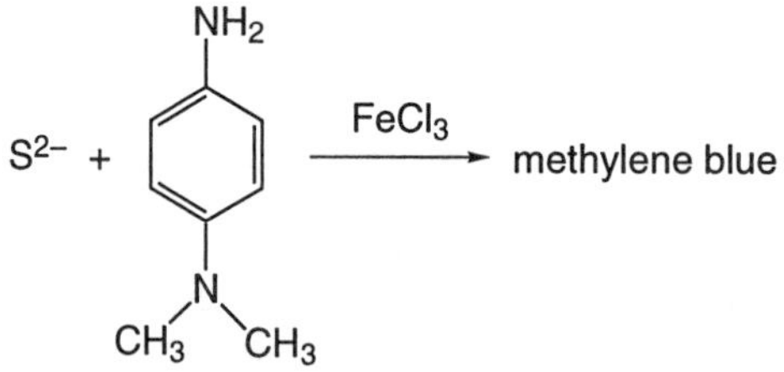

S^{2-} + NH_2 — C_6H_4 — $N(CH_3)_2$ $\xrightarrow{FeCl_3}$ methylene blue

The intensity of color developed is proportional to the concentration of sulfide in the sample. Color comparison may be done visually with methylene blue standard (sulfide equivalent). Alternatively, a spectrophotometer or a filter photometer may be used and the concentration of sulfide determined from a standard calibration curve. The color is measured at a wavelength maximum of 625 nm.

Procedure

Mark two matched test tubes as A and B and transfer 7.5 mL of sample to each. To tube A, add 0.5 mL amine-sulfuric acid reagent and 3 drops of $FeCl_3$ solution. Invert it once and mix the solution. If sulfide is present, the color of the solution develops to blue within a few minutes. Allow the solution to stand for 5 min. After this, add 2 mL of diammonium hydrogen phosphate solution.

To test tube B containing an equal amount of sample, add 0.5 mL 1:1 H_2SO_4, 3 drops of $FeCl_3$ solution, and invert once. After 5 min, add 2 mL of $(NH_4)_2HPO_4$ solution. Instead of amine-sulfuric acid, 1:1 acid is added to the sample in tube B.

Therefore, in this case, without the amine reactant, no methylene blue formation would occur. Add $(NH_4)_2HPO_4$ after 5 min. For photometric color comparison, use a 1 cm cell for 0.1 to 2.0 mg/L and a 10 cm cell for 2.0 to 20.0 mg/L sulfide concentration. Zero the instrument with a sample portion from tube B and read the absorbance of tube A. Plot a calibration curve between sulfide concentration and absorbance (see below) and determine the concentration of sulfide in the sample from the graph.

If the color comparison is done visually, proceed as described below.

If the sulfide concentration in the sample is expected to be high or found to be high from qualitative testing, use methylene blue standard solution I. Otherwise, for a low concentration, use a diluted solution of this standard, methylene blue solution II. Add methylene blue solution(s) dropwise to tube B until the color matches to that developed in tube A. For high concentrations, start with dropwise addition of solution I. When the color is close to matching, add solution II dropwise.

$$\text{Concentration of sulfide, mg/L} = \left(A + \frac{B}{10}\right) \times C$$

where A = number of drops of methylene blue solution I,
B = number of drops of methylene blue solution II, and
C = conc. sulfide (mg/L) equivalent to one drop of methylene.

METHYLENE BLUE SOLUTION: PREPARATION AND STANDARDIZATION

Solution I is made by dissolving 1.0 g powder having a dye content certified as 84% or more in 1 L of distilled water. This solution should be standardized against sulfide solution of known strength to determine mg/L sulfide that would react with 1 drop (0.05 mL) of the solution. Solution II is 1:10 dilution of solution I.

Standardization

Dissolve an accurately weighed 15.0 g sodium sulfide ($Na_2S \cdot 9\ H_2O$) in 50 to 60 mL distilled water in a 100-mL volumetric flask. Dilute to the mark. Add one drop of this solution (0.05 mL) to 1 L of distilled water in a volumetric flask. Shake well. That makes the concentration of the solution as 1 mg S^{2-}/L (1 ppm). Prepare five calibration standards in the range 1 to 10 mg S^{2-}/L using 1 to 10 drops of the stock solution. To a 7.5-mL aliquot of each standard, add the reagents and follow the procedure described above under colorimetric method. Read the absorbance of the color developed at 625 nm and draw a standard calibration curve plotting absorbance vs. concentration of sulfide.

If color comparison is made visually, standardize the methylene blue solution I or II as described below.

Match the color developed in one of the sulfide standard aliquots (after adding the reagents) to that of methylene blue solution, either by diluting the latter, or by adding more dye. For example, select the 1 ppm sulfide standard and match the color developed in it with the color of the prepared methylene blue solution I.

7.5 mL of 1 mg/L sulfide standard ≡ 7.5 μg S^{2-} which would react with 63.8 μg *N,N*-dimethyl-*p*-phenylenediamine in presence of $FeCl_3$ to produce 75 μg methylene blue ($C_{16}H_{18}N_3SCl$). Medicinal grade methylene blue is a trihydrate, $C_{16}H_{18}N_3SCl \cdot 3\ H_2O$ containing 84% methylene blue and 16% water. If the above grade methylene blue is used, 1 drop of 1 g/L solution ≡ 42 μg $C_{16}H_{18}N_3SCl \cdot 3\ H_2O$. Therefore, 1 drop of 1 mg/L S^{2-} standard solution should be theoretically equivalent to 1.8 drops of methylene blue, or 1 drop of this methylene blue solution I ≡ 0.56 mg/L of sulfide. If the zinc salt of methylene blue dye $C_{16}H_{18}N_3SCl_2 \cdot ZnCl_2 \cdot H_2O$ is used, then 1 drop of 1 g/L solution ≡ 40 μg methylene blue, $C_{16}H_{18}N_3SCl$ ≡ 1.9 drops of 1 mg/L S^{2-} standard, or 1 drop of methylene blue zinc salt ≡ 0.53 mg/L sulfide.

Because some amount of S^{2-} may escape out as H_2S during standard preparation and a small proportion may be oxidized to sulfate over a few hours, and also because methylene blue formation reaction may not thermodynamically go to completion, it is, therefore, always recommended that sulfide concentration should be determined by titrimetric iodide procedure and an average percent error of the methylene blue procedure be compared against this titrimetric procedure.

Reagents

- Aminosulfuric acid: 27 g *N,N*-dimethyl-*p*-phenylenediamine oxalate in 100 mL 1:1 H_2SO_4 solution in a volumetric flask. Store in dark. 25 mL of this solution is diluted to 1 L with 1:1 H_2SO_4. The reagent should be stored in a dark bottle.
- Ferric chloride solution: 100 g $FeCl_3 \cdot 6\ H_2O$ in 40 mL distilled water.
- Diammonium hydrogen phosphate solution: 400 g $(NH_4)_2HPO_4$ in 800 mL distilled water.

2.31 SULFITE

Sulfite, SO_3^{2-}, is a bivalent polyanion. It occurs in boiler feed water that is treated with sulfite for controlling dissolved oxygen. It also occurs in waters subjected to SO_2 treatment for dechlorination purpose. Sulfite forms sulfurus acid, H_2SO_3, which gradually oxidizes to sulfuric acid. Excess sulfite in boiler waters can cause corrosion. Sulfite is toxic to aquatic life.

Sulfite in water may be analyzed by the following methods:

1. Ion chromatography
2. Iodometric method
3. Phenanthroline colorimetric method

Method 1 is rapid and accurate and can be applied to analyze several anions including sulfite (see Section 1.8). The minimum detection limit for sulfite by iodometric method is 2 mg/L, while the detection limit is 0.01 mg/L for colorimetric method.

IODOMETRIC METHOD

Sample containing sulfite is first acidified and then titrated against a standard solution of potassium iodide-iodate to blue end point, using starch indicator. The sequence of reactions is as follows:

Potassium iodate and potassium iodide react in acid medium, liberating iodine.

$$IO_3^- + 5\,I^- + 6\,H^+ \Rightarrow 3\,I_2 + 3\,H_2O$$

$$(\text{or, } KIO_3 + 5\,KI + 6\,HCl \Rightarrow 6\,KCl + 3\,I_2 + 3\,H_2O)$$

The iodine produced reacts with SO_3^{2-} oxidizing sulfite to sulfate, and itself reduced to HI, as shown below:

$$SO_3^{2-} + I_2 + H_2O \Rightarrow SO_4^{2-} + 2\,HI$$

When all sulfite in the sample is consumed, the excess iodine reacts with starch to form blue coloration. Thus, at the end point of the titration, when no sulfite is left, iodine produced in situ from the addition of a single drop of standard iodide-iodate titrant into the acid solution forms the blue complex with starch previously added to the sample.

Interference

Presence of oxidizable substances in the sample would interfere in the test, thus giving high results. These include S^{2-}, $S_2O_3^{2-}$, and certain metal ions such as Fe^{2+} in lower oxidation state. Sulfide should be removed by adding 0.5 g zinc acetate, allowing the zinc sulfide precipitate to settle and drawing out the supernatant liquid for analysis. If thiosulfate is present, determine its concentration in an aliquot of sample by iodometric titration using iodine standard. Subract the concentration of thiosulfate from the iodometric sulfite results to calculate the true value of SO_3^{2-}.

Presence of Cu^{2+} ion can catalyze oxidation of SO_3^{2-} to SO_4^{2-}, which would give a low result. This is prevented by adding EDTA which complexes with Cu^{2+}. This also promotes oxidation of Fe^{2+} to Fe^{3+}. Nitrite, which reacts with sulfite, is destroyed by adding sulfamic acid.

Procedure

Place 1 mL of HCl in an Erlenmeyer flask. Add 0.1 g sulfamic acid. Add 100 mL sample. Add 1 mL of starch indicator solution (or 0.1 g solid). Titrate with potassium iodide-iodate standard solution until a faint blue color develops. Run a blank using distilled water instead of sample.

Calculation

$$\text{mg/L } SO_3^{2-} = \frac{(A - B) \times N \times 40,000}{\text{mL sample}}$$

where A = mL titrant for sample,
B = mL titrant for blank, and
N = Normality of KI-KIO_3 titrant.

The equivalent weight of SO_3^{2-} is $\frac{\text{formula wt}}{2} = \frac{80}{2}$ or 40. The milligram equivalent, therefore, is 40,000.

Reagents

- Standard potassium iodide-iodate titrant solution, 0.0125 N: Dissolve 0.4458 g anhydrous KIO_3 (dried for several hours at 120°C) and 4.25 g KI and 0.310 g $NaHCO_3$ in distilled water and dilute to 1 L. The equivalent weight of KIO_3

is $\frac{\text{formula wt.}}{6}$ or 35.67; thus, 0.0125 *N* solution is prepared by dissolving (35.7 g × 0.0125) or 0.4458 g salt in 1 L distilled water. The iodate-iodide anions upon reaction lose total six electrons, forming iodine. KIO_3 is the limiting reagent that determines the stoichiometry of iodine formation, while KI is used in excess as it has additional role in starch-iodide complex formation (see Iodometric Titration).

1 L of titrant ≡ 0.0125 g equivalent SO_3^{2-} ≡ 0.0125 × 40 g SO_3^{2-} or 500 mg SO_3^{2-}. Thus, 1 mL of KIO_3-KI titrant solution made ≡ 500 µg SO_3^{2-}.

- EDTA reagent: 2.5 g disodium EDTA in 100 mL distilled water.

COLORIMETRIC METHOD

Sample containing sulfite on acidification produces sulfur dioxide:

$$Na_2SO_3 + HCl \Rightarrow 2\ NaCl + SO_2 + H_2O$$

The liberated SO_2 is purged with nitrogen and trapped in an absorbing solution that contains Fe^{3+} and 1,10-phenanthroline. SO_2 reduces Fe^{3+} to Fe^{2+} which reacts with 1,10-phenanthroline to produce an orange-red complex, *tris* (1,10-phenanthroline)iron(II) (also called "ferroin"), having the following structure:

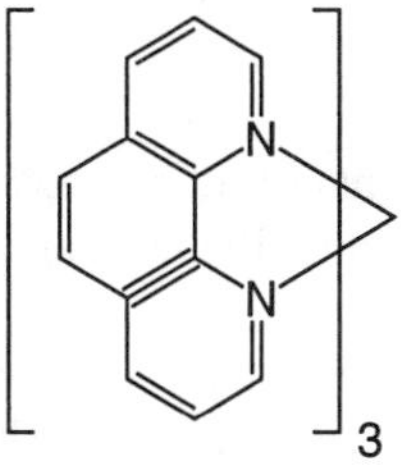

The intensity of the color developed is proportional to the amount of phenanthroline complex formed, which is proportional to the concentration of sulfite in the sample. After removing the excess ferric iron with ammonium bifluoride, the absorbance is measured at 520 nm. The concentration of SO_3^{2-} in the sample is calculated from a sulfite standard calibration curve. Because SO_3^{2-} solutions are unstable, the concentration of working standard is accurately determined by potassium iodide-iodate titration before colorimetric measurement.

Certain orthophenanthrolines may be used instead of 1,10-phenanthroline. These include 5-nitro-1,10-phenanthroline, Erioglaucin A, and *p*-ethoxychrysoidine which form iron-II complexes of violet-red, yellow-green, and red colors, respectively.

Procedure

Set up an apparatus to purge out SO_2 with N_2 from an acidified sample and trap the liberated gas in an absorbing solution, as shown below in Figure 2.31.1.

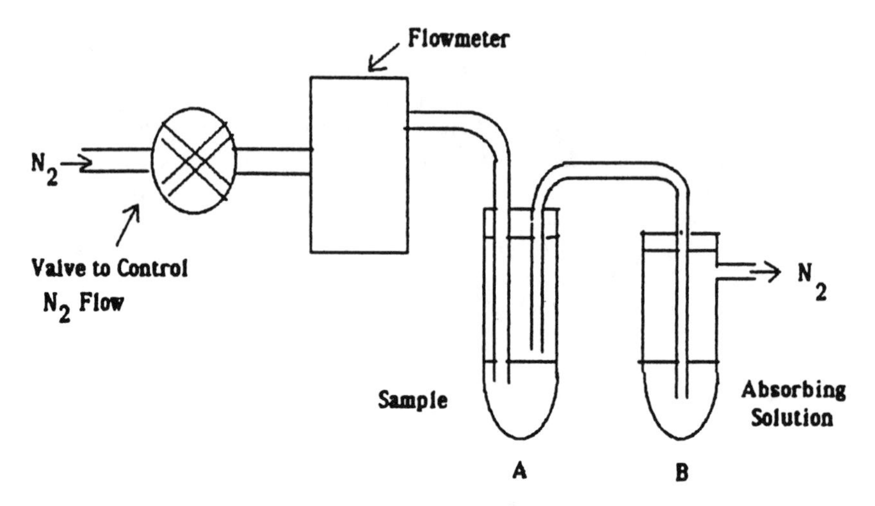

Figure 2.31.1 Diagram showing apparatus for purging out SO_2 with N_2.

The absorbing solution consists of 5 mL of 1,10-phenanthroline, 0.5 mL ferric ammonium sulfate, and 25 mL distilled water taken in a 100 mL glass tube B. A few drops of octyl alcohol is added to prevent foaming.

In tube A, take 1 mL sulfamic acid and 100 mL sample or a sample portion diluted to 100 mL. Add 10 mL 1:1 HCl and immediately connect tube A with tube B. Start nitrogen flow, adjust the flow between 0.5 and 1 L/minute, and purge for 1 h.

Disconnect the tubes after turning off the gas. Add 1 mL of NH_4HF_2 immediately to tube B. Dilute the contents of tube B to 50 mL. Allow it to stand for 5 min. Pipet 1 mL of this solution into a 1 cm cell and read the absorbance against distilled water at 510 nm.

Repeat the above steps exactly in the same manner and read absorbance for a distilled water blank and four SO_3^{2-} standards. The stock standard should be standardized by titrimetic method to determine the molarity of SO_3^{2-}. Prepare a calibration curve plotting absorbance vs. microgram sulfite. Run at least one standard for each batch of samples.

Calculation

$$\text{mg } SO_3^{2-}/\text{L} = \frac{\mu\text{g } SO_3^{2-} \text{ from calibration curve}}{\text{mL sample}} \times 50$$

If the sample volume is the same as the volume of calibration standards taken in the analysis, a calibration curve may be plotted in ppb concentration instead of microgram mass. In such case, read ppb concentration of SO_3^{2-} in the sample directly from the calibration curve.

Reagents

- Sulfite standards: Dissolve 1 g Na_2SO_3 in 1 L distilled water. Standardize this solution by titration using 0.0125 *N* potassium iodide-iodate titrant. (See the preceding section, Tritrimetric Method in this chapter for detailed procedure.) Determine the exact concentration of this stock solution.

  Pipet 1, 2, 5, 10, and 20 mL of the stock solution into 100-mL volumetric flasks and dilute to the mark with 0.04 *M* potassium tetrachloromercurate, K_2HgCl_4 solution. This gives a series of sulfite standards whose concentrations are listed in Table 2.31.1.

Table 2.31.1 Sulfite Standards

	Dilution	Calibration std, mg/L	Micrograms in 1 cm cell (1 mL std)
1 mL stock ⇒	100 mL	≡ 0.01 C_s	≡ 0.01 C_s
2 mL stock ⇒	100 mL	≡ 0.02 C_s	≡ 0.02 C_s
5 mL stock ⇒	100 mL	≡ 0.05 C_s	≡ 0.05 C_s
10 mL stock ⇒	100 mL	≡ 0.10 C_s	≡ 0.10 C_s
20 mL stock ⇒	100 mL	≡ 0.20 C_s	≡ 0.20 C_s

Note: C_s is the concentration of SO_3^{2-} in stock solution determined from titration.

- Standard potassium iodide-iodate: See Iodometric Method in the preceding section in this chapter.
- 1,10-Phenanthroline solution, 0.03 *M*: Dissolve 5.95 g in 100 mL of 95% ethanol and dilute to 1 L with distilled water.
- Ferric ammonium sulfate, $NH_4Fe(SO_4)_2 \cdot 12\ H_2O$, solution, 0.01 *M*: Add 1 mL conc. H_2SO_4 to 1 L distilled water and dissolve 4.82 g of the salt in this acidified water. Adjust the pH of the solution between 5 and 6.
- Ammonium bifluoride, 5% solution: Dissolve 5 g NH_2HF_2 in 100 mL distilled water. Store this reagent in a plastic bottle.
- Potassium tetrachloromercurate, 0.04 *M*: Dissolve 10.86 $HgCl_2$, 5.96 g KCl, and 0.066 g disodium EDTA in 1 L distilled water. Adjust the pH between 5.0 and 5.5. The solution is stable for 4 to 6 months.
- Sulfamic acid: 10 g in 100 mL distilled water.
- Disodium EDTA, 2.5%: 2.5 g in 100 mL distilled water.

2.32 SURFACTANT: ANIONIC

Many anionic surfactants can react with a cationic dye such as methylene blue to form strong ion pairs that can be extracted by a suitable organic solvent and can be determined using colorimetric techniques. The anionic surfactants that respond to the methylene blue test are primarily the sulfonate ($RSO_3^- Na^+$) and the sulfate ester ($ROSO_3^- Na^+$) type substances. On the other hand, soaps and the alkali salts of fatty acids (C-10 to C-20) used in certain detergents do not respond to the above test. The various anionic surfactants and their characteristic structural features are presented in Figure 2.32.1.

Such anionic surfactants that form ion pairs with methylene blue and that are extractable with chloroform are known as "Methylene Blue Active Substances" (MBAS). Other cationic dyes, such as crystal violet dye, may be used instead of methylene blue. Extraction of such an ion-pair complex into benzene has been reported (Hach, 1989). Detection Limit = 10 μg/L.

Reagents

- Methylene blue reagent: 30 mg methylene blue in 500 mL water; add 41 mL of 6 N H_2SO_4 and 50 g sodium dihydrogen phosphate monohydrate ($NaH_2PO_4 \cdot H_2O$); dilute to 1 L.
- Wash solution: 41 mL 6 N H_2SO_4 in 500 mL water; add 50 g $NaH_2PO_4 \cdot H_2O$; dilute to 1 L.
- Stock linear alkyl sulfonate (LAS) solution: Reference, LAS material may be obtained from the U.S. EPA (Environmental Monitoring Systems Laboratory, Cincinnati, OH). The average molecular weight of LAS should fall in the range of 310 to 340. LAS material in the above average molecular weight range may be specially ordered from commercial suppliers.

  Stock solution is made by dissolving 100 mg LAS in 1 L distilled water:

$$1 \text{ mL} = 100 \ \mu\text{g LAS}$$

A secondary standard solution is prepared from the above stock solution by diluting 1 mL to 100 mL:

$$1 \text{ mL} = 1 \ \mu\text{g LAS}$$

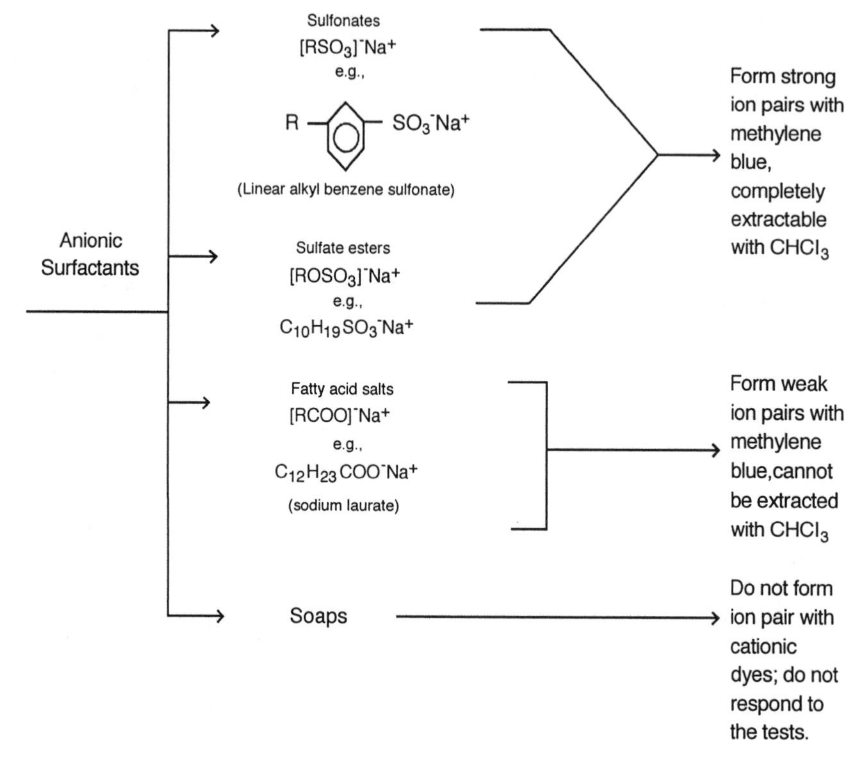

Figure 2.32.1 Schematic representation of the structural features of the various anionic surfactants.

Table 2.32.1 Series of Calibration Standard Solutions

Dilution of secondary standard	Concentration	Mass
1 mL diluted to 100 mL	10 μg/L	1 μg LAS
2 mL diluted to 100 mL	20 μg/L	2 μg LAS
5 mL diluted to 100 mL	50 μg/L	5 μg LAS
7 mL diluted to 100 mL	70 μg/L	7 μg LAS
10 mL diluted to 100 mL	100 μg/L	10 μg LAS
12 mL diluted to 100 mL	120 μg/L	12 μg LAS
15 mL diluted to 100 mL	150 μg/L	15 μg LAS

A series of calibration standard solutions are prepared from this secondary standard solution as seen in Table 2.32.1.

ANALYSIS

The determination of anionic surfactants in aqueous samples involves the following analytical steps:

1. Preparation of a standard calibration curve. A series of calibration standards are prepared from LAS. The concentrations of the calibration standards should range from 10 to 150 μg/L, which corresponds to a mass of 1 to 15 μg LAS, respectively, in 100 mL solution. Each LAS standard solution in 100 mL quantity should be treated with methylene blue reagent.
2. The standards are taken in separatory funnels. A few drops of phenolphthalein indicator followed by 1 *N* NaOH solution are added dropwise until a pink color forms. This is followed by the addition of 1 N H_2SO_4 dropwise until the pink colors decolorize.
3. Add 10 mL of chloroform followed by methylene blue reagent. The separatory funnel is shaken vigorously for 30 s. Methylene blue-surfactant ion pair separates into the bottom chloroform layer. The aqueous layer should be colorless. The above chloroform extraction is performed two more times. The extracts are combined and then washed repeatedly in another separatory funnel with the acid wash solution.
4. The absorbances of the chloroform extracts of the calibration standards are read at the wavelength 652 nm against a chloroform blank. A calibration curve is constructed plotting absorbance against microgram LAS.

Sample Analysis

A 100-mL sample aliquot is extracted as above with chloroform. The absorbance of the chloroform solution is read at 652 nm and the amount of anionic surfactants active to methylene blue is determined from the standard calibration curve.

Calculation:

$$\text{mg MBAS/L} = \frac{A}{\text{mL sample}}$$

where A is μg LAS read from the calibration curve.

Highly concentrated sample must be diluted such that MBAS concentration falls within the linear calibration range. Such concentrations, however, in routine environmental analyses are rarely encountered.

INTERFERENCE

The presence of chloride, nitrate, sulfide, organic sulfonates, cyanates, organic amines, and particulate matters may interfere in the test. The two major interferences, chloride and nitrate, however, may be removed in the acid backwash steps. Sulfide reacts with methylene blue to form a colorless complex. Interference from sulfide may be removed by treating the sample with a few drops of 30% H_2O_2, whereupon sulfide converted into sulfate.

2.33 THIOCYANATE

Thiocyanate is a monovalent polyanion that has the formula SCN^-. This anion is rarely found in wastewaters. However, cyanide-containing wastes or waters on contact with sulfides can form thiocyanates. On chlorination, thiocyanate could react with chlorine to form highly toxic cyanogen chloride, CNCl. Thiocyanate in water may be analyzed using 1. colorimetric method and 2. spot test. The latter method is a rapid spot test that can give semiquantitative results. Preserve samples at pH below 2 and refrigerate.

COLORIMETRIC METHOD

Thiocyanate reacts with ferric ion under acidic conditions to form ferric thiocyanate, which has an intense red color.

$$SCN^- + Fe^{3+} \Rightarrow Fe(SCN)_3$$

The intensity of color of the ferric thiocyanate formed is proportional to the concentration of thiocyanate ion in the sample. The absorbance or transmittance is measured at 460 nm using a spectrophotometer or a filter photometer. The concentration of SCN^- in the sample is determined from a standard calibration curve. The detection range of this method is 0.1 to 2.0 mg SCN^-/L. Dilute the samples if the concentration exceeds this range.

Interference

Reducing substances, hexavalent chromium and colored matter, interfere with the test. Reducing substances that may reduce Fe^{3+} to Fe^{2+} and thus prevent the formation of thiocyanate complex, may be destroyed by adding a few drops of H_2O_2. Add $FeSO_4$ to convert Cr^{6+} to Cr^{3+} at a pH below 2. Cr^{3+} and Fe^{3+} so formed are precipitated out when the pH is raised to 9 with NaOH.

Sample pretreatment is necessary if the sample contains color and various organic compounds. To remove such interference, pass the sample through a purified, solvent-washed adsorbent resin bed.

Procedure

To 100 mL sample or pretreated sample eluant eluted through the adsorbent resin, add a few drops of HNO_3 to bring down the pH to 2. Add 5 mL ferric nitrate solution and mix. Fill a 5 cm cell and measure absorbance against a reagent blank at 460 nm.

Prepare a calibration curve by plotting either mg SCN^- or mg/L SCN^- vs. absorbance using the standard solutions (see below). Determine the concentration of thiocyanate in the sample from the standard curve.

Calculation

$$\text{mg SCN}^- / \text{L} = \frac{\text{mg SCN}^- \text{ read from the calibration curve} \times 1000}{\text{mL sample}}$$

If the sample volume is same as the volumes of working standards, and if the calibration curve is plotted in mg/L concentration unit, then the concentration of thiocyanate as mg SCN^-/L can be directly read from the calibration curve. Multiply the result by dilution factor if the sample was diluted.

Reagents and Calibration Standards

- Ferric nitrate, $Fe(NO_3)_3 \cdot 9\,H_2O$ soln. 1 *M*: Dissolve 404 g in 800 mL distilled water, add 80 mL conc. HNO_3 and dilute to 1 L.
- Thiocyanate standard solns.: 1.673 g potassium thiocyanate, KSCN in distilled water, diluted to 1 L. This stock soln. = 1000 mg SCN^-/L or 1 mL = 1 mg SCN^-. Prepare a secondary standard from the stock solution by diluting 1 mL stock solution to 100 mL with distilled water. Secondary std = 10 mg SCN^-/L or 1 mL = 0.01 mg SCN^-. Prepare a series of calibration standards by diluting the secondary standard with distilled water as follows:

Secondary std	Diluted to:	SCN^- mass	SCN^- concn.
1 mL	100 mL	0.01 mg	0.1 mg/L (0.1 ppm)
2 mL	100 mL	0.02 mg	0.2 mg/L (0.2 ppm)
5 mL	100 mL	0.05 mg	0.5 mg/L (0.5 ppm)
10 mL	100 mL	0.10 mg	1.0 mg/L (1 ppm)
20 mL	100 mL	0.20 mg	2.0 mg/L (2 ppm)

SPOT TEST FOR THIOCYANATE

Thiocyanate reacts with chloramine-T on heating at pH below 8 to form cyanogen chloride, CNCl. The latter forms red color after adding pyridine-barbituric acid reagent. This test is similar to that of cyanide. Therefore, prior to chloramine treatment, the sample is treated with formaldehyde to mask the effect of cyanide. The addition of 4 drops of 37% formaldehyde solution should mask up to 5 to 10 mg CN^-/L if present in the sample.

Reagents are added after pretreatment of the sample. The spot test is performed in a porcelain spot plate with 6 to 12 cavities. The test is performed using a few drops of sample, standards, and reagents. The color developed is compared with thiocyanate standards. The test is semiquantitative and is suitable for screening SCN^- in water.

Procedure

Heat 25 mL sample in a water bath at 50°C. Add 4 drops of formaldehyde solution (37% pharmaceutical grade). Continue heating for another 10 min. Formaldehyde treatment would mask the effect of cyanide if present in the sample.

If the pH of the solution is above 10, add about 0.2 g Na_2CO_3 and mix. Add a drop of phenophthalein indicator which will turn the solution red (or pink) in alkaline medium. Add 1 *N* HCl dropwise till the color disappears. Place 3 drops of the above pretreated sample, 3 drops of distilled water, and 3 drops each of thiocyanate standards (0.05, 0.1, and 0.2 mg SCN^-/L) in the cavities of porcelain spot plate. Add 1 drop of chloramine-T solution to each cavity and mix with a clean glass rod. This is followed by the addition of 1 drop of pyridine-barbituric acid to each cavity. Again, mix the contents and allow it to stand for a minute. If thiocyanate is present, the sample spot will turn pink to red, depending on the concentration of SCN^- in the sample. If deep red coloration is produced, dilute the sample and repeat the test.

The above test would give SCN^- concentration in an estimated range. For greater accuracy, place more standards in the cavities of spot plate for color comparison. Alternatively, once the SCN^- concentration range in the sample is known from the above screening test, prepare several thiocyanate standards within that range and repeat the spot test for color comparison.

Reagents

- Chloramine-T solution: 1 g powder in 100 mL distilled water
- Pyridine-Barbituric acid: see under Cyanide, Total (Chapter 2.6)
- Thiocyanate standard solution: the stock standard is made from KSCN. The secondary and working standards are prepared from this. See the preceding section under Colorimetric Method.

3 SELECTED INDIVIDUAL COMPOUNDS

3.1 ACETALDEHYDE

Synonyms: ethanal, acetic aldehyde; Formula: CH_3CHO; MW 44.05; CAS [75-07-0]; used in the production of acetic acid, acetic anhydride, and many synthetic derivatives; found in water stored in plastic containers; colorless mobil liquid; fruity odor when diluted; boils at 20.8°C; solidifies at -121°C; highly volatile; vapor pressure 740 torr at 20°C; density 0.78 g/mL at 20°C; soluble in water, alcohol, acetone, ether, and benzene; highly flammable.

ANALYSIS OF AQUEOUS SAMPLE

Aqueous samples heated in a purging vessel under helium purge. Analyte trapped over an adsorbent (i.e., silica gel or Tenax) and then thermally desorbed out from the absorbent trap and transported onto a polar GC column for FID or GC/MS determination.

- Purging efficiency low because of high solubility of the analyte in water.

Aqueous samples buffered with citrate and pH adjusted to 3. Acidified sample derivatized with 2,4-dinitrophenylhydrazine (DNPH); derivative analyzed by GC-NPD or reverse phase HPLC with UV detection at 360 nm.

- For HPLC analysis, extract the derivative with methylene chloride and solvent exchange to acetonitrile.
- HPLC column: a C-18 reverse phase column such as Zorbax ODS or equivalent.
- GC column: a polar or an intermediate polar column (PEG-type phase) such as Carbowax 20 M, Supelcowax 10, DBWax, VOCOL, or equivalent; SPB-1, DB-1, or DB-5 columns are also suitable.
- Samples must be collected without headspace, refrigerated, and analyzed within 7 days.

AIR ANALYSIS

Air drawn through a midget impinger containing 0.05% DNPH reagent in dilute HCl and isooctane (immisicible); derivative formed partitions into iso-octane;

the organic solvent evaporated to dryness under a stream of N_2; residue dissolved in methanol and analyzed by HPLC (U.S. EPA Method TO5, 1988); recommended air flow rate 250 L/min; sample volume 25 L.

Alternatively, acidified DNPH coated silica gel or florisil use as adsorbent; derivative analyzed by HPLC (U.S. EPA Method TO11); recommended flow rate 500 mL/min; sample volume 100 L.

Air drawn through a solid sorbent tube containing 2-(hydroxymethyl) piperidine on XAD-2 (450 mg/250 mg); oxazolidine derivative of acetaldehyde formed; derivative desorbed into toluene under ultrasonic condition; analyzed by GC-FID (NIOSH Method 2538, 1989; OSHA Method 68, 1988).

(1 ppm acetaldehyde in air = 1.80 mg/m^3 at NTP)

3.2 ACETONE

Synonyms: 2-propanone, dimethyl ketone; Formula: CH_3COCH_3; MW 58.08; CAS [67-64-1]; used as a common solvent in many organic syntheses and in paint and varnish removers; colorless liquid; characteristic odor; sweetish taste; boils at 56.5°C; vapor pressure 180 torr at 20°C; freezes at –94°C; density 0.79 g/ML at 20°C; readily mixes with water and organic solvent; highly flammable.

ANALYSIS OF AQUEOUS SAMPLES

Aqueous samples purged by an inert gas; analyte trapped on a sorbent trap; transferred onto a GC column by heating the trap and backflushing with helium; determined by GC/MS.

- Purging efficiency of acetone is poor, because of its high solubility in water. The purging vessel should be heated.
- The characteristic masses for GC/MS identification: 43 and 58.
- GC column: Carbowax 20M, Carbopack, or a fused silica capillary column, such as DB-5, SPB-5, or equivalent.

Alternatively, an aliquot of aqueous sample directly injected onto an appropriate GC column for FID determination.

- The detection range on the FID by direct injection without sample concentration is several order higher than the purge and trap method (when purging vessel heated).

Trace acetone in water may be determined by a fast HPLC method (Takami et al., 1985); aqueous sample passed through a cartridge packed with a moderately sulfonated cation-exchange resin charged with 2,4-dinitrophenylhydrazine (DNPH); DNPH derivative eluted with acetonitrile and analyzed by HPLC with a 3-mm ODS column.

- Precision and accuracy data for all the three above methods not available.

AIR ANALYSIS

Air drawn through a midget impinger containing 10 mL of 2 *N* HCl/0.5% DNPH and 10 mL isooctane; the stable DNPH derivative formed partitions into isooctane layer; isooctane layer separated; aqueous layer further extracted with 10 mL 70/30 hexane/methylene chloride; the latter combined with isooctane; the combined organic layer evaporated under a steam of N_2; residue dissolved in methanol; the DNPH derivative determined by reversed phase HPLC using UV detector at 370 nm (U.S. EPA Method TO-5, 1988).

- Recommended air flow rate 0.1 L/min; sample volume 20 L.
- Isooctane must not completely evaporate out during sampling.
- Calibration standards are prepared in methanol from solid DNPH derivative of acetone.
- The molecular weight of DNPH derivative of acetone is 238.
- HPLC column: Zorbax ODS; mobile phase: 80/20 methanol/ water; flow rate 1 mL/min.

Alternatively, a measured volume of air drawn through a prepacked cartridge coated with acidified DNPH; the DNPH derivative of acetone eluted with acetonitrile and measured by HPLC-UV as above (U.S. EPA Method TO-11, 1988).

Alternatively, a measured volume of air drawn through a sorbent tube containing coconut shell charcoal (100 mg/50 mg); the analyte desorbed with CS_2 and analyzed by GC-FID (NIOSH Method 130D, 1984); recommended air flow rate 100 mL/min; sample volume 3 L.

- GC column: 10% SP-2100 or DB-1 fused silica capillary column or equivalent.

(1 ppm acetone in air = ~2.37 mg/m^3 at NTP)

3.3 ACETONITRILE

Synonyms: methyl cyanide, cyanomethane; Formula: CH_3CN; MW 41.06; CAS [75-05-8]; used as a solvent for polymers; boils at 81.6°C; solidifies at –45.7°C; vapor pressure 73 torr at 20°C; density 0.786 g/mL at 20°C; readily mixes with water and most organic solvents; immiscible with petroleum fractions and some saturated hydrocarbons; flammable.

ANALYSIS OF AQUEOUS AND NONAQUEOUS SAMPLES

Aqueous samples directly injected onto a GC column, and determined by FID.

Alternatively, sample subjected to purge and trap concentration; analyte determined by GC-FID, GC-NPD in N-mode, or GC/MS.

- Poor purging efficiency at room temperature due to high solubility of analyte in water; resulting in low recovery and high detection level; purging chamber must be heated.
- GC column: packed Porapak-QS (80/100 mesh) or Chromosorb 101 (60/80 mesh) or equivalent; capillary: –30 to 75 m length, 0.53 or 0.75 mm (or lower ID in conjunction with lower thickness) and 1.0 or 1.5 μm film VOCOL, DB-624, Rtx-502.2, DB-5, SPB-5, or equivalent.

Alternatively, sample analyzed by HPLC; UV detection recommended at 195 nm; HPLC column: Zorbax ODS or equivalent; mobile phase water, flow rate 1 to 2 mL/min.

Soils, sediments, or solid wastes mixed with water and subjected to purge and trap concentration; aqueous extract may directly be injected onto GC; sample/extract analyzed as above.

Alternatively, sample aliquot thermally desorbed under purge of an inert gas, analyte determined as above.

- Characteristic masses for GC/MS determination: 41, 50, and 39.
- Sample collected in glass vial without headspace, refrigerated, and analyzed within 14 days of collection.
- Detection level: in the range of 1 μg/L for a 25-mL sample size determined by purge and trap–GC-FID method.

AIR ANALYSIS

Air drawn through solid sorbent tube containing coconut shell charcoal (400/200 mg); analyte desorbed with benzene and determined by GC-FID (NIOSH Method 1606, 1984); recommended flow rate 100 mL/min; sample volume 10 L.

Alternately, air condensed at liquid argon temperature; contents of the trap swept by a carrier gas onto GC column for determination by FID or a mass spectrometer.

(1 ppm CH_3CN in air = ~1.68 mg/m^3 at NTP)

3.4 ACROLEIN

Synonyms: 2-propenal, acraldehyde, vinyl aldehyde; Formula: CH_2=CH–CHO an aldehyde; MW 56.07; CAS [107-02-8]; used to control aquatic weed, algae, and slime, and in leather tanning; colorless, volatile liquid; boils at 53°C; freezes at –87°C; density 0.843 g/mL at 20°C; soluble in water; mixes with alcohol and ether.

ANALYSIS OF AQUEOUS AND NONAQUEOUS SAMPLES

Aqueous samples extracted by purge and trap;volatile analyte thermally desorbed from the trap and swept onto a GC column for separation; detected by FID or GC/MS.

- Purging efficiency may be poor as the analyte is soluble in water; purging chamber should be heated.

Aqueous samples or aqueous extracts of nonaqueous samples analyzed by HPLC on a C-18 reverse phase column; analyte detected by UV at 195 nm; mobile phase, water; flow rate 2 mL/min; pressure 38 atm.

- Soils, solid wastes, or sludges mixed with water or methanol; the aqueous extract or a solution of methanol extract spiked into water, subjected to purge and trap concentration and analyzed as above.
- Characteristic masses for GC/MS identification: 56, 55, and 58.

AIR ANALYSIS

Air drawn through a solid sorbent tube containing 2-(hydroxymethyl)piperidine on XAD-2 (120mg/60mg); acrolein converted into 9-vinyl-1-aza-8-oxabicyclo[4.3.0]nonane; derivative desorbed with toluene (placed in ultrasonic bath) and measured by GC-NPD in N-specific mode (NIOSH 1984, Method 2501); recommended flow rate 50 mL/min; sample volume 10 L.

Air drawn through a midget impinger containing 10 mL of 2 *N* HCl/0.05% 2,4-dinitrophenyhydrazine (DNPH) and 10 mL isooctane; aqueous layer extracted with hexane/methylene chloride (70:30) and combined with iso-octane; combined extract evaporated to dryness, residue dissolved in methanol; derivative analyzed by reverse phase HPLC using UV detector at 370 nm (U.S. EPA 1988, Method TO5); recommended flow rate 200 mL/min; sample volume 20 L air.

Air drawn through a cartridge containing silica gel coated with acidified DNPH; derivative eluted with acetonitrile and determined using isocratic reverse phase HPLC with UV detection at 360 nm (U.S. EPA 1988, Method TO11); recommended flow rate 1 L/min; sample volume 100 L.

(1 ppm acrolein in air = ~2.3 mg/m^3 at NTP)

3.5 ACRYLONITRILE

Synonyms: 2-propenenitrile, vinyl cyanide; Formula: CH_2=CH–CN; MW 53.06; CAS [107-13-1]; a liquid at room temperature; boils at 77.5°C; density 0.806 g/mL; readily soluble in alcohol and ether; solubility in water ~7%; flammable and toxic.

ANALYSIS OF AQUEOUS AND NONAQUEOUS SAMPLES

Aqueous samples extracted by purge and trap; volatile analyte thermally desorbed from the trap and swept onto a GC column for separation; detected by FID, NPD, or by a mass spectrometer (MS).

- Because of moderate solubility in water, purging efficiency may be poor, resulting in low recovery and a high detection level; purging chamber should be heated.

Aqueous samples or aqueous extracts of nonaqueous samples directly injected onto a C-18 reverse phase column for HPLC analysis with UV detection at 195 nm (U.S. EPA Method 8316, 1992); mobile phase, water; flow rate 2 mL/min; pressure 38 atm.

Soils, solid wastes, or sludges mixed with water or methanol; the aqueous extract or a solution of methanol extract spiked into water; subjected to purge and trap concentration and analyzed as above.

Alternatively, aqueous and nonaqueous samples microextracted with tert-butyl ether; extract analyzed by GC or GC/MS.

- Characteristic masses for GC/MS identification: 53, 52, and 51.
- Limit of detection: in the range 0.1 µg/L when detected by HECD for a 5-mL sample aliquot (subject to matrix interference).
- GC column-packed: Porapak-QS (80/100 mesh), Chromosorb 101 (60/80 mesh), or equivalent; capillary: VOCOL, DB-624, Rtx-502.2, DB-5, SPB-5, or equivalent fused silica capillary column.
- Detection level: in the range 1 µg/L for purge and trap concentration (for a 5-mL sample aliquot) and FID determination.
- Sample collected in glass vial without headspace, refrigerated, and analyzed within 14 days of collection.

AIR ANALYSIS

Adsorbed over coconut shell charcoal (100 mg/50 mg); desorbed with acetone-CS_2 mixture (2:98); analyzed by GC-FID (NIOSH 1984, Method 1604); recommended flow rate 100 mL/min; sample volume 10 L.

Alternatively, air drawn over carbon molecular sieve in a cartridge; cartridge heated at 350°C under He purge; analyte desorbed and collected in a cryogenic trap and then flash evaporated onto a precooled GC column (–70°C) temperature programmed; determined by GC-FID or GC/MS; (U.S. EPA 1988, Method TO2); recommended flow rate 0.5 L/min; sample volume 100 L.

Alternatively, sample collected in a liquid argon trap; trap heated; sample vapors swept onto a precooled GC column for determination by FID, PID, or NPD (U.S. EPA 1988, Method TO3).

(1 ppm acrylonitrile in air at NTP = ~2.16 mg/m^3)

3.6 ANILINE

DESCRIPTION

Synonyms: aminobenzene, phenylamine, benzeneamine; Formula: $C_6H_5NH_2$; Structure:

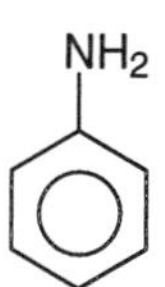

MW 93.14; CAS [62-53-3]; used in the manufacture of dyes, varnishes, resins, pharmaceuticals, and photographic chemicals; colorless oily liquid with characteristic odor; boils at 184.5°C; freezes at –6°C; vapor pressure 0.3 torr; density 1.02 g/mL at 20°C; moderately soluble in water 3.5% at 20°C; readily mixes with most organic solvents; weakly alkaline.

ANALYSIS OF AQUEOUS AND NONAQUEOUS SAMPLES

Aqueous samples made alkaline with NaOH to pH >12; repeatedly extracted with methylene chloride; analyte partitions into methylene chloride which is separated, concentrated, and analyzed by GC/NPD or GC/MS.

- GC column. Packed –3% SP-2250 on Supelcoport or equivalent; fused silica capillary column such as DB-5, PTE-5, or equivalent.
- Characteristic masses for GC/MS determination: 93, 66, 65, and 39; peak intensities ratios: m/z 93:66:65 = 100:32:16.
- Surrogates/IS: aniline-d_5 and nitrobenzene-d_5.

Aniline at low ppm level may be directly analyzed by HPLC; detector, UV at 220 nm; mobile phase: acetonitrile-K_3PO_4 buffer at pH 6.5; column: Suplex pKb-100 or equivalent (Supelco, 1995).

Soils, sediments, and solid samples mixed with anhydrous Na_2SO_4 and then methylene chloride; sonicated; extract cleaned by acid-base partitioning and analyzed as above.

- Sample refrigerated and extracted within 7 days of collection and extract analyzed within 40 days.

AIR ANALYSIS

Air drawn through a solid sorbent tube containing silica gel (150/15 mg); analyte desorbed into 95% ethanol in an ultrasonic bath and determined by GC-FID or GC-NPD (NIOSH Method 2002, 1985); recommended flow rate 200 mL/min; sample volume 20 L.

- High humidity reduces adsorption efficiency of silica gel.
- Aniline stable on silica gel at least for one week.
- Sensitivity is much greater on GC-NPD.
- GC column: Chromosorb 103 (80/100 mesh) or equivalent.

(1 ppm aniline in air = 3.8 mg/m^3 at NTP)

3.7 ARSINE

Synonyms: Arsenic hydride, arsenic trihydride, hydrogen arsenide; Formula: AsH_3; MW 77.95; CAS [7784-42-1]; used as a doping agent for solid-state electronic components; exposure risk arises from reaction of arsenic compounds with acids; colorless gas; disagreeable garlic odor; liquefies at –62°C; freezes at –117°C; gas density 2.70 g/L; slightly soluble in water (0.7g/L); highly toxic.

AIR ANALYSIS

Air drawn through a cellulose ester membrane (to remove any arsenic particulate) in front of a charcoal tube (100 mg/50 mg); arsine desorbed into 1 mL 0.01 *M* HNO_3 (in an ultrasonic bath for more than 30 min contact time); arsenic analyzed by graphite furnace-atomic absorption spectrophotometer at the wavelength 193.7 nm (NIOSH Method 6001, 1985); recommended air flow rate 0.1 L/min; sample volume 5 L.

- Multiply the concentration of arsenic by the stoichiometric factor 1.05 to calculate the concentration of arsine.
- Use background correction for molecular absorption and a matrix modifier (Ni^{2+} soln.).
- The working range reported in the method is 0.001 to 0.2 mg/m^3 for a 10 L air sample.

(1 ppm arsine in air = 3.19 mg/m^3 at NTP)

3.8 ASBESTOS

Asbestos constitutes several types of hydrated silicate mineral fibers. The types of asbestos, their chemical compositions, and CAS Numbers are presented in Table 3.8.1. These substances occur in nature in rocks, silicate minerals, fibrous stones, and underground mines. This class of substances exhibits unique properties of noncombustibility, high resistance to acids, and high tensile strength for which they were widely used in many products, including floor and roofing tiles, cement, textiles, ropes, wallboards, and papers. Because of the health hazards associated with excessive exposure to asbestos, the use of these substances is currently banned.

ANALYSIS

Asbestos can be determined by several analytical techniques, including optical microscopy, electron microscopy, X-ray diffraction (XRD), light scattering, laser microprobe mass analysis, and thermal analysis. It can also be characterized by chemical analysis of metals by atomic absorption, X-ray fluorescence, or neutron activation techniques. Electron microscopy methods are, however, commonly applied for the analysis of asbestos in environmental matrices.

Table 3.8.1 Asbestos Types and Their Compositions

Asbestos	CAS no.	Composition
Actinolite	[13768-00-8]	$2CaO.4MgO.FeO.8SiO_2.H_2O$
Anthophyllite	[17068-78-9]	$7MgO.8SiO_2.H_2O$
Amosite	[12172-73-5]	$11FeO.3MgO.16SiO_2.2H_2O$
Crocidolite	[12001-28-4]	$Na_2O.Fe_2O_3.3FeO.8SiO_2H_2O$
Chrysotile	[12001-29-5]	$3MgO.2SiO_2.2H_2O$
Tremolite	[14567-73-8]	$2CaO.5MgO.8SiO_2.H_2O$

ANALYSIS OF AQUEOUS SAMPLES

A measured volume of sample filtered through a membrane filter; the filter placed on a petri dish and dried on a bed of desiccant in a desiccator; a section of filter from any quadrant of sample and blank filters "prepared" and transferred to a transmission electron microscopy (TEM) grid; the fibers are identified and counted by TEM at 15,000 to 20,000 magnification.

- Samples containing high concentration of asbestos should be diluted appropriately with particle-free water.
- Use mixed cellulose ester or polycarbonate filter in the analysis.
- The term "sample or blank preparation" refers to a series of steps pertinent to TEM analysis. These include filter fusing, plasma etching, carbon coating, and specimen washing. Some of these steps may not be required if polycarbonate filters are used.
- The grids should be dry before placing in TEM.
- The instrument must be calibrated regularly.
- Perform a blank analysis for each batch using particle-free water, same reagents, and same type of filter.
- Do not use polypropylene bottles for sampling.
- Aqueous samples must be analyzed within 48 h of collection.
- Certain minerals can interfere in the test. They include halloysite, palygorskite, antigorite, hornblende, and others.
- The results are expressed as million fibers per liter (MFL) or million structures per liter (MSL).

AIR ANALYSIS

Air drawn through a 0.8 to 1.2 mm cellulose ester membrane filter; asbestos fibers counted by positive phase contrast microscopy technique; sample prepared by acetone/triacetin method (NIOSH Method 7400, 1985).

- The sampling flow rate and time should be adjusted to produce a fiber density of 100 to 1300 fibers/mm^2.
- A flow rate of 1 to 4 L/min for 8 h is appropriate for air containing less than 0.1 fiber/mL. Sample volume should be lower in dusty atmosphere.
- The method does not measure asbestos fibers less than 0.25-mm diameter.

Alternatively, asbestos fibers collected on the filter counted by TEM, following the modified Jaffe wick technique for sample preparation (NIOSH Method 7402, 1987).

- The method can measure the asbestos fibers of smallest diameter (<0.05 mm).

Percent chrysotile asbestos in bulk samples may be measured by X-ray powder diffraction (NIOSH Method 9000, 1984); sample dust grinded under liquid N_2; wet sieved through 10-mm sieve; sieved material treated with 2-propanol; agitated in an ultrasonic bath; filtered on a silver filter; measured by XRD, using a Cu target X-ray tube.

3.9 BENZENE

Synonyms: benzol, cyclohexatriene; Formula: C_6H_6; Structure:

MW 78.12; CAS [71-43-2]; occurs in gasoline and coal-tar distillation products; used as a solvent and in organic synthesis; colorless liquid with characteristic odor; boils at 80.1°C; freezes at 5.5°C; vapor pressure 76 torr at 20°C; density 0.88 g/L; slightly soluble in water (1.78 g/L); readily miscible with organic solvents; toxic and carcinogenic; flammable.

ANALYSIS OF AQUEOUS AND NONAQUEOUS SAMPLES

Purge and trap method; a measured volume of sample purged with helium in a purging vessel; benzene collected on the sorbent trap thermally desorbed from the trap and swept by an inert gas onto a GC column for separation from other volatile compounds and detection by PID, FID or a mass spectrometer (MS).

Solid samples mixed with water or methanol; an aliquot of aqueous extract or a portion of methanol extract spiked into water and measured into the purging vessel; subjected to purge and trap concentration and analyzed as above.

Alternatively, benzene thermally desorbed from the solid matrix under He purge (without any solvent treatment) and analyzed by GC or GC/MS as above.

- Characteristic masses for GC/MS identification: 78, 77, 52, and 39 (relative intensities of peaks m/z 78:77:52 = 100:20:19).
- Limit of detection: in the range 0.5 µg/L when detected by PID for a 5-mL sample aliquot (subject to matrix interference).
- Recommended surrogate/IS: benzene-d_6 and fluorobenzene.
- Samples collected in glass containers without headspace, refrigerated, and analyzed within 7 days; sample preserved with 1:1 HCl (0.2 mL acid per 40 mL sample) may be analyzed in 14 days.

AIR ANALYSIS

Air collected in a Tedlar bag; 1 mL air directly injected onto GC (portable) for PID detection; flow rate 20 to 50 mL air/min (NIOSH Method 3700, 1987).

- Analysis must be done within 4 h of sample collection.

Air drawn through a sorbent tube containing coconut shell charcoal (100 mg/50mg); analyte desorbed into CS_2 (on 30 min standing) and analyzed by GC-FID (NIOSH Method 1501, 1984); recommended flow rate 100 mL/min; sample volume 10 L.

- GC column: packed column—10% OV-275 on 100/120 mesh Chromosorb W-AW, Porapak P 50/80 mesh, or equivalent.

Air drawn through a cartridge containing Tenax (~2g); cartridge heated under He purge; benzene transferred into a cold trap and then to the front of a GC column at –70°C; column heated; analyte determined by GC/MS (U.S. EPA Method TO1); recommended flow rate 100 mL/min; sample volume 10 L.

Adsorbed over carbon molecular sieve (~400 mg) in a cartridge; heated at 350°C under He purge; transferred into a cryogenic trap and flash evaporated onto a capillary column; determined by GC/MS (U.S. EPA Method TO2); recommended flow rate 1 L/min; sample volume 100 L.

Collected in a trap in liquid argon; the trap removed and heated; sample transferred to a precooled GC column for PID or FID detection (U.S. EPA Method TO-3).

Collected in a SUMMA passivated canister under pressure using an additional pump or at subatmospheric pressure by initially evacuating the canister; the air transferred into cryogenically cooled trap attached to a GC column; the trap heated; benzene determined by GC/MS (U.S. EPA Method TO-14).

(1 ppm benzene in air = ~3.2 mg/m^3 at NTP)

3.10 BENZIDINE

Synonyms: 4,4′-biphenyldiamine; 4,4′-diamino-1,1-biphenyl, *p,p′*-dianiline; Formula: $C_{12}H_{12}N_2$; MW 184.26; Structure:

H2N—⟨benzene⟩—⟨benzene⟩—NH2

CAS [92-87-5]; used in the past in the manufacture of dyes; no longer currently used in dyes because of its cancer-causing effect; white crystalline solid; melts at 120°C; boils at 400°C; slightly soluble in cold water (0.04%); moderately soluble in hot water, alcohol, and ether.

ANALYSIS OF AQUEOUS AND NONAQUEOUS SAMPLES

Aqueous samples extracted with chloroform; chloroform extract further extracted with 1 *M* H_2SO_4; the acid extract buffered with 0.4 *M* sodium tribasic phosphate and then pH adjusted between 6 and 7 using NaOH; the neutralized extract then treated with chloroform; organic layer separated and concentrated; benzidine analyzed by HPLC using an electrochemical detector (i.e., glassy carbon electrode).

- Analyte, if found, must be confirmed either by HPLC, setting electrochemical conditions at a second potential, or by GC/MS.
- HPLC column: Lichrosorb RP-2 or equivalent, particle size 5 μm; mobile phase: acetonitrile/0.1 *M* acetate buffer (pH 4.7) (50:50).
- Characteristic masses for GC/MS identification: 184, 92, and 185.
- GC column: a packed column, such as 3% SP-2250 on Supelcoport or equivalent; or a fused silica capillary column such as DB-5, SPB-5, Rtx-5, or equivalent.
- MDL: in the range 0.1 μg/L by HPLC, when 1-L sample concentrated down to 1 mL, while ~50 μg/L by GC/MS with similar sample concentration.
- Soils, sediments, and solid wastes samples mixed with anhydrous Na_2SO_4 and extracted with chloroform by sonication; chloroform extract treated with acid, pH adjusted, and reextracted as above; analysis performed as above.

- All samples stored at 4°C in glass containers; protected from light and air oxidation; extracted within 7 days of collection and analyzed within 7 days of extraction; residual chlorine if present in aqueous samples must be removed by adding $Na_2S_2O_3$ (~100 mg/L sample).

3.11 BENZYL CHLORIDE

DESCRIPTION

Synonyms: (chloromethyl)benzene, tolyl chloride, α-chlorotoluene; Formula: C_7H_7Cl; Structure:

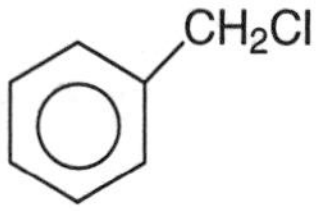

CH_2Cl

MW 126.59; CAS [100-44-7]; used in dyes, resins, perfumes, and lubricants; colorless liquid with pungent aromatic odor; boils at 179°C; freezes at –43°C; density 1.10 g/mL at 20°C; insoluble in water; miscible with most organic solvents.

ANALYSIS OF AQUEOUS AND NONAQUEOUS SAMPLES

Aqueous samples extracted by purge and trap; analyte thermally desorbed from the trap and swept onto a GC column for separation; detected by HECD, ECD, PID or by a mass spectrometer.

Alternatively, aqueous samples extracted with methylene chloride by LLE; extract concentrated and analyzed by GC using FID, PID, or ECD or by GC/MS.

Soils, solid wastes, or sludges mixed with water or methanol; the aqueous extract or a solution of methanol extract spiked into water, subjected to purge and trap concentration and analyzed as above.

- Methanol extract may be directly injected onto GC instead of purge and trap concentration.
- Solid matrices extracted with methylene chloride by sonication or Soxhlett extraction: extract concentrated and analyzed as above.
- Characteristic masses for GC/MS identification: 91 and 126.
- Limit of detection: in the range 0.1 μg/L when detected by HECD for a 5-mL sample aliquot (subject to matrix interference).

- Recommended surrogate/IS: 4-bromofluorobenzene and fluorobenzene.
- Sample collected in glass/plastic containers without headspace, pH adjusted to <2 with 1:1 HCl (if not analyzed immediately), refrigerated, and analyzed within 14 days.

See "Halogenated Hydrocarbons," Chapter 2.9 for GC columns and conditions and a detailed discussion.

AIR ANALYSIS

Adsorbed over coconut charcoal (100 mg/50 mg); desorbed with CS_2 and analyzed by GC-FID; recommended flow rate 200 mL/min; sample volume 20 L.

Adsorbed over Tenax (1 to 2 g) in a cartridge; analyte desorbed with an inert gas under heating and transferred onto a cold trap and then onto the front of a precooled GC column at 70°C; column heated; analyte eluted and detected by GC/MS, ECD, or FID; recommended flow rate 0.5 L/min; sample volume 100 L.

Collected in a trap in liquid argon; the trap removed and heated; sample transferred to a precooled GC column for ECD or FID detection, (U.S. EPA Method TO-3).

Collected in a SUMMA passivated canister under pressure using an additional pump or at subatmospheric pressure by initially evacuating the canister; the air transferred into cryogenically cooled trap attached to a GC column; the trap heated; the analyte separated on the column for detection by ECD, FID, or GC/MS (U.S. EPA Method TO-14).

(1 ppm benzyl chloride in air = ~5.2 mg/m^3 at NTP)

3.12 1,3-BUTADIENE

Synonyms: bivinyl, vinylethylene, biethylene; Formula: C_4H_6; Structure: CH_2=CH–CH=CH_2; MW 54.10; CAS [106-99-0]; petroleum product; used to produce synthetic rubber, elastomers, and food-wrapping materials; colorless gas, mild aromatic odor; heavier than air; gas density 1.865 (air = 1); 2.212 g/L at NTP; liquefies at –4.5°C; solidifies at –109°C; slightly soluble in water, 500 mg/L, soluble in organic solvents; carcinogenic; flammable.

AIR ANALYSIS

Air drawn through a solid sorbent tube packed with coconut charcoal (400 mg/200 mg); 1,3-butadiene desorbed with 4 mL methylene chloride (more than 30 min standing); analyzed by GC-FID (NIOSH Method 1024, 1987); recommended air flow rate 0.2L/min; sample volume 20 L.

- The upper limit of sampler is 100 ppm.
- Presence of other hydrocarbons may interfere in the test. High humidity can affect sampling efficiency.
- GC column: 10% FFAP on 80/100 mesh Chromosorb WAW. This column may not give adequate resolution from other hydrocarbons, C_4 to C_6 compounds. A 50 m × 0.32 mm ID fused silica porous-layer open tubular column coated with AL_2O_3/KCI gives better separation. Degradation of Al_2O_3 coating may be prevented by using a backflushable precolumn.
- The sample (when analyte remains adsorbed onto the trap) is stable for 3 weeks if stored at –4°C. At ambient temperature the average loss was found to be 1.5% per day for 26 mg of 1,3-butadiene loaded on the adsorbent (Lunsford and Gagnon, 1988).

(1 ppm 1,3-butadiene in air = 2.21 mg/m^3 at NTP)

3.13 CARBON DISULFIDE

DESCRIPTION

Synonym: dithiocarbonic anhydride; Formula: CS_2; MW 76.13; CAS [75-15-0]; used in the manufacture of rayon and soil disinfectants and as a solvent; colorless liquid with a strong foul odor; boils at 46.5°C; freezes at −111.6°C; density 1.263 g/mL at 20°C; slightly soluble in water (0.22% at 20°C); miscible with alcohol, ether, benzene, and chloroform; highly toxic and flammable (Patnaik, 1992).

ANALYSIS OF AQUEOUS AND NONAQUEOUS SAMPLES

Aqueous samples concentrated by purge and trap method; analyte determined by GC using a FPD or by GC/MS.

Solid matrices extracted with methanol; methanol extract spiked to reagent grade water and the aqueous solution subjected to purge and trap concentration and analyzed as above.

Alternatively, CS_2 thermally desorbed under purge of an inert gas and analyzed as above.

AIR ANALYSIS

Air drawn through a solid sorbent tube containing coconut shell charcoal (100 mg/50 mg) and anhydrous Na_2SO_4 (270 mg); analyte desorbed with benzene and determined by GC-FPD (NIOSH Method 1600, 1984); flow rate 0.01 to 0.2 L/min; sample volume 3 to 25 L.

- GC column: 5% OV-17 on GasChrom Q or 5% OV-210 on Chromosorb G-HP or equivalent.
- Water vapor interferes in sampling; it is removed by Na_2SO_4 in the sorbent tube.

(1 ppm CS_2 in air = 3.11 mg/m^3 at NTP)

3.14 CARBON MONOXIDE

Formula: CO; MW 28.01; CAS [630-08-0]; produced when substances burn in insufficient air; occurs in automobile exhaust gases and closed confinements; high risk of exposure under fire conditions; a colorless gas with no odor; liquefies at –191.5°C; solidifies at –205°C; slightly soluble in water; soluble in alcohol, acetone, chloroform, and acetic acid; highly toxic.

AIR ANALYSIS

Air collected in Tedlar bag or glass bulb and directly injected onto a GC column for TCD detection.

- GC-TCD sensitivity low; a detection level of 500 ppm may be achieved by injecting 1 mL air.
- GC column (Packed): Molecular Sieve 5A or Carboxen 1004 micropacked or equivalent in stainless steel column.

Air drawn through a vacuum pump into the gas cuvette of a nondispersive infrared spectrophotometer; IR absorption by CO is measured using two parallel IR beams through sample and reference cell and a selective detector; detector signal amplified; concentration of analyte determined from a calibration curve prepared from standard calibration gases (ASTM Method D 3162-91, 1993).

- Water vapor interferes in the test; air should be dried by passing through silica gel.
- Many nondispersive IR air analyzers (such as Myran and other models) are commercially available for continuous measurement of CO in the atmosphere.
- Detection limit in IR method is much lower than GC-TCD; detection range 0.5 to 100 ppm.

Diffusion type colorimetric dosimeters (such as Vapor Gard) give a direct readout of CO level without chemical analysis; a color forming reagent impregnated on purified silica gel; color changes from pale yellow to gray black and dosage exposure measured from the length of stain.

(1 ppm carbon monoxide in air = 1.14 mg/m^3 at NTP)

3.15 CARBON TETRACHLORIDE

Synonyms: tetrachloromethane, tetrachlorocarbon; Formula: CCl_4; MW 153.81; CAS [56-23-5]; used as a solvent; colorless liquid with characteristic odor; boils at 76.7°C; freezes at –23°C; vapor pressure 89.5 torr at 20°C; density 1.59 g/mL at 20°C; slightly soluble in water (~800 mg/L); miscible with organic solvents.

ANALYSIS OF AQUEOUS AND NONAQUEOUS SAMPLES

Aqueous samples subjected to purge and trap concentration; CCl_4 thermally desorbed and swept onto a GC column for separation from other volatile compounds and detection by HECD, ECD or MSD.

Alternatively, microextraction with hexane and analysis by GC-ECD; precision and accuracy of the method not established.

Soils, solid wastes, or sludges mixed with water or methanol; the aqueous extract or a solution of methanol extract spiked into water, subjected to purge and trap concentration and analyzed as above.

- Characteristic masses for GC/MS identification: 117, 119, and 121.
- Limit of detection: in the range 0.1 μg/L when detected by HECD for a 5-mL sample aliquot (subject to matrix interference).
- Recommended surrogate/IS: bromochloromethane and 1,2-dichloroethane-d_4.
- Samples collected in glass/plastic containers without headspace, refrigerated, and analyzed within 14 days.

See Halogenated Hydrocarbons, Chapter 2.9 for GC columns and conditions and further discussion.

AIR ANALYSIS

Adsorbed over coconut charcoal (100 mg/50 mg); desorbed with CS_2 and analyzed by GC-FID; recommended flow rate 100 mL/min; sample volume 15 L (NIOSH Method 1003).

Adsorbed overTenax (~2 g) in a cartridge; cartridge then heated and purged with He; compound transferred into a cold trap and then to the front of a GC column at –70°; column heated; CCl_4 detected by ECD or GC/MS; recommended flow rate 100 mL/min; sample volume 10 L (U.S. EPA Method TO-1).

Adsorbed over carbon molecular sieve (~400 mg) in a cartridge; heated at 350°C under He purge; transferred into a cryogenic trap and flash evaporated onto a capillary column GC/MS; recommended flow rate 1 L/min; sample volume 100 L (U.S. EPA Method TO-2).

Collected in a trap in liquid argon; the trap removed and heated; sample transferred to a precooled GC column for ECD or FID detection; (U.S. EPA Method TO-3).

Collected in a SUMMA passivated canister under pressure using an additional pump or at subatmospheric pressure by initially evacuating the canister; the air transferred into cryogenically cooled trap attached to a GC column; the trap heated; the analyte separated on the column for detection by ECD or GC/MS (U.S. EPA Method TO-14).

(1 ppm CCl_4 in air = ~6.3 mg/m^3 at NTP)

3.16 CAPTAN

Synonyms: N-(trichloromethylmercapto)-4-cyclohexene-1,2-dicarboximide
Formula: $C_9H_8Cl_3NO_2S$; Structure:

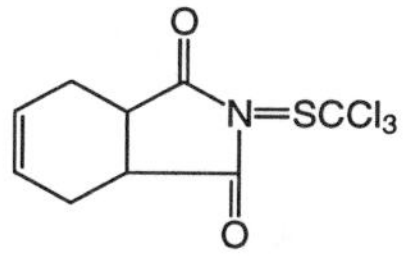

MW 300.57; CAS [133-06-2]; crystalline solid; melts at 173°C; insoluble in water; slightly soluble in alcohol, ether, and hydrocarbons; soluble in chloroform and other halogenated solvents; used as a fungicide.

ANALYSIS OF AQUEOUS AND NONAQUEOUS SAMPLES

Aqueous samples extracted with methylene chloride; solvent extract exchanged to hexane; extract cleaned up on a florisil column; concentrated and analyzed by GC-ECD.

Solid samples extracted with acetonitrile; extract diluted with water; the resulting solution mixed with methylene chloride-petroleum ether mixture (20:80) (A) and shaken; analyte partitions into (A); solvent layer (A) repeatedly washed with saturated NaCl solution; the extract then cleaned up on a florisil column (first eluted with 200 ml solution A and then with a mixture of methylene chloride 50% and 1.5% acetonitrile in petroleum ether; eluant concentrated and diluted to desired volume with petroleum ether; analyzed by GC-ECD (Pomerantz et al., 1970).

- Detection level: in the range 0.1 mg/kg.

3.17 CHLOROACETIC ACID

Synonyms: monochloroacetic acid, chloroethanoic acid; Formula $ClCH_2COOH$; MW 94.50; CAS [79-11-8]; used as a herbicide and in the manufacture of dyes; colorless or white crystals; occurring in three allotropic modifications: α, β, and γ; melts between 55 and 63°C; boils at 189°C; very soluble in water; moderately soluble in alcohol, benzene, ether, and chloroform.

ANALYSIS OF AQUEOUS SAMPLES

A 100-mL sample aliquot adjusted to pH 11.5; sample extracted with methyl *tert*-butyl ether (MTBE); chloroacetic acid partitions into aqueous phase; basic and neutral compounds in MTBE phase discarded; the aqueous phase now adjusted to pH 0.5 and extracted again with MTBE; the MTBE extract dried and concentrated; chloroacetic acid in the MTBE extract esterified with diazomethane; the methyl ester determined by capillary GC on an ECD (U.S. EPA Method 552, 1990).

Alternatively, a 30 mL sample portion microextracted with a 3-mL aliquot of MTBE; the extract esterified with diazomethane; the methyl ester of chloroacetic acid analyzed on GC-ECD.

- An alternate column must be used to confirm the presence of analyte. A 30 m DB-1701 and DB-210 having 0.32 mm ID and 0.25 to 0.50 mm film thickness were used in development of U.S. EPA method.
- The calibrations standards must be esterified prior to their injections into the GC.
- 1,2,3-Trichloropropane was found to be a suitable internal standard.
- The MTBE extract is concentrated down to 2 mL either at room temperature under a stream of dry N_2 or on a water bath at 35°C.
- Chloroacetic acid is a strong organic acid that may react with alkaline substances and may be lost during sample preparation. Glasswares and glass wool must be acid rinsed with HCI (1 + 9); Na_2SO_4 must be acidified with H_2SO_4.
- Many chlorinated compounds and organic acids may interfere in GC-ECD analysis. Some basic and neutral interference can be removed by acid–base partitioning sample cleanup.

- Unreacted diazomethane left after esterification should be removed by adding 0.2 g silica gel. Its presence is indicated from a persistent yellow color and also effervescence due to N_2.

Alternatively, a 100 mL sample portion pH adjusted to 5.0; chloroacetic acid separated on an anion exchange column and eluted with small aliquots of acidic methanol; a small volume of MTBE then added as cosolvent, resulting in esterification of the analyte; methyl ester partitions into the MTBE phase; the ester analyzed by GC-ECD (U.S. EPA Method 552.1, 1992).

- Use liquid-solid extraction cartridge or disk as anion exchange column. Add AG-1-X8 resin solution dropwise.
- Condition the column by passing 10-mL aliquots of methanol, reagent water, 1 *M* HCl/methanol, reagent water, 1 *M* NaOH and reagent water, respectively, through the resin under vacuum at a rate of 2 mL/min. The resin bed must not be allowed to dry.
- Sulfate, chloride, and other anions at high concentrations interfere in the test. Dilute the sample prior to analysis (1:5 or 1:10 dilution).
- Organic acids and phenols also interfere in the test. Perform a methanol wash after the acid analyte is adsorbed on the column.
- Samples dechlorinated with NH_4Cl and refrigerated at 4°C may be stable for 28 days.

AIR ANALYSIS

Air drawn through a solid sorbent tube containing silica gel (100 mg/50 mg) with glass wool plugs; analyte desorbed with 2 mL deionized water; chloroacetate ion determined by ion chromatography with conductivity detection (NIOSH Method 2008, 1987); recommended air flow rate 0.1 L/min; air sample volume 25 L.

- IC column: anion guard; anion separators in tandem packed with low capacity anion exchange resin; anion suppressor packed with high capacity cation exchange resin. Other IC column resins may be used; eluent: 1.5 mm $NaHCO_3$.
- Standard solutions of chloroacetic acid in deionized water are to be used for calibration.
- Particulate salts of acids and chloroacetyl chloride interfere in the test.
- Sample is stable for a week at room temepratures and 30 days if refrigerated.
- Working range for a 3 L air sample is 0.3 to 30 mg/m^3.

(1 ppm chloroacetic acid in air = 3.86 mg/m^3 at NTP)

3.18 CHLOROBENZENE

Synonyms: phenyl chloride, benzenechloride; Formula: C_6H_5Cl; Structure:

MW 112.56; CAS [108-90-7]; used in heat transfer medium and as a solvent in paint; colorless liquid with a faint almond odor; boils at 131°C; solidifies at –55°C; density 1.1 g/mL at 20°C; insoluble in water, miscible with organic solvents.

ANALYSIS OF AQUEOUS AND NONAQUEOUS SAMPLES

Aqueous samples extracted by purge and trap; volatile analyte thermally desorbed and swept onto a GC column for separation; detected by HECD, ECD, FID or GC/MS.

Soils, solid wastes, or sludges mixed with water or methanol; the aqueous extract or a solution of methanol extract spiked into water, subjected to purge and trap concentration and analyzed as above.

- Characteristic masses for GC/MS identification: 112, 77, and 114.
- Limit of detection: in the range 0.02 μg/L when detected by PID or HECD for a 5-mL sample aliquot (subject to matrix interference).
- Recommended surrogate/IS: 4-bromofluorobenzene and fluorobenzene.
- Sample collected in glass containers without headspace, refrigerated, and analyzed within 14 days.

See Halogenated Hydrocarbons, Chapter 2.9, three spacings for GC columns and conditions and a detailed discussion.

AIR ANALYSIS

Adsorbed over coconut charcoal (100 mg/50 mg); desorbed with CS_2 and analyzed by GC-FID; recommended flow rate: 200 mL/min; sample volume 20 L.

Adsorbed over Tenax (1 to 2 g) in a cartridge; cartridge heated under He purge; analyte desorbed and transferred onto a cold trap and then onto a precooled GC column at –70°C; column heated; analyte eluted and detected by GC/MS, ECD, or FID; recommended flow rate 0.5 L/min; sample volume 100 L (U.S. EPA Method TO-1).

Collected in a trap in liquid argon; the trap removed and heated; sample transferred to a precooled GC column for ECD or FID detection (U.S. EPA Method TO-3).

Collected in a SUMMA passivated canister under pressure using an additional pump or at subatmospheric pressure by initially evacuating the canister; the air transferred into cryogenically cooled trap attached to a GC column; the trap heated; the analyte separated on the column for detection by ECD, FID, or GC/MS (U.S. EPA Method TO-14).

(1 ppm chlorobenzene in air = ~4.6 mg/m^3 at NTP)

3.19 CHLOROFORM

Synonyms: trichloromethane, methenyl trichloride; Formula: $CHCl_3$; MW 119.37; CAS [67-66-3]; a volatile halogenated solvent; boils at 61.2°C; vapor pressure 158 torr at 20°C; heavier than water, density 1.484g/mL at 20°C; solubility in water very low (1.2%), miscible in organic solvents; nonflammable.

ANALYSIS OF AQUEOUS AND NONAQUEOUS SAMPLES

Aqueous samples subjected to purge and trap concentration; an aliquot of sample purged under helium flow chloroform and other volatile analyte thermally desorbed out from the sorbent trap and swept onto a GC column for separation; detected by HECD, ECD, or MSD.

Aqueous samples extracted with hexane; extract injected onto GC column for ECD detection; precision and accuracy of the method found to be low (Wirth and Patnaik, 1993).

Soils, solid wastes, or sludges mixed with water or methanol; the aqueous extract or a solution of methanol extract spiked into water, subjected to purge and trap concentration and analyzed as above.

- Characteristic masses for GC/MS identification: 83 and 85.
- Limit of detection: in the range 0.05 μg/L when detected by HECD for a 5-mL sample aliquot (subject to matrix interference).
- Recommended surrogate/IS: bromochloromethane and 1,2-dichloroethane-d_4.
- Samples collected in glass/plastic containers without headspace, refrigerated, and analyzed within 14 days.

See Halogenated Hydrocarbons, Chapter 2.9 for GC columns and conditions and for a detailed analysis.

AIR ANALYSIS

Adsorbed over coconut charcoal (100 mg/50 mg); desorbed with CS_2 and analyzed by GC-FID; recommended flow rate 100 mL/min; sample volume 15 L (NIOSH Method 1001).

Adsorbed over Tenax (~2 g) in a cartridge which is then heated and purged with He; compound transferred into a cold trap and then to the front of a GC column at –70°C; column heated; $CHCl_3$ detected by ECD or GC/MS; recommended flow rate 100 mL/min; volume 10 L (U.S. EPA Method TO-1).

Adsorbed over carbon molecular sieve (~400 mg) in a cartridge; heated at 350°C under He purge; transferred into a cryogenic trap and flash evaporated onto a capillary column GC/MS; recommended flow rate 2 L/min; sample volume 100 L (U.S. EPA Method TO-2).

Collected in a trap in liquid argon; the trap removed and heated; sample transferred to a precooled GC column for ECD or FID detection (U.S. EPA Method TO-3).

Collected in a SUMMA passivated canister under pressure using an additional pump or at subatmospheric pressure by initially evacuating the canister; the air transferred into a cryogenically cooled trap attached to a GC column; the trap heated; the analyte separated on the column for detection by ECD, FID, or GC/MS (U.S. EPA Method TO-3).

(1 ppm chloroform in air = 4.9 mg/m^3 at NTP)

3.20 2-CHLOROPHENOL

Formula: C_6H_5OCl; Structure:

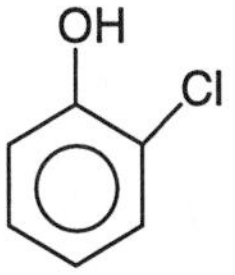

MW 128.56; CAS [95-57-8]; crystalline solid melting at 33.5°C; boils at 214°C; slightly soluble in cold water; soluble in organic solvents.

ANALYSIS OF AQUEOUS AND NONAQUEOUS SAMPLES

Aqueous samples pH adjusted below 2; extracted with methylene chloride; extract concentrated and analyzed by GC-FID or GC/MS.

- Alternatively, extract exchanged into 2-propanol; derivatized with pentafluorobenzyl bromide, and determined by GC-ECD.
- Aqueous samples may be analyzed by HPLC on an underivatized polystyrene-divinylbenzene column such as PolyRPCO (Alltech 1995) or a C-18 reverse phase column; gradient: acetonitrile and 0.01 *M* K_3PO_4 at pH 7 (55:45) and the analyte detected by UV at 254 nm.

Soils, solid wastes, and sludges extracted with methylene chloride by sonication, Soxhlett or supercritical fluid extraction; extract may be acid washed, concentrated, and analyzed by GC-FID, GC-ECD, or GC/MS as above.

- Characteristic masses for GC/MS determination 128, 64, and 130 (electron impact ionization); 129, 131, and 157 (chemical ionization).
- Limit of detection: in the range 5 to 10 μg/L on FID or GC/MS and below 1 μ/L on ECD (for aqueous samples concentrated by 1000 times).
- Recommended surrogate/IS: pentafluorophenol and 2-perfluoromethyl phenol.

- Samples collected in glass containers, refrigerated, and extracted within 7 days of collection and analyzed within 40 days of extraction.

See Phenols, Chapter 2.23 for GC columns and conditions.

AIR ANALYSIS

Recommended Method: Airborne particles collected on a sorbent cartridge containing polyurethane foam; extracted with 5% diethyl ether in hexane; extract analyzed by GC-ECD; recommended air flow 5 L/min; sample volume 1000 L.

3.21 CUMENE

Synonyms: isopropylbenzene, (1-methyl ethyl) benzene, 2-phenylpropane; Formula: C_9H_{12}; Structure:

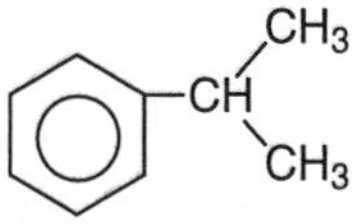

MW 120.21; CAS [98-82-8]; used as a solvent and in organic synthesis; colorless liquid with an aromatic odor; boils at 152.5°C; vapor pressure 8 torr at 20°C; freezes at –96°C; density 0.869 g/mL at 20°C; insoluble in water, miscible with organic solvents.

ANALYSIS OF AQUEOUS AND NONAQUEOUS SAMPLES

Purge and trap method; a measured volume of sample purged with helium; vapors of cumene collected on a sorbent trap; cumene thermally desorbed from the sorbent trap and backflushed with helium onto a GC column for separation from other volatile compounds; determined by PID, FID or a mass spectrometer.

Solid samples mixed with methanol or acetone; an aliquot of solvent extract spiked into a measured volume of water (25 mL) in the purging vessel; subjected to purge and trap concentration and analyzed as above.

Alternatively, cumene thermally desorbed from the solid matrix under He purge (without any solvent treatment) onto the GC column and analyzed on a suitable detector.

- Characteristic masses for GC/MS identification: 105 (primary ion) and 120.
- GC column: a nonpolar fused silica capillary column, such as DB-5, VOCOL DB-624, or equivalent; packed column: 1% SP-1000 on Carbopack B (60/80mesh) or equivalent.
- Limit of detection: in the range 0.2 mg/L for 25 mL sample analyzed by purge and trap GC/MS method.

- Recommended surrogates/IS: ethylbenzene-d_5 (m/z 111) and pentafluorobenzene (m/z 168).
- Samples collected in glass containers without headspace, refrigerated, and analyzed within 7 days of collection. Samples preserved with 1:1 HCl (0.2 mL acid/40 mL sample) may be analyzed in 14 days.

AIR ANALYSIS

Air drawn through a sorbent tube packed with coconut shell charcoal (100 mg/50 mg); analyte desorbed into CS_2 (more than 30 min standing); CS_2 extract analyzed by GC-FID (NIOSH Method 1501, 1984); recommended air flow rate 100 mL/min; sample volume 20 L.

- GC column: packed column 10% OV-275 on 100/120 mesh Chromosorb W-AW, Porapak P (50/80 mesh), or equivalent.

(Suggested method): Alternatively air drawn through a cartridge packed with carbon molecular sieve (0.5 g); cartridge heated at 350°C under He purge; analyte transported to the front of a precooled GC column, temperature programmed; cumene separated on the column and determined by a PID, FID, or a mass spectrometer; recommended air flow rate 0.5 L/min; sample volume 50 L.

- No precision and accuracy data available for this method.

(1 ppm cumene in air = 4.92 mg/m^3 at NTP)

3.22 CYANOGEN

Synonyms: ethanedinitrile, oxalonitrile, oxalyl cyanide, dicyan; Formula: C_2N_2; Structure: N≡C–C≡N; MW 52.04; CAS [460-19-5]; used as a fuel gas, fumigant, and propellant; occurs in blast furnace gases; colorless gas; almond-like odor, becoming pungent at high concentration; liquefies at –21°C; highly soluble in water, alcohol, and ether; highly toxic and highly flammable.

AIR ANALYSIS (SUGGESTED METHOD)

Air drawn through a 0.8-mm cellulose ester membrane (to separate cyanogen from particulate cyanide) followed by 0.1 *N* KOH bubbler soln.; cyanide ion analyzed by cyanide ion-selective electrode; recommended flow rate 200 mL/min; sample volume 10 L.

- No precision or accuracy data is available for the above method.
- HCN may interfere in the test.

Alternatively, air bubbled through a measured volume of methanol; the bubbler solution analyzed by GC-FID; peak areas compared against calibration standards for quantitation

- No precision or accuracy data are available for the above method.

(1 ppm cyanogen = 2.13 mg/m^3 at NTP)

3.23 CYANURIC ACID

Synonyms: trihydroxycyanidine; 2,4,6-trihydroxy-1,3,5-triazine; Formula: $C_3H_3N_3O_3$; Structure:

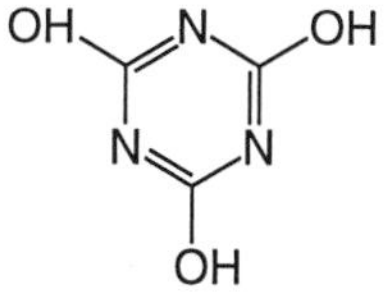

MW 129.02; CAS [108-80-5] solid crystals which may occur as dihydrate; loses water of crystallization on exposure to air; evolves cyanic acid on heating; does not melt; slightly soluble in cold water (0.5%) but much more soluble in hot water; soluble in pyridine, hot alcohols, HCl, H_2SO_4, and caustic soda solution; insoluble in cold methanol, ether, benzene, acetone, and chloroform.

ANALYSIS OF AQUEOUS AND NONAQUEOUS SAMPLES

Aqueous samples buffered with Na_2HPO_4 and analyzed by HPLC-UV at 225 nm (see under Air Analysis below).

Aqueous samples may be directly injected into a GC column for FID determination; separation from other triazines, however, may be difficult.

- Suggested GC column: DB-1701 or equivalent capillary column.
- No precision and accuracy data available.

Solid samples may be extracted with pyridine; extract analyzed by GC/MS.

- Characteristic masses for GC/MS identification: 129, 43, 44, 86, and 70; peak ratios: 129:43:44:86=100:59:56:15.

Solid samples extracted with water; aqueous extract analyzed by HPLC-UV at 225 nm (see under Air Analysis below).

AIR ANALYSIS

Air drawn through 5-μm PVC membrane filter; analyte extracted with a mixture of 0.005 *M* Na_2HPO_4 in 5% methanol (95% water) at pH 7 in ultrasonic bath for 10 min; analyte detected by HPLC using an UV detector (at 225 nm) (NIOSH Method 5030, 1989).

- HPLC column: μ-Bondapak C-18, 10 μm particle size; mobile phase: Na_2HPO_4 (0.005 *M*) in methanol/water (5:95).

3.24 DIAZOMETHANE

Synonym: azimethylene; Formula: CH_2N_2; Structure: $CH_2=N^+=N^-$; MW 42.04; CAS [334-88-3]; used as a methylating reagent in organic synthesis; yellow gas; liquefies at –23°C; solidifies at –145°C; decomposes on solid surfaces; soluble in ether and dioxane; solution unstable; readily breaks down to N_2; explodes on heating, rough surfaces, or in presence of impurities or when concentrated; highly toxic.

AIR ANALYSIS

Air passed through a solid sorbent tube containing octanoic acid-coated XAD-2 resin (100 mg/50 mg); diazomethane converted to methyl octanoate; the product desorbed with CS_2 (more than 30 min contact time) and analyzed by GC-FID (NIOSH Method 2515, 1985); recommended flow rate 0.2 L/min; sample volume 20 L.

- Multiply the concentration of methyl octanoate by the stoichiometric factor 0.266 to calculate the concentration of diazomethane.
- GC column: SP-1000 on 100/120 mesh Chromosorb WHP, preceded by a precolumn packed with 10% Carbowax 20M and 1.2% NaOH on 80/100 mesh Gas ChromQ.
- Working range: 0.1 to 0.6 ppm for a 10 L air sample.

(1 ppm diazomethane in air = 1.72 mg/m^3 at NTP)

3.25 DIBORANE

Synonyms: boron hydride, boroethane; Formula: B_2H_6; Structure:

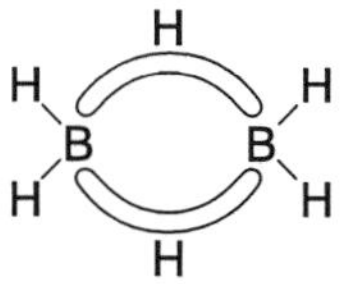

MW 27.69; CAS [19287-45-7]; used as a rocket propellant, a polymerization catalyst, a reducing agent, and in the vulcanization of rubber; colorless gas with a sweet, repulsive odor; liquefies at –92.5°C; freezes at –165°C; gas density 1.15 g/L; decomposes in water; soluble in CS_2; highly toxic and flammable; ignites in moist air.

AIR ANALYSIS

Air drawn through a PTFE filter and then through a solid sorbent tube containing oxidizer-impregnated charcoal (100 mg/50 mg); adsorbent after sampling, transferred into 10 mL 3% H_2O_2; allowed to stand for 30 min and then placed in an ultrasonic bath for 20 min; diborane oxidized to boron; the latter measured at the wavelength 249.8 nm by plasma emission spectrometer using a DC plasma emission source (NIOSH Method 6006, 1987); recommended air flow rate 0.5 to 1 L/min; sample volume 100 to 250 L.

- Use aliquots of 1% (v/v) diborane adsorbed on impregnated charcoal for calibration.
- Multiply the concentration of boron as determined in the analysis by the stoichiometric factor 1.28 to calculate the concentration of diborane.
- Boron may also be measured by inductively coupled plasma-emission spectrometry (NIOSH Method 7300, 1989).
- PTFE filter is used to remove boron-containing particulates. Use of mixed cellulose ester membrane prefilters is not recommended.

- A blank analysis is performed on the media blanks for background correction.
- The working range for a 250 L air sample is 0.1 to 0.5 ppm.
- Avoid borosilicate glassware during sample preparation. Use plastic bottles, pipettes, and flasks.

(1 ppm diborane = 1.131 mg/m^3 at NTP)

3.26 DICHLOROBENZENE

Formula: $C_6H_4C_{12}$; MW 147.00; Structures of three isomeric forms as follows:

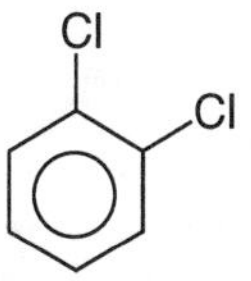

1,2-dichlorobenzene	1,3-dichlorobenzene	1,4-dichlorobenzene
(*o*-chlorobenzene)	(*m*-chlorobenzene)	(*ρ*-chlorobenzene)
[95-50-1]	[541-73-1]	[106-46-7]
(1)	(2)	(3)

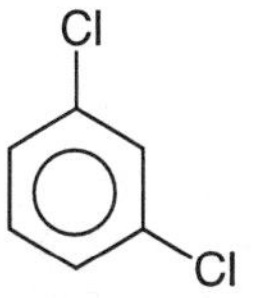

1. Colorless liquid with faint odor, density 1.28 g/mL, boils at 179°C; freezes at –17°C.
2. Colorless liquid, density 1.29 g/mL; boils at 172°C; freezes at 124°C.
3. Colorless crystalline solid, sublimes at ambient temperture; melts at 54°C; boils at 174°C.

All these isomers are insoluble in water, soluble in organic solvents.

ANALYSIS OF AQUEOUS AND NONAQUEOUS SAMPLES

Aqueous samples extracted by purge and trap; sample purged with helium or nitrogen analyte thermally desorbed out of the trap and swept onto a GC column for separation; detected by HECD, ECD, PID or a mass spectrometer.

- Alternatively, extracted with methylene chloride by LLE; extract concentrated and analyzed by a GC using a PID, FID, ECD, or GC/MS.

Soils, solid wastes, or sludges mixed with water or methanol; the aqueous extract or a solution of methanol extract spiked into water; subjected to purge and trap concentration and analyzed as above.

- Alternatively, extracted with methylene chloride by sonication or Soxhlett extraction; extract concentrated and analyzed.
- Characteristic masses for GC/MS identification: 146, 111, and 148.
- Limit of detection: in the range 0.1 μg/L when detected by HECD for a 5-mL sample aliquot (subject to matrix interference).
- Elution pattern: 1,3-isomer followed by 1,4- and then 1,2-isomer, on a fused silica capillary column, e.g., VOCOL.
- Recommended surrogate/IS: 4-bromofluorobenzene and fluorobenzene.
- Sample collected in glass containers without headspace, refrigerated, and analyzed within 14 days.

See Halogenated Hydrocarbons, Chapter 2.9 for GC columns and conditions and a detailed discussion.

AIR ANALYSIS

Adsorbed over Tenax (1 to 2 g) in a cartridge; cartridge heated under He purge; analyte desorbed and transferred onto a cold trap and then onto a precooled GC column at –70°C; column heated; analyte eluted and determined by GC/MS, ECD, or FID; recommended flow rate 0.5 L/min; sample volume 100 L (U.S. EPA Method TO-1).

Adsorbed over coconut charcoal (100 mg/50 mg); desorbed with CS_2 and analyzed by GC-FID; recommended flow rate: 200 mL/min; sample volume 20 L.

Collected in a trap in liquid argon; the trap removed and heated; sample transferred to a precooled GC column for ECD or FID detection (U.S. EPA Method TO-3).

Collected in a SUMMA passivated canister under pressure using an additional pump or at subatmospheric pressure by initially evacuating the canister; the air transferred into cryogenically cooled trap attached to a GC column; the trap heated; the analyte separated on the column for detection by ECD, FID, or GC/MS (U.S. EPA Method TO-14).

(1 ppm dichlorobenzene in air = ~6 mg/m^3 at NTP)

3.27 1,1-DICHLORETHYLENE

Synonym: vinylidene chloride; Formula: $C_2H_2Cl_2$; Structure:

$$Cl_2C{=}CH_2$$

MW 96.94; CAS[75-35-4]; colorless liquid; boils at 31.7°C; liquefies at −122.5°C; density 1.21 g/mL at 20°C; slightly soluble in water; miscible with organic solvents.

ANALYSIS OF AQUEOUS AND NONAQUEOUS SAMPLES

Analyte in aqueous samples extracted by purge and trap; a measured volume of sample purged with helium; volatile analytes transferred into the vapor phase and trapped on a sorbent trap; analyte thermally desorbed and swept onto a GC column for separation from other volatile compounds; detected by HECD, ECD, or MSD.

Soils, solid wastes, or sludges mixed with water or methanol; the aqueous extract or a solution of methanol extract spiked into water; subjected to purge and trap concentration and analyzed as above.

- Characteristic masses for GC/MS identification: 96, 61, and 98.
- Limit of detection: in the range 0.15 µg/L when detected by HECD for a 5-mL sample aliquot (subject to matrix interference).
- Recommended surrogate/IS: bromochloromethane and 1,2-dichloroethane-d_4.
- Samples collected in glass/plastic containers without headspace, refrigerated, and analyzed within 14 days.

See Halogenated Hydrocarbons in Chapter 2.9 for GC columns and conditions and a detailed discussion.

AIR ANALYSIS

Adsorbed over coconut charcoal (100 mg/50 mg); desorbed with CS_2 and analyzed by GC-FID; recommended flow rate 100 mL/min; sample volume 2 L.

Adsorbed over carbon molecular sieve (~400 mg) in a cartridge; heated at 350°C under He purge; transferred into a cryogenic trap and flash evaporated onto a capillary column GC/MS; recommended flow rate 2 L/min; sample volume 100 L (U.S. EPA Method TO-2).

Collected in a trap in liquid argon; the trap removed and heated; sample transferred to a precooled GC column for ECD or FID detection (U.S. EPA Method TO-3).

Collected in a trap in liquid argon; the trap removed and heated; sample transferred to a precooled GC column for ECD or FID detection (U.S. EPA Method TO-3).

Collected in a SUMMA passivated canister under pressure using an additional pump or at subatmospheric pressure by initially evacuating the canister; the air transferred into a cryogenically cooled trap attached to a GC column; the trap heated; the analyte separated on the column for detection by ECD, FID, or GC/MS (U.S. EPA Method TO-14).

3.28 2,4-DICHLOROPHENOL

Formula: $C_6H_4OCl_2$; Structure:

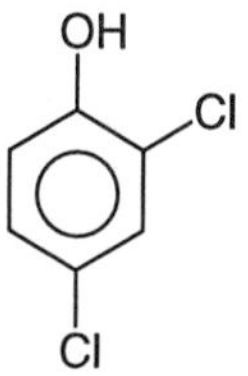

MW 163.01; CAS [120-83-2]; crystalline solid melts at 45°C; boils at 210°C; insoluble in water; soluble in organic solvents.

ANALYSIS OF AQUEOUS AND NONAQUEOUS SAMPLES

Aqueous samples pH adjusted below 2; extracted with methylene chloride; extract concentrated and analyzed by GC-FID or GC/MS.

- Alternatively, extract exchanged into 2-propanol; derivatized with pentafluorobenzyl bromide, and determined by GC-ECD.

Aqueous samples may be analyzed by HPLC on an underivatized polystyrene-divinylbenzene column such as PolyRPCO (Alltech 1995) or a C-18 reverse phase column; gradient: acetonitrile and 0.01 *M* K_3PO_4 at pH 7 (55:45) and the analyte detected by UV at 254 nm.

Soils, solid wastes, and sludges extracted with methylene chloride by sonication, Soxhlett, or supercritical fluid extraction; extract may be acid washed, concentrated, and analyzed by GC-FID, GC-ECD, or GC/MS as above.

- Characteristic masses for GC/MS determination: 162, 164, and 98 (electron impact ionization); 163, 165, and 167 (chemical ionization).
- Limit of detection: in the range 5 to 10 μg/L on FID or GC/MS and below 1 μg/L on ECD (for aqueous samples concentrated by 1000 times).
- Recommended surrogate/IS: pentafluorophenol and 2-perfluoromethyl phenol.

- Samples collected in glass containers, refrigerated, and extracted within 7 days of collection and analyzed within 40 days of extraction.

See Phenols, Chapter 2.23 for GC columns and conditions.

AIR ANALYSIS

Recommended method: Airborne particles collected on a sorbent cartridge containing polyurethane foam; analyte deposited on the cartridge extracted with 5% diethyl ether in hexane; extract analyzed by GC-ECD; recommended air flow 5 L/min; sample volume 1000 L.

3.29 DIETHYL ETHER

Synonyms: ethyl ether, solvent ether, diethyl oxide; Formula $(C_2H_5)_2O$; Structure: $H_3C–CH_2–O–CH_2–CH_3$; MW 74.14; CAS [60-29-7]; used as a solvent; used in organic synthesis and as an anesthetic; colorless liquid; pungent odor; sweet burning taste; boils at 34.6°C; vapor pressure 439 torr at 20°C; freezes at −116°C; density 0.71 g/mL at 20°C; miscible with organic solvents; solubility in water 6 g/100 mL; forms azeotrope with water (1.3%); extremely flammable; narcotic.

ANALYSIS OF AQUEOUS SAMPLES

Analyte purged out from sample matrix by purge and trap extraction; desorbed from the sorbent trap by heating and backflushing with He; transferred into a capillary GC column interfaced to a mass spectrometer; identified by mass spectra and retention time (U.S. EPA Method 524.2, 1992).

- GC column: a fused silica capillary column such as VOCOL, DB-5, DB-624, or equivalent.
- Characteristic masses for GC/MS identification: 74 (primary ion), 59, 45, and 73.
- Detection level: in the range of 0.1 mg/L.

Aqueous samples directly injected onto an appropriate GC column for FID detection.

- Poor detection level; sensitivity much lower than the purge and trap method.

AIR ANALYSIS

Air drawn through a solid sorbent tube containing coconut shell charcoal (100 mg/50 mg); diethyl ether desorbed with 1 mL ethyl acetate (30 min contact time) and analyzed by GC-FID (NIOSH Method 1610, 1985); recommended flow rate 0.1 L/min; sample volume 1 L.

- GC column: stainless steel packed with Porapak Q (50/80 mesh). Use a capillary column to prevent coelution of CS_2, hexane, and low molecular weight ketones
- Working range: 100 to 2700 mg/m^3 for a 3 L air sample.

(1 ppm diethyl ether = 3.03 mg/m^3 at NTP)

3.30 2,4-DINITROTOLUENE

Formula: $C_7H_6O_4N_2$; MW 182.14; Structure:

CH3, NO2, NO2

CAS [121-14-2]; crystalline solid; melts between 67 and 70°C; insoluble in water; soluble in organic solvents.

ANALYSIS OF AQUEOUS AND NONAQUEOUS SAMPLES

Aqueous samples repeatedly extracted with methylene chloride; extracts combined and concentrated by evaporation of methylene chloride solvent exchanged to hexane; florisil cleanup (for removal of interferences); extract analyzed on GC-ECD or GC/MS.

Soils, sediments, or solid wastes extracted with methylene chloride by sonication or Soxhlett extraction; extract concentrated; exchanged to hexane (if florisil cleanup required); analyzed by GC-ECD or GC/MS.

- For florisil cleanup, column eluted with methylene chloride/hexane (1:9) and then with acetone/methylene chloride (1:9) mixture; analyte eluted into the latter fraction.
- Column packed: 1.95% QF-1/1.5% OV-17 on Gas-Chrom Q (80/100 mesh) or 3% SP-2100 on Supelcoport (100/120 mesh) or equivalent; fused silica capillary column, such as PTE-5, SPB-5, DB-5, Rtx-5, or equivalent.
- 2,6-isomer elutes before the 2,4-isomer.
- Characteristic masses for GC/MS identification: 165, 63, and 182 (electron impact ionization); 183, 211, and 223 (chemical ionization).
- Suggested surrogate/IS: aniline d_5, 4-fluoroaniline, nitrobenzene d_5, and 1-fluoronaphthalene.

- Detection level: in the range 0.02 μg/L may be achieved when determined by GC-ECD (for aqueous samples concentrated by 1000 times).
- Samples collected in glass containers, refrigerated, extracted within 7 days of collection and analyzed within 40 days of extraction.

3.31 2,6-DINITROTOLUENE

Formula: $C_7H_6O_4N_2$; MW 182.14; Structure:

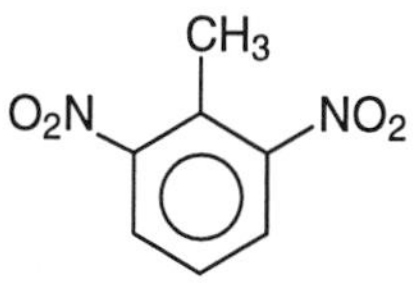

O_2N CH_3 NO_2

CAS [606-20-2]; crystalline solid; melts ≈ 5°C; insoluble in water; soluble in most organic solvents.

ANALYSIS OF AQUEOUS AND NONAQUEOUS SAMPLES

Aqueous samples extracted with methylene chloride; extract concentrated and solvent exchanged to hexane; florisil cleanup (for removal of interferences); extract analyzed on GC-ECD or GC/MS.

Soils, sediments, or solid wastes extracted with methylene chloride by sonication or Soxhlett extraction: extract concentrated; exchanged to hexane (if florisil cleanup required); analyzed by GC-ECD or GC/MS.

- For florisil cleanup, column eluted with methylene chloride/hexane (1:9) and then with acetone/methylene chloride (1:9) mixture; analyte eluted into the latter fraction.
- Column-packed: 1.95% QF-1/1.5% OV-17 on Gas-Chrom Q (80/100 mesh) or 3% SP-2100 on Supelcoport (100/120 mesh) or equivalent; capillary: fused silica capillary column, such as PTE-5, SPB-5, DB-5, Rtx-5, or equivalent.
- 2,6-isomer elutes before the 2,4-isomer.
- Characteristic masses for GC/MS identification: 165, 89, and 121 (electron impact ionization); 183, 211, and 223 (chemical ionization).
- Suggested surrogate/IS: aniline d_5, 4-fluoroaniline, nitrobenzene d_5, and 1-fluoronaphthalene.

- A detection level: in the range 0.02 μg/L may be achieved for aqueous samples concentrated up by 1000 times and determined by GC-ECD.
- Samples collected in glass containers, refrigerated, extracted within 7 days of collection and analyzed within 40 days of extraction.

3.32 EPICHLOROHYDRIN

Synonyms: 1,2-epoxy-3-chloropropane; 2-(chloromethyl)oxirane; 2,3-epoxypropyl chloride; Formula: C_3H_5ClO; Structure:

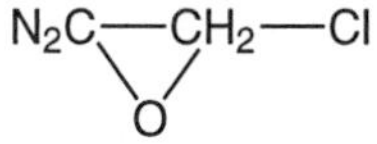

N_2C—CH_2—Cl
O

MW 92.53; CAS [106-89-8]; used to make epoxy resins, adhesives, surfactants, and plasticizers, and also as a solvent for gums, resins, and paints; colorless liquid; chloroform-like odor; boils at 116°C; solidifies at –57°C; vapor pressure 12 torr at 20°C; density 1.18 g/mL at 20°C; moderately soluble in water (6.6% at 20°C), soluble in most organic solvents; moderately toxic, carcinogenic, and a strong irritant (Patnaik, 1992).

ANALYSIS OF AQUEOUS AND NONAQUEOUS SAMPLES

Aqueous samples subjected to purge and trap concentration; analyte thermally desorbed from the trap and swept onto a GC clumn for separation from other volatile substances and detection by HECD, ECD or a mass spectrometer.

- Purging efficiency low because of its moderate solubility.
- Purging efficiency may be enhanced by heating the purging vessel.
- Analyte unstable; polymerizes in presence of acids, alkalies, and catalysts.
- Characteristic masses for GC/MS identification: 57, 27, 29, and 31; peak ratios: 57:27:29:31 = 100:96:71:39.
- Samples collected in glass vials without headspace, refrigerated, and analyzed within 14 days.

AIR ANALYSIS

Adsorbed over coconut shell charcoal (100 mg/50 mg); desorption with CS_2; epichlorohydrin desorbed into CS_2 then determined by GC-FID (NIOSH Method 1010, 1984); recommended flow rate 100 mL/min; sample volume 20 L.

- GC column: Chromosorb 101 or equivalent.

(1 ppm epichlorohydrin in air = ~3.78 mg/m^3 at NTP)

3.33 ETHYLBENZENE

Synonym: phenylethane; Formula C_8H_{10}; Structure:

CH_2-CH_3 (attached to benzene ring)

MW 106.18; CAS [100-41-4]; used as a solvent and an intermediate to produce styrene monomer; colorless liquid; characteristic aromatic odor; boils at 136°C; vapor pressure 7.1 torr at 21°C; freezes at –95°C; density 0.86 g/mL at 20°C; solubility in water 0.015 g/100 g; readily miscible with organic solvents.

ANALYSIS OF AQUEOUS AND NONAQUEOUS SAMPLES

Purge and trap method; ethylbenzene transferred from aqueous to vapor phase under helium purge; analyte adsorbed on a sorbent trap; thermally desorbed out from the sorbent trap backflushed with He onto a GC column for separation from other volatile compounds; determined by PID, FID or a mass spectrometer.

Solid samples mixed with methanol; a portion of methanol extract spiked into a measured volume of water (25 mL) in the purging vessel; subjected to purge and trap concentration and analyzed as above.

Alternatively, ethylbenzene thermally desorbed from the solid sample under He purge (without any solvent treatment) onto the GC column and analyzed on a suitable detector or by a mass spectrometer as above.

- Characteristic masses for GC/MS identification: 106 and 91.
- Limit of detection: in the range 0.5 mg/L for a 5 mL sample volume, when detected by PID.
- GC column: a nonpolar fused silica capillary column, such as DB-5, SPB-5, VOCOL, DB-624, or equivalent; packed column: 1% SP-1000 on Carbopack B (60/80 mesh) or equivalent.

- Recommended surrogate/IS: ethylbenzene-d_5 (m/z 111) and 1,4-difluorobenzene (m/z 114, 63,88).
- Samples collected in glass containers without headspace, refrigerated, and analyzed within 7 days. Samples preserved with 1:1 HCl (0.2 mL acid/40 mL sample) may be analyzed in 14 days.

AIR ANALYSIS

Air drawn through a sorbent tube packed with coconut shell charcoal (100 mg/50 mg); analyte desorbed into CS_2 (more than 30 min standing); CS_2 extract analyzed by GC-FID (NIOSH Method 1501, 1984); recommended air flow rate 100 mL/min; sample volume 20 L.

- GC column: packed column 10% OV-275 on 100/120 mesh Chromosorb W-AW, Porapak P (50/80 mesh), or equivalent.

Alternatively, air collected in a SUMMA passivated stainless steel canister either by pressurizing the canister using a sample pump or by preevacuating; canister then connected to an analytical system; air transferred to a cryogenically cooled trap; cryogen removed and temperature raised; analyte revolatilized; separated on a GC column; determined by PID, FID, or a mass spectrometer (U.S. EPA Method TO-14, 1988).

Alternatively, air drawn through a cartridge packed with either Tenax (1 g) or carbon molecular sieve (0.5 g); cartridge heated at 350°C under helium purge; analyte transported to the front of a precooled GC column, temperature programmed; ethylbenzene determined on a PID, FID, or a mass spectrometer; recommended air flow rate 0.5 L/min; sample volume 50 L.

- No precision and accuracy data available for this method.

(1 ppm ethylbenzene in air = 4.34 mg/m^3 at NTP)

3.34 ETHYL CHLORIDE

Synonym: chloroethane; Formula: C_2H_5Cl; MW 64.52; CAS [75-00-3]; colorless gas at room temperature with ether-like odor; liquefies at 12.5°C; slightly soluble in water, miscible with organic solvents.

ANALYSIS OF AQUEOUS SAMPLES

Aqueous samples subjected to purge and trap extraction; volatile analyte thermally desorbed out from the trap on heating and swept onto a GC column for separation; detected by HECD, ECD, or MDS; low retention time.

- Characteristic masses for GC/MS identification: 64 and 66.
- Limit of detection: in the range 0.1 μg/L when detected by HECD for a 5-mL sample aliquot (subject to matrix interference).
- Recommended surrogate/IS: bromochloromethane and 1,2-dichloroethane-d_4.
- Samples collected in glass/plastic containers without headspace, refrigerated, and analyzed within 14 days.

See Halogenated Hydrocarbons, Chapter 2.9, for GC columns and conditions and a detailed discussion.

AIR ANALYSIS

The following methods are recommended:

Adsorbed over carbon molecular sieve (~400 mg); desorbed at 350°C into a cryogenically cooled trap; flash evaporated onto a capillary column GC/MS system; recommended sample volume 10 L; flow rate 100 mL/min.

Collected in SUMMA passivated canister or a liquid argon trap; transferred onto a precooled GC column; determined by ECD or MSD.

The precision and accuracy of the above methods for this compound are not established.

(1 ppm ethyl chloride in air = ~2.2 mg/m^3 at NTP)

3.35 ETHYLENE CHLOROHYDRIN

Synonyms: 2-chloroethanol, 2-chloroethyl alcohol; Formula: C_2H_5OCl; Structure: Cl–CH_2–CH_2–OH; MW 80.52; CAS [107-07-3]; used as a solvent for cellulose esters and in making ethylene glycol and ethylene oxide; colorless liquid with a faint ether odor; boils at 129°C; freezes at –67°C; density 1.197 g/mL at 20°C; soluble in water, alcohol, and ether; highly toxic.

ANALYSIS OF AQUEOUS SAMPLE

Purge and trap concentration; sample purged with He under heating; analyte desorbed from the trap by heating and backflushing with He; transferred onto a GC column for separation; determined by GC-FID or a mass spectrometer.

- No precision and accuracy data are available.
- Purging efficiency is low because of high solubility of the analyte in water. Purging vessel should be heated under He purge.

AIR ANALYSIS

Air drawn through a solid sorbent tube containing petroleum charcoal (100 mg/50 mg); analyte desorbed with 1 mL 5% 2-propanol in CS_2 (more than 30 min standing) and analyzed by GC-FID (NIOSH Method 2513, 1985); recommended flow rate 0.1 L/min; sample volume 20 L.

- GC column: 10% FFAP on 80/100 mesh Chromosorb WHP or equivalent.
- Working range: 0.5 to 15 ppm for a 20 L air sample.

(1 ppm ethylene chlorohydrin in air = 3.29 mg/m^3 at NTP)

3.36 ETHYLENE DIBROMIDE

Synonyms: 1,2-dibromoethane, EDB; Formula: $BrCH_2CH_2Br$; MW 187.88; CAS [106-93-4]; used in fumigant and antiknock gasolines; colorless heavy liquid; chloroform odor; boils at 131°C; freezes at 10°C; vapor pressure 11 torr at 20°C; density 2.7 g/mL at 25°C; slightly soluble in water (0.4%); miscible with alcohol and ether; irritant and toxic.

ANALYSIS OF AQUEOUS AND NONAQUEOUS SAMPLES

Purge and trap method; aqueous samples purged with He; EDB thermally desorbed from the sorbent trap and swept by an inert gas onto a GC column for separation from other volatile compounds; analyzed by a halogen-specific detector or a mass spectrometer.

Solid samples mixed with methanol; methanol extract spiked into water; aqueous solution subjected to purge and trap extraction and analyzed as above.

- Characteristic masses for GC/MS identification: 107, 109, and 188.
- GC column: a fused silica capillary column such as DB-5, SPB-5, Rtx-5, or equivalent.
- Limit of detection: the range 0.1 μg/L for a 5-mL sample aliquot purged.
- Samples collected in glass containers without headspace, refrigerated, and analyzed within 7 days.

AIR ANALYSIS

Air drawn through a sorbent tube containing coconut shell charcoal (100 mg/50 mg); EDB desorbed from the charcoal by treatment with 10 mL 99:1 benzene-methanol (allowed to stand for 1 h); the solvent extract analyzed by GC-ECD (NIOSH Method 1008, 1987); recommended flow rate 100 mL/min; sample volume 10 L.

- GC column: 3% OV 210 on Gas Chrom Q (80/100 mesh).
- IS: 1,2-dibromopropane or 1,1,2,2-tetrachloroethane.

Air collected in a SUMMA passivated canister under pressure using an additional pump or at subatmospheric pressure by initially evacuating the canister; sample transferred into a cryogenically cooled trap attached to a GC column; the trap heated; EDB separated on the GC column and determined by a mass spectrometer (U.S. EPA Method TO-14).

(1 ppm EDB in air = 7.68 mg/m^3 at NTP)

3.37 ETHYLENE GLYCOL

Synonym: 1,2-ethanediol; Formula: $HOCH_2CH_2OH$; MW 62.07; CAS [107-21-1]; used as antifreeze in cooling and heating systems and in hydraulic brake fluids; colorless liquid; sweet taste; hygroscopic; density 1.11 g/mL; boils at 197.5°C; freezes at –13°C; highly soluble in water, lower alcohols, acetone, and pyridine; toxic.

ANALYSIS OF AQUEOUS AND NONAQUEOUS SAMPLES (SUGGESTED METHOD)

Aqueous samples directly injected onto a GC column and determined by an FID.

- GC column: a polar column (polar PEG type phase), such as Supelcowax 10, Nukol, or equivalent.
- No precision or accuracy data are available.

10 g soil, sediment, or solid wastes sonicated with 50 mL 2% isopropanol in water; extract injected onto GC column; determined on FID.

- No precision or accuracy data are available.
- Detection level: in the range 50 mg/kg.

AIR ANALYSIS

Air drawn through a glass fiber filter and then through a sorbent tube containing silica gel (520 mg/260 mg), ethylene glycol desorbed out from silica gel with 2% isopropanol in water (on 5-min standing); the analyte in eluant determined by GC-FID (NIOSH Method 5500, 1984); recommended air flow rate 0.2 L/min; sample volume 30 L.

- GC column: packed glass column 3% Carbowax 20M on 80/100 mesh Chromosorb 101. Any appropriate polar column may be used.
- The working range for a 3 L air sample was found to be 7 to 330 mg/m^3.

Alternatively, air bubbled through water; aqueous soln. of the analyte oxidized with periodic acid into formaldehyde; the latter analyzed by chromotropic acid colorimetric method (Tucker and Deye, 1981).

(1 ppm ethylene glycol in air = 2.54 mg/m^3 at NTP)

3.38 ETHYLENE OXIDE

Synonyms: 1,2-epoxyethane, oxirane; Formula: C_2H_4O; Structure:

$$H_2C\text{—}CH_2 \text{ (bridged by O)}$$

MW 44.06; unstable, ring cleaves readily; CAS [75-21-8]; used as a fumigant and sterilizing agent, and in the manufacture of many glycol ethers and ethanolamines; colorless gas with ether-like odor; liquefies at 10.4°C; density 0.88 g/mL at 10°C; vapor pressure 1095 torr at 20°C; soluble in water and most organic solvents; highly flammable, toxic, and severe irritant.

AIR ANALYSIS

Air drawn through a solid sorbent tube containing HBr-coated petroleum charcoal; bromo derivative of analyte formed desorbed with dimethylformamide (DMF) and analyzed by GC-ECD (NIOSH Method 1614, 1987); recommended air flow 100 mL/min; sample volume 10 L.

- Calibration standard solutions: 2-bromoethanol in DMF.
- GC column: 10% SP-1000 on Chromosorb WHP or equivalent.

Alternatively, air bubbled through impingers contain 0.1 *N* H_2SO_4; ethylene oxide converted into ethylene glycol; latter measured by GC-FID (Romano and Renner, 1979).

- Activated charcoal, impregnated with H_2SO_4 may alternatively, be employed in air sampling; glycol formed desorbed with water or a suitable organic solvent analyzed by GC-FID.

Ethylene oxide in air may be measured directly *in situ* by a rapid colorimetric technique (Pritts et al., 1982). Air is drawn through a multipart detector tube consisting of three reactor tubes: containing periodic acid, xylene, and conc. H_2SO_4, respectively. Ethylene oxide is oxidized by periodic acid to formaldehyde which then reacts with xylenes to form diaryl methylene compounds. The latter products are oxidized by H_2SO_4, impregnated on silica gel to red p-quinoidal compounds. Thus, the color in the detector tube changes from white to reddish brown and the concentration of the analyte in the range 5 to 1000 ppm may be monitored from the length of stain formed in the detector tube.

(1 ppm ethylene oxide in air = 1.80 mg/m^3 at NTP)

3.39 FORMALDEHYDE

Synonyms: methanal, methylene oxide, oxymethane; Formula: HCHO; MW 30.03; CAS[50-00-0]; constitutes about 50% of all aldehydes present in air; released in trace quantities from pressed wood products, burning wood, and synthetic polymers; and automobiles; colorless gas at ambient conditions; pungent suffocating odor; liquefies at –19.5°C; solidifies at –92°C; density 1.07 (air = 1); very soluble in water, soluble in organic solvents; readily polymerizes; flammable, toxic, and carcinogenic (Patnaik, 1992).

WATER ANALYSIS

Aqueous samples analyzed by HPLC using a postcolumn reaction detector; formaldehyde separated on a reversed phase C-18 column; derivatized with 3-methyl-2-benzothiazolinone hydrazone and detected at 640 nm (Igawa et al., 1989). (The method was developed for cloud and fogwater analysis.)

AIR ANALYSIS

Formaldehyde may be analyzed by several techniques involving GC, colorimetry, polarography, HPLC, and GC/MS.

Air drawn through a solid sorbent tube containing 10% (2-hydroxymethyl) piperidine on XAD-2 (120 mg/60 mg); oxazolidine derivative of formaldehyde formed desorbed into toluene under ultrasonic conditions (60 min); analyzed by GC-FID (NIOSH Method 2541, 1989); recommended air flow 200 mL/min; sample volume 15 L.

- Calibration standards prepared from formaldehyde (formalin) stock solutions; each standard injected into 120 mg portions coated-adsorbent; derivative desorbed into toluene and analyzed by GC-FID as above; calibration curve constructed (plotting area/height response against concentrations) for quantitation.
- GC column: fused silica capillary column such as DBWax or equivalent.

Alternatively, air passed through a solid sorbent tube containing 2-(benzylamino)ethanol on Chromosorb 102 or XAD-2; the derivative, 2-benzyloxazolidine desorbed with isooctane and analyzed by GC-FID (NIOSH Method 2502, 1984).

- 2-Benzyloxazolidine peak often masked under interferences from decomposition and/or polymerization products from derivatizing agent; isooctane solution may be readily analyzed by GC/MS using a DB-5 capillary column (Patnaik, 1991).

Alternatively, air drawn through a PTFE membrane followed by sodium bisulfite solution in impingers; impinger solution treated with chromotropic acid and H_2SO_4; color developed due to formation of a derivative of formaldehyde; absorbance measured by a spectrophotometer at 580 nm; a standard calibration curve prepared from formaldehyde standard solutions for quantitation (NIOSH Method 3500, 1989); recommended air flow 500 mL/min; sample volume 50 L.

- Phenols, alcohols, olefins, and aromatics interfere in the test to a small extent.

Air drawn through a midget bubbler containing 15 mL Girard T reagent; Girard T derivative of formaldehyde analyzed by polarography; a calibration curve constructed (plotting concentration formaldehyde/15 mL Girard T solution vs. diffusion current) for quantitation (NIOSH Method 3501, 1989); recommended air flow rate 100 mL/min; sample volume 10 L.

- The half-wave potential for the derivative vs. saturated calomel electrode is –0.99 V.

Alternatively, air drawn through Florisil or silica gel adsorbent coated with acidified 2,4-dinitrophenylhydrazine (DNPH); the DNPH derivative of formaldehyde desorbed with isooctane, and solvent exchanged to methanol or acetonitrile; the solution analyzed by reverse phase HPLC (U.S. EPA Method TO-11, 1988); recommended air flow rate 500 mL/min; sample volume 100 L.

Alternatively, a measured volume of air drawn through an impinger containing ammonium acetate and 2,4-pentanedione; formaldehyde forms a fluorescence derivative, 3,5-diacetyl-1,4-dihydrolutidine; fluorescence of the solution measured by a filter fluorometer (Dong and Dasgupta, 1987).

(1 ppm formaldehyde in air = 1.23 mg/m^3 at NTP)

3.40 FREON 113

Synonyms: 1,1,2-trichloro-1,2,2-trifluoroethane, fluorocarbon-113; Formula $C_2O_3F_3$; Structure:

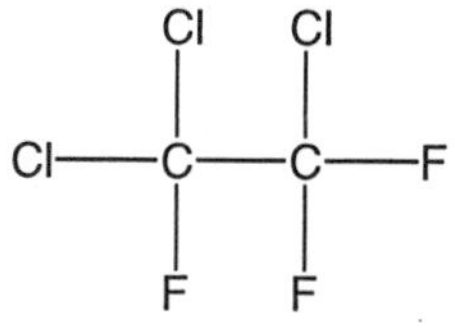

```
      Cl   Cl
      |    |
Cl----C----C----F
      |    |
      F    F
```

MW [37.37]; CAS [76-13-1]; colorless liquid with a characteristic odor; boils at 47.6°C; freezes at –35°C; vapor pressure 284 torr at 20°C; density 1.56 g/mL; insoluble in water; miscible with organic solvents; used as a refrigerant and dry cleaning solvent, and also in extracting oil, grease, and hydrocarbons in chemical analysis.

ANALYSIS OF AQUEOUS SAMPLES

An aliquot of sample is purged with He; the analyte adsorbed over a trap (Tenax or equivalent); the trap heated under He flow; the analyte desorbed from the trap and transported onto a GC column for separation from other volatile compounds and determination by a halogen specific detector (ECD or HECD) or a mass spectrometer.

- The characteristic ions for GC/MS identifications 101, 151, 103, 153, 85, and 66.
- Samples collected in glass containers without headspace, refrigerated, and analyzed within 14 days of collection.

AIR ANALYSIS

Air drawn through coconut shell charcoal (100 mg/50 mg); analyte desorbed into CS_2 (allowed to stand for 30 min) and analyzed by GC-FID (NIOSH Method 1020, 1987); flow rate 10 to 50 mL/min; sample volume 1 to 2 L.

Air collected in a cryogenic trap under liquid argon; trap warmed; sample transferred onto a precooled GC column; temperature programmed; separated on the column and determined by FID, halogen specific detector, or GC/MS.

(1 ppm freon-113 in air = 7.66 mg/m^3 at NTP)

3.41 HYDROGEN CYANIDE

Formula: HCN; MW 27.03; CAS [74-90-8]; occurs in the root of certain plants, beet sugar residues, coke oven gas, and tobacco smoke; released during combustion of wool, polyurethane foam, and nylon; produced when metal cyanides react with dilute mineral acids; colorless or pale liquid or a gas; odor of bitter almond; boils at 25.6°C; solidifies at –13.4°C; density of liquid 0.69 g/mL at 20°C and gas 0.95 (air = 1) at 31°C; soluble in water and alcohol, very weakly acidic; dangerously toxic and highly flammable (Patnaik, 1992).

ANALYSIS OF AQUEOUS SAMPLES

- Aqueous samples distilled directly without any acid treatment; HCN liberated and collected in NaOH solution; cyanide analyzed by colorimetric or by ion selective electrode method (See Cyanide, Chapter 2.6).
- Avoid acid treatment of samples, as it converts metal cyanides into HCN.

Test papers treated with p-nitrobenzaldehyde and K_2CO_3 produces reddish-purple stain due to HCN.

- Method suitable for detecting 10 ppm HCN.

Presence of HCN indicated by color change of 4-(2-pyridylazo) resorcinol-palladium in carbonate-bicarbonate buffer solution from intense red to yellow (Carducci et al., 1982).

- The method originally developed for determination of plants, tissue, and toxicological substances; no precision and accuracy data available for environmental samples of aqueous matrices.

AIR ANALYSIS

Air drawn through a solid sorbent tube containing soda lime (600 mb/200 mg); cyanide complex desorbed into deionized water; solution treated with N-chlorosuccinimide-succinimide oxidizing reagent; after several minutes standing barbituric acid-pyridine coupling reagent added; color development measured at 580 nm by a spectrophotometer; concentration determined from a calibration standard (NIOSH Method 6010, 1989); recommended flow rate 100 mL/min; sample volume 50 L.

- Oxidizing agent: To 10 g succinimide in 200 mL water, add 1 g N-chlorosuccinimide; stirred to dissolve; volume adjusted to 1 L; stable for 6 months if refrigerated.
- Coupling agent: To 6 g barbituric acid in 30 mL water, add 30 mL pyridine slowly with stirring; adjust volume to 100 mL; stable for 2 months if refrigerated.

Alternatively, 50 L air bubbled through 10 mL 0.1 *N* KOH solution at a rate of 0.5 L/min; the solution analyzed for CN^- by ion-selective electrode method (NIOSH Method 7904, 1984).

- $$\text{HCN} = \frac{(W_b - B_b) \times 1.04}{V} \text{ mg/m}^3$$

  where W_b and B_b are µg CN^- in sample bubbler and media blank bubbler, respectively, and V, the air volume (L) sampled (and HCN to CN^- stoichiometric factor 1.04).
- Metals interfere in the test, forming strong complexes with CN^-; such interference is eliminated by special treatment with EDTA.

Alternatively, air bubbled through an impinger containing 15 mL 0.2 *N* NaOH solution; HCN converted into NaCN; content of impinger transferred into a vial, rinsed with 2 mL 0.2 *N* NaOH (total volume 17 mL); sealed and heated overnight in an oven; NaCN hydrolyzed into sodium formate; contents cooled and diluted to 50 mL; formate ion analyzed by ion chromatography (Dolzine et al., 1982).

- Concentration of formate determined from a formate calibration standards in 0.068 *N* NaOH.
- µg HCN in sample = µg/mL formate ion × 0.60 × 50
  (factor for converting formate ion to HCN 0.60, sample solution 50 mL)
- mg HCN/m^3 air = µg HCN in sample/liters of air sampled.
- Separator column: strong base anion exchange resin; suppressor column: strong acid resin in hydrogen form; eluting solvent 0.005 *M* sodium borate.

3.42 HYDROQUINONE

Synonyms: 1,4-benzenediol, hydroquinol, *p*-dihydroxybenzene; Formula: $C_6H_6O_2$; Structure:

MW 110.11; CAS [123-31-9]; crystalline solid; melts at 170°C; boils at 286°C; low solubility in water (7%); dissolves readily in alcohol and ether; oxidizes slowly in air.

ANALYSIS OF AQUEOUS AND NONAQUEOUS SAMPLES

Aqueous samples serially extracted with methylene chloride; the extract concentrated and analyzed by GC/MS.

- The characteristic masses for GC/MS identification: 110 (primary ion), 81, 53, and 55.
- GC column: a fused silica capillary column, such as DB-5 or equivalent.

Soils, sediments, and solid wastes mixed with anhydrous Na_2SO_4; sonicated or Soxhlett extracted with methylene chloride; extract concentrated and analyzed by GC/MS.

AIR ANALYSIS

A measured volume of air drawn through a 0.8-mm cellulose ester membrane filter; the particulates of hydroquinone extracted with 1% acetic acid in water;

extract analyzed by HPLC using an UV detector at 290 nm (NIOSH Method 5004, 1984); recommended air flow rate 2 L/min; sample volume 100 L.

- HPLC column: M-Bondapak C-18 or equivalent (e.g., 25 cm × 4.6 mm ID Partisil 10-ODS) at ambient temperature, 400 to 600 psi; mobile phase 1% acetic acid in water, 1 mL/min; calibration standard: same as above.
- Sample stable for 2 weeks at ambient temperature.
- Acetic acid stabilizes the sample, preventing air oxidation.
- Working range is 0.5 to 10 mg/m^3 for a 100 L air sample.
- Hydroquinone could occur in air at ambient temperature as particulates and not vapor, as its vapor pressure is too low.

3.43 HYDROGEN SULFIDE

Formula: H_2S; MW 34.08; CAS [7783-06-4]; ocurs in natural gas and sewer gas; formed when metal sulfides react with dilute mineral acids; colorless gas with rotten egg odor; liquefies at –60.2°C; solidifies at 85.5°C; slightly soluble in water (4000 mg/L at 20°C); aqueous solution unstable, absorbs oxygen and decomposes to sulfur; highly toxic and flammable.

AIR ANALYSIS

Atmospheric monitoring of H_2S may be performed by commercially available photorateometric analyzer.

- Paper tape H_2S analyzers use paper strips treated with lead acetate; colored stains of PbS detected by photocells (Kimbell, 1982); color ranges from light yellow to silvery black; extremely high signal amplification required to measure stains far too light in color for rapid and continuous measurement of H_2S at low ppb level.

H_2S exposure may be monitored by diffusion type colorimetric dosimeters (such as Vapor Gard); color changes from white to brown-black; dosage exposure measured from the length of stain in the indicator tube.

(1 ppm H_2S in air = 1.39 mg/m^3 at NTP)

3.44 ISOPHORONE

Synonyms: 1,1,3-trimethyl-3-cyclohexene-5-one, isoacetophorone; Formula: $C_9H_{14}O$; Structure:

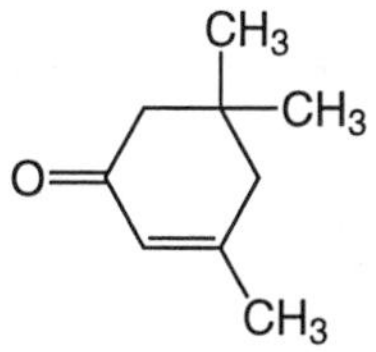

MW 138.20; CAS [78-59-1]; used as a solvent for vinyl resins and cellulose esters; boils at 215°C; solidifies at –8°C; density 0.92 g/mL at 20°C; slightly soluble in water; readily miscible with alcohol, ether, and acetone.

ANALYSIS OF AQUEOUS AND NONAQUEOUS SAMPLES

Aqueous samples extracted with methylene chloride; extract concentrated and solvent exchanged to hexane; florisil cleanup (for removal of interferences); extract analyzed on GC-FID or GC/MS.

Soils, sediments, or solid wastes extracted with methylene chloride by sonication or Soxhlett extraction; extract concentrated; exchanged to hexane (if florisil cleanup required); analyzed by GC-FID or GC/MS.

- For florisil cleanup, column eluted with methylene chloride/hexane (1:9) and then with acetone/methylene chloride (1:9) mixture; analyte eluted into the latter fraction.
- Column-packed: 1.95% QF-1/1.5% OV-17 on Gas-Chrom Q (80/100 mesh) or 3% SP-2100 on Supelcoport (100/120 mesh) or equivalent; capillary: fused silica capillary column, such as PTE-5, SPB-5, DB-5, Rtx-5, or equivalent.
- Characteristic masses for GC/MS identification: 82, 95, and 138.
- Suggested surrogate/IS: aniline d_5, 4-fluoroaniline, nitrobenzene d_5, and 1-fluoronaphthalene.
- Samples collected in glass containers, refrigerated, extracted within 7 days of collection and analyzed within 40 days of extraction.

3.45 METHANE

Synonym: marsh gas; Formula CH_4; MW 16.04; CAS [74-82-8]; prime constituent of natural gas; formed from petroleum cracking, decay of animal and plant remains, and anaerobic fermentation of municipality landfill contents; occurs in marshy pools, landfill gas, and leachate from the landfill; colorless and odorless gas; lighter than air; gas density 0.717 g/L; liquefies at –161.4°C; soluble in organic solvents, slightly soluble in water (25 mg/L); flammable gas.

ANALYSIS OF AQUEOUS SAMPLES AND SLUDGE DIGESTER GAS

Samples from wells are collected using a pump sufficiently submerged. Sludge digester gas samples are collected in sealed containers, such as glass sampling bulbs with three-way stopcocks. The containers are flushed with the digester gas or the gases from the aqueous samples to purge out air from the containers, prior to sample collection. Analytical methodologies are similar for both the aqueous and the sludge digester gas samples. Gas chromatographic method, discussed separately below under "Air Analysis," is also applicable.

A measured volume of sample transferred into an Orsat-type gas-analysis apparatus; sample equilibrated to atmospheric pressure by adjusting leveling bulb; CO_2 removed from sample by passing through KOH soln.; O_2 removed by passing through alkaline pyrogallol; H_2 removed by passing over heated CuO; CH_4 in the sample then oxidized to CO_2 and H_2O by passing through a catalytic oxidation assembly or a slow combustion pipet assembly by controlled electrical heating using a platinum filament; volume of CO_2 formed during combustion determined to measure the fraction of methane originally present.

Methane in aqueous sample determined by direct readout combustible gas detector, available commercially.

- The method determines the partial pressure of methane in the gas phase above the solution (Henry's law). Methane catalytically oxidizes on a heated platinum filament, that is part of a Wheatstone bridge. The heat generated increases the electrical resistance of the filament which is measured and compared against calibrated standards.

- Ethane, hydrogen, and other combustible gases interfere in the test. H_2S interference can be reduced by adding solid NaOH to the container before sampling.

AIR ANALYSIS

Air collected in Tedlar bag; an aliquot of sample injected onto the GC column at ambient temperature; determined by TCD or FID.

- Use TCD if the methane concentration is expected to be more than 2000 ppm. Also, the sample volume for packed column injection should be greater (~500 μL air). Detector response (1 to 2 mg).
- A lower detection level may be obtained using an FID.
- GC column: molecular sieve 5A, 13X Chromosorb 102 (80/100 mesh), Carboxen 1004, Haye Sep Q, Carbosieve, Carbosphere, silica gel, activated alumina, or any other equivalent material. Fused silica nonpolar capillary column may be used for low sample volume.

Alternatively, sample may be analyzed by GC/MS (using a capillary column) under cryogenic conditions.

- Characteristic mass: 15.

(1 ppm methane in air = 0.66 mg/m^3 at NTP)

3.46 METHYL BROMIDE

Synonym: bromomethane; Formula; (CH_3Br; MW 94.95; CAS [74-83-9]; colorless gas with a chloroform-like smell at high concentrations; liquefies at 3.5°C; slightly soluble in water; miscible with organic solvents.

ANALYSIS OF AQUEOUS SAMPLES

Aqueous samples subjected to purge and trap extraction; volatile analyte adsorbed on the sorbent trap thermally desorbed and swept with helium onto a GC column for separation; detected by HECD, ECD, FID or MSD; low retention time.

- Characteristic masses for GC/MS identification: 94 and 96.
- Limit of detection: in the range of 0.1 μg/L when detected by HECD for a 5-mL sample aliquot (subject to matrix interference).
- Recommended surrogate/IS: bromochloromethane and 1,2-dichloroethane d_4.
- Samples collected in glass/plastic containers without headspace, refrigerated, and analyzed within 14 days.

See Halogenated Hydrocarbons, Chapter 2.9 for GC columns and conditions and for a detailed discussion.

AIR ANALYSIS

The following methods are recommended:

Carbon molecular sieve adsorption; desorption at 350°C into a cryogenically cooled troop; flash evaporated onto a capillary column GC/MS system; recommended sample volume 10 L; flow rate 100 mL/min.

Collected in SUMMA passivated canister; transferred into a cryogenically cooled trap attached to a GC column; determined by ECD or MSD.

Precision and accuracy of the above methods not established.

(1 ppm methyl bromide in air = ~3.9 mg/m^3 at NTP)

3.47 METHYL CHLORIDE

Synonym: chloromethane; Formula: CH_3Cl; MW 50.49; CAS [74-87-3]; colorless gas with a faint sweet odor; freezes at –32.7°C; slightly soluble in water; miscible with organic solvents.

ANALYSIS OF AQUEOUS AND NONAQUEOUS SAMPLES

Aqueous samples subjected to purge and trap extraction; sample aliquot purged under helium flow; highly volatile methyl chloride transferred from aqueous matrix to vapor phase; absorbed on a sorbent trap; analyte thermally desorbed and swept onto a GC column for separation; detected by HECD, ECD, FID or MSD; low retention time.

Soils, solid waste, or sludges mixed with water or methanol; the aqueous extract or a solution of methanol extract spiked into water; subjected to purge and trap concentration and analyzed as above.

- Alternatively, sample thermally desorbed under helium flow and analyzed as above.
- Characteristic masses for GC/MS identification: 50 and 52.
- Limit of detection: in the range 0.1 μg/L when detected by HECD for a 5-mL sample aliquot (subject to matrix interference).
- Recommended surrogate/IS: bromochloromethane and 1,2-dichloroethane-d_4.
- Samples collected in glass/plastic containers without headspace, refrigerated and analyzed within 14 days.

See under Halogenated Hydrocarbons, Chapter 2.9 for GC columns and conditions and for a detailed discussion.

AIR ANALYSIS

Collected in a SUMMA passivated canister under pressure using an additional pump or at subatmospheric pressure by initially evacuating the canister; the air transferred into a cryogenically cooled trap attached to a GC column; the trap heated; the analyte separated on the column for detection by ECD, FID, or GC/MS (U.S. EPA Method TO-14).

Adsorbed over carbon molecular sieve (~400 mg); heated to 350°C under helium purge; transferred into a cryogenic trap and flash evaporated onto a capillary column GC/MS; recommended sample volume 10 L; flow rate 100 mL/min; the precision and accuracy data for the compound not established.

(1 ppm methyl chloride in air = ~2 mg/m^3 at NTP)

3.48 METHYLENE CHLORIDE

Synonyms: dichloromethane; Formula: CH_2Cl_2; MW 84.94; CAS [75-09-2]; a volatile halogenated hydrocarbon; widely used as a solvent; boils at 40°C; vapor pressure 349 torr at 20°C; density 1.323 g/mL at 20°C; solubility in water, very low (1.3%) miscible in organic solvents; nonflammable.

ANALYSIS OF AQUEOUS AND NONAQUEOUS SAMPLES

Aqueous samples extracted by purge and trap method; sample purged under helium flow; volatile analyte thermally desorbed out from sorbent trap and swept onto a GC column for separation; detected by HECD, ECD, FID or MSD.

Soils, solid wastes, or sludges mixed with water or methanol; the aqueous extract or a solution of methanol extract spiked into water; subjected to purge and trap concentration and analyzed as above.

- Alternatively, analyte thermally desorbed under helium flow and analyzed as above.
- Characteristic masses for GC/MS identification: 84, 49, 51, and 86.
- Limit of detection: in the range 0.3 μg/L when detected by HECD for a 5-mL sample aliquot (subject to matrix interference).
- Recommended surrogate/IS: bromochloromethane and 1,2-dichloroethane-d_4.
- Samples collected in glass/plastic containers without headspace, refrigerated, and analyzed within 14 days.

See Halogenated Hydrocarbons, Chapter 2.9 for GC columns and conditions and a detailed discussion.

AIR ANALYSIS

Air analysis may be performed by NIOSH, OSHA, or EPA methods. The latter give a lower range of detection.

Adsorbed over coconut charcoal (100 mg/50 mg); desorbed with CS_2 and analyzed by GC-FID; flow rate 10 to 200 mL/min; sample volume 0.5 to 2.5 L (NIOSH 1987); sample volume 10 L for 350 mg sorbent (OSHA Method 59).

Adsorbed over carbon molecular sieve (~400 mg) in a cartridge; heated at 350°C under helium purge; transferred into a cryogenic trap and flash evaporated onto a capillary column GC/MS; recommended flow rate 1 L/min; sample volume 60 L (U.S. EPA Method TO-2).

Collected in a trap in liquid argon; the trap removed and heated; sample transferred to a precooled GC column for ECD or FID detection (U.S. EPA Method TO-3).

Collected in a SUMMA passivated canister under pressure using an additional pump or at subatmospheric pressure by initially evacuating the canister; the air transferred into a cryogenically cooled trap attached to a GC column; the trap heated; the analyte separated on the column for detection by ECD, FID, or GC/MS (U.S. EPA Method TO-14).

(1 ppm methylene chloride in air = 3.47 mg/m^3 at NTP)

3.49 METHYL IODIDE

Synonym: iodomethane; Formula CH_3I; MW 142.94; CAS [74-88-4]; used as a methylating agent and in microscopy; colorless liquid, turns yellow or brown on exposure to light or moisture; boils at 42.5°C; vapor pressure 375 torr at 20°C; freezes at –66.5°C; decomposes at 270°C; density 2.28 g/mL; low solubility in water (2%); soluble in alcohol and ether; toxic and carcinogenic.

ANALYSIS OF AQUEOUS SAMPLES

A measured volume of sample purged by an inert gas; analyte trapped on a sorbent trap; transferred onto a GC column by heating the trap and backflushing with He; determined by GC/MS.

- The characteristic masses for GC/MS identification: 142 and 127.
- GC column: a fused silica capillary column such as DB-624, VOCOL, DB-5, or equivalent.
- A halogen specific detector such as HECD may be used in the GC system.
- Detection limit: in the range 0.02 mg/L for 25 mL sample volume.

AIR ANALYSIS

Air drawn through a solid sorbent tube containing coconut shell charcoal (100 mg/50 mg); the analyte desorbed with 1 mL toluene (more than 8 h standing); toluene extract analyzed by GC-FID (NIOSH Method 1014, 1985); recommended air flow rate 0.5 L/min; sample volume 30 L.

- GC column: Chromosorb 101 or equivalent.
- For a 50 L air sample, the working range of this method is 10 to 100 mg/m^3.

Suggested Method: Alternatively, air sample collected in a SUMMA passivated stainless steel canister pressurized using a pump; canister attached to the analytical system; analyte concentrated by collection in a cryogenically cooled trap; cryogen removed; temperature of trap raised; analyte revolatilized and separated on a GC column; determined by a mass spectrometer.

- No precision or accuracy data are available.
- A fused silica capillary column such as VOCOL, DB-5, or equivalent should give satisfactory performance.
- Methyl iodide may be identified from its retention time and the characteristic masses 142 and 127.

(1 ppm methyl iodide in air = 5.80 mg/m^3 at NTP)

3.50 METHYL ISOBUTYL KETONE

Synonyms: MIBK, hexone, isopropyl acetone, 4-methyl-2-pentanone; Formula $C_6H_{12}O$; Structure:

$$CH_3-\overset{\overset{\displaystyle O}{\|}}{C}-CH_2-\overset{\overset{\displaystyle CH_3}{|}}{CH}-CH_3$$

MW 100.16; CAS [108-10-1]; used as a solvent for gums, resins, oils, and waxes; colorless liquid; faint camphor-like odor; boils at 117°C; vapor pressure 7.5 torr at 25°C; freezes at –87.4°C; density 0.80 g/mL; solubility in water 19.1 g/L, soluble in most organic solvents; flammable.

ANALYSIS OF AQUEOUS AND NONAQUEOUS SAMPLES

Purge and trap extraction; MIBK purged out from aqueous matrix under He flow; thermally desorbed out from the sorbent trap and backflushed with He onto a GC column for separation from other volatile compounds; detected by a FID or a mass spectrometer.

- Characteristic masses for GC/MS identification: 43 (primary ion), 58 and 85.
- GC column: a 75m × 0.53 mm ID x 3 mm film thickness fused silica capillary column such as DB-624 or equivalent.
- Detection limit: in the range 0.25 mg/L for a 25 mL sample volume, detected by GC/MS.
- Samples collected in a glass container without headspace, refrigerated, and analyzed in 14 days.

Alternatively, a 100-mL aliquot of aqueous sample pH adjusted to 3; treated with 2,4-dinitrophenylhydrazine (DNPH); heated at 40°C for an hour under gentle swirling; DNPH derivative extracted with methylene chloride by liquid-liquid extraction; extract solvent exchanged to acetonitrile; determined by HPLC-UV at 360 nm.

- HPLC column: C-18 reverse phase column, such as Zorbax ODS or equivalent; mobile phase acetonitrile/water (70:30); flow rate 1 mL/min.
- The DNPH derivative may alternatively be extracted by solid-phase extraction on a sorbent cartridge, conditioned with 10 mL dilute 1 *M* citrate buffer (1:25 dilution) and 10 mL saturated NaCl solution. The extract loaded on the cartridge, and the derivative eluted with acetonitrile for HPLC analysis.
- The DNPH derivative of MIBK may be determined by GC instead of HPLC.

A measured amount of solid sample extracted with methanol; an aliquot of methanol extract spiked into a measured volume of reagent water (25 mL) and subjected to purge and trap extraction and GC/MS determination as above.

Alternatively, MIBK thermally desorbed out from the solid matrix under He purge (without any solvent treatment) and analyzed by GC or GC/MS.

AIR ANALYSIS

Air drawn through a solid sorbent tube packed with coconut shell charcoal (100 mg/50 mg); MIBK desorbed with 1 mL CS_2 (more than 30 min standing); analyzed by GC- FID (NIOSH Method 1300, 1989); recommended flow rate 0.1 L/min; sample volume 5 L.

- GC column: 10% SP-2100 and 0.1% Carbowax 1500 on Chromosorb WHP or a fused silica capillary such as DB-1, DB-5 or equivalent.

Alternatively, air drawn through an impinger solution containing 2 *N* HCl solution of 0.05% DNPH reagent and isooctane; the DNPH derivative of MIBK partitions into isooctane; aqueous layer extracted with hexane/methylene chloride mixture (70:30); extract combined with isooctane; the three solvent mixture evaporated to dryness under N_2 stream; residue dissolved in methanol; analyzed by HPLC-UV as above (U.S. EPA Method TO-5).

- Recommended air flow rate 200 mL/min; sample volume 20 L.
- Isooctane volume: 10 to 20 mL.

Alternatively, DNPH coated silica gel or florisil adsorbent may be used instead of impinger solution; derivative eluted with acetonitrile for HPLC determination.

(1 ppm MIBK in air = 4.10 mg/m^3 at NTP)

3.51 METHYL ISOCYANATE

Synonyms: isocyanic acid methyl ester, isocyanatomethane, MIC; Formula: CH_3NCO; MW 57.05; CAS [624-83-9]; used in the manufacture of carbamate pesticides; colorless liquid with an unpleasant odor; boils at 39°C; freezes at –80°C; vapor pressure 400 torr at 20°C; density 0.96 g/mL at 20°C; vapor density 1.97 (air = 1); decomposes in water; soluble in most organic solvents; highly toxic and flammable.

AIR ANALYSIS

Air drawn through a glass tube containing ion exchange resin XAD-2; MIC desorbed from the resin into a solution of fluorescamine (Fluram) in tetrahydrofuran; an intense fluorescent derivative formed; derivative analyzed by HPLC using a multiwavelength fluorescent detector (Vincent and Ketcham, 1980); recommended air flow rate 200 mL/min; sample volume 15 L.

- The detection limit 0.02 ppm.
- Chemical name for fluorescamine is 4-phenylspiro[furan-2(3H),1-phthalan]3,3′-dione.
- Silica gel may, alternatively, be used as adsorbent; the method sensitivity, however, is greater with Amberlite XAD-2.
- Monomethylamine interferes in the test; interference is removed by passing air through an impinger containing 0.5% $CuCl_2$ solution before drawing the air through the solid adsorbent.
- A reverse phase column, such as Varian CH-10 or equivalent, is suitable for separation.

(1 ppm MIC in air = 2.33 mg/m^3 at NTP)

3.52 METHYL METHACRYLATE

Synonym: methacrylic acid methyl ester; Formula: $C_5H_8O_2$; Structure:

$$H_2C{=}C(CH_3){-}C({=}O){-}OCH_3$$

MW 100.12; CAS [80-62-6]; used in the manufacture of plastics and resins; colorless liquid; polymerizes; boils at 100°C; vapor pressure 35 torr at 20°C; density 0.944 g/mL; soluble in benzene, chloroform, tetrahydrofuran, and methyl ethyl ketone.

ANALYSIS OF AQUEOUS SAMPLES

A measured volume of sample purged with an inert gas; analyte adsorbed on a sorbent trap; trap heated and backflushed with He to desorb the analyte into a capillary GC column interfaced to a mass spectrometer (MS); determined on the MS.

- Characteristic masses for GC/MS identification: 69 (primary ion) and 99.
- GC column: a fused silica capillary column, such as VOCOL, DB-624, or equivalent.
- Detection limit: in the range 0.5 mg/L when the sample volume is 25 mL.
- Methylmethacrylate polymerizes quickly. The sample must be collected without headspace, refrigerated, and analyzed immediately.

AIR ANALYSIS

Air drawn through a solid sorbent tube packed with XAD-2 (400 mg/200 mg); analyte desorbed with 2 mL CS_2; extract analyzed by GC-FID (NIOSH Method 2537, 1989); recommended air flow rate 0.05 L/min; sample volume 5 L.

- GC column: a fused silica capillary column; DB-1, 30 m × 0.25 mm × 1 mm or equivalent; packed column: 10% FFAP on Supelcoport (120/100 mesh) or equivalent.
- Analyte adsorbed on the trap is stable for 30 days, if refrigerated; stable for 7 days at 25°C.
- The working range is 10 to 1000 mg/m^3 for a 3 L air sample.

(1 ppm methylmethacrylate in air = 4.10 mg/m^3 at NTP)

3.53 NITROBENZENE

Synonyms: nitrobenzol, mirbane oil; Formula: $C_6H_5NO_2$; Structure:

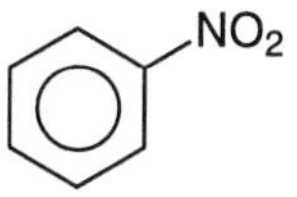

MW 123.12; CAS [98-95-3]; a pale yellow oily liquid; boils at 210°C; solidifies at –6°C; density 1.205 g/mL at 15°C; miscible in most organic solvents; slightly soluble in water (0.2%); toxic.

ANALYSIS OF AQUEOUS AND NONAQUEOUS SAMPLES

Aqueous samples extracted with methylene chloride; extract concentrated and solvent exchanged to hexane; florisil cleanup (for dirty samples); extract analyzed on GC using FID or NPD in N-mode or by GC/MS.

Soils, sediments, or solid waste samples extracted with methylene chloride by sonication or Soxhlett extraction; extract concentrated, exchanged to hexane (for florisil cleanup), and determined as above.

- Column-packed: 1.95% QF-1/1.5% OV-17 on Gas-Chrom Q (80/100 mesh), 3% OV-101 on Gas-Chrom Q (80/100 mesh), or 3% SP-2250 on Supelcoport; capillary: PTE-5, SPB-5, DB-5, Rtx-5 or equivalent.
- FID response: ~10 ng.
- Characteristic masses for GC/MS identification: 77, 123, and 65.
- Suggested surrogate/IS: aniline d_5, 4-fluoroaniline, and 1-fluoronaphthalene.
- Samples collected in glass containers, refrigerated, extracted within 7 days of collection and analyzed within 40 days of extraction.

AIR ANALYSIS

Analyte adsorbed over silica gel in a solid sorbent tube; desorbed with methanol in an ultrasonic bath for 1 h; analyzed by GC-FID using a suitable column (i.e.,

10% FFAP on Chromosorb W-HP) (NIOSH Method 2005, 1984); recommended flow rate 0.05 to 1 L/min; sample volume 50 to 150 L.

Adsorbed over Tenax (2 g) in a cartridge; cartridge heated under He purge; analyte transferred successively onto a cold trap and then to a precooled GC column; column heated; analyte eluted and determined by GC-FID or GC/MS; recommended flow rate 0.5 L/min; sample volume 100 L.

Collected in a trap in liquid argon; the trap removed and heated; sample transferred onto a precooled GC column for FID detection.

(1 ppm nitrobenzene in air = 5 mg/m^3 at NTP)

3.54 NITROGEN DIOXIDE

Synonyms: nitrogen peroxide; Formula: NO_2; MW 46.01; CAS [10102-44-0]; occurs in the exhausts of automobiles and in cigarette smoke; produced by the reaction of nitric acid with metals and decomposition of nitrates or during fire; reddish-brown fuming liquid or gas; sharp pungent odor; liquefies at 21°C; solidifies at –9.3°C; density of liquid 1.45 at 20°C; vapor 1.58 (air = 1); reacts with water to form nitric acid and nitrogen oxide; reacts with alkalies to form nitrates and nitrites; highly toxic.

AIR ANALYSIS

Analysis may be performed by passive indicator tubes, various passive samplers, and electrochemical instruments.

Air drawn through Palmes tube with three triethanolamine (TEA)-treated screens; analyte converted into nitrite ion (NO_2^-); NO_2^- treated with an aqueous solution of a reagent mixture containing sulfanilamide, H_3PO_4, and *N*-1-naphthylethylenediamine dihydrochloride; color develops; absorbance measured at 540 nm by a spectrophotometer; concentration determined from a standard calibration curve made from $NaNO_2$ (NIOSH Method 6700, 1984).

- Alternatively, a bubbler may be used instead of Palmes tube; NO_2^- measured by colorimetry as above.

Concentration of NO_2 may be measured directly by passive colorimetric dosimeter tubes (i.e., Vapor Guard); NO_2 in air diffuses into the tube; color changes (white to brown in Vapor Guard); concentration determined from the length of stain.

- Temperature and relative humidity may affect measurement; tubes should be used between 15 and 40°C.

(1 ppm NO_2 = 1.88 mg/m^3 at NTP)

3.55 PENTACHLOROPHENOL

Synonym: pentachlorophenate; Formula: C_6Cl_5OH; Structure:

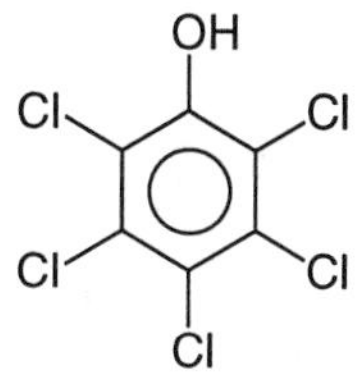

MW 166.32; CAS [87-86-5]; crystalline solid melts at 190°C; insoluble in water (14 ppm at 20°C), soluble in organic solvents.

ANALYSIS OF AQUEOUS AND NONAQUEOUS SAMPLES

Aqueous samples pH adjusted below 2; extracted with methylene chloride; extract concentrated and analyzed by GC-FID or GC/MS.

Alternatively, extract exchanged into 2-propanol; derivatized with pentafluorobenzyl bromide; derivative dissolved in hexane; fractionated over silica gel; eluted with 15% toluene in hexane and determined by GC-ECD.

Aqueous samples may be analyzed by HPLC using a C-18 derivatized reverse phase column or on an underivatized polystyrene-divinylbenzene column such as Poly-RP CO (Alltech 1995); gradient: acetonitrile and 0.01 *M* K_3PO_4 at pH 7 (55:45) and the analyte detected by UV at 254 nm.

Soils, solid wastes, and sludges extracted with methylene chloride by sonication, Soxhlett or supercritical fluid extraction; extract may be acid washed, concentrated, and analyzed by GC-FID, GC-ECD, or GC/MS as above.

- Characteristic masses for GC/MS determination: 266, 264, and 268 (electron impact ionization); 267, 265, and 269 (chemical ionization).
- Limit of detection: in the range 5 to 10 μg/L on FID or GC/MS and below 1 μg/L on ECD (for aqueous samples concentrated by 1000 times).

- Recommended surrogate/IS: pentafluorophenol and 2-perfluoromethyl phenol.
- Samples collected in glass containers, refrigerated, and extracted within 7 days of collection and analyzed within 40 days of extraction.

See Phenols, Chapter 2.23 for GC columns and conditions.

AIR ANALYSIS

Not likely to occur in ambient air in vapor state; the vapor pressure too low (0.00017 torr at 20°C).

Airborne particles collected on a sorbent cartridge containing polyurethane foam; extracted with 5% diethyl ether in hexane; extract analyzed by GC-ECD; recommended air flow 5 L/min; sample volume 1000 L (U.S. EPA Method TO-10).

3.56 PHOSGENE

Synonyms: carbonyl chloride, carbon oxychloride, chloroformyl chloride; Formula: $COCl_2$; Structure:

$$\mathrm{Cl_2C{=}O}$$

MW 98.91; CAS [75-44-5]; formed in air in trace amounts by photodecomposition of chlorinated solvent vapors; used in the synthesis of dyes and pesticides, and as a war gas; colorless; suffocating odor at low ppm concentration; sweet hay-like odor at low ppb concentration; liquefies at 8°C; solidifies at –118°C; density 1.43 g/mL at 0°C; slightly soluble in water (reacting slowly), soluble in hexane, benzene, and glacial acetic acid; highly poisonous.

AIR ANALYSIS

Air drawn through a midget impinger containing 10 mL of 2% aniline in toluene (by volume); phosgene reacts with aniline forming carbanilide (1,3-diphenylurea); solvent evaporated at 60°C under N_2 flow; residue dissolved in 1 mL acetonitrile; carbanilide analyzed by reverse-phase HPLC with an UV detector set at 254 nm (EPA Method T06); recommended flow rate 200 mL/min; sample volume 20 L.

- A C-18 reverse-phase column such as Zorbax ODS or equivalent may be used.
- Chloroformates and acidic substances (at high concentrations) may interfere in the test.
- Detection limit: in the range 0.1 ppb.

Air drawn through an impinger containing 4,4′-nitrobenzyl pyridine in diethyl phthalate; phosgene forms a colored derivative; absorbance measured by a spectrophotometer.

- Detection limit: in the range 10 ppb.

Air drawn through a midget impinger containing a 10% solution of equal parts of ρ-dimethylaminobenzaldehyde and diphenylamine in carbon tetrachloride; phosgene reacts forming a deep orange derivative; absorbance measured by a spectrophotometer and compared against a standard calibration curve.

- The method applicable to measure low ppm concentration.

(1 ppm phosgene in air = 4 mg/m^3 at NTP)

3.57 PYRIDINE

Synonyms: azabenzene, azine; Formula: C_5H_5N; Structure:

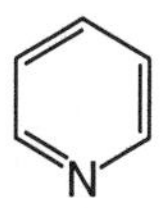

CAS [110-86-1]; used as a solvent and an intermediate in organic synthesis; colorless liquid; characteristic disagreeable odor; boils at 115°C; freezes at –41.5°C; density 0.98 g/ml; miscible with water and most organic solvents; weak base; toxic and flammable.

ANALYSIS OF AQUEOUS AND NONAQUEOUS SAMPLES

A measured volume of aqueous sample pH adjusted to >11 and then mixed and shaken repeatedly with methylene chloride; pyridine being basic partitions into the organic layer while acidic compounds partition into the basic aqueous phase; aqueous layer discarded; organic solvent extract concentrated and separated on a GC column; determined by a mass spectrometer, a NPD or a FID.

A measured quantity of soil, sediment, or solid waste sample extracted with a measured volume of water; aqueous solution pH adjusted to >11 and serially extracted with methylene chloride; pyridine partitions into the organic phase, which is then concentrated and analyzed as above.

- Characteristic mass for GC/MS identification: 79.
- GC column: silicone-coated fused silica capillary column, such as DB-5 or equivalent.
- Precision, accuracy, and the detection limit for pyridine have not been determined.

AIR ANALYSIS

Air drawn through a solid sorbent tube packed with coconut shell charcoal (100 mg/50 mg); pyridine desorbed with 1 ml methylene chloride (more than 30 min standing); eluant analyzed by GC-FID (NIOSH method 1613, 1987); recommended air flow rate 0.5 L/min; sample volume 100 L.

- GC column: 3 m × 3 mm stainless steel packed with 5% Carbowax 20M on 80/100 mesh acid-washed DMCS Chromosorb W or a silicon coated fused silica.
- Capillary column (e.g., DB 1) may be used. The working range for this method was determined to be 1 to 14 ppm for a 100 L air sample.

(1 ppm pyridine in air = 3.23 mg/m^3 at NTP)

3.58 PYROCATECHOL

Synonyms: 2-hydroxyphenol, *o*-diphenol, 1,2-benzenediol, *o*-dihydroxybenzene; Formula: $C_6H_5O_2$; Structure:

An isomer of resorcinal; MW 110.12; CAS [120-80-9]; used in photography and dyeing fur; crystalline solid; melts at 105°C; soluble in water and most organic solvents.

ANALYSIS OF AQUEOUS AND NONAQUEOUS SAMPLES

Recommended Method:

Aqueous samples pH adjusted below 2; extracted with methylene chloride; extract concentrated and analyzed by GC-FID or GC/MS.

- Alternatively, extract exchanged into 2-propanol; derivatized with pentafluorobenzyl bromide, and determined by GC-ECD.

Soils, solid wastes, and sludges mixed with anhydrous Na_2SO_4 and extracted with methylene chloride by sonication or Soxhlett extraction; extract may be acid washed, concentrated, and analyzed by GC-FID, GC-ECD, or GC/MS as above.

- Characteristic masses for GC/MS determination: 110, 81, and 82.
- Limit of detection: in the range 100 μg/L for aqueous samples concentrated by 1000 times; GPC cleaned and determined by GC/MS.
- Recommended surrogate/IS: 2-fluorophenol and pentafluorophenol.
- Samples collected in glass containers, refrigerated, and extracted within 7 days of collection and analyzed within 40 days of extraction.

See Phenols, Chapter 2.23 for GC columns and conditions.

AIR ANALYSIS

Recommended Method: airborne particles collected on a sorbent cartridge containing polyurethane foam; particulates deposited on the filter extracted with methylene chloride; extract analyzed by GC-FID or GC/MS; recommended air flow 5 L/min; sample volume 1000 L.

3.59 PYROGALLOL

Synonyms: 1,2,3-trihydroxybenzene; pyrogallic acid; 1,2,3-benzenetriol; Formula: $C_6H_5O_3$; Structure:

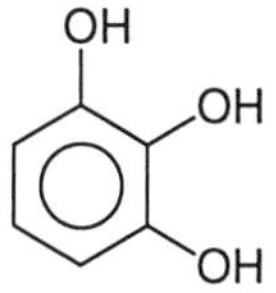

MW 126.12; CAS [87-66-1]; white crystalline solid; melts at 132°C; soluble in water and most organic solvents.

ANALYSIS OF AQUEOUS AND NONAQUEOUS SAMPLES

Aqueous samples pH adjusted below 2; extracted with methylene chloride; extract concentrated and analyzed by GC-FID or GC/MS.

- Alternatively, extract exchanged into 2-propanol; derivatized with pentafluorobenzyl bromide and determined by GC-ECD.
- Aqueous samples directly injected onto the GC column for FID determination.

Soils, solid wastes, and sludges extracted with methylene chloride by sonication, Soxhlett, or supercritical fluid extraction; extract may be acid washed, concentrated, and analyzed by GC-FID, GC-ECD, or GC/MS as above.

- Recommended surrogate/IS: 2-fluorphenol and pentafluorophenol.
- Samples collected in glass containers, refrigerated, and extracted within 7 days of collection and analyzed within 40 days of extraction.

See Phenols, Chapter 2.23 for GC columns and conditions.

AIR ANALYSIS

Not likely to occur in ambient air in vapor state; vapor pressure too low.

Recommended Method to analyze airborne particles: Air drawn through a sorbent cartridge containing polyurethane foam; deposited particles extracted with methylene chloride; extract analyzed by GC-FID or GC/MS; recommended air flow 5 L/min; sample volume 1000 L.

3.60 RESORCINOL

Synonyms: *m*-hydroxyphenol; 3-hydroxyphenol; 1,3-benzenediol; 1,3-dihydroxybenzene, Formula: $C_6H_6O_2$; Structure:

OH

OH

An iosomer of pyrocatechol; MW 110.12; CAS [108-46-3]; used in the manufacture of resins, dyes, and explosives; white crystalline solid; melts at 110°C; boils at 214°C; soluble in water, alcohol, and ether; slightly soluble in chloroform.

ANALYSIS OF AQUEOUS AND NONAQUEOUS SAMPLES

Aqueous samples pH adjusted below 2; extracted with methylene chloride; extract concentrated and analyzed by GC-FID or GC/MS.

- Alternatively, extract exchanged into 2-propanol; derivatized with pentafluorobenzyl bromide and determined by GC-ECD.

Soils, solid wastes, and sludges mixed with anhydrous Na_2SO_4, extracted with methylene chloride by sonication or Soxhlett extraction; extract may be acid washed, concentrated, and analyzed by GC-FID, GC-ECD, or GC/MS as above.

- Characteristic masses for GC/MS determination: 110, 81, 82, 53, and 69
- Limit of detection: in the range 100 µg/L for aqueous sample concentrated by 1000 times; extract GPC cleaned and determined by GC/MS.
- Recommended surrogate/IS: 2-fluorophenol and pentaflurophenol.
- Samples collected in glass containers, refrigerated, and extracted within 7 days of collection and analyzed within 40 days of extraction.

See Phenols, Chapter 2.23 for GC columns and conditions.

AIR ANALYSIS

Recommended Method: airborne particles collected on a sorbent cartridge containing polyurethane foam; deposited particles extracted with methylene chloride; extract analyzed by GC-FID or GC/MS; recommended air flow 5 L/min; sample volume 1000 L.

3.61 STIBINE

Synonyms: antimony hydride, hydrogen antimonide; Formula SbH_3; MW 124.78; CAS [7803- 52-3]; used as a fumigating agent; colorless gas with a disagreeable odor; decomposes slowly at ambient temperature; liquefies at −18°C; gas density 5.515 g/L; slightly soluble in water, dissolves in organic solvents; highly toxic; flammable gas.

AIR ANALYSIS

Air drawn through a solid sorbent tube containing $HgCl_2$-coated silica gel (1000 mg/500 mg); stibine extracted out from the sorbent trap with 15 ml conc. HCl (over 30 min swirling); acid extract diluted to a measured volume; an aliquot of the sample extract treated with ceric sulfate, followed by isopropyl ether and water; mixture transferred into a separatory funnel and shaken; aqueous layer discarded; ether extract treated with Rhodamine B soln. and shaken; allow phases to separate; an aliquot of organic phase centrifuged; color developed due to antimony measured by a spectrophotometer at 552 nm (NIOSH method 6008, 1987); recommended air flow rate 0.1 L /min; sample volume 20 L.

- Concentration of antimony is read from a standard calibration curve.
- Antimony in HCl extract may also be measured by atomic absorption or ICP emission spectrometry.
- Certain metals such as gold, iron(III), thallium (I), and tin(II) at high concentrations interfere in the above colorimetric analysis (iron (III) >30,000 mg, while the other metals >1000 mg).
- Color forming reagent, Rhodamine B soln. should have a strength of 0.01% in 0.5 *M* HCl.
- The working range is 0.02 to 0.7 ppm for a 20 L air sample.
- Sample is stable at least for 7 days at 25°C.
- Use a cellulose ester membrane if particulate antimony compounds are suspected to be present in the air.
- Activated charcoal absorbent showed a poor collection efficiency.

(1 ppm stibine in air = 5.10 mg/m^3 at NTP)

3.62 STRYCHNINE

Synonym: strychnidin-10-one; Formula: $C_{21}H_{22}N_2O_2$; Structure:

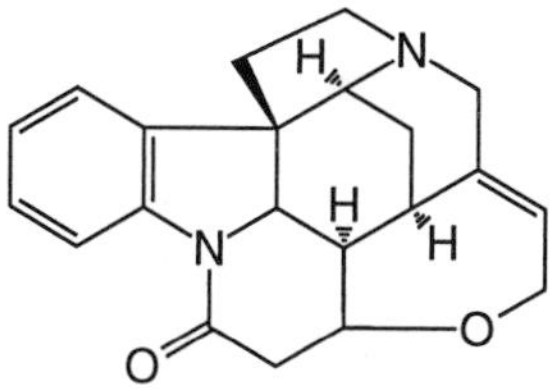

MW 334.40; CAS [57-24-9]; used as a rodent poison; white crystalline powder; highly bitter, melts at 268°C; solubility in water (0.16 g/L); moderately soluble in hot alcohol; dissolves readily in chloroform; highly toxic.

ANALYSIS OF AQUEOUS AND NONAQUEOUS SAMPLES

A measured volume of aqueous sample acidified to pH <2 and serially extracted with methylene chloride; strychnine being basic partitions into aqueous phase; organic layer discarded; aqueous phase pH adjusted to >10 and serially extracted with methylene chloride; solvent extract concentrated; analyzed by GC/MS.

- Characteristic masses for GC/MS identification: 334, 335, and 333.
- GC column: a fused silica capillary column, such as DB-5, SPB-5, or equivalent; the compound eluted relatively at a long retention time.

A measured quantity of solid sample mixed with anhydrous Na_2SO_4; the mixture sonicated or Soxhlett extracted with methylene chloride or chloroform; solvent extract concentrated; cleanup (if required) by acid-base partitioning; analyzed as above by GC/MS using an appropriate column.

Alternatively, a measured quantity of solid sample extracted with aqueous 1-heptane sulfonic acid and acetonitrile; the solution analyzed by HPLC using an UV detector at 254 mm (HPLC column and conditions given below under "Air Analysis").

- No precision and accuracy study data are available for the above method.

AIR ANALYSIS

Air drawn through a 37-mm glass fiber filter; strychnine deposited on the filter, desorbed with 5 mL mobile phase for HPLC-UV detection at 254 mm (NIOSH method 5016, 1985); recommended air flow rate 2 L/min; sample volume 500 L.

- The mobile phase for HPLC analysis, as well as the solvent media for desorption-aqueous 1-heptane sulfonic acid and acetonitrite mixture; pH 3.5; flow rate 1 mL/min at ambient temperature.
- HPLC column: packed with M-Bondapak C-18, 10-mm particle size.
- The working range for the above method is 0.05 to 10 mg/m^3 for a 200-L air sample.

3.63 STYRENE

Synonyms: vinylbenzene, phenylethylene, ethenylbenzene; Formula: C_8H_8; Structure:

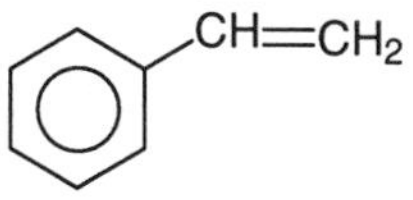

C_6H_5–CH=CH$_2$

MW 104.14; CAS [100-42-5]; used for manufacturing plastics, resins, and synthetic rubber; colorless to yellowish oily liquid; penetrating odor; slowly polymerizes on exposure to light and air; boils at 145°C; vapor pressure 6.1 torr at 25°C; freezes at –31°C; density 0.906 g/ml at 20°C; sparingly soluble in water, miscible with most organic solvents; forms peroxide.

ANALYSIS OF AQUEOUS AND NONAQUEOUS SAMPLES

A measured volume of sample purged with an inert gas; styrene purged out from water; absorbed on a sorbent trap; trap heated and back flushed with He; analyte transported into the GC column; separated from other volatiles; determined on a PID, FID or a mass spectrometer.

Solid samples mixed with methanol; an aliquot of methanol extract spiked into a measured volume of water (25 mL) in the purging vessel; subjected to purge and trap concentration and analyzed as above.

- GC column: a fused silica capillary column, such as VOCOL, DB-624, or equivalent; packed column: 1% SP-1000 on Carbopack B (60/80 mesh) or equivalent.
- Characteristic masses for GC/MS identification: 104, 103, 78, and 51.
- Recommended surrogate/IS: ethyl benzene-d_5 (m/z 114, 63, 88).
- Limit of detection: in the range 0.5 mg/L for a 5-mL sample volume, when detected by PID.
- Samples collected in glass containers without headspace, refrigerated, and analyzed immediately. Because styrene undergoes polymerization, oxidation, and addition reactions, exposure to sunlight or air should be avoided,

and analysis should be done immediately. Preservation of sample with HCI is not recommended due to its formation of an additional product.

AIR ANALYSIS

Air drawn through a sorbent tube packed with coconut shell charcoal (100 mg/50 mg); analyte desorbed into CS_2 (more than 30 min standing); the eluant analyzed by GC-FID (NIOSH method 1501, 1984); recommended air flow rate 100 mL/min; sample volume 20 L.

- GC column: packed with 10% OV-275 on 100/120 mesh Chromosorb W-AW, Porapak P (50/80 mesh), or equivalent.

Alternatively, air drawn through a cartridge packed with carbon molecular sieve (0.5 g); cartridge heated at 350°C under He purge; analyte transferred to the front of a precooled GC column; temperature programmed; styrene determined on a PID, FID, or a mass spectrometer; recommended air flow rate 0.5 L/min; sample volume 50 L.

- No precision and accuracy data available for this method.

(1 ppm styrene in air = 4.26 mg/m^3 at NTP)

3.64 SULFUR DIOXIDE

Formula: SO_2; MW 64.06; CAS [7446-09-5]; a major air pollutant; produced when soft coals, oils, and automobile fuels burn; used as a fumigating and bleaching agent; colorless gas with a strong suffocating odor; liquefies at –10°C; solidifies at –72°C; soluble in water (8.5% at 25°C), alcohol, ether, and chloroform; highly toxic and a strong irritant.

AIR ANALYSIS

Air drawn through two 0.8 μm cellulose ester membranes: (1) the front filter to collect any interfering particulate sulfate and sulfite in the air, and (2) the back filter, treated with KOH to trap SO_2; SO_2 converted to K_2SO_3 on the back filter; this filter treated with 10 mL mixture solution of 3 mmol $NaHCO_3$ and 2.4 mmol Na_2CO_3; K_2SO_3 dissociates into SO_3^{2-} anion, which is then determined by ion chromatography (NIOSH Method 6004, 1989); recommended flow rate 1 L/min; sample volume 100 L.

- Use SO_3^{2-} calibration standard to determine the mass of SO_3^{2-}; a blank analysis performed on the media blanks for background correction.
- Multiply SO_3^{2-} concentration by the stoichiometric factor 0.8 to calculate the concentration of SO_2.
- Bicarbonate-carbonate eluant of the back filter may also be oxidized with a drop of 30% H_2O_2 to SO_4^{2-}; SO_4^{2-} formed measured by ion chromatography; to calculate the concentration of SO_2, multiply the mass of SO_4^{2-} found in the analysis, multiplied by the stoichiometric factor 0.667.

Alternatively, air drawn through an impinger solution containing 0.3 *N* H_2O_2; SO_2 converted to SO_4^{2-} which is measured by titration with NaOH or barium perchlorate (NIOSH Method S308, 1978).

Alternatively, air drawn though a solid sorbent tube containing molecular sieve 5A; SO_2 adsorbed on this trap desorbed on heating under He purge; transported onto GC column and determined by GC/MS.

- Characteristic mass for SO_2 is 64.

A passive colorimetric dosimeter tube may also be used for direct measurement; SO_2 diffuses into the tube; color changes (red-purple to yellow in Vapor Guard tube); concentration determined from the length of stain.

- Temperature and relative humidity may affect measurement; tubes should be used between 15 and 40°C.

(1 ppm SO_2 = 2.62 mg/m^3 at NTP)

3.65 TETRACHLORETHYLENE

Synonyms: perchloroethylene, tetrachloroethene, ethylene tetrachloride; Formula: $Cl_2C{=}CCl_2$; MW 165.82; CAS [127-18-4]; used in dry cleaning and metal degreasing; colorless liquid with ether-like odor; boils at 121°C; freezes at –22°C; vapor pressure 19 torr at 25°C; density 1.62 g/mL at 20°C; insoluble in water; miscible with organic solvents.

ANALYSIS OF AQUEOUS AND NONAQUEOUS SAMPLES

Aqueous samples extracted by purge and trap method; analyte thermally desorbed out of trap and swept onto a GC column for separation from other volatile compounds; detected by HECD, ECD, FID, or MSD.

Soils, solid wastes, or sludges mixed with water or methanol; the aqueous extract or a solution of methanol extract spiked into water; subjected to purge and trap concentration and analyzed as above.

Methanol or hexane extract may be directly injected for GC (ECD or FID) or GC/MS determination.

- Characteristic masses for GC/MS identification: 164, 129, 131, and 166.
- Limit of detection: in the range 0.05 µg/L when detected by HECD for a 5-mL sample aliquot (subject to matrix interference).
- Recommended surrogate/IS: 2-bromo-1-chloropropane and 1,4-dichlorobutane.
- Samples collected in glass/plastic containers without headspace, refrigerated, and analyzed within 14 days.

See Halogenated Hydrocarbons, Chapter 2.9 for GC columns and conditions and a further discussion.

AIR ANALYSIS

Adsorbed over coconut charcoal (100 mg/50 mg); desorbed with CS_2 and analyzed by GC-FID (NIOSH Method 1003, 1987); recommended flow rate 100 mL/min; sample volume 3 L.

Adsorbed over Tenax (~2 g) in a cartridge; desorbed by heating under He purge; transferred into a cold trap and then to the front of a GC column at –70°C; column heated; analyte detected by ECD or MSD (U.S. EPA Method TO-1); recommended flow rate 100 mL/min; sample volume 10 L.

Adsorbed over carbon molecular sieve (~400 mg) in a cartridge; heated at 350°C under He purge; transferred into a cryogenic trap and flash evaporated onto a capillary column GC/MS (U.S. EPA Method TO-2); recommended flow rate 1 L/min; sample volume 100 L.

Collected in a trap in liquid argon; the trap removed and heated; sample transferred to a precooled GC column for ECD or FID detection (U.S. EPA Method TO-3).

Collected in a SUMMA passivated canister under pressure using an additional pump or at subatmospheric pressure by initially evacuating the canister; the air transferred into cryogenically cooled trap attached to a GC column; the trap heated; the analyte separated on the column for detection by ECD or GC/MS (U.S. EPA Method TO-14).

(1 ppm tetrachloroethylene in air = ~6.8 mg/m^3 at NTP)

3.66 TETRAETHYLLEAD

Synonym: tetraethylplumbane, lead tetraethyl; Formula: $Pb(C_2H_5)_4$; MW 323.47; CAS [78-00-2]; used in motor gasoline as an additive to prevent "knocking"; such an application, however, is currently curtailed because of environmental pollution; boils at 200°C; vapor pressure 0.2 torr at 20°C; density 1.653 g/mL at 20°C; insoluble in water; slightly soluble in alcohols; dissolves in benzene, toluene, hexane, petroleum ether, and gasoline; highly toxic (Patnaik, 1992).

ANALYSIS OF AQUEOUS AND NONAQUEOUS SAMPLES

Aqueous samples may be microextracted with toluene, benzene, or hexane and the extract analyzed by GC/MS.

- The m/z of primary characteristic ion 237; the secondary ions for compound identification 295, 208, 235, and 266 (electron impact ionization); peak intensity ratios for m/z 237:295:208 = 100:73:61 (Hites, 1992).
- GC column: a fused silica capillary column such as DB-5, SPB-5, Rtx-5, or equivalent.

Soils, sediments, or other nonaqueous matrices mixed with anhydrous Na_2SO_4 and extracted by sonication; the solvent extract analyzed as above by GC/MS.

- If lead analysis is performed by AA spectrophotometry, following acid digestion of samples, then the stoichiometric calculation for tetraethyllead (TEL) may be done as follows:

$$\text{conc. of TEL} = \text{conc. of Pb} \times 1.56$$

(assuming all Pb in sample occurs as TEL; such an assumption, however, could be erroneous.)

3.67 TETRAETHYL PYROPHOSPHATE

Synonyms: pyrophosphoric acid tetraethyl ester, TEPP; Formula: $C_8H_{20}O_7P_2$; Structure:

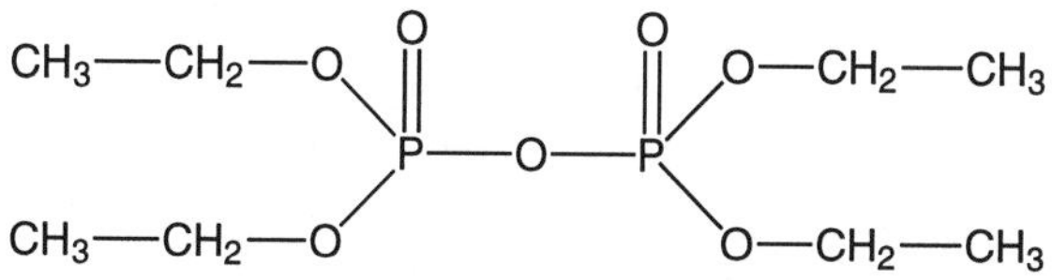

$$\begin{array}{c} CH_3\text{—}CH_2\text{—}O \quad \overset{O}{\underset{}{\|}} \qquad \overset{O}{\underset{}{\|}} \quad O\text{—}CH_2\text{—}CH_3 \\ P\text{—}O\text{—}P \\ CH_3\text{—}CH_2\text{—}O \qquad\qquad\qquad O\text{—}CH_2\text{—}CH_3 \end{array}$$

MW 290.19; CAS [107-49-3]; an organophosphorus pesticide; colorless liquid; hygroscopic; boiling point 124°C at 1 mm Hg; vapor pressure 0.00047 torr at 20°C; decomposes at 170 to 213°C; density 1.185 g/ml at 20°C; miscible with most organic solvents; miscible with water but rapidly hydrolyzed; highly toxic.

ANALYSIS OF AQUEOUS AND NONAQUEOUS SAMPLES

Aqueous samples (TEPP unstable) extracted serially with methylene chloride; the extract concentrated and analyzed by GC/MS.

- Characteristic masses for GC/MS identifications: 99, 155, 127, 81, and 109.
- GC column: a fused silica capillary column, such as DB-5 or equivalent.

Soil, sediment, or solid waste sample mixed with anhydrous Na_2SO_4; sonicated or Soxhlett extracted with methylene chloride; extract concentrated and analyzed by GC/MS as above.

AIR ANALYSIS

Air drawn through a solid sorbent tube containing Chromosorb 102 (100 mg/50 mg); analyte desorbed with 1 ml toluene (on 60 min standing); toluene extract analyzed by GC-FPD (NIOSH method 2504, 1984); recommended flow rate 0.1 L/min; sample volume 20 L.

- GC column: Super-Pak 20M or equivalent.
- Calibration standards are made from TEPP in toluene.
- The working range of the method was determined to be 0.025 to 0.15 mg/m^3 for a 40 L sample.

(1 ppm TEPP in air = 11.86 mg/m^3 at NTP)

3.68 TETRAHYDROFURAN

Synonyms: oxolane, 1,4-epoxybutane, oxacyclopentane; Formula: C_4H_8O; Structure:

$$\begin{array}{ccc} H_2C & \text{---} & CH_2 \\ | & & | \\ H_2C & \diagdown O \diagup & CH_2 \end{array}$$

MW 72.12; CAS [109-99-9]; used as a solvent; colorless liquid; ether-like odor; boils at 66°C; vapor pressure 145 torr at 20°C; freezes at –188.5°C; density 0.89 g/ml at 20°C; miscible with water and most organic solvents; highly flammable.

ANALYSIS OF AQUEOUS SAMPLES

Aqueous samples extracted by purge and trap method; the sorbent trap heated and back flushed with He to transfer the analyte onto a capillary GC column interfaced to a mass spectrometer; determined by GC/MS.

- Because of high solubility of tetrahydrofuran in water, its purging efficiency is low. The purging vessel should be heated.
- The characteristic masses for GC/MS identification: 71 (primary), 72, and 42.
- GC column: a fused silica capillary column such as DB-624, 75 m × 0.53 mm × 3 mm, or equivalent.

Alternatively, aqueous samples directly injected onto the GC column for FID determination.

- The detection range on FID without sample concentration is several order higher than the purge and trap method (under heating or purging vessel).
- No precision and accuracy data are available for both the above methods.

AIR ANALYSIS

Air drawn through a solid sorbent tube containing coconut shell charcoal (100 mg/ 50 mg); analyte desorbed into 0.5 mL CS_2 (on 30 min contact time); the CS_2 extract analyzed by GC-FID (NIOSH method 1609, 1985); recommended air flow rate 0.1 L/min; sample volume 5 L.

- GC column: stainless steel column packed with Porapak Q (50/80 mesh). A fused silica capillary column such as DBWAX or DB-624 may also be used.
- High humidity may affect the measurement.
- For a 5 L air sample, working range of the method was determined to be 100 to 2600 mg/m^3.

(1 ppm tetrahydrofuran in air = 2.95 mg/m^3 at NTP)

3.69 TOLUENE

Synonyms: methylbenzene, phenylmethane. Formula: C_7H_8; Structure:

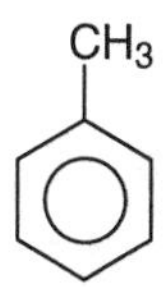

MW 92.15; CAS [108-88-3]; occurs in gasoline, petroleum solvents, and coal-tar distillates; used as a solvent and in many organic syntheses; colorless liquid with a characteristic aromatic odor; boils at 110.7°C; freezes at –95°C; vapor pressure 22 torr at 20°C; density 0.866 g/mL at 20°C; slightly soluble in water (0.63 g/L at 25°C); readily miscible with organic solvents.

ANALYSIS OF AQUEOUS AND NONAQUEOUS SAMPLES

Purge and trap method; toluene bound onto the trap, thermally desorbed from the sorbent trap, and swept by an inert gas onto a GC column for separation from other volatile compounds; detected by PID, FID, or a mass spectrometer.

Solid samples mixed with water or methanol; the aqueous extract or a portion of methanol extract spiked into water; subjected to purge and trap concentration and analyzed as above.

Alternatively, toluene thermally desorbed from the solid sample under He purge (without any solvent treatment) and analyzed by GC or GC/MS as above.

- Characteristic masses for GC/MS identification: 91, 92, 39, 65; peak ratios 91:92:39:65 = 100:73:20:14 (Hites, 1992).
- Limit of detection: in the range 0.5 μg/L when detected by PID for a 5-mL sample volume (subject to matrix interference).
- GC column: a nonpolar fused silica capillary column, such as DB-5, SPB-5, VOCOL, or equivalent.
- Recommended surrogate/IS: toluene d_7 and fluorobenzene.

- Samples collected in glass containers without headspace, refrigerated, and analyzed within 7 days; samples preserved with 1:1 HCl (0.2 mL acid/40 mL sample) may be analyzed in 14 days.

Air drawn through a sorbent tube containing coconut shell charcoal (100 mg/50 mg); analyte desorbed into CS_2 over 30 min. standing; analyzed by GC-FID (NIOSH Method 1501, 1984); recommended flow rate 200 mL/min; sample volume 20 L.

- GC column: packed column 10% OV-275 on Chromosorb W-AW (100/120 mesh), Porapak P 50/80 mesh, or equivalent.

Air drawn through a cartridge containing Tenax (~2 g); cartridge heated under He purge; toluene transported into a cold trap and then to the front of a GC column at –70°C; column heated; toluene determined by GC/MS (U.S. EPA Method TO-1); recommended flow rate 100 mL/min; sample volume 10 L.

Air passed through a cartridge containing carbon molecular sieve (~400 mg); cartridge heated at 350°C under He purge; analyte carried into a cryogenic trap and flash evaporated onto a capillary column; determined by GC/MS (U.S. EPA Method TO-2); recommended flow rate 1 L/min; sample volume 100 L.

Collected in a trap in liquid argon; the trap removed and heated; sample transferred to a precooled GC column for PID or FID detection (U.S. EPA Method TO-3).

(1 ppm toluene in air = 3.76 mg/m^3 at NTP)

3.70 TOLUENE-2,4-DIISOCYANATE

Synonyms: 2,4-diisocyanatotoluene, TDI, isocyanic acid 4-methyl-*m*-phenylene ester; Formula: $CH_3C_6H_3(NCO)_2$; Structure:

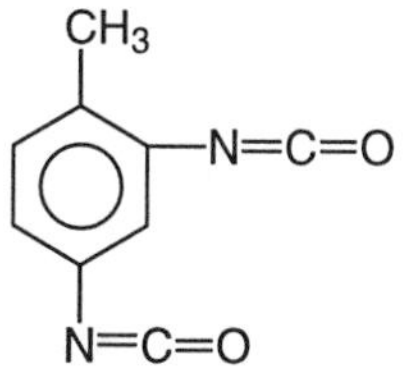

MW 174.16; CAS [584-84-9]; used in the production of urethane foams, elastomers, and coatings; also commercially available as a mixture of 2,4- and 2,6-isomers; colorless liquid or solid; turns dark on exposure to light; pungent fruity odor; boils at 238°C; melts at 20.5°C (80:20 mixture of the isomers melts at 13°C); density 1.22 g/mL; decomposes with water and alcohol; soluble in most organic solvents; highly toxic.

ANALYSIS OF AQUEOUS AND NONAQUEOUS SAMPLES

Sample extracted with methylene chloride; extract concentrated and analyzed by GC/MS.

- The characteristic masses for compound identification: 174, 145, 173, 146, 132, and 91.

Alternatively, sample extracted with toluene; extract treated with *N*-4-nitrobenzyl-*N*-propylamine (derivatizing agent) in toluene; TDI forms an urea derivative; the derivative extracted with acetonitrile and the extract solution then analyzed by HPLC-UV at 272 nm.

- The method originally developed for air analysis (Tiesler and Eben, 1985); no data available on precision and accuracy study for aqueous or solid matrices.

- HPLC column and conditions: column packing: RP-8 or RP-18 (10 mm); temperature: ambient; mobile phase: acetonitrile-water; gradient program: from 30% acetonitrile to 60% in 15 min; 5 min isocratic; flow rate 1.5 mL/min.
- For quantitation, prepare a calibration standard, plotting the concentration of urea derivative of each reference standard against peak areas.
- Derivatizing agent prepared as follows: 50 g 4-nitrobenzyl chloride in 250 mL toluene heated under reflux; 36 g *n*-propylamine added dropwise in 30 min; refluzed for 5 h; solvent removed at 50°C on a rotary evaporator; residue treated with 80 mL water followed by slow addition of 30 mL NaOH 40%; the alkaline solution extracted with 100 mL toluene; extract treated with 2 g charcoal and then filtered. The derivatizing agent prepared above is stable for 2 to 3 days. It may be converted into its hydrochloride as follows (which may be stable for several months):

  Excess solvent and *n*-propylamine removed on a rotary evaporator; product dissolved in 50 mL acetone; 35 mL conc. HCl added; mixture dried at 50°C; hydrochloride repeatedly washed with acetone-toluene (1:1); the salt desiccated in the vacuum oven.

  Hydrochloride may be converted back to the free amine derivatizing agent prior to use. This may be done by dissolving the salt in water, then adding 1 *N* NaOH soln. and extracting the free amine into toluene. The toluene phase is dried with Na_2SO_4 and then diluted to 250 mL. The concentration of this amine solution is 0.002 *M* (Soln. A). This may be tenfold diluted to give 0.0002 *M* amine solution (Soln. B).

AIR ANALYSIS

Air drawn through glass wool coated with *N*-[(4-nitropheny)methylpropylamine; derivative desorbed into 2 mL methanol in ultrasonic bath; analyzed by HPLC using an UV detector at 254 nm (NIOSH Method 2535, 1987); recommended flow rate 0.5 L/min; sample volume 50 L.

- Derivative stable for 14 days if protected from light.
- HPLC column: Octadecylsilylated silica (5 μm).

(1 ppm TDI = 7.12 mg/m^3 at NTP)

3.71 *o*-TOLUIDINE

Synonyms: 2-methylaniline, 2-methylbenzenamine, 2-aminotoluene, *o*-tolylamine; Formula: $C_7H_7NH_2$; Structure:

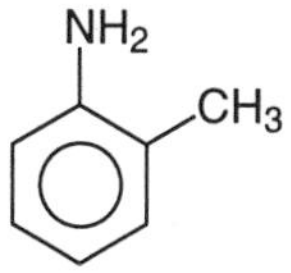

MW 107.17; CAS [95-53-4]; used in the manufacture of various dyes and as an intermediate in rubber chemicals and pharmaceuticals; colorless liquid, becoming yellowish to reddish-brown on exposure to air or light; boils at 200°C; solidifies at −16°C; vapor pressure 0.1 torr at 20°C; slightly soluble in water (1.5% at 20°C), readily miscible in organic solvents; carcinogenic and toxic at high doses (Patnaik, 1992).

ANALYSIS OF AQUEOUS AND NONAQUEOUS SAMPLES

Aqueous samples made alkaline with NaOH to pH >12 and extracted with methylene chloride; analyte partitions into the organic phase; the organic extract concentrated and analyzed by GC/NPD or GC/MS.

- GC column: Packed –3% SP-2250 on Supelcoport, Chromosorb-103, or equivalent; capillary: fused silica capillary column such as DB-5, PTE-5, or other equivalent.
- Characteristic masses for GC/MS identification: 107 and 106.

Solid matrices mixed with anhydrous Na_2SO_4 followed by methylene chloride; sonicated; extract cleaned by acid-based partitioning and analyzed by GC-NPD, GC-FID, or GC/MS.

- Aqueous and nonaqueous samples should be stored at 4°C, extracted within 7 days of collection, and analyzed within 40 days from extraction.

AIR ANALYSIS

Analyte adsorbed over silica gel (150/15 mg); desorbed into 95% ethanol in an ultrasonic bath and analyzed by GC-FID or GC-NPD (NIOSH Method 2002, 1985); recommended flow rate 200 mL/min; sample volume 20 to 30 L.

- Adsorption efficiency of silica gel decreases under high humidity.
- *o*-Toluidine stable on silica gel at least for one week.
- Sensitivity is much greater on GC-NPD.

(1 ppm *o*-toluidine in air = ~4.4 mg/m^3 at NTP)

3.72 1,1,1-TRICHLOROETHANE

Synonyms: methyl chloroform, strobane; Formula: $C_2H_3Cl_3$; Structure: $H_3C–CCl_3$; MW 133.40; CAS [71-55-6]; used as a cleaning solvent; colorless liquid with a mild chloroform-like odor; boils at 74°C; vapor pressure 100 torr at 20°C; freezes at –32.5°C; density 1.34 g/mL at 20°C; very slightly soluble in water (~70 mg/L); soluble in organic solvents.

ANALYSIS OF AQUEOUS AND NONAQUEOUS SAMPLES

Analyte in aqueous samples extracted by purge and trap method; volatile compounds thermally desorbed and swept onto a GC column for separation; detected by HECD, ECD, or MSD.

Soils, solid wastes, or sludges mixed with water or methanol; the aqueous extract or a solution of methanol extract spiked into water; subjected to purge and trap concentration and analyzed as above.

- Characteristic masses for GC/MS identification: 97, 99, 117, and 119.
- Limit of detection: in the range 0.05 µg/L when detected by HECD for a 5-mL sample aliquot (subject to matrix interference).
- Recommended surrogate/IS: bromochloromethane and 1,2-dichloroethane d_4.
- Samples collected in glass/plastic containers without headspace, refrigerated, and analyzed within 14 days.

See Halogenated Hydrocarbons, Chapter 2.9 for GC columns and conditions and a detailed discussion.

AIR ANALYSIS

Suggested method: adsorbed over coconut charcoal (100 mg/50 mg); desorbed with CS_2 and analyzed by GC-FID; recommended flow rate 100 mL/min; sample volume 10 L; precision and accuracy of the method not established.

Suggested method: adsorbed over carbon molecular sieve (~400 mg) in a cartridge; heated at 350°C under He purge; transferred into a cryogenic trap and flash evaporated onto a capillary column GC/MS; recommended flow rate 2 L/min; sample volume 100 L; precision and accuracy of the method not known.

Collected in a SUMMA passivated canister under pressure using an additional pump or at subatmospheric pressure by initially evacuating the canister; the air transferred into cryogenically cooled trap attached to a GC column; the trap heated; the analyte separated on the column for detection by ECD or GC/MS (U.S. EPA Method TO-14).

3.73 TRICHLOROETHYLENE

Synonyms: trichloroethene, ethylene trichloride; Formula C_2HCl_3; Structure:

MW 131.38; CAS [79-01-6]; used in dry cleaning and degreasing; colorless liquid with chloroform-like odor; boils at 87°C; freezes at –85°C; vapor pressure 58 torr at 20°C; density 1.46 g/mL at 20°C; very slightly soluble in water (about 1000 mg/L), miscible with organic solvents.

ANALYSIS OF AQUEOUS AND NONAQUEOUS SAMPLES

Aqueous samples extracted by purge and trap method; analyte thermally desorbed and swept onto a GC column for separation from other volatile compounds; detected by HECD, ECD, FID, or MSD.

Soils, solid wastes, or sludges mixed with water or methanol; the aqueous extract or a solution of methanol extract spiked into water; subjected to purge and trap concentration and analyzed as above.

Methanol extract may be directly injected for GC or GC/MS determination.

- Characteristic masses for GC/MS identification: 95, 97, 130, and 132.
- Limit of detection: in the range 0.1 μg/L when detected by HECD for a 5-mL sample aliquot (subject to matrix interference).
- Recommended surrogate/IS: 1,2-dichloroethane d_4 and 2-bromo-1-chloropropane.
- Samples collected in glass/plastic containers without headspace, refrigerated, and analyzed within 14 days.

See Halogenated Hydrocarbons, Chapter 2.9 for GC columns and conditions and further discussion.

AIR ANALYSIS

Adsorbed over coconut charcoal (100 mg/50 mg); desorbed with CS_2; and analyzed by GC-FID; recommended flow rate 100 mL/min; sample volume 5 L.

Adsorbed over carbon molecular sieve (~400 mg) in a cartridge heated at 350°C under He purge; transferred into a cryogenic trap and flash evaporated onto a capillary column GC/MS (U.S. EPA Method TO-2); 1 L/min recommended flow rate; sample volume 100 L.

Collected in a trap in liquid argon; the trap removed and heated; sample transferred to a precooled GC column for ECD or FID detection (U.S. EPA Method TO-3).

Collected in a SUMMA passivated canister under pressure using an additional pump or at subatmospheric pressure by initially evacuating the canister; the air transferred into cryogenically cooled trap attached to a GC column; the trap heated; the analyte separated on the column for detection by ECD or GC/MS (U.S. EPA Method TO-14).

(1 ppm trichloroethylene in air = 5.36 mg/m^3 at NTP)

3.74 2,4,6-TRICHLOROPHENOL

Formula: $C_6H_2Cl_3OH$; Structure:

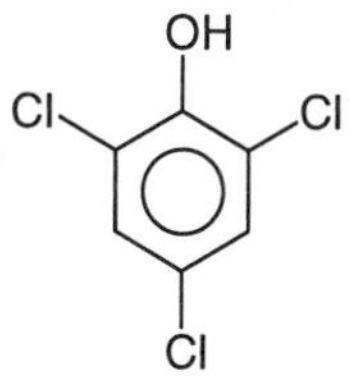

MW 197.46; CAS [88-06-2]; crystalline solid with strong phenolic odor; melts at 69°C; boils at 246°C; insoluble in water; soluble in most organic solvents.

ANALYSIS OF AQUEOUS AND NONAQUEOUS SAMPLES

Aqueous samples pH adjusted below 2; extracted with methylene chloride; extract concentrated and analyzed by GC-FID or GC/MS.

- Alternatively, extract exchanged into 2-propanol; derivatized with pentafluorobenzyl bromide, and determined by GC-ECD.
- Aqueous samples may be analyzed by HPLC on an underivatized polystyrene-divinylbenzene column such as Poly-RP CO (Alltech 1995) or a C-18 reverse-phase column; gradient: acetonitrile and 0.01 *M* K_3PO_4 at pH 7 (55:45) and the analyte detected by UV at 254 nm.

Soils, solid wastes, and sludges extracted with methylene chloride by sonication, Soxhlett, or supercritical fluid extraction; extract may be acid washed, concentrated, and analyzed by GC-FID, GC-ECD, or GC/MS as above.

- Characteristic masses for GC/MS determination: 196, 198, and 200 (electron impact ionization); 197, 199, and 201 (chemical ionization).
- Limit of detection: in the range 5 to 10 μg/L on FID or GC/MS and below 1 μg/L on ECD (for aqueous samples concentrated by 1000 times).
- Recommended surrogate/IS: pentafluorophenol and 2-perfluoromethyl phenol.

- Sample collected in glass containers, refrigerated, and extracted within 7 days of collection and analyzed within 40 days of extraction.

See Phenols, Chapter 2.23 for GC columns and conditions.

AIR ANALYSIS

Not likely to occur in ambient air in vapor state; the vapor pressure too low.

Recommended Method: Airborne particles collected on a sorbent cartridge containing polyurethane foam; deposited particles extracted with 5% diethyl ether in hexane; extract analyzed by GC-ECD; recommended air flow 5 L/min; sample volume 1000 L.

3.75 VINYL CHLORIDE

Synonyms: chloroethene, monochloroethylene; Formula: CH_2=CHCl; MW 62.50; CAS [75-01-4]; used as a monomer to make PVC resins and plastics; a colorless gas; liquefies at –13.4°C; slightly soluble in water; miscible in organic solvents.

ANALYSIS OF AQUEOUS SAMPLES

Aqueous samples extracted by purge and trap method; analyte thermally desorbed and swept onto a GC column for separation; detected by HECD, ECD, and MSD.

- Characteristic masses for GC/MS identification: 62 and 64.
- Limit of detection: in the range 0.1 μg/L when detected by HECD for a 5-mL sample aliquot (subject to matrix interference).
- Recommended surrogate/IS: bromochloromethane and 1,2-dichloroethane d_4.
- Samples collected in glass/plastic containers without headspace, refrigerated, and analyzed within 14 days.

See Halogenated Hydrocarbons, Chapter 2.9 for GC columns and conditions and for a detailed discussion.

AIR ANALYSIS

Adsorbed over carbon molecular sieve (~400 mg) in cartridge; heated at 350°C under He purge; transferred into a cryogenic trap and flash evaporated onto a capillary column GC/MS; recommended flow rate 100 mL/min; sample volume 10 L (U.S. EPA Method TO-2).

Collected in a trap in liquid argon; the trap removed and heated; sample transferred to a precooled GC column for ECD detection (U.S. EPA Method TO-3).

Collected in a SUMMA passivated canister under pressure using an additional pump or at subatmospheric pressure by initially evacuating the canister; the air transferred into cryogenically cooled trap attached to a GC column; the trap heated; the analyte separated on the column for detection by ECD or GC/MS (U.S. EPA Method TO-14).

(1 ppm vinyl chloride in air = 2.5 mg/m^3 at NTP).

3.76 XYLENE

Synonym: dimethyl benzene; Formula: C_8H_{10}; Structures: Xylene occurs in three isomeric forms as ortho-, meta-, and para isomers.

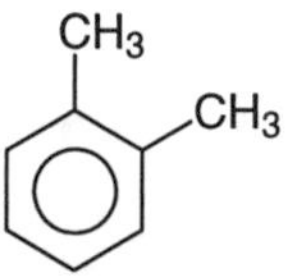

(*o*-Xylene) 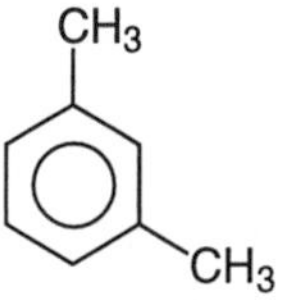 (*m*-Xylene)  (*p*-Xylene)

MW 106.18; CAS [1330-20-7]; [95-47-6] for *o*-xylene, [108-38-3] for *m*-xylene and [106-42-3] for *p*-xylene; used as a solvent and in the manufacture of dyes and drugs; occurs in gasoline and petroleum solvents; colorless liquid with characteristic odor; boiling point for *o*-, *m*-, and *p*- isomers 144.4, 139.1, and 138.4°C, respectively; their vapor pressures at 20°C are 6.7, 8.4, and 8.8 torr, respectively; density 0.880, 0.864, and 0.861 g/ml respectively; practically insoluble in water; readily miscible with organic solvents.

ANALYSIS OF AQUEOUS AND NONAQUEOUS SAMPLES

Purge and trap extraction; xylene thermally desorbed from the sorbent trap and backflushed with an inert gas onto a GC column for separation from other volatile compounds; detected by PID, FID, or a mass spectrometer.

Solid samples mixed with methanol; an aliquot of methanol extract spiked into a measured volume of water in purging vessel and subjected to purge and trap concentration and analyzed as above.

Alternatively, xylene thermally desorbed out from the solid matrix under He purge (without any solvent treatment) and analyzed by GC or GC/MS.

- The characteristic masses for GC/MS identification: 106 and 91; all three isomers of xylene and ethyl benzene produce the same characteristic masses. These isomers and ethyl benzene should, therefore, be identified from their retention times, which are in the following order of elution:

ethyl benzene < *p*-xylene ~ *m*-xylene < *o*-xylene

In most columns, two peaks will be observed, coeluting *m*- and *p*-xylene, followed by *o*-xylene; *m*- and *p*-isomers mostly coelute. Separation of these two isomers may be achieved to some degree by using a narrow bore capillary column of 100 m length. Under such conditions, *p*-isomer should elute before *m*-xylene; *o*-xylene shows a longer retention time and may be readily separated from its other isomers.

- GC column: a fused silica capillary column such as DB-5, DB-624, VOCOL, or equivalent.
- Recommended surrogate/IS: ethyl benzene-d_5 (m/z 111) and 1,4-difluorobenzene (w/z 114,63,88).
- Samples collected in glass containers without headspace, refrigerated, and analyzed within 7 days. Samples preserved with 1:1 HCI (a few drops/40 mL sample) may be analyzed in 14 days.

AIR ANALYSIS

Air drawn through a sorbent tube packed with coconut shell charcoal (100 mg/50 mg); xylene desorbed into CS_2 (more than 30 min standing); CS_2 extract analyzed by GC-FID (NIOSH method 1501, 1984); recommended air flow rate 100 mL/min; sample volume 20 L.

- GC column: 10% OV-275 on 100/120 mesh Chromosorb W-AW or equivalent. Xylene elutes as two peak, *p*- and *m*-xylene coeluting, followed by *o*-xylene.

Air drawn through a cartridge packed with Tenax (2 g); cartridge heated under He purge; analyte transported into a cold trap and then to the front of a GC column at –70°C; column temperature programmed; xylene determined by GC/MS (U.S. EPA method TO-1); recommended flow rate 100 ml/min; sample volume 10 L.

Air drawn through a cartridge packed with carbon molecular sieve (400 mg); cartridge heated at 350°C under He purge; analyte transported into a cryogenic trap; flash evaporated onto a capillary column; determined by GC/MS (U.S. EPA method TO-2); recommended flow rate 2 L/min; sample volume 100 L.

Air collected in a SUMMA passivated stainless steel canister either by pressurizing the canister using a sample pump or by repeated preevacuation; canister then attached to an analytical system; air sample collected transferred to a cryogenically cooled trap; cryogen removed and temperature raised; analyte revolatilized; separated on a GC column; determined by PID, FID, or a mass spectrometer (U.S. EPA method TO-14).

(1 ppm xylene in air = 4.34 mg/m^3 at NTP)

BIBLIOGRAPHY

BIBLIOGRAPHY

Alltech Corporation, Product catalog and literature, Deerfield, IL, 1995.

American Public Health Association, American Water Works Association and Water Environment Federation, *Standard Methods for the Examination of Water and Wastewater,* 18th ed., Washington, DC, 1992.

American Society for Testing and Materials, Atmospheric analysis, in *Annual Book of ASTM Standards,* Vol. 11.03, Philadelphia, PA, 1993.

American Society for Testing and Materials, Water (1) and (2), in *Annual Book of ASTM Standards,* Vol. 11.01-02, Philadelphia, PA, 1993.

Cavanaugh, M. and Patnaik, P., Decompostion of chlorinated pesticides on cleanup treament of extract with sulfuric acid, unpublished data, Rancocas Environmental Laboratory, Delanco, NJ, 1995.

Dolzine, T. W., Esposito, G. G., and Rinehart, D. S., Determination of hydrogen cyanide in air by ion chromatography, in *Toxic Materials in the Atmosphere: Sampling and Analysis,* ASTM special technical publication 786, American Society for Testing and Materials, Philadelphia, PA, 1982, 142.

Dong, S. and Dasgupta, P. K., Fast fluorometric flow injection analysis of formaldehyde in atmospheric water, *Environ. Sci. Technol.*, 21(6), 581, 1987.

Hach Company, *Water Analysis Handbook,* Loveland, CO, 1989.

Hewlett Packard, HP Environmental Solutions Catalog, Palo Alto, CA, 1995–1996.

Hites, R. A., *Handbook of Mass Spectra of Environmental Contaminants,* 2nd ed., CRC/Lewis, Boca Raton, FL, 1992.

Igawa, M., Munger, W. J., and Hoffmann, M. R., Analysis of aldehydes in clound- and fogwater samples by HPLC with a post-column reaction detector, *Environ. Sci. Technol.*, 23(5), 556, 1989.

J. & W. Scientific Corporation, Product literature and catalog, Folsom, CA 1995.

Kimbell, C. L., Atmospheric monitoring for hydrogen sulfide by photorateometric analysis, in *Toxic Materials in the Atmosphere: Sampling and Analysis,* ASTM special technical publication 786, American Society for Testing and Materials, Philadelphia, PA, 1982, 60.

McConnaughey, P. W., McKee, E. S., and Pritts, I. M., Passive colorimetric dosimeter tubes for ammonia, carbon monoxide, carbon dioxide, hydrogen sulfide, nitrogen dioxide, and sulfur dioxide, in *Toxic Materials in the Atmosphere: Sampling and Analysis,* ASTM special technical publication 786, American Society for Testing and Materials, Philadelphia, PA, 1982, 113.

Milke, R. and Patnaik, P., Soil pH: effect of soil to water ratio, unpublished data, Rancocas Environmental Laboratory, Delanco, NJ, 1995.

Millipore Corporation, Product bulletins on EnviroGard immunoassay kits, Millipore, Bedford, MA, 1994.

National Institute for Occupational Safety and Health (NIOSH), *Manual of Analytical Methods*, 3rd ed., Cincinnati, OH, 1984–1989.

Ohmicron Corporation, *Rapid Assay*, literature on the environmental immunoassay, Ohmicron, Newtown, PA, 1994.

Orion Research Corporation, Handbook of Electrode Technology, Cambridge, MA, 1982.

Patnaik, P., *A Comprehensive Guide to the Hazardous Properties of Chemical Substances*, Van Nostrand Reinhold, New York, 1992.

Patnaik, P., Derivatization of formaldehyde and its analysis by GC/MS, Unpublished data, 1989.

Pomerantz, I. H., Miller, L. J., and Kava, G., Analysis of pesticide residues: Captan, Folpet and Difolatan, *J. Am. Oil Assoc.*, 53, 154, 1970.

Pritts, I. M., McConnaughey, P. W., Roberts, C. C., and McKee, E. S., Ethylene oxide detection and protection, in *Toxic Materials in the Atmosphere: Sampling and Analysis*, ASTM special technical publication 786, American Society for Testing and Materials, Philadelphia, PA, 1982, 14.

Raisglid, M. and Burke, M. F., Optimizing solid phase extraction for oil and grease and particulate laden samples. Paper presented at EPA's 19th annual conference on Analysis of Pollutants in the Environment, Norfolk, VA, May 15–16, 1996.

Restek Corporation, Chromatography products catalog and literature, Bellefonte, PA, 1995.

Romano, S. and Renner, V. A., Analysis of ethylene oxide-worker exprosure, *Am. Ind. Hygiene Assoc. J.*, 40, 1979.

Skoog, D. A., West, D. M., and Holler, F. J., *Fundamentals of Analytical Chemistry*, 6th ed., Saunders College Publishing, New York, 1992.

Supelco Inc., Chromatography products catalog and literature, Bellefonte, PA, 1995.

Tiesler, A. and Eben, A., Hexamethylene diisocyanate, 2,4- and 2,6-toluylene diisocyanate, in *Analysis of Hazardous Substances in Air*, Vol. 1, Kettrup, A. and Henschler, D., Eds., VCH Verlagsgesellschaft mbH, Weinheim, Germany, 1991, 71.

U.S. EPA, *Methods for the Determination of Organic compounds in Drinking Water*, Reproduced by US Department of Commerce, National Technical Inforamtion Service, Springfield, VA, 1992.

U.S. EPA, Methods for organic chemical analysis of municipal and industrial wastewater, in *Code of Federal Regulations*, title 40, part 136, Office of the Federal Register, Washington, DC, 1992.

U.S. EPA, *Methods for the Determination of Inorganic Compounds in Drinking Water*, Reproduced by US Department of Commerce, National Technical Information Service, Springfield, VA, 1992.

U.S. EPA, Test Methods for Evaluating Solid Waste, Physical/Chemical Methods, SW-846, 3rd ed., Office of Solid Wastes, Washington DC, 1995.

U.S. EPA, *Compendium of Methods for the Determination of Toxic Organic Compounds in Ambient Air*, Reproduced by US Department of Commerce, National Technical Information Service, Springfield, VA, 1988.

Wirth, S. and Patnaik, P., Determination of chloroform on electron capture detector, unpublished data, Rancocas Environmental Laboratory, Delanco, NJ, 1993.

APPENDICES

APPENDIX A
SOME COMMON QC FORMULAS AND STATISTICS

MEASUREMENT OF PRECISION

Standard Deviation

$$s = \sqrt{\frac{\sum x^2 - \frac{\left(\sum x\right)^2}{n}}{n-1}}$$

where s = standard deviation,
Σx^2 = sum of squares of individual measurements,
Σx = sum of individual measurements, and
n = number of individual measurements.

An estimate of standard deviation can be calculated from the following formula:

$$s = \sqrt{\frac{\sum (x - \bar{x})^2}{n-1}}$$

Relative Standard Deviation (RSD)

(Also known as coefficient of variance)

$$\text{RSD} = \frac{s}{\bar{x}} \times 100\%$$

where $\bar{x}$ is the average of test results.

Standard Error of Mean (M)

$$M = \frac{s}{\sqrt{n}}$$

where n is the number of measurements.

Relative Percent Difference (RPD)

$$\text{RPD} = \frac{\text{Difference of two measurements}}{\text{Average of the two measurements}} \times 100\%$$

MEASUREMENT OF LINEARITY OF DATA POINTS

Correlation Coefficient (γ)

The correlation coefficient (γ) is a measure of linear relationship between two sets of data. It can attain a value which may vary between 0 and ±1. A value of +1 (or −1, when the slope is negative) indicates the maximum possible linearity; on the other hand, a zero γ indicates there is absolutely no link between the data. In environmental analysis, especially in spectrophotometric methods, γ is calculated to determine the linearity of the standard calibration curve. γ may be calculated from one of the following equations.

$$\gamma = \frac{s_x}{s_y} \times b$$

where s_x and s_y are the standard deviations in x and y sets of data, respectively, and b is the slope of the line. The above relationship, however, is susceptible to error, because b can change depending on how we manipulate to construct the line of best fit. The correlation coefficient, however can be best determined from one of the following two equations.

$$\gamma = \frac{\sum xy - \frac{\sum x \sum y}{n}}{\sqrt{\left[\sum x^2 - \frac{\left(\sum x\right)^2}{n}\right]\left[\sum y^2 - \frac{\left(\sum y\right)^2}{n}\right]}}$$

or

$$\gamma = \frac{\sum xy - n\bar{x}\bar{y}}{\sqrt{\left[\sum x^2 - n(\bar{x})^2\right]\left[\sum y^2 - n(\bar{y})^2\right]}}$$

Example

Determine the correlation coefficient of the standard calibration curve for residual chlorine from the following sets of data.

Concentration (mg/L)	Absorbance
(*x*)	(*y*)
0.10	0.023
0.15	0.030
0.30	0.066
0.35	0.075
0.45	0.094
0.55	0.112

$$\gamma = \frac{\sum xy - n\bar{x}\bar{y}}{\sqrt{\left[\sum x^2 - n\bar{x}^2\right]\left[\sum y^2 - n\bar{y}^2\right]}}$$

$\sum xy = 0.3167,\ n = 6$

$\bar{x} = 0.3167$ $\qquad \bar{y} = 0.0667$

$(\bar{x})^2 = 0.1003$ $\qquad (\bar{y})^2 = 0.00444$

$\sum x^2 = 0.75$ $\qquad \sum y^2 = 0.0328$

$$\gamma = \frac{0.1567 - 0.1267}{\sqrt{(0.7500 - 0.6017)(0.03828 - 0.0267)}}$$

$$= \frac{0.0300}{\sqrt{(0.1483)(0.0062)}}$$

$$= \frac{0.0300}{0.0303}$$

$$= 0.9901$$

The above value of γ is very close to +1.0000. Therefore, the standard calibration curve is highly linear.

MEASUREMENT OF ACCURACY

Percent Spike Recovery by U.S. EPA Formula

$$\% \text{ Recovery} = \frac{X_s - X_u}{K} \times 100\%$$

where X_s = measured value for the spiked sample,
X_u = measured value for the unspiked sample adjusted for dilution of the spike, and
K = known value of the spike in the sample.

(This is a U.S. EPA percent spike recovery formula.)

Percent Spike Recovery by Alternate Method

$$\% \text{ Recovery} = \frac{\text{measured concentration}}{\text{true concentration}} \times 100\%$$

The above formula may be directly used without any volume correction, if the volume of spike added is very small, i.e., <1% sample volume. However, if the volume of the spike solution added to the sample is large (>5% of the volume of the sample), the true (or expected) concentration in the above formula may be determined as follows:

$$\text{true concentration} = \frac{(C_u \times V_u)}{(V_u + V_s)} + \frac{(C_s \times V_s)}{(V_u + V_s)}$$

where C_u = measured concentration of the analyte in the sample,
C_s = concentration of the analyte in the spike standard,
V_u = volume of the sample, and
V_s = volume of the spike standard added.

(See Chapter 1.2 for solved examples.)

METHOD DETECTION LIMITS FOR ORGANIC POLLUTANTS IN AQUEOUS SAMPLES

Method Detection Limit (MDL)

$$\text{MDL} = t \times s$$

where t is the students' t value for $(n-1)$ measurements and s is the standard deviation of the replicate analyses.

Note that t-statistics should be followed when the sample size is small, i.e., <30. In the MDL measurements, the number of replicate analyses are well below 30, generally 7. For example, if the number of replicate analyses are 7, then the degrees of freedom, i.e., the $(n-1)$ is 6, and, therefore, the t value for 6 should be used in the above calculation. MDL must be determined at the 99% confidence level. When analyses are performed by GC or GC/MS methods, the concentrations of the analytes to be spiked into the seven aliquots of the reagent grade water for the MDL determination should be either at the levels of their IDL (instrument detection limit) or five times the background noise levels (the noise backgrounds) at or near their respective retention times.

***t* Values at the 99% Confidence Level**

Number of replicate measurements	Degrees of freedom $(n-1)$	t Values
7	6	3.143
8	7	2.998
9	8	2.896
10	9	2.821
11	10	2.764
16	15	2.602
21	20	2.528
26	25	2.485
31	30	2.457

APPENDIX B
SAMPLE CONTAINERS, PRESERVATIONS, AND HOLDING TIMES

Analyte	Container	Preservation	Maximum holding time
	Inorganics and microbial tests		
Acidity	P,G	Cool, 4°C	14 d
Alkalinity	P,G	Cool, 4°C	14 d
Bacterias, Coliform (total and fecal)	P,G	Cool, 4°C; add 0.008% $Na_2S_2O_3$ if residual chlorine is present	6 h
Biochemical oxygen demand	P,G	Cool, 4°C	48 h
Bromide	P,G	None required	28 d
Chloride	P,G	None required	28 d
Chlorine, residual	P,G	None required	Analyze immediately
Chemical oxygen demand	P,G	Cool, 4°C, H_2SO_4 to pH<2	28 d
Color	P,G	Cool, 4°C	48 h
Cyanide	P,G	Cool, 4°C, pH>12, 0.6 g ascorbic acid	14 d
Fluoride	P	None required	28 d
Hardness	P,G	pH<2 with HNO_3 or H_2SO_4	6 mon
Iodine	P,G	None required	Analyze immediately
Kjeldahl nitrogen	P,G	Cool, 4°C, pH<2 with H_2SO_4	28 d
Metals (except chromium-VI, boron and mercury)	P,G	HNO_3 to pH<2	6 mon
Chromium-VI	P,G	Cool, 4°C	24 h
Mercury	P,G	HNO_3 to pH<2	28 d
Boron	P	HNO_3 to pH<2	28 d
Nitrate	P,G	Cool, 4°C, H_2SO_4 to pH<2	28 d
Nitrite	P,G	Cool, 4°C	48 h
Odor	G	None required	Analyze immediately
Oil and grease	G	Cool, 4°C, H_2SO_4 or HCl to pH<2	28 d

Analyte	Container	Preservation	Maximum holding time
		Inorganics and microbial tests	
Oxygen, dissolved	G (BOD bottle)	None required	Analyze immediately
pH	P,G	None required	Analyze immediately
Phenolics	G	Cool 4°C, H_2SO_4 to pH<2	28 d
Phosphorus			
elemental	G	Cool, 4°C	48 h
orthophosphate	P,G	Cool, 4°C	48 h
total	P,G	Cool, 4°C, H_2SO_4 to pH<2	28 d
Residue			
Total	P,G	Cool, 4°C	7 d
Filterable	P,G	Cool, 4°C	7 d
Nonfilterable (TSS)	P,G	Cool, 4°C	7 d
Settleable	P,G	Cool, 4°C	48 h
Volatile	P,G	Cool, 4°C	7 d
Silica	P	Cool, 4°C	28 d
Specific conductance	P,G	Cool, 4°C	28 d
Sulfate	P,G	Cool, 4°C	28 d
Sulfide	P,G	Cool, 4°C, zinc acetate plus NaOH to pH >9	7 d
Sulfite	P,G	None required	Analyze immediately
Surfactants	P,G	Cool, 4°C	48 h
Taste	G	Cool, 4°C	24 h
Temperature	P,G	None required	Analyze
Total organic carbon	G	Cool, 4°C, HCl or H_2SO_4 to pH<2	28 d
Total organic halogen	G (amber bottles)	Cool, 4°C, store in dark, HNO_3 to pH<2, add Na_2SO_3 if residual chlorine present	14 d
Turbidity	P,G	Cool, 4°C	48 h
		Organics tests	
Purgeable halocarbons	G (Teflon-lined septum)	Cool, 4°C, no headspace (add 0.008% $Na_2S_2O_3$ if residual chlorine is present)	14 d
Purgeable aromatics	G (Teflon-lined septum)	Cool, 4°C, no headspace (add 0.008% $Na_2S_2O_3$ if residual chlorine is present), HCl to pH<2	14 d
Pesticides, chlorinated	G (Teflon-lined cap)	Cool, 4°C, pH 5-9	7 days until extraction; 40 days after extraction
PCBs	G (Teflon-lined cap)	Cool, 4°/C	7 days until extraction; 40 days after extraction
Phthalate esters	G (Teflon-lined cap)	Cool, 4°/C	7 days until extraction; 40 days after extraction

Parameter	Container	Preservation	Maximum holding time
Nitroaromatics	G (Teflon-lined cap)	Cool, 4°C (add 0.008% $Na_2S_2O_3$, if residual chlorine present), store in dark	7 days until extraction; 40 days after extraction
Nitrosamines	G (Teflon-lined cap)	Cool, 4°C (add 0.008% $Na_2S_2O_3$, if residual chlorine present), store in dark	7 days until extraction; 40 days after extraction
Polynuclear aromatic hydrocarbons	G (Teflon-lined cap)	Cool, 4°C (add 0.008% $Na_2S_2O_3$, if residual chlorine present), store in dark	7 days until extraction; 40 days after extraction
Haloethers	G (Teflon-lined cap)	Cool, 4°C (add 0.008% $Na_2S_2O_3$ if residual chlorine present)	7 days until extraction; 40 days after extraction
Phenols	G (Teflon-lined cap)	Cool, 4°C (add 0.008% $Na_2S_2O_3$ if residual chlorine present)	7 days until extraction; 40 days after extraction
Dioxins and dibenzofurans	G (Teflon-lined cap)	Cool, 4°C (add 0.008% $Na_2S_2O_3$ if residual chlorine present)	7 days until extraction; 40 days after extraction

Note: P, polyethylene; G, glass. If there is no residual chlorine in the sample, the addition of $Na_2S_2O_3$ may be omitted.

APPENDIX C PREPARATION OF MOLAR AND NORMAL SOLUTIONS OF SOME COMMON REAGENTS

Concentrations of reagents and titrants in wet analysis are commonly expressed in terms of molarity (*M*) or normality (*N*). One mole (molecular or formula weight expressed in grams) of a substance dissolved in 1 L of its aqueous solution produces 1 *M*.

$$\text{Molarity, } M = \frac{\text{number of moles of the solute (or reagent)}}{\text{liter of its solution}}$$

For example, 158.10 g sodium thiosulfate ($Na_2S_2O_3$) in 1 L of its solution is 1 *M*; or 1.581g $Na_2S_2O_3$ per liter is 0.01 *M*. The amount of a substance to be dissolved in water and made up to any specific volume of its solution for obtaining a concentration of any specific molarity is determined as follows:

EXAMPLE 1

How may grams of potassium hydrogen phthalate ($KHC_8H_4O_4$) must be added to water to produce 250 mL of 0.015 *M* solution?

$$M = 0.015, \text{ volume of solution} = 0.250 \text{ L}$$

Thus,

$$0.015 = \frac{\text{number of moles of } KHC_8H_4O_4}{0.250}$$

Therefore, the number of moles of $KHC_8H_4O_4 = 0.015 \times 0.250 = 0.00375$ mol, which is equal to:

$$0.00375 \text{ mol } KHC_8H_4O_4 \times \frac{204.23 \text{ g } KHC_8H_4O_4}{1 \text{ mol } KHC_8H_4O_4} = 0.766 \text{ g } KHC_8H_4O_4$$

That is, 0.766 g $KHC_8H_4O_4$ is to be dissolved in water to a volume of 250 mL to produce 0.015 *M* solution.

Normality (*N*), on the other hand, is the number of gram equivalent of a substance per liter of solution; or the number of milligram equivalent per mL of solution. Gram equivalent is the equivalent weight of the compound expressed in grams. Often, normality of a reagent is the same as its molarity. Also, the normality of a substance may differ widely from its molarity, depending on the reactions. Such a wide difference may conspicuously be noted, especially in the case of oxidizing and reducing agents in redox reactions. In such reactions, equivalent weight is determined from the change in oxidation number on the element during titration. The molarity (gram molecular weights/liter) and normality (gram equivalent weight/liter) of some common reagents in wet analysis are presented below.

Compound	Formula	1.00 *M* (g/L)	1.00 *N*[(a)] (g/L)
Arsenic trioxide	As_2O_3	197.85	49.455
Barium hydroxide	$Ba(OH)_2$	171.26	85.68
Ferrous ammonium sulfate (hexahydrate)	$Fe(NH_4)_2(SO_4)_2 \cdot 6H_2O$	392.14	—
Hydrochloric acid	HCl	36.46	36.46
Iodine	I_2	253.80	126.90
Mercuric nitrate	$Hg(NO_3)_2$	324.61	162.30
Nitric acid	HNO_3	63.01	63.01
Potassium biiodate	$KH(IO_3)_2$	389.92	32.49
Potassium bromate	$KBrO_3$	163.01	27.17
Potassium chlorate	$KClO_3$	122.55	20.4
Potassium chloride	KCl	74.56	74.56
Potassium dichromate	$K_2Cr_2O_7$	294.19	49.04
Potassium hydrogen phthalate	$KHC_8H_4O_4$	204.23	204.23
Potassium hydroxide	KOH	56.11	56.11
Potassium iodate	KIO_3	214.00	42.8[(b)]
Potassium oxalate	$K_2C_2O_4$	166.19	83.09
Potassium permanganate	$KMnO_4$	158.04	31.61[(c)]
Potassium persulfate	$K_2S_2O_8$	270.33	135.16
Silver nitrate	$AgNO_3$	169.87	169.87
Sodium carbonate	Na_2CO_3	105.99	53.00
Sodium chloride	$NaCl$	58.44	58.44
Sodium thiosulfate	$Na_2S_2O_3$	158.10	158.10
Sulfuric acid	H_2SO_4	98.08	49.04

[(a)]The equivalent weight is highly reaction-specific, which can change with pH and redox conditions.

[(b)]Under strong acid conditions, the reaction is $IO_3^- + 6\ H^+ + 5\ e^- \rightarrow \frac{1}{2}\ I_2 + 3\ H_2O$

[(c)]Equivalent weight of $KMnO_4$ could be 52.68, when the redox reaction is

$MnO_4^- + 3\ e^- + 2\ H_2O \rightarrow MnO_{2(s)} + 4\ OH^-$

APPENDIX D
TOTAL DISSOLVED SOLIDS AND SPECIFIC CONDUCTANCE: THEORETICAL CALCULATIONS

TOTAL DISSOLVED SOLIDS (TDS)

The total dissolved solids (TDS) in aqueous samples is determined by filtering an aliquot of the sample, evaporating the filtrate to dryness in a weighted dish, and then measuring the weight of the dried residue to a constant weight. The residue is dried at 180°C at least for an hour.

TDS is essentially attributed to many soluble inorganic salts that are commonly found in the surface and groundwaters. The ions that are often found in significant concentrations, contributing to the TDS include the metal ions (e.g., Na^+, K^+, Mg^{2+} and Ca^{2+}) and anions (e.g., Cl^-, F^-, SO_4^{2-}, CO_3^{2-}, NO_3^{2-}, and SiO_3^{2-}). In addition, certain metal ions, such as Al^{3+}, Cr^{3+}, Fe^{3+}, Mn^{2+} also often occur in many waters. The latter, however, occur as insoluble oxides and hydroxides and, therefore, do not contribute to TDS at the ambient temperature and the pH conditions.

Total dissolved solids may be calculated as follows by summing up the concentrations of the common soluble ions found in the water:

$$\text{TDS, mg/L} =$$

$$\left[Na^+ + K^+ + Ca^{2+} + Mg^{2+} + Cl^- + SO_4^{2-} + NO_3^- + SiO_3^{2-} + F^- + 0.6\ (\text{alkalinity, as } CaCO_3)\right] \text{mg/L}$$

In the above calculation, alkalinity (as $CaCO_3$) is multiplied by the factor 0.6 to account for the concentrations of CO_3^{2-}, HCO_3^-, and OH^-. Carbonate constitutes 60% of $CaCO_3$ (MW 100). Since there might be other soluble ions present in the sample as well, the TDS, calculated as above cannot be greater than the measured TDS. In other words, measured TDS should be either equal or slightly greater than the calculated TDS.

CONDUCTIVITY

Specific conductance, or the conductivity of a solution, is attributed to the ionic species (cations and anions) present in the solution. The conductivity to TDS ratio should be between 1.4 and 1.8, i.e.,

$$\frac{\text{Conductivity (expressed as } \mu\text{mhos/cm)}}{\text{TDS (mg/L)}} = 1.4 \text{ to } 1.8$$

or

$$\frac{\text{TDS}}{\text{conductivity}} = 0.55 \text{ to } 0.70$$

The conductivity of a solution may be accurately determined by a conductivity meter. If the sample resistance of the sample is measured, conductivity at 25°C is calculated as follows:

$$k = \frac{(1{,}000{,}000)(C)}{R[1 + 0.019(t - 25)]}$$

where k = conductivity, μmhos/cm,
C = cell constant, cm^{-1},
R = resistance of the sample in ohm, as measured, and
t = temperature of the sample.

C is determined using a 0.01 *M* KCl solution

$$C\left(cm^{-1}\right) = (0.001412)\left(R_{KCl}\right)[1 + 0.019\ (t - 25)]$$

where R_{KCl} is resistance, in ohms of 0.01 *M* KCl solution.

The units for *k* are as follows:

Conductivity of a solution is usually expressed as μmhos/cm. The SI units for conductivity and their conversion are as follows:

1 μmhos/cm = 1 μS/cm (microsiemens/centimeter)

= 0.1 mS/m (millisiemens/meter)

or

1 mS/m = 10 μmhos/cm

where $1 \text{ S (siemens)} = \frac{1}{\text{ohm}}$

Conductivity of a solution (a standard solution) may be estimated using the following equation;

$$k = \frac{\lambda c}{0.001}$$

where the conductivity k is expressed as μmhos/cm, λ is the equivalent conductivity in mho-cm²/equivalent, and c is the gram equivalent/L. Such calculation, however, does not give an accurate determination of k. It applies to a solution of infinity dilution and, therefore, is susceptible to appreciable departure from the true value at higher concentrations. This is illustrated below in the following example:

Example 1

Calculate the conductivity of 0.01 M KCl solution using the above equation.

$$KCl_{(aq)} \rightarrow K^+_{(aq)} + Cl^-_{(aq)}$$

Thus, 0.01 M KCl solution is 0.01 M K^+ and 0.01 M Cl^- which is also 0.01 equivalent/L for both K^+ and Cl^-.

$$\text{Thus, } k \text{ due to } K^+ = \frac{\lambda c}{0.001} = \frac{73.5 \times 0.01}{0.001} = 735$$

$$\text{and } k \text{ due to } Cl^- = \frac{\lambda c}{0.001} = \frac{76.4 \times 0.01}{0.001} = 764$$

The λ values for K^+ and Cl^- at the infinite dilution are 73.5 and 76.4 mho-cm²/equivalent, respectively (Table 1). Therefore, the conductivity of 0.01 M (or 0.01 N) KCl solution is estimated to be 735 + 764 or 1499 μmhos/cm, which is a distinct deviation from the measured value of 1412 μmhos/cm. This problem may be overcome by using ion activity coefficient as shown below in the following equations and examples.

The conductivity of a sample may be theoretically calculated if the concentrations of the ions present in the sample are known. Such calculations may be performed, using a series of equations, as follows:

$$k_{calc} = k^\circ \times y^2$$

where k_{calc} = conductivity calculated
k° = conductivity at infinity dilution
y = the monovalent ion activity coefficient.

y can be determined at the ambient temperature (ranging between 20 and 30°C) for any solution that is less than 0.5 *M* in concentration, using Davies equation as follows:

$$y = 10^{-0.5\left[\left\{(a)^{0.5} \div \left(1+(a)^{0.5}\right)\right\} - 0.3 \times a\right]}$$

where a is the ionic strength in molar units; a can be calculated from the following equation:

$$a = \frac{\sum z^2 c}{2000}$$

where z = charge on the ion, and
c = concentration of the ion, expressed as millimole (m*M*).

Finally, $k°$ is calculated from the following equation:

$$k° = z \times \lambda° \times c$$

where $\lambda°$ is the equivalent conductance, expressed as mho/cm^2.

The value of $\lambda°$ for ions in water at 25°C has been measured. The $\lambda°$ values for some common ions are tabulated below:

The theoretical determination of conductivity is illustrated in the following two examples.

Example 2

Calculate the conductivity of 0.01 *M* KCl solution at 25°C.

KCl is a strong elecrolyte that readily dissociates into K^+ and Cl^- ions when dissolved in water:

$$KCl_{(aq)} \rightarrow K^+_{(aq)} + Cl^-_{(aq)}$$

As we know, the conductivity, $K_{calc} = k° \times y^2$. Now we have to determine $k°$ and y.

$$K° = z\lambda°c$$

Both K^+ and Cl^- are monovalent ions, having a charge of +1 and –1, respectively. Thus the magnitude of z for each ion is 1 unit. The equivalent conductance's, $\lambda°$ for K^+ and Cl^- from Table 1 are 73.5 and 76.4 mho-cm^2/equivalent, respectively. The millimolar (m*M*) concentration of each of these ions is $\left(0.01\ M \times \frac{1000\ mM}{1\ M}\right)$ or 10 m*M*.

Table 1 Equivalent Conductance, λ° for Common Ions in Water at 25°C

Ions	Equivalent conductance λ°(mho-cm²/equivalent)
H^+	350.0
NH_4^+	73.5
1/3 Al^{3+}	61.0
Na^+	50.1
K^+	73.5
1/2 Ca^{2+}	59.5
1/2 Mg^{2+}	53.1
1/2 Mn^{2+}	53.5
1/2 Fe^{2+}	54.0
1/3 Fe^{3+}	68.0
1/2 Ni^{2+}	50.0
1/2 Cu^{2+}	53.6
1/2 Pb^{2+}	71.0
Ag^+	61.9
OH^-	198.6
F^-	54.4
HCO_3^-	44.5
1/2 CO_3^{2-}	72.0
1/2 SO_4^{2-}	80.0
Ac^- [a]	40.9
Cl^-	76.4
NO_3^-	71.4
NO_2^-	71.8
$H_2PO_4^-$	33.0
1/2 HPO_4^{2-}	57.0
Br^-	78.1
I^-	76.8
ClO_2^-	52.0
ClO_3^-	64.6
ClO_4^-	67.3
Hs^-	65.0
HSO_3^-	50.0
HSO_4^-	50.0
CN^-	78.0
CNO^-	64.6
OCN^-	64.6
SCN^-	54.9
MnO_4^-	61.3
1/2 CrO_4^{2-}	85.0

[a] Acetate ion (CH_3COO^-).

Now $k°$ for KCl = $k°$ for K^+ + $k°$ for Cl^-.

Thus, the $k°$ for $K^+ = 1 \times 73.5 \times 10 = 735$ and $Cl^- = 1 \times 76.4 \times 10 = 764$. Therefore, the $k°$ for KCl = 735 + 764 = 1499.

Now we have to calculate the value for y.

$$y = 10^{-0.5\left[\left\{(a)^{0.5} \div \left(1+(a)^{0.5}\right)\right\}-0.3a\right]}$$

The ionic strength,

$$a = \frac{\sum z^2 c}{2000} M \text{ or } \frac{z^2 c \text{ for K}^+ + z^2 c \text{ for Cl}^-}{2000}$$

and c = 10 mM for both the K^+ and Cl^-, respectively.

Therefore,

$$a = \frac{1^2 \times 10 \text{ (for K}^+) + 1^2 \times 10 \text{ (for Cl}^-) \ M}{2000} = \frac{20}{2000} \text{ or } 0.001 \ M$$

Therefore, $y = 10^{-0.5\left[\left\{0.001^{0.5} \div \left(1 + 0.001^{0.5}\right)\right\} - 0.3 \times 0.001\right]}$

$$\text{or } 10^{-0.0138} = 0.9687$$

Substituting the values for y and $k°$, we can determine the theoretical conductivity as:

$$k_{calc} = 1499 \times (0.9687)^2 \text{ or } 1407 \ \mu\text{mhos/cm}$$

This closely agrees with the experimental value of 1412 µmhos/cm.

Example 3

Analysis of a aqueous sample showed the following results: Cl^- 207 mg/L, SO_4^{2-} 560 mg/L, NO_3^- 22.5 mg/L, Na^+ 415 mg/L, and Ca^{2+} 54 mg/L. The pH of the sample was found to 3.8. Calculate the conductivity of the sample.

At first, we convert the concentration of these ions into their mM concentrations (c).

$$Cl^- = \frac{207 \text{ mg}}{1 \text{ L}} \times \frac{1 \text{ mmol}}{35.45 \text{ mg}} \times \frac{1 \text{ L}}{1 \text{ mmol}} = 5.84 \text{ m}M$$

$$SO_4^{2-} = \frac{560 \text{ mg}}{1 \text{ L}} \times \frac{1 \text{ mmol}}{96 \text{ mg}} \times \frac{1 \text{ L}}{1 \text{ mmol}} = 5.83 \text{ m}M$$

$$NO_3^- = \frac{22.5 \text{ mg}}{1 \text{ L}} \times \frac{1 \text{ mmol}}{62 \text{ mg}} \times \frac{1 \text{ L}}{1 \text{ mmol}} = 0.363 \text{ m}M$$

$$Na^+ = \frac{415 \text{ mg}}{1 \text{ L}} \times \frac{1 \text{ mmol}}{23 \text{ mg}} \times \frac{1 \text{ L}}{1 \text{ mmol}} = 18.04 \text{ m}M$$

$$Ca^{2+} = \frac{54 \text{ mg}}{1 \text{ L}} \times \frac{1 \text{ mmol}}{40 \text{ mg}} \times \frac{1 \text{ L}}{1 \text{ mmol}} = 1.35 \text{ m}M$$

$$\text{pH} = 3.8; \; C_{H^+} = 10^{-\text{pH}} = 10^{-3.8}$$

$$= 0.00158\ M$$

$$H^+ = 0.158 \text{ m}M$$

Now we calculate $k°$ for each of these ions, using the formula

$k° = z \times \lambda° \times c$ ($\lambda°$ values are presented in Table 1)

$k°$ for Cl^- $= 1 \times 76.4 \times 5.84 = 446.2$

$k°$ for SO_4^{2-} $= 2 \times 80.0 \times 5.83 = 932.8$

$k°$ for NO_3^- $= 1 \times 71.4 \times 0.363 = 25.9$

$k°$ for Na^+ $= 1 \times 50.1 \times 18.04 = 903.8$

$k°$ for Ca^{2+} $= 2 \times 59.5 \times 1.35 = 160.6$

$k°$ for H^+ $= 1 \times 350.1 \times 0.158 = 55.3$

$k°$ $= 446.2 + 932.8 + 25.9 + 903.8 + 160.6 + 55.3 =$

2524.6

Next, we calculate the ionic strength, a, for the sample using the equation:

$$a = \frac{\sum z^2 c}{2000}$$

$$= \frac{(1 \times 5.84) + (4 \times 5.83) + (1 \times 0.363) + (1 \times 18.04) + (4 \times 1.35) + (1 \times 0.158)}{2000}$$

$$= 0.02656\ M$$

From the value of a, we now calculate y, the monovalent ion activity, from the equation:

$$y = 10^{-0.5\left[\left\{(a)^{0.5} \div \left(1+(a)^{0.5}\right)\right\} - 0.3a\right]}$$

$$= 10^{-0.06607}$$

$$= 0.8589$$

Therefore, $k_{calc} = 2524.6 \times (0.8580)^2 = 1862$ µmhos/cm

Thus, the conductivity of any aqueous sample may be precisely calculated, as we see in the above two examples, if we know the concentrations of the metal ions and the anions in the sample. The presence of such metal ions and the anions and their concentrations may be simultaneously measured by ICP atomic emission spectrophotometer and ion chromatograph, respectively.

APPENDIX E

Characteristic Masses of Miscellaneous Organic Pollutants (Not Listed in Text) for GC/MS Identification[a]

CAS no.	Compounds	Primary ion (m/z)	Secondary ions (m/z)
75-86-5	Acetone cyanohydrin	43	58,27
123-54-6	Acetylacetone	43	85, 100
260-94-6	Acridine	179	89, 178, 180
79-06-1	Acrylamide	44	71, 55, 27
15972-60-8	Alachlor	45	160, 188, 237, 146
116-06-3	Aldicarb	58	86, 41, 89, 144, 100
2032-59-9	Aminocarb	151	150, 136, 208
834-12-8	Ametryne	227	212, 58, 170, 185, 98
61-82-5	Amitrole	84	29, 57, 44
84-65-1	Anthraquinone	208	108, 152, 76, 126
1912-24-9	Atrazine	200	58, 215, 173, 202, 69
103-33-3	Azobenzene	77	51, 182, 105, 152
275-51-4	Azulene	128	51, 102
101-27-9	Barban	222	87, 51, 104, 143, 257
22781-23-3	Bendiocarb	151	166, 126,58, 223
1861-40-1	Benfluralin	292	41, 264, 160, 105, 206
65-85-0	Benzoic acid	105	122, 77, 51
100-47-0	Benzonitrile	103	76, 50, 104
119-61-9	Benzophenone	105	77, 182, 51
98-88-4	Benzoyl chloride	105	77, 51, 50
100-44-7	Benzyl chloride	91	126, 65, 92
17804-35-2	Benomyl	191	159, 105, 40, 132, 146
314-40-9	Bromacil	205	207, 41, 69, 163, 233, 262
90-11-9	1-Bromonaphthalene	206	127, 208, 126
18181-80-1	Bromopropylate	43	75, 185, 104, 155, 341
23184-66-9	Butachlor	57	176, 160, 188, 237
123-86-4	Butyl acetate	43	86, 73
75-64-9	*tert*-Butylamine	58	41, 42
111-76-2	Butyl cellosolve	57	45, 87, 41, 75
15271-41-7	Butyllate	58	39, 184, 148, 118, 96
76-22-2	Camphor	95	81, 152, 108, 69
63-25-2	Carbaryl	144	115, 116, 145, 201
86-74-8	9H-Carbazole	167	166, 139, 168, 44
1563-66-2	Carbofuran	164	149, 122, 123, 201
118-75-2	Chloranil	87	246, 209, 244, 211, 218

Characteristic Masses of Miscellaneous Organic Pollutants (Not Listed in Text) for GC/MS Identification[a] (Continued)

CAS no.	Compounds	Primary ion (m/z)	Secondary ions (m/z)
78-95-5	Chloroacetone	49	42, 92, 77, 51
510-15-6	Chlorobenzilate	251	139, 253, 111, 141
118-91-2	2-Chlorobenzoic acid	139	156, 111, 75, 50, 121
563-47-3	2-Chloro-3-methylpropene	55	39, 90, 54, 75
90-13-1	Chloronaphthalene	162	127, 164, 63
1897-45-6	Chlorothalonil	266	264, 268, 109, 124
191-07-1	Coronene	300	150, 149, 301
123-73-9	Crotonaldehyde	70	41, 39, 69
98-82-8	Cumene	105	120, 77, 51, 79
110-82-7	Cyclohexane	56	84, 41, 55, 69
108-93-0	Cyclohexanol	57	44, 67, 82
108-91-8	Cyclohexylamine	56	43, 99, 70
94-75-7	2,4-D	162	220, 164, 133, 175, 63
94-82-6	2,4-DB	162	164, 87, 63
96-12-8	DBCP	157	75, 155, 77, 49, 62
1861-32-1	DCPA	301	299, 303, 332, 221
75-99-0	Dalapon	28	36, 62, 43, 97, 106
2303-16-4	Diallate	86	234, 70, 128, 109, 58
1918-00-9	Dicamba	173	220, 191, 97, 149, 175
117-80-6	Dichlone	191	226, 163, 228, 135, 99
91-94-1	3,3′-Dichlorobenzidine	252	254, 126, 154
75-71-8	Dichlorodifluoromethane	85	87, 50, 101
120-36-5	Dichlorprop	162	164, 234, 189, 98
542-75-6	1,3-Dichloropropene	75	39, 77, 49, 110, 112
97-23-4	Dichlorophene	128	141, 268, 215, 152
115-32-2	Dicofol	139	111, 141, 250, 75
111-42-2	Diethanolamine	74	30, 56, 42
109-89-7	Diethylamine	30	27, 58, 44, 73
88-85-7	Dinoseb	211	163, 147, 117, 240
122-39-4	Diphenylamine	169	168, 77, 51
330-54-1	Diuron	72	232, 234, 44
145-73-3	Endothall	68	100, 69, 82, 140
75-08-1	Ethanethiol	62	47, 29, 34, 45
100-41-4	Ethylbenzene	91	106, 51, 39
110-77-0	2-(Ethylthio)ethanol	75	47, 106, 45, 61
101-42-8	Fenuron	72	164, 44, 70, 77
14484-64-1	Ferbam	88	296, 44, 175, 120, 416
944-22-9	Fonofos	109	137, 246, 110, 81
98-01-1	Furfural	39	96, 95, 38, 29
118-74-1	Hexachlorobenzene	284	286, 282, 142, 249
67-72-1	Hexachloroethane	201	117, 119, 203, 166
110-19-0	Isobutyl acetate	43	56, 73
314-42-1	Isocil	204	206, 163, 161, 120, 70, 246
465-73-6	Isodrin	193	195, 66, 263, 147
148-24-3	Isoquinoline	129	102, 51, 128
566-61-6	Isothiocyanatomethane	73	72, 45, 70
143-50-0	Kepone	272	274, 270, 237, 355
330-55-2	Linuron	61	46, 248, 160
2032-65-7	Methiocarb	168	153, 109, 225, 91

16752-77-5	Methomyl	54	105, 87, 42, 28
99-76-3	Methylparaben	121	152, 93, 65
51218-45-2	Metolachlor	162	45, 238, 146, 91
1929-82-4	Metribuzin	198	41, 57, 103, 74, 144, 214
315-18-4	Mexacarbate	165	150, 134, 222
2385-85-5	Mirex	272	274, 270, 237, 332
150-68-5	Monuron	72	198, 40, 28
110-91-8	Morpholine	57	29, 87, 56, 86
134- 32-7	1-Naphthylamine	143	115, 116, 89, 63
91-59-8	2-Naphthylamine	143	115, 116, 89, 63
555-37-3	Neburon	40	114, 187, 274, 276
54-11-5	Nicotine	84	162, 133, 161
1836-75-5	Nitrofen	283	285, 202, 139, 162
930-55-2	*N*-Nitrosopyrrolidine	41	100, 42, 69, 43
25154-52-3	Nonyl phenol	149	107, 121, 55, 77
23135-22-0	Oxamyl	72	44, 162, 115, 88, 145
101-84-8	Phenyl ether	170	141, 77, 51
84-62-8	Phenyl phthalate	65	44, 66, 91, 105, 120
1918-02-1	Picloram	196	198, 161, 163, 86, 240
2631-37-0	Promecarb	135	150, 91, 58
7287-19-6	Prometryne	241	58, 184, 226, 106
1918-16-7	Propachlor	120	77, 93, 176, 211
709-98-8	Propanil	161	163, 57, 29, 217
122-42-9	Propham	43	93, 179, 137, 120
114-26-1	Propoxur	110	152, 81, 58, 209
91-22-5	Quinoline	129	102, 51, 128
83-79-4	Rotenone	192	394, 191, 177, 28
93-72-1	Silvex	196	198, 97, 270, 268, 167
1982-49-6	Siduron	93	55, 56, 94, 232
122-34-9	Simazine	44	201, 186, 68, 173
1014-70-6	Simetryne	213	68, 170, 155, 71
57-24-9	Strychnine	334	335, 120, 36, 162
100-42-5	Styrene	104	103, 78, 51, 77
95-06-7	Sulfallate	188	72, 88, 60, 148, 223
93-76-5	2,4,5-T	196	198, 254, 256, 167, 97
5902-51-2	Terbacil	161	160, 162, 56, 116, 216
53555-64-9	Tetrachloronaphthalene	266	264, 268, 194
119-64-2	Tetralin	104	132, 91, 41
137-26-8	Thiram	88	120, 240, 44
119-93-7	2-Tolidine	212	106, 196, 180
76-03-9	Trichloroacetic acid	44	83, 85, 36, 117
87-61-6	Trichlorobenzene	180	182, 145, 184, 109
75-69-4	Trichlorofluoromethane	101	103, 66, 105, 152
55720-37-1	Trichloronaphthalene	230	232, 160, 234
102-71-6	Triethanolamine	118	56, 45, 42
121-44-8	Triethylamine	86	30, 58, 101
1582-09-8	Trifluralin	306	264, 43, 290, 335
75-50-3	Trimethylamine	58	59, 42, 30
51-79-6	Urethan	31	44, 45, 62, 74
108-05-4	Vinyl acetate	43	86, 42, 44
1330-20-7	Xylene	91	106, 105, 77, 51

[a] Electron impact ionization at 70 V (nominal).

APPENDIX F
VOLATILITY OF SOME ADDITIONAL ORGANIC SUBSTANCES (NOT LISTED IN TEXT) FOR PURGE AND TRAP ANALYSIS

Volatility data are presented in the following table for some additional organic compounds. These substances may be analyzed in aqueous samples using purge and trap technique. For purge and trap extraction of an organic compound from aqueous matrice, the compound should ideally have a high vapor pressure, low boiling point, and low solubility in water. An inert gas or a gas that does not react with the analyte is bubbled through a measured voume of sample aliquot in a purgin vessel. The compound is swept onto a sorbent trap, filled with tenax, activated charcoal, and silica gel (or equivalent materials), where it is adsorbed. It is then thermally desorbed from the trap and then transported under the flow of a carrier gas onto a termperature programmed GC column for separation from other volatile substances. The compounds are detected either by a mass spectrometer or by a FID. Halogenated organics may, alternatively, be determined by an ECD or other halogen specific detector.

CAS no.	Compound	Boiling point (°C)	Vapor pressure at 20°C (torr)
67-64-1	Acetone[(a)]	56.2	105[(c)] at 8°C
506-96-7	Acetyl bromide	76.0	80[(c)]
75-36-5	Acetyl chloride	50.9	175[(c)]
107-18-6	Allyl alcohol[(a)]	97.1	10 at 10°C
300-57-2	Allyl benzene[(b)]	156	5[(c)]
106-95-6	Allylbromide	70	150[(c)]
107-05-1	Allyl chloride (3-Chloropropene)	45	400 at 27.5°C
557-40-4	Allyl ether	94	20[(c)]
557-31-3	Allyl ethyl ether	66	150[(c)]
870-23-5	Allyl mercaptan	67	100 at 15°C
627-40-7	Allyl methyl ether	55	300[(c)] at 25°C
1471-03-0	Allyl propyl ether	91	25[(c)]

Volatility of Some Additional Organic Substances (Not Listed in the Text) for Purge and Trap Analysis (Continued)

CAS no.	Compound	Boiling point (°C)	Vapor pressure at 20°C (torr)
100-52-7	Benzaldehyde[b]	178	1 at 26°C
109-65-9	1-Bromobutane	101.6	40 at 25°C
29576-14-5	1-Bromo-2-butene	98	70[c]
74-97-5	Bromochloromethane	68	100[c]
762-49-2	1-Bromo-2-fluoroethane	72	110[c]
107-82-4	1-Bromo-3-methylbutane	120.5	12 at 15°C
106-94-5	1-Bromopropane	71	100 at 18°C
75-26-3	2-Bromopropane	60	175[c]
75-62-7	Bromotrichloromethane	105	50[c]
109-79-5	*n*-Butylmercaptan	98	3.1
123-86-4	Butyl acetate	125	15[c]
71-36-3	*n*-Butyl alcohol[a]	117.2	4.4
78-92-2	*sec*-Butyl alcohol [a]	99.5	12
75-65-0	*tert*-Butyl alcohol[a]	83	31
109-69-3	*n*-Butyl chloride	78.4	80
78-86-4	*sec*-Butyl chloride	68	100 at 14°C
507-20-0	*tert*-Butyl chloride	51	375[c] at 30°C
628-81-9	Butyl ethyl ether	91.5	25[c]
123-72-8	Butyraldehyde[a]	75.7	71
141-75-3	Butyryl chloride	102	40[c]
142-96-1	*n*-Butyl ether[b]	141	4.8
75-87-6	Chloral (Trichloroacetaldehyde)	97.8	39
107-20-0	Chloroacetaldehyde	90	50[c]
78-95-5	Chloroacetone[a]	119	25[c]
107-14-2	Chloroacetonitrile	127	79-04-9
79-04-9	Chloroacetyl chloride	107	35[c]
78-86-4	2-Chlorobutane	68.2	85[c]
928-51-8	4-Chloro-1-butanol	84.5	60[c]
4091-39-8	3-Chloro-2-butanone	116	30[c]
563-52-0	3-Chloro-1-butene	64	200[c]
1120-57-6	Chlorocyclobutane	83	140[c]
542-18-7	Chlorocyclohexane	143	10[c]
930-28-9	Chlorocyclopentane	114	60[c]
544-10-5	1-Chlorohexane[b]	134.5	20[c]
543-59-9	1-Chloropentane	108	75[c]
75-29-6	2-Chloropropane	35	450[c]
540-54-5	1-Chloropropane	46.6	375[c]
590-21-6	1-Chloropropene	37	400 at 18°C
557-98-2	2-Chloropropene	22.5	>700
95-49-8	2-Chlorotoluene[b] (α-Tolylchloride)	159	5[c]
123-73-9	Crotonaldehyde[a] (2-Butenal)	104	19
108-93-0	Cyclohexanol[b]	161	1
78-75-1	1,2-Dibromopropane[b]	141.5	7
109-64-8	1,3-Dibromopropane[b]	167	3[c]
594-16-1	2,2-Dibromopropane[b]	120	10[c]
79-02-7	Dichloroacetaldehyde	90.5	45[c]
513-88-2	1,1-Dichloroacetone	120	20[c]
616-21-7	1,2-Dichlorobutane	123.5	22[c]
541-33-3	1,1-Dichlorobutane	115	16[c]
7581-97-7	2,3-Dichlorobutane	116	17[c]

1190-22-3	1,3-Dichlorobutane[b]	134	10[c]
79-36-7	Dichloroacetyl chloride	109	15[c]
594-37-6	1,2-Dichloro-2-methylpropane	108	15[c]
78-99-9	1,1-Dichloropropane	88	45[c]
78-87-5	1,2-Dichloropropane	97	41
142-28-9	1,3-Dichloropropane	120.4	40[c]
594-20-7	2,2-Dichloropropane	69.3	60[c]
542-75-6	1.3-Dichloro-1-propene	108	39[c]
78-88-6	2,3-Dichloro-1-propene	84	53 at 25°
60-29-7	Diethyl ether[a]	34.5	442
352-93-2	Diethyl sulfide	92	50[c]
108-20-3	Diisopropyl ether	69	130
75-18-3	Dimethyl sulfide	38	420
141-78-6	Ethyl acetate[a]	77	73
140-88-5	Ethyl acrylate[a]	100	29
97-95-0	2-Ethyl-1-butanol[b]	150	1.8
105-54-4	Ethyl butyrate	121	11.3
107-07-3	Ethylene chlorohydrin	128	4.9
109-94-4	Ethyl formate[a]	54	192
75-08-1	Ethyl mercaptan	36	440
97-63-2	Ethyl methacrylate	117	15[c]
110-43-0	2-Heptanone[b]	150	2.6
106-35-4	3-Heptanone[b]	148.5	1.4 at 25°C
123-19-3	4-Heptanone[b]	144	1.2 at 25°C
78-84-2	Isobutyraldehyde[a]	61.5	170
98-82-8	Isopropylbenzene[b] (Cumene)	152.5	3.2
75-29-6	Isopropyl chloride	35.7	450[c]
108-21-4	Isopropyl acetate[a]	90	47.5
108-83-8	Isovalerone[b] (Diisobutylketone)	165	1.7
513-36-0	Isobutyl chloride	69	100 at 16°C
590-86-3	Isovaleraldehyde	90	70[c]
126-98-7	Methacrylonitrile[a]	90.3	65 at 25°C
100-66-3	Methoxybenzene[b] (Anisole)	155	5[c]
96-33-3	Methyl acrylate[a]	80	70
96-34-4	Methyl chloroacetate	130	10[c]
74-95-3	Methylene bromide	98	36
78-93-3	Methyl ethyl ketone[a]	79.6	77.5
74-88-4	Methyl iodide	42.5	400 at 25°C
563-80-4	Methyl isopropyl ketone	97.5	42 at 25°C
108-10-1	Methyl isobutyl ketone[a]	119	6
80-62-6	Methyl methacrylate	101	38[c] at 25°C
557-17-5	Methyl propyl ether	38.5	400 at 22.5°C
107-87-9	2-Pentanone[a]	101	13[c]
96-22-0	3-Pentanone[a]	102	13
109-60-4	*n*-Propyl acetate[a]	102	25
103-65-1	*n*-Propylbenzene[b]	159	2.5
107-19-7	Propargyl alcohol	115	
123-38-6	Propionaldehyde[a]	49	235
107-12-0	Propionitrile	97	40 at 22°C
107-03-9	*n*-Propyl mercaptan	68	150[c]
79-03-8	Propionyl chloride	80	85[c]
918-00-3	1,1,1-Trichloroacetone[b]	149	5[c]
76-02-8	Trichloroacetyl chloride	118	10[c]
10403-60-8	2,2,3-Trichlorobutane[b]	144	5[c]
96-18-4	1,2,3-Trichloropropane[b]	156	2

Volatility of Some Additional Organic Substances (Not Listed in the Text) for Purge and Trap Analysis (Continued)

CAS no.	Compound	Boiling point (°C)	Vapor pressure at 20°C (torr)
2567-14-8	1,1,2-Trichloropropylene	118	20[c]
354-58-5	1,1,1-Trichloro-2,2,2-trifluoroethane	46	275[c]
76-13-1	1,1,2-Trichloro-1,2,2-trifluoroethane	48	270
110-62-3	Valeraldehyde	103	50 at 25°C
108-05-4	Vinyl acetate[a]	72	83
593-60-2	Vinyl bromide	15.8	>700
109-93-3	Vinyl ether	28	>600[c]

[a] Poor purging efficiency because of moderate solubility in water.
[b] Compounds show high retention times on GC chromatogram.
[c] Vapor pressure estimated.

APPENDIX G
ANALYSIS OF ELEMENTS BY ATOMIC SPECTROSCOPY: AN OVERVIEW

Technique	Principle	Comparison/comments
Flame-AA	Atomic absorption: sample is vaporized and atomized in high temperature flame. Atoms of the analyte element absorb light of a specific wavelength from a hollow cathode lamp, passing through the flame. Amount of energy absorbed by these atoms is measured, which is proportional to the number of atoms in the light path. Components: lamp, flame, monochromator, and detector.	Single element determination; detection limits relatively higher than other techniques.
Graphite furnace-AA	Atomic absorption: flame is replaced by an electrically heated graphite tube into which sample is directly introduced. All of the analyte is atomized. This significantly enhances the sensitivity and detection limit. The general principle of this technique, otherwise, is same as flame-AA.	Single element determination; capability limited to fewer elements; analysis time longer than flame; the sensitivity and detection limits, however, are much greater than flame technique, and significantly better than ICP techniques.
Inductively coupled plasma (ICP)	Atomic emission: Atoms or ions of the element present in the sample absorb energy, causing excitation of their electrons to unstable energy states. When these atoms or ions return to their stable ground-state configuration, they emit light of specific wavelengths characteristic of the elements. ICP is an argon plasma that can heat the sample to a very high temperature in the range 5500 to 8000°C. Sample is injected as an aerosol, and transported onto the ICP by argon, which is also used as a carrier gas. The emitted light is separated according to its wavelength and measured.	Multielement determination (sequential or simultaneous); faster analysis time; minimal chemical interaction; detection limits and sensitivity fall in between that of flame and graphite furnace measurements.

Analysis of Elements by Atomic Spectroscopy: An Overview (Continued)

Technique	Principle	Comparison/comments
ICP-Mass Spectrometry (ICP-MS)	The method combines an inductively coupled plasma with a quadrupole mass spectrometer. High energy ICP generates singly charged ions from the atoms of the elements present in the sample. Such ions are now directed onto the mass spectrometer, separated, and measured according to their mass-to-charge ratio.	Multielement determination; sensitivity and detection limits exceptionally good (over 100 times greater than furnace techniques for some metals); isotopes also may be measured; also has the capability to determine nonmetals (at a much lower sensitivity); broad linear-working range; high cost.

APPENDIX H
ANALYSIS OF TRACE ELEMENTS BY INDUCTIVELY COUPLED PLASMA-MASS SPECTROMETRY

Multi-element determination of dissolved metals at ultratrace level may be performed by Inductively Coupled Plasma-Mass Spectrometry (ICP-MS). U.S. EPA's Methods 200.8 and 1638 present a methodology for measuring trace elements in waters and wastes by the above technique. Sample is acid digested and the solution is introduced by pneumatic nebulization into a radio-frequency plasma. The elements in the compounds are atomized and ionized. The ions are extracted from the plasma through a differentially pumped vacuum interface and separated by a quadrupole mass spectrometer by their mass to charge ratios. The mass spectrometer must have a resolution capability of 1 amu peak width at 5% peak height.

ICP-MS technique is susceptible to interference that may arise from the presence of isobar elements (atoms of different elements having same masses) and isobaric polyatomic ions, buildup of sample material in the plasma torch, transport and sample conversion processes in the plasma, quadrupole operating pressure, and ion energy. Such interferences may be minimized by using alternative analytical isotopes, correcting for elemental equation in data calculation, adjusting rinse times, diluting the sample extracts to reduce dissolved solids concentrations to less than 0.2% (w/v), and using appropriate internal standards similar to the elements being determined.

Recommended analytical masses, elemental equations, interference effects, and internal standards are summarized below in the following tables.

Table 1 Recommended Analytical Masses for Element Detection

Element	Primary isotope	Additional isotope(s)
Aluminum	27	—
Antimony	121	123
Arsenic	75	—
Barium	137	135
Beryllium	9	—
Cadmium	111	106, 108, 114
Chromium	52	53
Cobalt	59	—
Copper	63	65
Krypton	83	—
Lead	206	207, 208
Manganese	55	—
Molybdenum	98	95, 97
Nickel	60	62
Palladium	105	—
Ruthenium	99	—
Selenium	82	77, 78, 80
Silver	107	109
Thallium	205	203
Tin	118	—
Thorium	232	—
Uranium	238	—
Vanadium	51	—
Zinc	66	67, 68

Table 2 Mass Ions for Internal Standards

Internal standard	Mass
Lithium	6
Scandium	45
Yttrium	89
Rhodium	103
Indium	115
Terbium	159
Holmium	165
Lutetium	175
Bismuth	209

Table 3 Interference from Isobars and Matrix Molecular Ions

Element	Mass	Isobar/matrix molecular ion
Arsenic	75	$Ar^{35}Cl^+$
Chromium	52	$ClOH^+$, ArO^+, ArC^+
	53	$^{37}ClO^+$
	54	$^{37}ClOH^+$, ArN^+
Cadmium	106	ZrO
	108	MoO, ZrO
	111	MoO, ZrO
Copper	63	PO_2^+, $ArNa^+$, TiO
	65	TiO
Indium	115	^{115}Sn
Manganese	55	$ArNH^+$
Nickel	62	TiO
Scandium	45	$COOH^+$
Selenium	77	$Ar^{37}C1^+$
	78	$^{40}Ar^{38}Ar^+$
	80	$^{40}Ar_2^+$
Silver	107	ZrO
	109	ZrO, MoO
Tin	115	^{115}In
Vanadium	51	$^{35}ClO^+$, $^{34}SOH^+$
Zinc	64	SO_2^+, S_2^+, TiO
	66	TiO

Table 4 Recommended Elemental Equations for Calculations

Element	Elemental equation
Aluminum	$(1.000)\ (^{27}C)$
Antimony	$(1.000)\ (^{121}C)$
Arsenic	$(1.000)\ (^{75}C) - (3.127)\ \{(^{77}C) - (0.815)\ (^{82}C)\}$ [a]
Barium	$(1.000)\ (^{137}C)$
Beryllium	$(1.000)\ (^{9}C)$
Cadmium	$(1.000)\ (^{111}C) - (1.073)\ \{(^{108}C) - (0.712)\ (^{106}C)\}$ [b]
Chromium	$(1.000)\ (^{52}C)$ [c]
Cobalt	$(1.000)\ (^{59}C)$
Copper	$(1.000)\ (^{63}C)$
Lead	$(1.000)\ (^{206}C) + (1.000)\ (^{207}C) + (1.000)\ (^{208}C)$ [d]
Manganese	$(1.000)\ (^{55}c)$
Molybdenum	$(1.000)\ (^{98}C) - (0.146)\ (^{99}c)$ [e]
Nickel	$(1.000)\ (^{60}C)$
Selenium	$(1.000)\ (^{82}C)$ [f]
Silver	$(1.000)\ (^{107}C)$
Thalium	$(1.000)\ (^{205}C)$
Thorium	$(1.000)\ (^{232}C)$
Uranium	$(1.000)\ (^{238}C)$
Vanadium	$(1.000)\ (^{51}C) - (3.127)\ \{(^{53}C) - (0.113)\ (^{52}C)\}$ [g]
Zinc	$(1.000)\ (^{66}C)$

Table 4 Recommended Elemental Equations for Calculations (Continued)

Element	Elemental equation
Internal Standard	
Bismuth	(1.000) (^{209}C)
Indium	(1.000) (^{115}C) – (0.016) (^{118}C) [h]
Scandium	(1.000) (^{45}C)
Terbium	(1.000) (^{159}C)
Ytterbium	(1.000) (^{89}C)

Note: C, counts at specified mass (calibration blank subtracted).

[a] Correction for chloride interference with adjustment for ^{77}Se.
[b] Correction for MoO interference.
[c] There may be contribution from ClOH. This can be estimated from the reagent blank.
[d] For variable isotopes of lead.
[e] Isobaric elemental correction for ruthenium.
[f] Krypton might be present as an impurity in some argon supplies. Correction for ^{82}Kr is done by background subtraction.
[g] Correction for chloride interference with adjustment for ^{53}Cr.
[h] Isobaric elemental correction for tin.

APPENDIX I
OIL AND GREASE ANALYSIS: AN OVERVIEW

The term "oil and grease" refers to a broad class of organic substances recovered from the sample matrices by extraction with an appropriate solvent. Such recovery, therefore, is characteristic of certain physical properties of the compounds, primarily the volatility of the compounds and their solubility in the extraction solvent. The solvent must be immiscible in water and volatile, as well as readily distilled on a water bath. Many solvents or mixed-solvent systems should be suitable for the extraction of oil and grease in aqueous and nonaqueous samples. These include petroleum ether, *n*-hexane, methylene chloride, methyl *tert*-butyl ether, and trichlorotrifluoroethan (freon). These solvents are listed in Table 1.

Among these solvents, freon has been extensively used in the oil and grease extraction. However, because of its ozone depletion action, the manufacture and use of Freon is currently being curtailed, and other solvents are now being used.

These substances that are classified under the above definition of oil and grease belong to both the biological lipids and the petroleum hydrocarbons and include straight chain and branched hydrocarbons, fatty acids, and the esters of fatty acids. Certain organic dyes, sulfur compounds, and chlorophyll are also extracted, and contribute to the measurement.

OIL AND GREASE ANALYSIS

Oil and grease may be analyzed by (1) gravimetric method and (2) infrared method.

Gravimetric Method

The aqueous sample is acidified to pH 2 or lower with 1:1 HCl. A measured volume of sample (1 L) is transferred to a separatory funnel. A 30-mL portion of extracting solvent is added to the sample. The content are shaken vigorously for 1 to 2 min. The organic layer separated is collected in a distillation flask. The

Table 1 Extraction Solvents for Oil and Grease

Extraction Solvent	Not applicable to substances volatilizing below	Solvent layer in separatory funnel
Trichlorotrifluoroethan (Freon)	70°C	Lower
n-Hexane	85°C	Upper
n-Hexane/methyl tertbutyl ether	85°C	Upper
Methylene chloride	70°C	Lower
Petroleum ether	70°C	Upper

aqueous layer is reextracted two more times using 30 mL portion of solvent each time. The extracts are combined together in tared (or weighted) distillation flask. The solvent is distilled from the flask on a water bath. After all the solvent distills out, the residue in the flask is dried under air flow and vacuum. The distillation flask containing the oil and grease residue is cooled in a desiccator and weighed to a constant weight.

Infrared Method

The extraction of oil and grease by infrared method is same as above for the gravimetric method. The only difference, however, is that the solvent extract is not evaporated nor is the solvent distilled out. The absorbance of the solvent extract is measured at 2930 cm using a 1 cm path-length cell and compared against the calibration standards solutions.

Thus, oil and grease measured by both the methods are susceptible to show variation. While gravimetric method measures "all" substances that are solvent extractables and "nonvolatile" under the conditions of distillation and drying, infrared method measures the absorbance of carbon-hydrogen bond of substances extracted. Also, compounds boiling below the distillation temperature of the extraction solvent may occur in the extract and contribute to oil and grease measured by the infrared method.

U.S. EPA Method 1664

This is a performance-based method that avoids the use of chlorofluorocarbon solvent. The method is applicable to aqueous matrices, using *n*-hexane as the extraction solvent and gravimetry as the determinative technique. Because hexane is a hydrocarbon solvent, and if this solvent is employed for extraction, the method performance cannot be evaluated by IR measurement. The substances that may be determined by this method are relatively nonvolatile hydrocarbons, vegetable oils, greases, waxes, animal fats, and related materials. The method permits the use of other extraction solvents also, provided that the QC criteria are met.

Technique

A 1-L sample is acidified to pH 2 and serially extracted three times with *n*-hexane in a separatory funnel. The extract is dried over Na_2SO_4. The solvent is evaporated from the extract and the oil and grease is weighed. Substances that volatilize below 85°C cannot be measured by this method. This includes gasoline and many components of #2 fuel oil. Also some crude oils and heavy fuel oils are not completely soluble in hexane. Therefore, the recovery will be low.

Hydrocarbons constituting oil and grease can be determined by such gravimetry technique, whether the extraction solvent is Freon or hexane. The oil and grease residue after weighing is redissolved in the extraction solvent. Silica gel, which has the ability to absorb polar materials, is added to the solution. Polar fatty acids are retained on the silica gel. The mixture is filtered. The filtrate should contain dissolved hydrocarbons which may be determined directly by infrared measurement or by gravimetry following distillation and drying of the residue. If only hydrocarbons need to be determined, the solvent extract of the sample may be treated with silica gel, filtered, and the filtrate analyzed as above either by IR or gravimetry.

Sludges and solid wastes are acidified and combined with $MgSO_4 \cdot H_2O$. This can dry the sample by forming $MgSO_4 \cdot 7\ H_20$. Samples should not be heated for drying because heating the acidified sample can decompose many component compounds. The dried sample is then serially extracted with an appropriate extraction solvent and analyzed as above.

Solid phase extraction (SPE) is being currently investigated as an alternative to liquid-liquid extraction (LLE). The SPE approach may offer certain advantages over the LLE, such as low solvent requirement and elimination of fromation of emulsion (Raisglid and Burke, 1996). The proper choice of solvent and sorbent may enable the determination of specific classes of target analytes.

APPENDIX J
NIOSH METHODS FOR AIR ANALYSIS

CAS no.	Analyte	NIOSH method
[75-07-0]	Acetaldehyde	2538, 3500, 3507
[64-19-7]	Acetic acid	1603, 3501
[108-24-5]	Acetic anhydride	3506
[75-86-5]	Acetone cyanohydrin	2506
[75-05-8]	Acetonitrile	1606
—	Acids, inorganic	7903
[10035-10-6]	Hydrobromic acid	7903
[7664-38-2]	Hydrofluoric acid	7903
[7664-38-2]	Phosphoric acid	7903
[7647-01-0]	Hydrochloric acid	7903
[7697-37-2]	Nitric acid	7903
[7664-93-9]	Sulfuric acid	7903
[107-02-8]	Acrolein	2501
[107-13-1]	Acrylonitrile	1604, 2505
—	Alcohols	1400, 1401, 1402
[107-18-6]	Allyl alcohol	1402
[71-36-3]	*n*-Bytul alcohol	1401
[78-92-2]	*sec*-Butyl alcohol	1401
[75-65-0]	*tert*-Butyl alcohol	1400
[108-93-0]	Cyclohexanol	1402
[123-42-2]	Diacetone alcohol	1402
[64-17-5]	Ethanol	1400
[123-51-3]	Isoamyl alcohol	1402
[78-83-1]	Isobutyl alcohol	1401
[67-63-0]	Isopropyl alcohol	1400
[105-30-6]	Methyl isobutyl carbinol	1402
[71-23-8]	*n*-Propyl alcohol	1401
—	Aldehydes, screening	2539, 2531
—	Alkaline dusts	7401
[107-05-1]	Allyl chloride	1000
[7429-90-5]	Aluminum	7013
—	Amines, Alphatic	2010, 6010
[109-89-7]	Diethylamine	2010, 6010
[124-40-3]	Dimethylamine	2010, 6010
—	Amines, aromatic	2002
[62-53-3]	Aniline	2002

NIOSH Methods for Air Analysis (Continued)

CAS no.	Analyte	NIOSH method
[121-69-7]	*N,N*-Dimethylaniline	2002
[99-97-8]	*N,N*-Dimethyl-*p*-toluidine	2002
[95-68-1]	α-Toluidine	2002
[95-68-1]	2,4-Xylidine	2002
—	Aminoethanols	2007/3509
[141-43-5]	2-Aminoethanol (monoethanolamine)	2007/3509
[102-81-8]	2-(Dibutyl)aminoethanol	2007
[111-42-2]	Diethanolamine	3509, 5521
[100-37-8]	2-Diethylaminoethanol	2007
[102-71-6]	Triethanolamine	3509, 5521
[7664-41-7]	Ammonia	6701
[29191-52-4]	Anisidine	2514
[90-04-0]	*o*-Anisidine	2514
[104-94-9]	*p*-Anisidine	2514
[7440-38-2]	Arsenic	7900
—	Arsenic, organo	5022
[98-50-0]	*p*-Aminophenylarsonic acid	5022
—	Dimethylarsonic acid	5022
[124-58-3]	Methylarsonic acid	5022
[1237-53-3]	Arsenic trioxide	7901
[7784-42-1]	Arsine	6001
—	Asbestos, bulk	9002, 1300
—	Asbestos, fibers	7402, 6009 [a]
[123-99-9]	Azelaic acid	5019
—	Barium, soluble compounds	7056
[92-87-5]	Benzidine	5509
[94-36-0]	Benzoyl peroxide	5009
[7440-41-7]	Beryllium	7102
[92-52-4]	Biphenyl	2530
[12069-32-8]	Boron carbide	7506
[75-63-8]	Bromotrifluoromethane	1017
[1689-84-5]	Bromoxynil	5010
[1689-99-2]	Bromoxynil octanoate	5010
[106-99-0]	1,3-Butadiene	1024
[109-79-5]	1-Butanethiol	2525, 2537
[78-93-3]	2-Butanone	2500
[7440-43-9]	Cadmium	7048
[7440-70-2]	Calcium	7020
[63-25-2]	Carbaryl	5006
[1333-86-4]	Carbon black	5000
[75-15-0]	Carbon disulfide	1600
[55720-99-5]	Chlorinated diphenyl ether	5025
—	Chlorinated terphenyl	5014
[79-11-8]	Chloroacetic acid	2008
[126-99-8]	Chloroprene	1002
[7440-47-3]	Chromium	7024
—	Chromium, hexavalent	7600, 7604
[65996-93-2]	Coal tar pitch volatiles	5023
[7440-48-4]	Cobalt	7027
[7440-50-8]	Copper (dust and fumes)	7029
[1319-77-3]	Cresol	2001

—	Cyanides (aerosol and gas)	7904
[108-80-5]	Cyanuric acid	5030
[542-92-7]	1,3-Cyclopentadiene	2523
[8065-48-3]	Demeton	5514
[334-88-3]	Diazomethane	2515
[19287-45-7]	Diborane	6006
[75-61-6]	Dibromodifluoromethane	1012
[107-66-4]	Dibutyl phosphate	5017
[91-94-1]	3,3′-Dichlorobenzidine	5509
[75-71-8]	Dichlorodifluoromethane	1018
[111-44-4]	sym-Dichloroethyl ether	1004
[75-43-4]	Dichlorofluoromethane	2516
[594-72-9]	1,1-Dichloro-1-nitroethane	1601
[76-14-2]	1,2-Dichlorotetrafluoroethane	1018
[127-19-5]	Dimethylacetamide	2004
[68-12-2]	Dimethylformamide	2004
[77-78-1]	Dimethyl sulfate	2524
[123-91-1]	Dioxane	1602
—	Dyes	5013
[90-04-0]	*o*-Anisidine	5013, 2514
[92-87-5]	Benzidine	5013, 2514
[119-93-7]	α-Tolidine	5013, 2514
[106-89-8]	Epichlorohydrin	1010
[2104-64-5]	EPN	5012
—	Esters	1450
[628-63-7]	*n*-Amyl acetate	1450
[626-38-0]	*sec*-Amyl acetate	1450
[123-86-4]	*n*-Butyl acetate	1450
[105-46-4]	*sec*-Butyl acetate	1450
[540-88-5]	*tert*-Butyl acetate	1450
[111-15-9]	2-Ethoxyethyl acetate	1450
[140-88-5]	Ethyl acrylate	1450
[123-92-2]	Isoamyl acetate	1450
[110-19-0]	Isobutyl acetate	1450
[108-84-9]	Methyl isoamyl acetate	1450
[109-60-4]	*n*-Propyl acetate	1450
—	Ethylene amines	2540
[111-40-0]	Diethylenetriamine	2540
[107-15-3]	Ethylenediamine	2540
[280-57-9]	Triethylenetetramine	2540
[107-07-3]	Ethylene chlorohydrin	2513
[107-21-1]	Etylene glycol	5500
[75-21-8]	Ethylene oxide	1614, 3702
[96-45-7]	Ethylene thiourea	5011
[60-29-7]	Ethyl ether	1610
—	Fibers	7400
[50-00-0]	Formaldehyde	2541, 3500, 3501
[98-01-1]	Furfural	2529
[98-00-0]	Furfuryl alcohol	2505
[111-30-8]	Glutaraldehyde	2531
[556-52-5]	Glycidol	1608
—	Glycol ethers	1403
[111-76-2]	2-Butoxyethanol	1403
[110-80-5]	2-Ethoxyethanol	1403

NIOSH Methods for Air Analysis (Continued)

CAS no.	Analyte	NIOSH method
[109-86-4]	2-Methoxyethanol	1403
[118-74-1]	Hexachlorobenzene	—
—	Hydrocarbons, alphatic and naphthenic	1500
[110-82-7]	Cyclohexane	1500
[110-83-8]	Cyclohexene	1500
[142-82-5	*n*-Heptane	1500
[110-54-3]	*n*-Hexane	1500
[111-65-9]	*n*-Pentane	1500
—	Hydrocarbons, aromatic	1501
[71-43-2]	Benzene	1501, 3700
[98-82-8]	*p-tert*-Butyltoluene	1501
[98-82-8]	Cumene	1501
[100-41-4]	Ethylbenzene	1501
[98-83-9]	α-Methylstyrene	1501
[91-20-3]	Napththalene	1501, 5506, 5515
[100-42-5]	Styrene	1501
[108-88-3]	Toluene	1501, 4000, 1501
[25013-15-4]	Vinyltoluene	1501
[1330-20-1]	Xylene	1501
—	Hydrocarbons, halogenated	1003
[100-44-7]	Benzyl chloride	1003
[75-25-2]	Bromoform	1003
[56-23-5]	Carbon tetrachloride	1003
[108-90-7]	Chlorobenzene	1003
[74-97-5]	Chlorobromomethane	1003
[67-66-3]	Chloroform	1003
[90-50-1]	*o*-Dichlorobenzene	1003
[106-46-1]	*p*-Dichlorobenzene	1003
[75-34-3]	1,1-Dichlorobenzene	1003
[107-06-2]	1,2-Dichloroethylene	1003
[107-06-2]	Ethylene dichloride	1003
[67-72-1]	Hexachloroethane	1003
[71-55-6]	Methylchloroform	1003
[127-18-4]	Tetrachloroethylene	1003
[79-00-5]	1,1,2-Trichloroethane	1003
[96-18-4]	1,2,3-Trichloropropane	1003
—	Hydrocarbons, halogenated (miscellaneous)	—
[107-05-1]	Allyl chloride	1000
[75-63-8]	Bromotrifluoromethane	1017
[75-61-6]	Dibromofifluoromethane	6006
[78-87-5]	1,2-Dichloropropane	1013
[74-96-4]	Ethyl bromide	1011
[75-00-3]	Ethyl chloride	2519
[106-93-4]	Ethylene dibromide	1008
[74-87-9]	Methyl bromide	2520
[74-87-3]	Methyl chloride	1001
[75-09-2]	Methylene chloride	1005
[74-88-4]	Methyl iodide	1014
[76-01-7]	Pentachloroethane	2517
[79-27-6]	1,1,2,2-Tetrabromoethane	2003
[79-34-5]	1,1,2,2-Tetrachloroethane	1019
[79-01-5]	Trichloroethylene	1022, 3701

[593-60-2]	Vinyl bromide	1009
[75-01-4]	Vinyl chloride	1007
[75-35-4]	Vinylidene chloride	1015
[74-09-8]	Hydrogen cyanide	6010
[123-31-9]	Hydroquinone (1,4-Benzenediol)	5004
[7553-56-2]	Iodine	6005
—	Isocyanate	5521
[584-89-9]	Toluene-2,4-diisocyanate	5521, 2535
[101-68-8]	4,4′-Methylenediphenyl isocyanate	5521
[822-06-0]	Hexamethylene diisocyanate	5521
[78-59-1]	Isophorone	2508
[143-50-0]	Kepone	5508
—	Ketones	—
[67-64-1]	Acetone	1300
[76-22-2]	Camphor	1301
[108-94-1]	Cyclohexanone	1300
[108-83-8]	Diisobutylketone	1300
[106-68-3]	Ethyl amyl ketone	1301
[106-35-4]	Ethyl butyl ketone	1301
[591-78-6]	2-Hexanone	1300
[141-79-7]	Mesityl oxide	1301
[583-60-8]	2-Methyl isobutyl ketone	1300
[110-43-0]	Methyl-*n*-amyl ketone	1301
[107-87-9]	2-Pentanone	1300
[7439-92-1]	Lead	7082
[1314-87-0]	Lead sulfide	7505
[121-75-5]	Malathion	5012
[7439-97-6]	Mercury	6009
[67-56-1]	Methanol	2000
[109-87-5]	Methylal	1611
[1338-23-4]	Methyl ethyl ketone peroxide	3508
[80-62-6]	Methyl methacrylate	2537
[7786-34-7]	Mevinphos	2503
[8012-95-1]	Mineral oil mist	5026
—	Naphthas	1550
—	Coal tar naphtha	1550
[8008-20-6]	Kerosene	1550
—	Mineral spirits	1550
[8032-32-4]	Petroleum ether	1550
[8030-30-6]	Petroleum naphtha	1550
—	Rubber solvent	1550
[8052-41-3]	Stoddart solvent	1550
—	Naphthylamines	5518
[134-32-7]	α-Naphthylamine	5518
[91-59-8]	β-Naphthylamine	5518
[13463-39-3]	Nickel carbonyl	6007
—	Nitrobenzenes	2005
[100-00-5]	4-Chloronitrobenzene	2005
[98-95-3]	Nitrobenzene	2005
[88-72-2]	2-Nitrotoluene	2005
[99-08-1]	3-Nitrotoluene	2005
[99-99-0]	4-Nitrotoluene	2005
[79-24-3]	Nitroethane	2527
[10102-44-0]	Nitrogen dioxide	6700
[55-63-0]	Nitroglycerine	2507

NIOSH Methods for Air Analysis (Continued)

CAS no.	Analyte	NIOSH method
[75-52-5]	Nitromethane	2527
[79-46-9]	2-Nitropropane	2528
—	Nitrosamines	2522
[924-16-3]	*N*-Nitrosodibutylamine	2522
[55-18-5]	*N*-Nitrosodiethylamine	2522
[62-75-9]	*N*-Nitrosodimethylamine	2522
[621-64-7]	*N*-Nitrosodipropylamine	2522
[59-89-2]	*N*-Nitrosomorpholine	2522
[100-75-4]	*N*-Nitrosopiperidine	2522
[930-55-2]	*N*-Nitrosopyrrolidine	2522
[10024-97-2]	Nitrous oxide	6600
—	Nuisance dust	—
—	Respirable	0600
—	Total	0500
[111-88-6]	1-Octanethiol	2510
—	Organotin compounds	5504
[7782-44-7]	Oxygen	6601
[4685-14-7]	Paraquat, tetrahydrate	5003
[56-38-2]	Parathion	5012
—	Pesticides and herbicides, chlorinated	—
[309-00-2]	Aldrin	5502
[57-74-9]	Chlordane	5510
[94-75-7	2,4-D	5001
[72-20-8]	Endrin	5519
[58-89-9]	Lindane	5502
[93-76-5]	2,4,5-T	5001
[608-93-5]	Pentachlorobenzene	5517
[87-86-5]	Pentachlorophenol	5512
[108-95-2]	Phenol	3502
[7723-14-0]	Phosphorus	7905
[7719-12-2]	Phosphorus trichloride	6402
—	Polychlorinated biphenyls	5503
—	Polynuclear aromatic hydrocarbons	5506 (GC) or 5515 (HPLC)
[83-32-9]	Acenaphthene	5506 (GC) or 5515 (HPLC)
[208-96-8]	Acenaphthylene	5506 (GC) or 5515 (HPLC)
[120-12-7]	Anthracene	5506 (GC) or 5515 (HPLC)
[56-55-3]	Benz[*a*]anthracene	5506 (GC) or 5515 (HPLC)
[205-99-2]	Benzo[*b*]fluoranthene	5506 (GC) or 5515 (HPLC)
[207-08-9]	Benzo[*k*]fluoranthene	5506 (GC) or 5515 (HPLC)
[191-24-2]	Benzo[*ghi*]perylene	5506 (GC) or 5515 (HPLC)
[192-97-2]	Benzo[*e*]pyrene	5506 (GC) or 5515 (HPLC)
[218-01-9]	Chrysene	5506 (GC) or 5515 (HPLC)
[53-70-3]	Dibenz[*a,h*]anthracene	5506 (GC) or 5515 (HPLC)
[206-44-0]	Fluoranthene	5506 (GC) or 5515 (HPLC)
[86-73-7]	Fluorene	5506 (GC) or 5515 (HPLC)
[193-39-5]	Indeno[1,2,3-*cd*]pyrene	5506 (GC) or 5515 (HPLC)
[91-20-3]	Naphthalene	5506 (GC) or 5515 (HPLC)
[85-10-8]	Phenanthrene	5506 (GC) or 5515 (HPLC)
[129-00-0]	Pyrene	5506 (GC) or 5515 (HPLC)
[75-56-9]	Propylene oxide	1612
[8003-34-7]	Pyrethrum	5008
[110-86-1]	Pyridine	1613

[14808-60-7]	Quartz (in coal mine dust)	7609
—	Ribaviron	5027
[83-79-4]	Rotenone	5007
—	Silica	—
—	Amorphous	7501
—	Crystalline (XRD)	7500
—	Crystalline (color)	7601
—	Crystalline (IR)	7602
[7803-52-3]	Stibine	6008
[57-24-9]	Strychnine	5016
[7446-09-5]	Sulfur dioxide	6004
[84-15-1]	α-Terphenyl	5021
[634-66-2]	1,2,3,4-Tetrachlorobenzene	—
[78-00-2]	Tetraethyl lead	2533
[107-49-3]	Tetraethyl pyrophosphate	2504
[109-99-9]	Tetrahydrofuran	1609
[75-74-1]	Tetramethyl lead	2534
[2782-91-4]	Tetramethyl thiourea	3505
[137-26-8]	Thiram	5005
[95-80-7]	2,4-Toluenediamine	5516
[823-40-5]	2,6-Toluenediamine	5516
[584-84-9]	Toluene-2,4-diisocyanate	2535
[120-82-1]	1,2,4-Trichlorobenzene	5517
[129-79-3]	2,4,7-Trinitrofluoren-9-one	7074
[7440-33-7]	Tungsten	7074
[8006-64-2]	Turpentine	1551
[110-62-3]	Valeraldehyde	2536
—	Vanadium oxides	7504
[81-81-2]	Warfarin	5002
—	Welding and brazing fumes (metals)	7200
[7440-66-6]	Zinc	7030
[1314-13-2]	Zinc oxide	7502

APPENDIX K
U.S. EPA METHODS FOR AIR ANALYSIS

There are 14 analytical methods developed by U.S. EPA for measuring common organic pollutants in air. These analytes include aldehydes and ketones, chlorinated pesticides, polynuclear aromatic hydrocarbons, and many volatile organic compounds. These methods may also be applied to analyze other similar substances. All these methods are numbered from TO-1 to TO-14 and based on GC, GC/MS, and HPLC analytical techniques. Method numbers, sampling and analytical techniques, and the types of pollutants are outlined in Table 1, while individual substances are listed in Table 2.

Table 1 Description of U.S. EPA Methods for Air Analysis

Method no.	Sampling and analytical techniques	Types of pollutants
TO-1	Tenax adsorption, GC/MS analysis	Volatile, nonpolar organics in the b.p. range 80° to 200°C
TO-2	Carbon molecular sieve adsorption, GC/MS analysis	Highly volatile, nonpolar organics in the b.p. range –15° to +120°C
TO-3	Cryogenic trapping, GC-FID, or ECD analysis	Volatile, nonpolar organics, in the b.p. range –10°C to +200°C
TO-4	High volume Polyurethane foam sampling, GC/ECD analysis	Chlorinated pesticides and PCBs
TO-5	Derivatization with Dinitrophenylhydrazine in impinger solution, HPLC-UV analysis	Aldehydes and ketones
TO-6	Bubbled through aniline solution; carbanilide formed; analyzed by HPLC	Phosgene
TO-7	Thermosorb/N adsorption; desorbed into methylene chloride; GC/MS analysis	*N*-Nitrosodimethylamine
TO-8	Sodium hydroxide impinger solution, HPLC analysis	Cresol and phenol
TO-9	High volume Polyurethane foam sampling with High Resolution GC/High Resolution MS (HRGC/HRMS) analysis	Polychlorinated dibenzo-*p*-dioxin
TO-10	Low volume Polyurethane foam sampling, GC-ECD analysis	Chlorinated pesticides
TO-11	DNPH-coated adsorbent cartridge; elution with acetonitrile; HPLC analysis	Formaldehyde

Table 1 Description of U.S. EPA Methods for Air Analysis (Continued)

Method no.	Sampling and analytical techniques	Types of pollutants
TO-12	Cryogenic preconcentration and direct flame ionization detection	Nonmethane organic compounds
TO-13	Polyurethane foam/XAD-2 adsorption; GC and HPLC detection	Polynuclear aromatic hydrocarbons
TO-14	SUMMA Passivated Canister sampling; GC analysis using FID, ECD, NPD, PID, or GC/MS	Different types of organics including chlorinated and aromatic compounds

Table 2 Individual Organic Pollutants in Air: U.S. EPA Methods

CAS no.	Aldehydes and ketones	Method(s)
[75-07-0]	Acetaldehyde	TO-5, TO-11
[67-64-1]	Acetone	TO-5, TO-11
[107-02-8]	Acrolein	TO-5, TO-11
[100-52-7]	Benzaldehyde	TO-5, TO-11
[123-72-8]	Butyraldehyde	TO-5, TO-11
[123-73-9]	Crotonaldehyde	TO-5, TO-11
[5779-94-2]	2,5-Dimethylbenzaldehyde	TO-11
[50-00-0]	Formaldehyde	TO-5, TO-11
[66-25-1]	Hexanal	TO-5, TO-11
[78-84-2]	Isobutyraldehyde	TO-5
[590-86-3]	Isovaleraldehyde	TO-5, TO-11
[78-93-3]	Methyl ethyl ketone	TO-5
[110-62-3]	Pentanal	TO-5, TO-11
[123-38-6]	Propanal	TO-5, TO-11
[620-23-5]	*m*-Tolualdehyde	TO-5, TO-11
[529-20-4]	*o*-Tolualdehyde	TO-5, TO-11
[104-87-0]	*p*-Tolualdehyde	TO-5, TO-11
[110-62-3]	Valeraldehyde	TO-5, TO-11
[95-48-7]	*o*-Cresol	TO-8
[108-39-4]	*m*-Cresol	TO-8
[106-44-5]	*p*-Cresol	TO-8
	Miscellaneous Pollutants	
[75-69-4]	Freon-11	TO-14
[75-75-8]	Freon-12	TO-14
[76-13-1]	Freon-113	TO-14
[76-14-1]	Freon-114	TO-14
[118-74-1]	Hexachlorobenzene	TO-10
[87-68-3]	Hexachlorobutadiene	TO-14
[319-84-6]	α-Hexachlorocyclohexane	TO-10
[77-47-4]	Hexachlorocyclopentadiene	TO-10
[98-95-3]	Nitrobenzene	TO-1, TO-3
[62-75-9]	*N*-Nitrosodimethylamine	TO-7
[108-95-2]	Phenol	TO-8
[75-44-5]	Phosgene	TO-6
	Pesticides, Chlorinated	
[309-00-2]	Aldrin	TO-4, TO-10

[133-06-2]	Captan	TO-10
[57-74-9]	Chlordane	TO-4
[1897-45-6]	Chlorothalonil	TO-10
[2921-88-2]	Chlorpyrifos	TO-10
[72-55-9]	4,4′-DDE	TO-4, TO-10
[50-29-3]	4,4′-DDT	TO-4, TO-10
[62-73-7]	Dichlorovos	TO-10
[115-32-2]	Dicofol	TO-10
[60-57-1]	Dieldrin	TO-10
[72-20-8]	Endrin	TO-10
[7421-93-4]	Endrin aldehyde	TO-10
[133-07-3]	Folpet	TO-10
[76-44-8]	Heptachlor	TO-10
[1024-57-3]	Heptachlor epoxide	TO-10
[58-89-9]	Lindane	TO-10
[72-43-5]	Methoxychlor	TO-10
[315-18-4]	Mexacarbate	TO-10
[2385-85-5]	Mirex	TO-10
[39765-80-5]	*trans*-Nonachlor	TO-10
—	Oxychlordane	TO-10
[299-84-3]	Ronnel	TO-10
[608-93-5]	Pentachlorobenzene	TO-10
[87-86-5]	Pentachlorophenol	TO-10
—	Polychlorinated Biphenyls	TO-4, TO-9
—	Polychlorinated dibenzo-p-dioxins	TO-9
[57653-85-7]	1,2,3,4,7,8-H_xCDD	TO-9
[3268-87-9]	Octachlorodibenzo-p-dioxin	TO-9
—	1,2,3,4-TCDD	TO-9
[1746-01-6]	2,3,7,8-TCDD	TO-9

Polynuclear Aromatic Hydrocarbons

[83-32-9]	Acenaphthene	TO-13
[208-96-8]	Acenaphthylene	TO-13
[120-12-7]	Anthracene	TO-13
[56-55-3]	Benz[*a*]anthracene	TO-13
[50-32-8]	Benzo[*a*]pyrene	TO-13
[205-99-2]	Benzo[*b*]fluoranthene	TO-13
[191-24-2]	Benzo[*g,h,i*]perylene	TO-13
[207-08-9]	Benzo[*k*]fluoranthene	TO-13
[218-01-9]	Chrysene	TO-13
[53-70-3]	Dibenz[*a,h*]anthracene	TO-13
[206-44-0]	Fluoranthene	TO-13
[86-73-7]	Fluorene	TO-13
[193-39-5]	Indeno[1,2,3-*cd*]pyrene	TO-13
[91-20-3]	Napththalene	TO-13
[85-01-8]	Phenanthrene	TO-13
[129-00-0]	Pyrene	TO-13

Volatile Organic Compounds TO-1, TO-2, TO-3, TO-14

[107-13-1]	Acrylonitrile	TO-2, TO-3
[107-05-1]	Allyl chloride	TO-2, TO-3
[71-43-2]	Benzene	TO-1, TO-2, TO-3, TO-4

Table 2 Individual Organic Pollutants in Air: U.S. EPA Methods (Continued)

CAS no.	Aldehydes and ketones	Method(s)
[56-23-5]	Carbon tetrachloride	TO-1, TO-2, TO-3, TO-4
[108-90-7]	Chlorobenzene	TO-1, TO-2, TO-14
[67-66-3]	Chloroform	TO-1, TO-2, TO-3, TO-14
[126-99-8]	Chloroprene (2-Chloro-1,3-butadiene)	TO-1, TO-3
[98-82-8]	Cumene	TO-1
[106-93-4]	1,2-Dibromoethane	TO-14
[95-50-1]	1,2-Dichlorobenzene	TO-14
[541-73-1]	1,3-Dichlorobenzene	TO-14
[106-46-7]	1,4-Dichlorobenzene	TO-1, TO-3, TO-14
[75-34-3]	1,1-Dichloroethane	TO-14
[107-06-2]	1,2-Dichloroethylene	TO-14
[78-87-5]	1,2-Dichloropropane	TO-14
[142-28-9]	1,3-Dichloropropane	TO-14
[100-41-4]	Ethyl benzene	TO-14
[75-00-3]	Ethyl chloride	TO-14
[106-93-4]	Ethylene dibromide	TO-1
[107-06-02]	Ethylene dichloride (1,2-Dichloroethane)	TO-1, TO-2, TO-3, TO-14
[622-96-8]	4-Ethyltoluene	TO-14
[74-87-3]	Methyl chloride (Chloromethane)	TO-14
[75-09-2]	Methylene chloride (Dichloromethane)	TO-2, TO-3, TO-14
[79-34-5]	1,1,2,2-Tetrachloroethane	TO-14
[127-18-4]	Tetrachloroethylene	TO-14
[108-88-3]	Toluene	TO-1, TO-2, TO-3, TO-14
[75-25-2]	Tribromomethane (Bromoform)	TO-1
[71-55-6]	1,1,1-Trichloroethane (Methyl chloroform)	TO-1, TO-2, TO-3, TO-14
[79-00-5]	1,1,2-Trichloroethane	TO-14
[79-01-5]	Trichloroethylene	TO-1, TO-2, TO-3, TO-14
[100-42-5]	Vinyl benzene (Styrene)	TO-14
[75-01-4]	Vinyl chloride	TO-2, TO-3, TO-14
[75-35-4]	Vinylidine chloride (1,1-Dichloroethylene)	TO-2, TO-3, TO-14
[95-47-6]	*o*-Xylene	TO-1, TO-2, TO-14
[108-38-3]	*m*-Xylene	TO-1, TO-2, TO-14
[106-42-3]	*p*-Xylene	TO-1, TO-2, TO-14

APPENDIX L
INORGANIC TEST PROCEDURES FOR ANALYSIS OF AQUEOUS SAMPLES: EPA, SM, AND ASTM REFERENCE METHODS

Analytes, types of analysis	U.S. EPA methods	Standard method (Vol. 18)	ASTM
Acidity, as mg $CaCO_3$/L			
Electrometric or	305.1	2310 B	D1067-88
Phenolphthalein titration	305.1	2310 B	—
Alkalinity, as mg $CaCO_3$/L			
Electrometric or colorimetric	310.1	2310 B	D1067-88
Manual titration	310.1	2320 B	—
Automate	310.2	—	—
Aluminum			
AA direct aspiration	202.1	3500-A1 B	—
AA furnace	202.2	3500-A1 B	—
ICP	200.7	3500-A1 C	D4190-88
DCP	—	—	—
Colorimetric (eriochrome cyanine R)	—	3500-A1 D	—
Ammonia (as N)			
Colorimetric Nesslerization	350.2	4500-NH_3C	D1426-79(A)
Titration	350.2	4500-NH_3E	—
Electrode	350.3	4500-NH_3F	D1426-79(D)
Automated phenate	350.1	4500-NH_3H	D1426-79(C)
Colorimetric phenate	—	4500-NH_3D	—
Antimony			
AA direct aspiration	204.1	3500-Sb B	—
AA furnace	204.2	3500-Sb B	—
ICP	200.7	3500-Sb C	—
Arsenic			
AA gaseous hydride	206.3	3500-As B	D2972-84(B)
AA furnace	206.2	3500-As B	—
ICP	200.7	3500-As D	—
Colorimetric (SDDC)	206.4	3500-As C	D2972-84(A)
Barium			
AA direct aspiration	208.1	3500-Ba B	—
AA furnace	208.2	3500-Ba B	—

Inorganic Test Procedures for Analysis of Aqueous Samples: EPA, SM, and ASTM Reference Materials (Continued)

Analytes, types of analysis	U.S. EPA methods	Standard method (Vol. 18)	ASTM
ICP	200.7	3500-Ba C	—
Beryllium			
AA direct aspiration	210.1	3500-Be B	D3645-84-88(A)
AA furnace	210.2	3500-Be B	—
ICP	200.7	3500-Be C	—
DCP		3500-Be D	D4190-88
Colorimetric (aluminon)			—
Biochemical Oxygen Demand (dissolved oxygen depletion)	405.1	5210 B	—
Boron			
Colorimetric (curcumin)	212.3	4500-B B	—
TCP	200.7	4500-B D	—
DCP	200.7	4500-B D	D4190-88
Bromide			
Colorimetric, phenol red	—	4500-Br B	—
Ion chromatography	—	4110	—
Titrimetric	320.1	—	D1246-82C
Cadmium			
AA direct aspiration	213.1	3500-Cd B	D3557-90(A,B)
AA furnace	213.2	3500-Cd B	
ICP	200.7	3500-Cd C	
DCP			D4190-90
Colorimetric (dithiozone)		3500-Cd D	
Calcium			
AA direct aspiration	215.1	3500-Ca B	D511-88(B)
ICP	200.7	3500-Ca C	
Titrimetric (EDTA)	215.2	3500-Ca D	D511-88(A)
Chemical oxygen demand			
Titrimetric	410.1	5220 C	D1252-88
Colorimetric	410.4	5220 D	—
Chloride			
Titrimetric (silver nitrate)	—	4500-Cl^- B	D512-89(B)
Titrimetric (mercuric nitrate)	325.3	4500-Cl^- C	D512-89(A)
Colorimetric, manual	—	—	D512-89(c)
Automated (ferricyanide)	325.1 325.2	4500-Cl^- E	—
Potentiometric titration	—	4500-Cl^- D	—
Ion chromatography	—	4110	—
Chlorine, residual			
Amperometric titration	330.1	4500-Cl E	D1253-76(A)
Iodometric direct titration	330.3	4500-Cl B	D1253-76(B)
Iodometric back titration	330.2	4500-Cl C	—
DPD-FAS titration	330.4	4500-Cl F	—
DPD-colorimetric	330.5	4500-Cl G	—
Chromium VI (dissolved)			
AA chelation extraction	218.4	3111 C	—
Colorimetric-diphenylcarbazide	—	3500-Cr D	—
Chromium, total			
AA direct aspiration	218.1	3500-Cr B	D1687-86(D)

AA chelation extraction	218.3	3111 C	—
AA furnace	218.2	3500-Cr B	—
ICP	200.7	3500-Cr C	—
DCP	—	—	D4190-88
Colorimetric-di-phenylcarbazide	—	3500-Cr D	D1687-86(A)
Cobalt			
AA direct aspiration	219.1	3500-Co B	D3558-90
AA furnace	219.2	3500-Co B	—
ICP	200.7	3500-Co C	—
DCP	—	—	D4190-88
Color			
Colorimetric (ADMI)	110.1	2120 E	—
Platinum-cobalt	110.2	2120 B	—
Spectrophotometric	110.3	2120 C	—
Copper			
AA direct aspiration	220.1	3500-Cu B	D1688-90 (A,B)
AA furnace	220.2	3500-Cu B	
ICP	200.7	3500-Cu C	
DCP	—	—	D4190-88
Colorimetric (neo cuproine)	—	3500-Cu D	D1688-84(88)(A)
Cyanide, total			
Titrimetric	—	4500-CN^- D	—
Spectrophotometric	335.2	4500-CN^- D	D2036-89(A)
Automated	335.3	—	D2036-89(A)
Electrode	—	4500-CN^- F	—
Cynanide amenable to chlorination	335.1	4500-CN^- G,H	D2036-89(B)
Fluoride			
Electrode	340.2	4500-F^- C	D1179-88(B)
Metric (SPADNS)	340.1	4500-F^- D	D1179-88(A)
Automated complexone	340.3	4500-F^- E	—
Ion chromatography	—	4110	—
Gold			
Direct aspiration	231.1	3500-Au B	
AA furnace	231.2	3500-Au B	
Hardness			
Automated colorimetric	130.1	3500-Ir B	—
EDTA titration or Ca & Mg by AA	130.2	3500-Ir B	D1126-86 (1990)
Iridium			
AA direct aspiration	235.1	3500-Ir B	—
AA furnace	232.2	3500-Ir B	—
Iron			
AA direct aspiration	236.1	3500-Fe B	D1068-90(A,B)
AA furnace	236.2	3500-Fe B	
ICP	200.7	3500-Fe C	
Colorimetric (phenanthroline)	—	3500-Fe D	D1068-90(D)
Kjeldahl nitrogen	—	4500-N_{org}	3590-84(A)
Titration	351.3	B or C	D3590-89(A)
Nesslerization	351.3	—	D3590-89(A)
Elecrode	351.3	—	—
Automated phenate	351.1	—	
Semi-automated block digester	351.2	—	D3590-89(B)
Potentiometric	351.4	—	D3590-89(A)
Lead			
AA direct aspiration	239.1	3500-Pb B	D3559-90(A,B)
AA furnace	239.2	3500-Pb B	—

Inorganic Test Procedures for Analysis of Aqueous Samples: EPA, SM, and ASTM Reference Materials (Continued)

Analytes, types of analysis	U.S. EPA methods	Standard method (Vol. 18)	ASTM
ICP	200.7	3500-Pb C	—
DCP	—	—	D4190-88
Voltametry	—	—	D3559-90(C)
Colorimetric (dithiozone)	—	3500-Pb D	—
Magnesium			
AA direct aspiration	242.1	3500-Mg B	D511-88(B)
ICP	200.7	3500-Mg C	
Gravimetric	—	3500-Mg D	D511-77(A)
Manganese			
AA direct aspiration	243.1	3500-Mn B	D858-90(A,B)
AA furnace	243.2	3500-Mn B	—
ICP	200.7	3500-Mn C	
DCP	—	—	D4190-88
Colorimetric, persulfate	—	3500-MnD	D858-84(A) (1988)
Mercury			
Cold vapor, manual	245.1	3500-Hg B	D3223-86
Automated	245.2	—	—
Molybdenum			
AA direct aspiration	246.1	3500-Mo B	—
AA furnace	246.2	3500-Mo B	
ICP	200.7	3500-MoC	
Nickel			
AA direct aspiration	249.1	3500-Ni B	D1886-90(A,B)
AA furnace	249.2	3500-Ni B	—
ICP	200.7	3500-Ni C	—
DCP	—	—	D4190-88
Nitrate (as N)			
Electrode method	—	4500-NO_3^- C	—
Ion chromatography	—	4110	—
Colorimetric	352.1	—	D992-71
Nitrate-nitrite (as N)			
Cd reduction, manual	353.3	4500-NO_3^- E	D3867-90(B)
or automated	353.2	4500-NO_3^- F	D3867-90(A)
Hydrazine reduction (automated)	353.1	4500-NO_3^- H	—
Nitrite (as N)			
Colorimetric, manual	354.1	4500-NO_2^- B	D1254-67
Oil and grease			
Gravimetric	413.1	5520 B	—
Infrared method	—	5520 C	—
Orthophosphate (as P)			
Ascorbic acid method	—	4500 P E	—
Automated	365.1	—	
Manual, single reagent	365.2	—	D515.88(A)
Manual, two reagent	365.3	—	—
Vanadomolyldophosphoric acid	—	4500-PC	—
Osmium			
AA direct aspiration	252.1	3500-Os B	—
AA furnace	252.2	3500-Os B	—
Oxygen, dissolved			

Winkler (azide modification)	360.2	4500-O C	D888-81 (C) (1988)
Electrode	360.1	4500-O G	—
Ozone, residual			
Colorimetric, indigo	—	4500-O_3	—
pH electrometric	150	4500-H^+	D1293-84(AB) (1990)
Palladium			
AA direct aspiration	253.1	3500-Pd B	—
AA furnace	253.2	3500-Pd B	—
Phenols	—	5530	—
Colorimetric, manual	420.1	—	D1783-80(A,B)
Automated	420.2	—	—
Phosphorus, total	—	4500 P	—
Colorimetric, manual	365.2	—	D515-88(A)
Automated, ascorbic acid	365.1	—	—
Semiautomated	365.4	—	—
Platinum			
AA direct aspiration	255.1	3500-Pt B	—
AA furnace	255.2	3500-Pt B	—
Potassium			
AA direct aspiration	258.1	3500-K B	—
ICP	200.7	3500-K C	—
Flame photometric	—	3599-K D	D1428-82(A)
Residue, total	160.3	2540 B	—
Filterable	160	2540 C	—
Nonfilterable (TSS)	160.2	2540 D	—
Settleable	160.5	2540 F	—
Volatile	160.4	2540 G	—
Rhodium			
AA direct aspiration	265.1	3500-Rh B	—
AA furnace	265.2	3500-Rh B	—
Ruthenium			
AA direct aspiration	267.1	3500-Ru B	—
AA furnace	267.2	3500-Ru B	—
AA gaseous hydride	270.3	3500-Se C	D3859-88(A)
Silica, dissolved			
Colorimetric	370.1	—	D859-88(B)
Silver			
AA direct aspiration	272.1	3500-Ag B	—
AA furnace	272.2	3500-Ag B	—
ICP	200.7	3500-Ag C	—
Sodium			
AA direct aspiration	273.1	3500-Na B	—
ICP	200.7	3500-Na C	—
Flame photometric	—	3500-Na D	D1428-82(A)
Specific conductance			
Wheatstone bridge	120.1	2510	D1125-82(A)
Sulfate			
Gravimetric	375.3	4500-SO_4^{2-} C,D	D516-82A (1988)
Turbidimetric	375.4	4500-SO_4^{2-} E	D516-88
Colorimetric, automated (barium chloranilate)	375.1	4500-SO_4^{2-} F	—
Ion chromatography	—	4110	—
Sulfide			
Iodometric titration	376.1	4500-S^{2-} E	—
Colorimetric (methylene blue)	376.2	4500-S^{2-} D	—

Inorganic Test Procedures for Analysis of Aqueous Samples: EPA, SM, and ASTM Reference Materials (Continued)

Analytes, types of analysis	U.S. EPA methods	Standard method (Vol. 18)	ASTM
Sulfite			
Titrimetric	377.1	4500-SO_3^{2-} B	D1339-84(C)
Surfactants			
Colorimetric (methylene blue)	425.1	5540 C	D2330-88
Temperature	170.1	2550	—
Thallium			
AA direct aspiration	279.1	3500-TLB	—
AA furnace	279.2	3500-TLB	—
ICP	200.7	3500-TLC	—
Tin			
AA direct aspiration	282.1	3500-Sn B	—
AA furnace	282.2	3500-Sn B	—
Titanium			
AA direct aspiration	283.1	3500-Ti B	—
AA furnace	283.2	3500-Ti B	—
Total organic carbon (TOC) combustion	415.1	—	D2579-85(A,B)
Turbidity			
Nephelometric	180.1	—	D1889-88(A)
Vanadium			
AA direct aspiration	286.1	3500-V B	—
AA furnace	286.2	3500-V B	—
ICP	200.7	3500-V C	—
DCP	—	—	D4190-88
Colorimetric (gallic acid)	—	3500-V D	D3373-84(A) (1988)
Zinc			
AA direct aspiration	289.1	3500-Zn B	D1691.90(A,B)
AA furnace	289.2	3500-Zn B	—
ICP	200.7	3500-Zn C	—
DCP	—	—	D4190-88
Colorimetric			
Dithizon	—	3500-Zn D,E	—
Zincon	—	3500-Zn F	—

APPENDIX M
THE U.S. EPA'S ANALYTICAL METHODS FOR ORGANIC POLLUTANTS IN AQUEOUS AND SOLID MATRICES

The U.S. EPA's major analytical methods for the determination of organic pollutants in aqueous and solid matrices are presented in the following table. Organic compounds, their CAS registry numbers, the types or classes in which they belong, the EPA method numbers, and the instrumental techniques for analysis are listed in the table. The sample extraction techniques, however, are not mentioned. This is excluded because most of these compounds can be extracted by more than one specific method. Readers should refer to the specific EPA methods for a full detailed procedures, including sample preparation, cleanup and chromatographic conditions.

The analytical methods in the series 500 address to potable waters, while the methods in the 600 series refer to analysis of wastewater. Methods in the 8000 series refer to the instrumental analysis of organic pollutants in several types of matrices, including groundwater, soils, sediments, and hazardous wastes. Methods for sample extractions for the 8000 series are written separately under the 3000 series, which should go in conjunction with the 8000 series. The extraction methods under the 3000 series are not presented in this appendix. All of these three parallel series of methods under 500, 600, and 8000 do not differ significantly from each other. Also, there are a few methods under 1600 that are similar to the above methods. These methods are not included in the table.

The instrumentation techniques are primarily GC, GC/MS, and HPLC. The term GC-HSD refer to the Halide Specific Detector, which includes Hall Electrolytic Conductivity Detector and Microcoulometric Detector. For such analysis, either of these detectors may be used. An Electron Capture Detector (ECD) may be effective to a lesser extent for GC-HSD measurement. NPD and FPD refer to Nitrogen-Phosphorus Detector and Flame Photometric Detector, respectively. GC-FID (Flame Ionization Detector) and GC/MS (Mass Spectrometer) are the most versatile instrumental techniques. HPLC methods primarily require the use of either a Fluorescence Detector (HPLC-FL) or an UV Detector (HPLC-UV).

The U.S. EPA's Analytical Methods for Organic Pollutants in Aqueous and Solid Matrices

Compounds	CAS no.	Type/class	U.S. EPA method #	Instrumental techniques
Acenaphthene	83-32-9	PAH	550.1	HPLC-UV
			610	HPLC-UV
			625	GC/MS
			8310	HPLC-UV
			8270	GC/MS
			OLMO1	GC/MS
Acenaphthylene	208-96-8	Polynuclear aromatic hydrocarbon	525	GC/MS
			550.1	HPLC-UV
			610	HPLC-UV
			625	GC/MS
			8310	HPLC-UV
			8270	GC/MS
			OLMO1	GC/MS
Acetamidofluorene	53-96-3	Polynuclearamide	8270	GC/MS
Acetone (2-propanone)	67-64-1	VOC(ketone)	OLMO1	GC/MS
Acetophenone (1-penylethanone)	98-86-2	Aromatic ketone	8270	GC/MS
1-Acetyl-2-thiourea	591-08-2	Thiourea	8270	GC/MS
Aciflurofen	50594-66-6	Pesticide, chlorinated	515.1	GC-ECD
Acrolein (2-propenal)	107-02-8	Aldehyde (VOC)	603	GC-FID
			624	GC/MS
Acrylonitrile (vinyl cyanide)	107-13-1	VOC (nitrate)	603	GC-FID
			624	GC/MS
Alachlor (Metachlor, Lasso)	15972-60-8	Pesticide, chlorinated	505	GC-ECD
			507	GC-NPD
			525	GC/MS
Aldicarb (Temik)	116-06-3	Pesticide, carbamate	531.1	HPLC-FL
Aldicarb sulfone (Aldoxycarb)	1646-88-4	Pesticide, carbamate	531.1	HPLC-FL
Aldicarb sulfoxide	1646-87-3	Pesticide, carbamate	531.1	HPLC-FL

Aldrin	309-00-2	Pesticide, organochlorine	505	GC-ECD
			508	GC-ECD
			525	GC/MS
			608	GC-ECD
			625	GC/MS
			8080	GC-ECD
			8270	GC/MS
			OLMO1	GC-ECD
Allyl chloride (3-chlor-1-propene)	107-05-1	Halogenated hydrocarbon (VOC)		
2-Aminoanthraquinone	117-79-3	Aromatic ketone	8270	GC/MS
4-Aminobiphenyl	92-67-1	Aromaticamine	8270	GC/MS
Anilazine (Dyrene)	101-05-3	Triazine	8270	GC/MS
Aniline (aminobenzene)	62-53-3	Aromaticamine	8270	GC/MS
o-Anisidine	90-04-0	Aromatic amine	8270	GC/MS
Anthracene	120-12-7	Polynuclear aromatic hydrocarbons	525	GC/MS
			550.1	HPLC-FL
			610	HPLC-FL
			625	GC/MS
			8310	HPLC-FL
			8270	GC/MS
			OLMO1	GC/MS
Aramite	140-57-8	Pesticide, organosulfur	8270	GC/MS
Aroclor-1016 (PCB-1016)	12674-11-2	Polychlorinated biphenyl	505	GC-ECD
			508	GC-ECD
			608	GC-ECD
			625	GC/MS
			8080	GC-ECD
			8270	GC/MS
			OLMO1	GC-ECD
Aroclor-1221 (PCB-1221)	11104-28-2	Polychlorinated biphenyl	505	GC-ECD
			508	GC-ECD
			608	GC-ECD

Compounds	CAS no.	Type/class	U.S. EPA method #	Instrumental techniques
Aroclor-1221 (PCB-1221) (continued)			625	GC/MS
			8080	GC-ECD
			8270	GC/MS
			OLMO1	GC-ECD
Aroclor-1232 (PCB-1232)	11141-16-5	Polychlorinated biphenyl	505	GC-ECD
			508	GC-ECD
			608	GC-ECD
			625	GC/MS
			8080	GC-ECD
			8270	GC/MS
			OLMO1	GC-ECD
Aroclor-1242 (PCB-1242)	53469-21-9	Polychlorinated biphenyl	505	GC-ECD
			508	GC-ECD
			608	GC-ECD
			625	GC/MS
			8080	GC-ECD
			8270	GC/MS
			OLMO1	GC-ECD
Aroclor-1248 (PCB-1248)	12672-29-6	Polychlorinated biphenyl	505	GC-ECD
			508	GC-ECD
			608	GC-ECD
			625	GC/MS
			8080	GC-ECD
			8270	GC/MS
			OLMO1	GC-ECD
Aroclor-1254	11097-69-1	Polychlorinated biphenyl	505	GC-ECD
			508	GC-ECD
			608	GC-ECD
			8080	GC-ECD

			625	GC/MS
			8270	GC/MS
			OLMO1	GC-ECD
Aroclor-1260 (PCB-1260)	11096-82-5	Polychlorinated biphenyl	505	GC-ECD
			508	GC-ECD
			608	GC-ECD
			625	GC/MS
			8080	GC-ECD
			8270	GC/MS
			OLMO1	GC-ECD
Atrazine	1912-24-9	Pesticide, triazine	505	GC-ECD
			507	GC-NPD
			525	GC/MS
Avadex	2303-16-4	Pesticide, carbamate	8270	GC/MS
Azinphos-methyl (Guthion)	86-50-0	Organophosphorus pesticide	8140	GC-FPD
			8270	GC/MS
Barban (Carbyn)	101-27-9	Pesticide, carbamate	8270	GC/MS
Basalin	33245-39-5	Pesticide, chlorinated	8270	GC/MS
Baygon (Propoxur)	114-26-1	Pesticide, carbamate	531.1	HPLC-FL
Benzal chloride (dichloromethyl benzene)	98-87-3	Halogenated hydrocarbon	8120	GC-ECD
Benzene	71-43-2	Aromatic	502.2	GC-PID
			524.2	GC/MS
			602	GC-PID
			624	GC/MS
			8260	GC/MS
			8020	GC-PID
			OLMO1	GC/MS
1,4-Benzenediamine (*p*-phenylenediamine)	106-50-3	Aromatic amine	8270	GC/MS
Benzidine	92-87-5	Aromaticamine	605	HPLC-EL
			625	GC/MS
			8270	GC/MS

Compounds	CAS no.	Type/class	U.S. EPA method #	Instrumental techniques
Benzo[*a*]anthracene	56-55-3	PAH	550.1	HPLC-FL
			525	GC/MS
			610	HPLC-FL
			625	GC/MS
			8310	HPLC-FL
			8270	GC/MS
			OLMO1	GC/MS
Benzo[*b*]fluoranthene	205-99-2	Polynuclear aromatic hydrocarbon	550.1	HPLC-FL
			610	HPLC-FL
			625	GC/MS
			8310	HPLC-FL
			8270	GC/MS
			OLMO1	GC/MS
Benzo[*j*]fluoranthene	205-82-3	Polynuclear aromatic hydrocarbon	525	GC/MS
Benzo[*k*]fluoranthene	207-08-9	Polynuclear aromatic hydrocarbon	525	GC/MS
			550.1	HPLC-FL
			610	HPLC-FL
			625	GC/MS
			8310	HPLC-FL
			8270	GC/MS
			OLMO1	GC/MS
Benzo[*g,h,i*]perylene	191-24-2	Polynuclear aromatic hydrocarbon	525	GC/MS
			550.1	HPLC-FL
			610	HPLC-FL
			625	GC/MS
			8310	HPLC-FL
			8270	GC/MS
			OLMO1	GC/MS
Benzo[*a*] pyrene	50-32-8	Polynuclear aromatic hydrocarbon	550.1	HPLC-FL

			525	GC/MS
			610	HPLC-FL
			625	GC/MS
			8310	HPLC-FL
			8270	GC/MS
			OLMO1	GC/MS
Benzoic acid (benzenecarboxylic acid)	65-85-0	Carboxylic acid	8270	GC/MS
Benzyl alcohol	100-51-6	Aromatic alcohol	8270	GC/MS
Benzyl chloride [(chloromethyl) benzene]	100-44-7	Chlorinated hydrocarbon	8120	GC-ECD
BHC (Hexachlorocyclohexane)	608-73-1	Pesticide, chlorinated	8120	GC-ECD
α-BHC	319-84-6	Pesticide, chlorinated	508	GC-ECD
			608	GC-ECD
			625	GC/MS
			8080	GC-ECD
			8270	GC/MS
			OLMO1	GC-ECD
β-BHC	319-85-7	Pesticide, chlorinated	508	GC-ECD
			608	GC-ECD
			625	GC/MS
			8080	GC-ECD
			8270	GC/MS
			OLMO1	GC-ECD
δ-BHC	319-86-8	Pesticide, chlorinated	508	GC-ECD
			608	GC-ECD
			625	GC/MS
			8080	GC-ECD
			8270	GC/MS
			OLMO1	GC-ECD
Bromacil	314-40-9	Pesticide, urea	507	GC-NPD
Bromobenzene	108-86-1	Halogenated hydrocarbon	502.2	GC-ELCD/PID
			524.2	GC/MS
			8260	GC/MS

Compounds	CAS no.	Type/class	U.S. EPA method #	Instrumental techniques
Bromochloromethane (chlorobromomethane)	74-97-5	VOC (halocarbon)	502.2	GC-ELCD
			524.2	GC/MS
			8260	GC/MS
Bromoform	75-25-2	VOC (halocarbon)	502.2	GC-ELCD
			524.2	GC/MS
			601	GC-HSD
			624	GC/MS
			8260	GC/MS
			OLMO1	GC/MS
4-Bromophenylphenylether (1-bromo-4-phenoxybenzene)	101-55-3	Haloether	611	GC-HSD
			625	GC/MS
			8270	GC/MS
			OLMO1	GC/MS
Bromoxynil	1689-84-5	Bromonitrile	8270	GC/MS
n-Butyl benzene	104-51-8	VOC (aromatic)	502.2	GC-PID
			524.2	GC/MS
			8260	GC/MS
sec-Butylbenzene	135-98-8	Aromatic	502.2	GC-PID
			524.2	GC/MS
			8260	GC/MS
tert-Butylbenzene	98-06-6	Aromatic	502.2	GC-ELCD
			524.2	GC/MS
			8260	GC/MS
Butyl benzyl phthalate	85-68-7	Phthalate ester	525	GC/MS
			606	GC-ECD
			625	GC/MS
			8060	GC-ECD/FID
			8270	GC/MS
			OLOM1	GC/MS

Captafol (Difolatan)	2425-06-1	Pesticide, thiocarboximide	8270	GC/MS
Captan (Ortholide-406)	133-06-2	Pesticide, carbamate	8270	GC/MS
Carbaryl	63-25-2	Pesticide, carbamate	531.1	HPLC-FL
			8270	GC/MS
Carbazole	86-74-8	Heteroclic nitrogen compound	OLMO1	GC/MS
Carbofuran (Furadan)	1563-66-2	Pesticide, carbamate	531.1	HPLC-FL
			8270	GC/MS
Carbon disulfide	75-15-0	Organic sulfide	OLMO1	GC/MS
Carbon tetrachloride	56-23-5	VOC (halogenated organic)	502.2	GC-ELCD
			524.2	GC/MS
			601	GC-HSD
			624	GC/MS
			8260	GC/MS
			OLMO1	GC/MS
CDEC (Sulfallate)	95-06-7	Pesticide, thiocarbamate	8270	GC/MS
Chlordane	57-74-9	Pesticide, chlorinated	505	GC-ECD
			508	GC-ECD
			608	GC-ECD
			625	GC/MS
			8080	GC-ECD
			8270	GC/MS
α-Chlordane	5103-71-9	Pesticide, chlorinated	505	GC-ECD
			508	GC-ECD
			525	GC/MS
			608	GC-ECD
			625	GC/MS
			OLMO1	GC-ECD
γ-Chlordane	5103-74-2	Pesticide, organochlorine	505	GC-ECD
			508	GC-ECD
			525	GC/MS
			608	GC-ECD
			625	GC/MS

Compounds	CAS no.	Type/class	U.S. EPA method #	Instrumental techniques
γ-Chlordane (continued)			OLMO1	GC-ECD
Chlorfenvinphos	470-90-6	Pesticide, organophosphorus	8270	GC/MS
4-Chloro-3-methylphenol	59-50-7	Phenol	604	GC-ECD/FID
			625	GC/MS
			8040	GC-ECD/FID
			8270	GC/MS
			OLMO1	GC/MS
4-Chloroaniline	106-47-8	Halogenated amine	8270	GC/MS
			OLMO1	GC/MS
Chlorobenzene	108-90-7	Halogenated hydrocarbon	502.2	GC-ELCD/PID
			524.2	GC/MS
			601	GC-HSD
			602	GC-PID
			624	GC/MS
			8260	GC/MS
			OLMO1	GC/MS
Chlorobenzilate (Acaraben)	510-15-6	Pesticide, chlorinated	508	GC-ECD
			8270	GC/MS
bis (2-Chloroethoxy) methane	111-91-1	Haloether	611	GC-HSD
			625	GC/MS
			8270	GC/MS
			OLMO1	GC/MS
2-Chloroethylvinyl ether	110-75-8	Haloether	601	GC-HSD
			624	GC/MS
Chloroform	67-66-3	VOC (halocarbon)	502.2	GC-ELCD
			524.2	GC/MS
			601	GC-HSD
			624	GC/MS
			8260	GC/MS

			OLMO1	GC/MS
bis (2-Chloroisopropyl) ether	108-60-1	Haloether	625	GC/MS
			8270	GC/MS
			OLMO1	GC/MS
3-Chloromethylpyridine hydrochloride	6959-48-4	Pyridine, substituted	8270	GC/MS
1-Chloronaphthalene	90-13-1	Chlorinated hydrocarbon	8270	GC/MS
2-Chloronaphthalene	91-58-7	Chlorinated hydrocarbon	612	GC-ECD
			625	GC/MS
			8120	GC-ECD
			8270	GC/MS
			OLMO1	GC/MS
5-Chloro-*o*-toluidine	95-79-4	Aromatic amine	8270	GC/MS
2-Chlorophenol	95-57-8	Phenol	604	GC-FID/ECD
			625	GC/MS
			8040	GC-ECD/FID
			8270	GC/MS
			OLMO1	GC/MS
4-Chlorophenyl phenyl ether (1-chloro-4-phenoxybenzene)	7005-72-3	Haloether	625	GC/MS
			8270	GC/MS
			OLMO1	GC/MS
Chlorothalonil (Daconil 2787)	1897-45-6	Pesticide, chlorinated	508	GC-ECD
o-Chlorotoluene	95-49-8	VOC (halocarbon)	502.2	GC-ELCD/PID
			524.2	GC/MS
			8260	GC/MS
p-Chlorotoluene	106-43-4	VOC (halogenated hydrocarbon)	502.2	GC-ELCD/PID
			524.2	GC/MS
			8260	GC/MS
Chrysene	218-01-9	Polynuclear aromatic hydrocarbon	525	GC/MS
			550.1	HPLC-FL
			610	HPLC-UV
			625	GC/MS
			8310	HPLC-FL

			8270	GC/MS

Compounds	CAS no.	Type/class	U.S. EPA method #	Instrumental techniques
Chrysene (continued)			OLMO1	GC/MS
Coumaphos	56-72-4	Pesticide, organophosphorus	8140	GC-FPD
			8270	GC/MS
m-Cresol (3-methylphenol)	108-39-4	Phenol	8270	GC/MS
o-Cresol (2-methylphenol)	95-48-7	Phenol	8270	GC/MS
			OLMO1	GC/MS
p-Cresol (4-methylphenol)	106-44-5	Phenol	8270	GC/MS
			OLMO1	GC/MS
Cresylic acid	1319-77-3	Phenol	8040	GC-FID
Crotoxyphos	7700-17-6	Pesticide, organophosphorus	8270	GC/MS
Cumene (isopropylbenzene)	98-82-8	VOC (aromatic)	502.2	GC-PID
			524.2	GC/MS
			8260	GC/MS
p-Cymene (*p*-isopropyl toluene)	99-87-6	VOC (aromatic)	502.2	GC-PID
			524.2	GC/MS
			8260	GC/MS
2,4-D	94-75-7	Chlorophenoxy acid herbicide	515.1	GC-ECD
			8150	GC-ECD
Dacthal (DCPA)	1861-32-1	Chlorinated aromatic	508	GC-ECD
Dalapon (2,2-dichloropropanoic acid)	75-99-0	Pesticide, chlorinated	515.1	GC-ECD
			8150	GC-ECD
4,4′-DDD	72-54-8	Chlorinated pesticide	508	GC-ECD
			608	GC-ECD
			625	GC/MS
			8080	GC-ECD
			8270	GC/MS
			OLMO1	GC-ECD
4,4′-DDE	72-55-9	Chlorinated pesticide	508	GC-ECD

			608	GC-ECD
			625	GC/MS
			8080	GC-ECD
			8270	GC/MS
			OLMO1	GC-ECD
4,4′-DDT	50-29-03	Chlorinated pesticide	508	GC-ECD
			608	GC-ECD
			625	GC/MS
			8080	GC-ECD
			8270	GC/MS
			OLMO1	GC-ECD
o,p-DDT	789-02-6	Pesticide, chlorinated	625	GC/MS
Dechlorane	2385-85-5	Pesticide, chlorinated	8270	GC/MS
2,4-Diaminotoluene	95-80-7	Aromatic amine	8270	GC/MS
Diazinon (Basudin)	333-41-5	Pesticide, organophosphorus	507	GC-NPD
			8140	GC-FPD
Dibenz[*a,j*]acridine	224-42-0	Polynuclear aromatic hydrocarbon	8270	GC/MS
Dibenz[*a,h*]anthracene	53-70-3	Polynuclear aromatic hydrocarbon	515.1	HPLC-FL
			525	GC/MS
			610	HPLC-FL
			625	GC/MS
			8310	HPLC-FL
			8270	GC/MS
			OLMO1	GC/MS
Dibenzofuran	132-64-9	Benzofuran	8270	GC/MS
			OLMO1	GC/MS
Dibenzo[*a,e*]pyrene	192.65.4	Polynuclear aromatic hydrocarbon	8270	GC/MS
Dibromochloromethane	124-48-1	Halogenated hydrocarbon (VOC)	502.2	GC-ELCD
			524.2	GC/MS
			601	GC-HSD
			624	GC/MS
			8260	GC/MS

Compounds	CAS no.	Type/class	U.S. EPA method #	Instrumental techniques
Dibromochloromethane (continued)			OLMO1	GC/MS
Dibromochloropropane	96-12-8	VOC (halocarbon)	502.2	GC-ELCD
			524.2	GC/MS
			8011	GC-ECD
			8260	GC/MS
tris-(2,3-Dibromopropyl)phosphate	126-72-7	Organic phosphate	8270	GC/MS
Dibutyl phthalate	84-74-2	Phthalate ester	525	GC/MS
			606	GC-ECD
			625	GC/MS
			8060	GC-ECD/FID
			8270	GC/MS
			OLMO1	GC/MS
Dicamba	1918-00-9	Herbicide, chlorophenoxy acid	515.1	GC-ECD
			8150	GC-ECD
Dichlone (Phygon)	117-80-6	Pesticide, chlorinated	8270	GC/MS
trans-1,2-Dichlons ethylene	156-60-5	Halogenated hydrocarbon (VOC)	502.2	GC/PID
			524.2	GC/MS
			601	GC/HSD
			624	GC/MS
			8260	GC/MS
1,2-Dichlorobenzene	95-50-1	Chlorinated hydrocarbon	502.2	GC-ELCD/PID
			524.2	GC/MS
			601	GC-HSD
			602	GC-PID
			612	GC-ECD
			624	GC/MS
			625	GC/MS
			8020	GC/PID
			8120	GC-ECD

			8260	GC/MS
			8270	GC/MS
			OLMO1	GC/MS
1,3-Dichlorobenzene	541-73-1	Halogenated hydrocarbon	502.2	GC-ELCD/PID
			524.2	GC/MS
			601	GC-HSD
			602	GC-PID
			612	GC-ECD
			624	GC/MS
			625	GC/MS
			8120	GC-ECD
			8020	GC-PID
			8260	GC/MS
			8270	GC/MS
			OLMO1	GC/MS
1,4-Dichlorobenzene	106-46-7	VOC (halogenated hydrocarbon)	524.2	GC/MS
			601	GC-HSD
			602	GC-PID
			612	GC-ECD
			624	GC/MS
			625	GC/MS
			8020A	GC-PID
			8120	GC-ECD
			8260	GC/MS
			8270	GC/MS
			OLMO1	GC/MS
3,3′-Dichlorobenzidine	91-94-1	Aromatic amine	605	HPLC-EL
			625	GC/MS
			8270	GC/MS
			OLMO1	GC/MS
Dichlorodifluoromethane (CFC-12)	75-71-8	VOC (halocarbon)	502.2	GC-ELCD
			524.2	GC/MS

Compounds	CAS no.	Type/class	U.S. EPA method #	Instrumental techniques
Dichlorodifluoromethane (CFC-12) (continued)			601	GC-HSD
			8260	GC/MS
1,1-Dichloroethane (ethyledene dichloride)	75-34-3	VOC (halocarbon)	502.2	GC-ELCD
			524.2	GC/MS
			601	GC-HSD
			624	GC/MS
			8260	GC/MS
			OLMO1	GC/MS
1,1-Dichloroethene (vinylidine chloride)	75-35-4	VOC (halocarbon)	502.2	GC-ELCD
			524.2	GC/MS
			601	GC-HSD
			624	GC/MS
			8260	GC/MS
			OLMO1	GC/MS
cis-1,2-Dichloroethylene	156-59-2	Halogenated hydrocarbon (VOC)	502.2	GC-ELCD
			524.2	GC/MS
			8260	GC/MS
Dichloroethyl ether	111-44-4	Haloether	611	GC-HSD
			625	GC/MS
			8270	GC/MS
			OLMO1	GC/MS
2,4-Dichlorophenol	120-83-2	Phenol	604	GC-ECD/FID
			625	GC/MS
			8040	GC-ECD/FID
			8270	GC/MS
			OLMO1	GC/MS
2,6-Dichlorophenol	87-65-0	Phenol	8040	GC-FID
			8270	GC/MS
1,2 Dichloropropane (propylene dichloride)	78-87-5	VOC (halocarbon)	502.2	GC-ELCD

			524.2	GC/MS
			601	GC-HSD
			624	GC/MS
			8260	GC/MS
			OLMO1	GC/MS
1,3-Dichloropropane	142-28-9	Halogenated hydrocarbon (VOC)	502.2	GC-ELCD
			524.2	GC/MS
			8260	GC/MS
2,2-Dichloropropane (*sec*-dichloropropane)	594-20-7	Halogenated hydrocarbon (VOC)	502.2	GC-ELCD
			524.2	GC/MS
			8260	GC/MS
1,1-Dichloropropene	563-58-6	Halogenated hydrocarbon (VOC)	502.2	GC-ELCD/PID
			524.2	GC/MS
			8260	GC/MS
cis-1,3-Dichloropropene	10061-01-5	Halogenated hydrocarbon (VOC)	502.2	GC-PID
			524.2	GC/MS
			601	GC-HSD
			624	GC/MS
			OLMO1	GC/MS
trans-1,3-Dichloropropene	10061-01-5	Halogenated hydrocarbon (VOC)	502.2	GC-PID
			524.2	GC/MS
			601	GC-HSD
			624	GC/MS
			OLMO1	GC/MS
Dichlorvos (Vapona)	62-73-7	Pesticide, organophosphorus	507	GC-NPD
			8270	GC/MS
Dicrotophos (Bidrin)	141-66-2	Pesticide, organophosphorus	8270	GC/MS
Dieldrin	60-57-1	Chlorinated pesticide	505	GC-ECD
			508	GC-ECD
			608	GC-ECD
			625	GC/MS
			8080	GC-ECD

Compounds	CAS no.	Type/class	U.S. EPA method #	Instrumental techniques
Dieldrin (continued)			8270	GC/MS
			OLMO1	GC-ECD
Diethyl phthalate	84-66-2	Phthalate ester	606	GC-ECD
			625	GC/MS
			8060	GC-ECD/FID
			8270	GC/MS
			OLMO1	GC/MS
Diethylstilbestrol	56-53-1	Phenol	8270	GC/MS
Diethyl sulfate	64-67-5	Sulfate ester	8270	GC/MS
Dilantin	57-41-0	Heterocyclic nitrogen compound	8270	GC/MS
Dimethoate (Cygon)	60-51-5	Pesticide, organophosphorus	8270	GC/MS
3,3-Dimethoxybenzidine	119-90-4	Aromatic amine	8270	GC/MS
7,12-Dimethylbenz[*a*]anthracene	57-97-6	PAH	8270	GC/MS
N,*N*-Dimethyl-4-(phenylazo) benzenamine	60-11-7		8270	GC/MS
Dimethyl phthalate	131-11-3	Phthalate ester	525	GC/MS
			606	GC-ECD
			625	GC/MS
			8060	GC-ECD/FID
			8270	GC/MS
			OLMO1	GC/MS
2,4-Dimethylphenol	105-67-9	Phenol	604	GC-FID/ECD
			625	GC/MS
			8040	GC-FID/ECD
			8270	GC/MS
			OLMO1	GC/MS
Dinex (2-cyclohexyl-4,6-dinitrophenol)	131-89-5	Nitrophenol	8040	GC-FID
			8270	GC/MS
1,2-Dinitrobenzene	512-56-1	Nitroaromatic	8270	GC/MS
1,3-Dinitrobenzene	99-65-0	Nitroaromatic	8270	GC/MS

1,4-Dinitrobenzene	100-25-4	Nitroaromatic	8270	GC/MS
2,4-Dinitrophenol	51-28-5	Phenol	604	GC-FID/ECD
			625	GC/MS
			8040	GC-ECD/FID
			8270	GC/MS
			OLMO1	GC/MS
2,4-Dinitrotoluene	121-14-2	Nitroaromatic	625	GC/MS
			8270	GC/MS
			OLMO1	GC/MS
2,6-Dinitrotoluene	606-20-2	Nitroaromatic	625	GC/MS
			8270	GC/MS
			OLMO1	GC/MS
Dinoseb	88-85-7	Phenol (nitrophenol pesticide)	515.1	GC-ECD
			8040	GC-FID
			8150	GC-ECD
			8270	GC/MS
Dioctyl phthalate	117-84-0	Phthalate ester	606	GC-ECD
			625	GC/MS
			8060	GC-ECD/FID
			8270	GC/MS
			OLMO1	GC/MS
Dioxin (2,3,7,8-TCDD)	1746-01-6	Dioxin	613	GC/MS
			8280	GC/MS
Diphenylamine (*N*-phenylbenzeneamine)	122-39-4	Aromatic amine	8270	GC/MS
1,2-Diphenyl hydrazine	122-66-7	Azobenzene	8270	GC/MS
Diquat dibromide (Aquacide, Reglone)	85-00-7	Triazine pesticide	549	HPLC-UV
Disulfoton (Disyston)	298-04-4	Pesticide, organophosphorus	507	GC-NPD
			8140	GC-FPD
			8270	GC/MS
Disulfoton sulfoxide (Oxydisulfoton)	2497-07-6	Pesticide, organophosphorus	507	GC-NPD
Dowicide 6 (2,3,4,6-tetrachlorophenol)	58-90-2	Phenol	8270	GC/MS
Dursban (Chlorpyrifos)	2921-88-2	Pesticide, organophosphorus	508	GC-ECD

Compounds	CAS no.	Type/class	U.S. EPA method #	Instrumental techniques
Dursban (chlorpyrifos) (continued)			8140	GC-FPD
Endosulfan-I	959-98-8	Pesticide, chlorinated	508	GC-ECD
			608	GC-ECD
			625	GC/MS
			8080	GC-ECD
			8270	GC/MS
			OLMO1	GC-ECD
Endosulfan-II (β-endosulfan, thiodan-II)	33213-65-9	Pesticide, chlorinated	508	GC-ECD
			608	GC-ECD
			625	GC/MS
			8080	GC-ECD
			8270	GC/MS
			OLMO1	GC-ECD
Endosulfan sulfate	1031-07-8	Pesticide, chlorinated	508	GC-ECD
			608	GC-ECD
			625	GC/MS
			8080	GC-ECD
			OLMO1	GC-ECD
Endrin	72-20-8	Pesticide, chlorinated	505	GC-ECD
			508	GC-ECD
			525	GC/MS
			608	GC-ECD
			625	GC/MS
			8080	GC-ECD
			8270	GC/MS
			OLMO1	GC-ECD
Endrin aldehyde	7421-93-4	Pesticide, chlorinated	508	GC-ECD
			608	GC-ECD
			625	GC/MS

			8080	GC-ECD
			8270	GC/MS
			OLMO1	GC-ECD
Endrin ketone	53494-70-5	Pesticide, chlorinated	8270	GC/MS
			OLMO1	GC-ECD
Ethion (Nialate)	563-12-2	Pesticide, organophosphorus	8270	GC/MS
Ethoprophos (Ethoprop)	13194-48-4	Pesticide, organophosphorus	507	GC-NPD
			8140	GC-FPD
Ethyl carbamate	51-79-6	Carbamate pesticide	8270	GC/MS
Ethyl chloride (chloroethane)	75-00-3	VOC (halocarbon)	502.2	GC-ELCD
			524.2	GC/MS
			601	GC-HSD
			624	GC/MS
			8260	GC/MS
			OLMO1	GC/MS
Ethylbenzene	100-41-4	VOC (aromatic)	502.2	GC-PID
			524.2	GC/MS
			602	GC-PID
			624	GC/MS
			8020	GC-PID
			8260	GC/MS
			OLMO1	GC/MS
Ethylene dibromide (1,2-dibromoethane, EDB)	106-93-4	VOC (halogenated hydrocarbon)	502.2	GC-ELCD
			524.2	GC/MS
			8011	GC-ECD
			8260	GC/MS
Ethylene dichloride (1,2-dichloroethane)	107-06-2	VOC (halogenated hydrocarbon)	502.2	GC-ELCD
			524.2	GC/MS
			601	GC-HSD
			624	GC/MS
			8260	GC/MS
			OLMO1	GC/MS

Compounds	CAS no.	Type/class	U.S. EPA method #	Instrumental techniques
bis (2-Ethylhexyl) adipate	103-23-1	Adipate ester	525	GC/MS
bis (2-Ethylhexyl) phthalate	117-81-7	Phthalate ester	525	GC/MS
			606	GC-ECD
			625	GC/MS
			8060	GC-ECD/FID
			8270	GC/MS
			OLMO1	GC/MS
Ethyl methanesulfonate	62-50-0	Sulfonate	8270	GC/MS
Famphur (Famophos)	52-85-7	Pesticide, organophosphorus	8270	GC/MS
Fenamiphos (Nemacur)	22224-92-6	Pesticide, organophosphorus	507	GC-NPD
Fensulfothion	115-90-2	Pesticide, organophosphorus	8140	GC-FPD
			8270	GC/MS
Fenthion (Baytex)	55-38-9	Pesticide, organophosphorus	8140	GC-FPD
			8270	GC/MS
Fluoranthene	206-44-0	Polynuclear aromatic hydrocarbon	550.1	HPLC-FL
			610	HPLC-FL
			625	GC/MS
			8310	HPLC-FL
			8270	GC/MS
			OLMO1	GC/MS
Fluorene (9H-fluorene)	86-73-7	Polynuclear aromatic hydrocarbon	525	GC/MS
			550.1	HPLC-UV
			610	HPLC-UV/GC-FID
			625	GC/MS
			8310	HPLC-UV
			8270	GC/MS
			OLMO1	GC/MS
o-Fluorophenol	367-12-4	Phenol	8270	GC/MS

Glyphosate	1071-83-6	Pesticide, organophosphorus	547	HPLC-FL
Heptachlor	76-44-8	Pesticide, chlorinated	505	GC-ECD
			508	GC-ECD
			525	GC/MS
			608	GC-ECD
			625	GC/MS
			8080	GC-ECD
			8270	GC/MS
			OLMO1	GC-ECD
Heptachlor epoxide	1024-57-3	Pesticide, chlorinated	505	GC-ECD
			508	GC-ECD
			608	GC-ECD
			625	GC/MS
			8080	GC-ECD
			8270	GC/MS
			OLMO1	GC/MS
1,2,3,4,6,7,8-Heptachlorodibenzo-*p*-dioxin	35822-46-9	Dioxin	8280	GC/MS
Hexachlorobutadiene	87-68-3	Halocarbon	502.2	GC-PID/ELCD
			524.2	GC/MS
			612	GC-ECD
			625	GC/MS
			8120	GC-ECD
			8260	GC/MS
			8270	GC/MS
			OLMO1	GC/MS
Hexachlorocyclopentadiene	77-47-4	Halogenated hydrocarbon	505	GC-ECD
			525	GC/MS
			612	GC-ECD
			625	GC/MS
			8120	GC-ECD
			8270	GC/MS
			OLMO1	GC/MS

Compounds	CAS no.	Type/class	U.S. EPA method #	Instrumental techniques
1,2,3,6,7,8-Hexachlorodibenzo-*p*-dioxin	57653-85-7	Dioxin	8280	GC/MS
1,2,3.4.7,8-Hexachlorodibenzofuran	70648-26-9	Dibenzofuran	8280	GC/MS
Hexachlorobenzene	118-74-1	Halogenated aromatic	505	GC-ECD
			508	GC-ECD
			525	GC/MS
			612	GC-ECD
			625	GC/MS
			8120	GC-ECD
			8270	GC/MS
			OLMO1	GC/MS
Hexachloroethane	67-72-11	Halocarbon	612	GC-ECD
			625	GC/MS
			8120	GC-ECD
			8270	GC/MS
			OLMO1	GC/MS
Hexachlorophene (Nabac)	70-30-4	Phenol	8270	GC/MS
Hexachloropropene	1888-71-7	Chlorinated hydrocarbon	8270	GC/MS
Hexamethylphosphoramide	680-31-9	Phosphoric amide	8270	GC/MS
2-Hexanone (methyl butyl ketone)	591-78-6	Ketone	OLMO1	GC/MS
Hydroquinone	123-31-9	Aromatic ketone	8270	GC/MS
Indeno (1,2,3-*cd*)pyrene	193-39-5	Polynuclear aromatic hydrocarbon	525	GC/MS
			550.1	HPLC-FL
			610	HPLC-FL
			625	GC/MS
			8310	HPLC-FL
			8270	GC/MS
			OLMO1	GC/MS
Isodrin	465-73-6	Chlorinated aromatic	8270	GC/MS
Isophorone	78-59-1	Ketone	609	GC-FID

			625	GC/MS
			8270	GC/MS
			OLMO1	GC/MS
Isosafrole	120-58-1	Benzodioxole	8270	GC/MS
Kepone (Chlordecone)	143-50-0	Pesticide, chlorinated	8270	GC/MS
Kerb	23950-58-5	Pesticide, chlorinated amide	507	GC-NPD
			8270	GC/MS
Leptophos (Phosvel)	21609-90-5	Pesticide, organophosphorus	8270	GC/MS
Lindane (γ-BHC)	58-89-9	Pesticide, chlorinated	505	GC-ECD
			508	GC-ECD
			525	GC/MS
			608	GC-ECD
			8080	GC-ECD
			8270	GC/MS
			OLMO1	GC-ECD
Malathion (Phosphothion)	121-75-5	Pesticide, organophosphorus	8270	GC/MS
Maleic anhydride	108-31-6	Anhydride	8270	GC/MS
Mesitylene	108-67-8	Aromatic	502.2	GC-PID
			524.2	GC/MS
			8260	GC/MS
Mestranol	72-33-3	Steroid	8270	GC/MS
Metaphos (methyl parathion)	298-00-0	Pesticide, organophosphorus	8140	GC-FPD
			8270	GC/MS
Methapyrilene	91-80-5	Pyridinyl amine	8270	GC/MS
Methiocarb (Mesurol)	2032-65-7	Pesticide, carbamate	531.1	HPLC-FL
Methomyl (Lannate)	16752-77-5	Pesticide, thiocarbamate	531.1	HPLC-FL
Methoxychlor	72-43-5	Pesticide, chlorinated	505	GC-ECD
			508	GC-ECD
			525	GC/MS
			608	GC-ECD
			8080	GC-ECD
			8270	GC/MS
			OLMO1	GC-ECD

Compounds	CAS no.	Type/class	U.S. EPA method #	Instrumental techniques
3-Methylcholanthrene	56-49-51	PAH	8270	GC/MS
Methyl bromide (bromomethane)	74-83-9	VOC (halocarbon)	502.2	GC-ELCD
			524.2	GC/MS
			601	GC-HSD
			624	GC/MS
			8260	GC/MS
			OLMO1	GC/MS
Methyl chloride (chloromethane)	74-87-3	VOC (halocarbon)	502.2	GC-ELCD
			524.2	GC/MS
			601	GC-HSD
			624	GC/MS
			8260	GC/MS
			OLMO1	GC/MS
2-Methyl-4,6-dinitrophenol	534-52-1	Phenol	604	GC-PID
			625	GC/MS
			8040	GC-ECD/FID
			8270	GC/MS
			OLMO1	GC/MS
Methyl ethyl ketone (2-butanone)	78-93-3	Ketone	OLMO1	GC/MS
Methyl isobutyl ketone (MIBK)	108-10-1	Ketone	OLMO1	GC/MS
Methyl methanesulfonate	66-27-3	Sulfonate ester	8270	GC/MS
2-Methylnaphthalene	91-57-6	PAH	8270	GC/MS
			OLMO1	GC/MS
N-Methyl-*N*-nitrosoethanamine	10595-95-6	Nitrosamine	8270	GC/MS
Methylene bromide (dibromomethane)	74-95-3	VOC (halocarbon)	502.2	GC-ELCD
			524.2	GC/MS
			8260	GC/MS
Methylene chloride (dichloromethane)	75-09-2	VOC (halocarbon)	502.2	GC-ELCD
			524.2	GC/MS

			601	GC-HSD
			624	GC/MS
			8260	GC/MS
			OLMO1	GC/MS
Metolachlor	51218-45-2	Pesticide, chlorinated	507	GC-NPD
Metribuzin	21087-64-9	Pesticide, triazine	507	GC-NPD
Mevinphos (Phosdrin)	7786-34-7	Pesticide, organophosphorus	507	GC-NPD
			8140	GC-FPD
			8270	GC/MS
Mexacarbate	315-18-4	Pesticide, carbamate	8270	GC/MS
MOCA (4,4′-Methylene-*bis* [2-chloroaniline])	101-14-4	Aromatic amine	8270	GC/MS
Monocrotophos	6923-22-4	Pesticide, organophosphorus	8270	GC/MS
Naled (Dibrom)	300-76-5	Pesticide, organophosphorus	8140	GC-FPD
			8270	GC/MS
Naphthalene	91-20-3	PAH	502.2	GC-PID
			524.2	GC/MS
			550.1	HPLC-UV
			610	HPLC-UV
			625	GC/MS
			8260	GC/MS
			8270	GC/MS
			8310	HPLC-UV
			OLMO1	GC/MS
1,4-Naphthoquinone (1,4-naphthalenedione)	130-15-4	Aromatic ketone	8270	GC/MS
1-Naphthylamine	134-32-7	Aromatic amine	8270	GC/MS
2-Naphthylamine	91-59-8	Aromatic amine	8270	GC/MS
Nicotine	54-11-5	Alkaloid	8270	GC/MS
5-Nitroacenaphthene	602-87-9	Nitroaromatic	8270	GC/MS
2-Nitroaniline (2-nitrobenzenamine)	88-74-4	Nitroaromatic	8270	GC/MS
			OLMO1	GC/MS
3-Nitroaniline	99-09-2	Nitroaromatic (aromatic amine)	8270	GC/MS

Compounds	CAS no.	Type/class	U.S. EPA method #	Instrumental techniques
			OLMO1	GC/MS
p-Nitroaniline	100-01-6	Aromatic amine (nitroaromatic)	8270	GC/MS
			OLMO1	GC/MS
5-Nitro-*o*-anisidine	99-59-2	Aromatic amine	8270	GC/MS
Nitrobenzene	98-95-3	Nitroaromatic	609	GC-FID
			625	GC/MS
			8270	GC/MS
			OLMO1	GC/MS
4-Nitrobiphenyl	92-93-3	Nitroaromatic	8270	GC/MS
2-Nitrophenol	88-75-5	Phenol	604	GC-FID/ECD
			625	GC/MS
			8040	GC-FID/ECD
			8270	GC/MS
			OLMO1	GC/MS
4-Nitrophenol	100-02-7	Phenol	515.1	GC-ECD
			604	GC-FID/ECD
			625	GC/MS
			8040	GC-ECD/FID
			8270	GC/MS
			OLMO1	GC/MS
4-Nitroquinoline-1-oxide	56-57-5	Heterocyclic nitrogen compound	8270	GC/MS
N-Nitrosodiethylamine	55-18-5	Nitrosamine	8270	GC/MS
N-Nitrosodi-*n*-butylamine	924-16-3	Nitrosamine	8270	GC/MS
N-Nitrosodimethylamine (*N*-methyl-*N*-nitrosomethanamine)	62-75-9	Nitrosamine	625	GC/MS
			8270	GC/MS
N-Nitrosodiphenylamine	86-30-6	Nitrosamine	607	GC-NPD
			625	GC/MS
			8270	GC/MS
			OLMO1	GC/MS

Nitrofen (TOK)	1836-75-5	Chlorinated nitroaromatic	8270	GC/MS
N-Nitrosodi-*n*-propylamine	621-64-7	Nitrosamine	625	GC/MS
			8270	GC/MS
			OLMO1	GC/MS
4-Nitrosomorpholine	59-89-2	Nitrosamine	8270	GC/MS
N-Nitrosopiperidine	100-75-4	Heterocyclic nitrogen compound	8270	GC/MS
N-Nitrosopyrrolidine	930-55-2	Nitrosamine	8270	GC/MS
5-Nitro-*o*-toluidine	99-55-8	Aromatic amine (nitroaromatic)	8270	GC/MS
cis-Nonachlor	5103-73-1	Pesticide, chlorinated	505	GC-ECD
trans-Nonachlor	39765-80-5	Pesticide, chlorinated	505	GC-ECD
			525	GC/MS
Octachlorodibenzo-*p*-dioxin (OCDD)	3268-87-9	Dioxin	8280	GC/MS
Octamethylpyrophosphoramide	152-16-9	Pesticide, organophosphorus	8270	GC/MS
4,4′-Oxydianiline (4,4′-diaminodiphenyl ether)	101-80-4	Aromatic amine	8270	GC/MS
Paraquat	1910-42-5	Pesticide, bipyridinum	549	HPLC-UV
Parathion (Niran)	56-38-2	Pesticide, rganophosphorus	8270	GC/MS
1,2,3,7,8,-Pentachlorodibenzo-*p*-dioxin	40321-76-4	Dioxin	8280	GC/MS
1,2,3,7,8-Pentachlorodibenzofuran	57117-41-6	Dibenzofuran	8280	GC/MS
Pentachlorobenzene	608-93-5	Pesticide, chlorinated	8270	GC/MS
Pentachloronitrobenzene	82-68-8	Chlorinated nitroaromatic	8270	GC/MS
Pentachlorophenol	87-86-5	Phenol	515.1	GC-ECD
			525	GC/MS
			604	GC-ECD/FID
			625	GC/MS
			8040	GC-ECD/FID
			8270	GC/MS
			OLMO1	GC/MS
Phenanthrene	85-01-8	Polynuclear aromatic hydrocarbon	525	GC/MS
			550.1	HPLC-FL
			610	HPLC-FL
			625	GC/MS
			8310	HPLC-FL
			8270	GC/MS

Compounds	CAS no.	Type/class	U.S. EPA method #	Instrumental techniques
Phenanthrene (continued)			OLMO1	GC/MS
Phenobarbital	50-06-6	Barbiturate	8270	GC/MS
Phenol	108-95-2	Phenol	604	GC-FID/ECD
			625	GC/MS
			8040	GC-ECD
			8270	GC/MS
			OLMO1	GC/MS
Phentermine (1,1-dimethyl-2-phenyl ethanamine)	122-09-8	Aromatic amine	8270	GC/MS
p-(Phenylazo) aniline (aminoazobenzene)	60-09-3	Aromatic amine	8270	GC/MS
Phorate (Thimet)	298-02-2	Pesticide, organophosphorus	8140	GC-FPD
			8270	GC/MS
Phosmet (Imidan)	732-11-6	Pesticide, organophosphorus	8270	GC/MS
Phosphamidon (Dimecron)	13171-21-6	Pesticide, organophosphorus	8270	GC/MS
Phthalic anhydride	85-44-9	Acid anhydride	8270	GC/MS
Picloram	1918-02-1	Pesticide, chlorinated	515.1	GC-ECD
2-Picoline (2-methylpyridine)	109-06-8	Pyridine	8270	GC-MS
Piperonyl sulfoxide	120-62-7	Sulfoxide	8270	GC/MS
Prometon	1610-18-0	Pesticide, triazine	507	GC-NPD
Pronamide	25057-89-0	Amide	515.1	GC-ECD
n-Propylbenzene	103-65-1	VOC (aromatic)	502.1	GC-PID
			524.2	GC/MS
			8260	GC/MS
Propylthiouracil	51-52-5	Organosulfur compound	8270	GC/MS
Pyrene	129-00-0	Polynuclear aromatic hydrocarbon	525	GC/MS
			550.1	HPLC-FL
			610	HPLC-FL
			625	GC/MS
			8310	HPLC-FL
			8270	GC/MS

			OLMO1	GC/MS
Pyridine	110-86-1	Pyridine	8270	GC/MS
Quinone	106-51-4	Aromatic ketone	8270	GC/MS
Resorcinol (1,3-benzenediol)	108-46-3	Phenol	8270	GC/MS
Safrole (1,2-methylendioxy-4-allyl benzene)	94-59-7	Aromatic	8270	GC/MS
Santox (EPN)	2104-64-5	Pesticide, organophosphorus	8270	GC/MS
Silvex (2,4,5-TP)	93-72-1	Chlorophenoxy acid herbicide	515.1	GC-ECD
			8150	GC-ECD
Simazine	122-34-9	Pesticide, triazine	505	GC-ECD
			507	GC-NPD
			525	GC/MS
Stirofos (Rabon, tetrachlorvinphos)	22248-79-9	Pesticide, organophosphorus	507	GC-NPD
			8140	GC-FPD
			8270	GC/MS
Strychnine sulfate	60-41-3	Alkaloid	8270	GC/MS
Styrene (vinylbenzene)	100-42-5	VOC (aromatic)	502.2	GC-PID
			524.2	GC/MS
			602	GC-PID
			624	GC/MS
			8020	GC-PID
			8260	GC/MS
			OLMO1	GC/MS
2,4,5-T	93-76-5	Chlorophenoxy acid herbicide	515.1	GC-ECD
			8150	GC-ECD
2,3,7,8-TCDF (2,3,7,8-tetrachlorodibenzofuran)	51207-31-9	Dibenzofuran	8280	GC/MS
Terbufos (Counter)	13071-79-9	Pesticide, organophosphorus	507	GC-NPD
			8270	GC/MS
1,2,4,5-Tetrachlorobenzene	95-94-3		8270	GC/MS
1,1,1,2-Tetrachloroethane	630-20-6	Halogenated hydrocarbon (VOC)	502.2	GC-ELCD
			524.2	GC/MS
			8260	GC/MS
1,1,2,2-Tetrachloroethane	79-34-5	VOC (halocarbon)	502.2	GC-ELCD

Compounds	CAS no.	Type/class	U.S. EPA method #	Instrumental techniques
1,1,2,2-Tetra-chloroethane (continued)			524.2	GC/MS
			601	GC-HSD
			624	GC/MS
			8260	GC/MS
			OLMO1	GC/MS
Tetrachloroethylene (perchloroethylene)	127-18-4	Halogenated hydrocarbon (VOC)	502.2	GC-ELCD
			524.2	GC/MS
			601	GC-HSD
			624	GC/MS
			8260	GC/MS
			OLMO1	GC/MS
Tetrachlorophenol	25167-83-3	Phenol	8040	GC-FID
Tetraethylpyrophosphate (TEPP)	107-49-3	Organic phosphate	8270	GC/MS
Thiophenol (benzenethiol, phenyl mercaptan)	108-98-5	Mercaptan	8270	GC/MS
Toluene	108-88-3	Aromatic	502.2	GC-PID
			524.2	GC/MS
			602	GC-PID
			624	GC/MS
			8020	GC-PID
			8260	GC/MS
Toluene-2,4-diisocyanate	584-84-9	Organic isocyanate	8270	GC/MS
o-Toluidine	95-53-4	Aromatic amine	8270	GC/MS
Toxaphene (Camphechlor)	8001-35-2	Pesticide, chlorinated	505	GC-ECD
			508	GC-ECD
			525	GC/MS
			608	GC-ECD
			625	GC/MS
			8080	GC-ECD
			8270	GC/MS

			OLMO1	GC/MS
1,2,3-Trichlorobenzene	87-61-6	VOC (halocarbon)	502.2	GC-ELCD/PID
			524.2	GC/MS
			8260	GC/MS
1,2,4-Trichlorobenzene	120-82-1	Halogenated, aromatic	502.2	GC-ELCD/PID
			524.2	GC/MS
			612	GC-ECD
			625	GC/MS
			8120	GC-ECD
			8260	GC/MS
			8270	GC/MS
			OLMO1	GC/MS
1,1,1-Trichloroethane (methyl chloroform)	71-55-6	Halocarbon	601	GC-HSD
			624	GC/MS
			8260	GC/MS
			OLMO1	GC/MS
1,1,2-Trichloroethane	79-00-5	VOC (halocarbon)	502.2	GC-ELCD
			524.2	GC/MS
			601	GC-HSD
			624	GC/MS
			8260	GC/MS
			OLMO1	GC/MS
Trichloroethene	79-01-6	VOC (halocarbon)	502.2	GC-ELCD\PID
			524.2	GC/MS
			601	GC-HSD
			624	GC/MS
			8260	GC/MS
			OLMO1	GC/MS
Trichlorofluoromethane (CFC-11)	75-69-4	VOC (halocarbon)	502.2	GC-ELCD
			524.2	GC/MS
			601	GC-HSD
			8260	GC/MS

Compounds	CAS no.	Type/class	U.S. EPA method #	Instrumental techniques
(Trichloromethyl) benzene	98-07-7	Chlorinated hydrocarbon	8120	GC-ECD
Trichloronate	327-98-0	Pesticide, organophosphorus	8140	GC-FPD
2,4,5-Trichlorophenol	95-95-4	Phenol	8270	GC/MS
			OLMO1	GC/MS
2,4,6-Trichlorophenol	88-06-2	Phenol	604	GC-ECD/FID
			625	GC/MS
			8040	GC-ECD/FID
			8270	GC/MS
			OLMO1	GC/MS
Trichlorophenol	25167-82-2	Phenol	8040	GC-FID
1,2,3-Trichloropropane	96-18-4	VOC (halocarbon)	502.2	GC-ELCD
			524.2	GC/MS
			8260	GC/MS
1,1,2-Trichloro-1,2,2-trifluoroethane (Freon-113)				
O,O,O-Triethylphosphorothioate	126-68-1	Pesticide, organophosphorus	8270	GC/MS
Trifluralin (Treflan)	1582-09-8	Pesticide, organofluorine	508	GC-ECD
			8270	GC/MS
2,4,5-Trimethylaniline	137-17-7	Aromatic amine	8270	GC/MS
1,2,4-Trimethylbenzene	95-63-6	Aromatic	502.2	GC-PID
			524.2	GC/MS
			8260	GC/MS
Trimethylphosphate	512-56-1	Organic phosphate	8270	GC/MS
1,3,5-Trinitrobenzene	99-35-4	Nitroaromatic	8270	GC/MS
Trithion (Carbofenothion)	786-19-6	Pesticide, organophosphorus	8270	GC/MS
Vancide-89	133-90-4	Pesticide, chlorinated	515.1	GC-ECD
Vinyl chloride (chloroethene)	75-01-4	VOC (halocarbon)	502.2	GC-PID/ELCD
			524.2	GC/MS
			601	GC-HSD
			624	GC/MS

			8260	GC/MS
			OLMO1	GC/MS
Vydate (Oxamyl)	23135-22-0	Pesticide, thiocarbamate	531.1	HPLC-FL
m-Xylene	108-38-3	Aromatic	502.2	GC-PID
			524.2	GC/MS
			8260	GC/MS
o-Xylene	95-47-6	Aromatic	502.2	GC-PID
			524.2	GC/MS
			8260	GC/MS
p-Xylene	106-42-3	VOC (aromatic)	502.2	GC-PID
			524.2	GC/MS
			8260	GC/MS
Xylenes (total xylene)	1330-20-7	Aromatic	8020	GC-PID
			OLMO1	GC/MS
Zinophos (Thionazin)	297-97-2	Pesticide, organophosphorus	8270	GC/MS

INDICES

CHEMICAL COMPOUNDS INDEX

Chemical Compounds	CAS Registry Number	Page
Abate	[3383-96-8]	218
Acaraben	[510-15-6]	508
Acconame	[21548-32-3]	217
Acenaphthene	[83-32-9]	166,486,491
Acenaphthylene	[208-96-8]	166,486,491
Acephate	[30560-19-1]	217
Acetaldehyde	[75-07-0]	108,285,481,490
Acetamidofluorene	[53-96-3]	500
Acetic acid	[64-19-7]	481
Acetic aldehyde	[75-07-0]	285
Acetic anhydride	[108-24-5]	481
Acetone	[67-64-1]	108,287,467,485,490
Acetone cyanohydrin	[75-86-5]	463,481
Acetonitrile	[75-05-8]	289,481
Acetophenone	[98-86-2]	500
Acetyl acetone	[123-54-6]	108,463
Acetyl bromide	[506-96-7]	467
Acetyl chloride	[75-36-5]	467
2-Acetylaminofluorene	[53-96-3]	500
1-Acetyl-2-thiourea	[591-08-2]	505
Aciflurofen	[50594-66-6]	500
Acraldehyde	[107-02-8]	291
Acridine	[260-94-6]	463
Acrolein	[107-02-8]	108,291,481,490
Acrylamide	[79-06-1]	463
Acrylonitrile	[107-13-1]	293,481,491
Akton	[1757-18-1]	217,218
Alachlor	[15972-60-8]	208,463
Aldicarb	[116-06-3]	201,463,500
Aldicarb sulfone	[1646-88-4]	201,500
Aldicarb sulfoxide	[1646-87-3]	201,500
Aldoxycarb	[1646-88-4]	201,500
Aldrin	[309-00-2]	208,486,490
Allyl alcohol	[107-18-6]	467,481
Allyl benzene	[300-57-2]	467
Allyl chloride	[107-05-1]	146,467,481,484,491
Allyl ether	[557-40-4]	467
Allyl ethyl ether	[557-31-3]	467

Allyl mercaptan	[870-23-5]	467
Allyl methyl ether	[627-40-7]	467
Allyl propyl ether	[1471-03-0]	467
Allylbromide	[106-95-6]	467
Aluminum	[7429-90-5]	481
Ametryne	[834-12-8]	205,463
2-Aminoanthraquinone	[117-79-3]	501
Aminoazobenzene	[60-09-3]	528
Aminobenzene	[62-53-3]	295
4-Aminobiphenyl	[92-67-1]	501
Aminocarb	[2032-59-9]	463
2-Aminoethanol	[141-43-5]	482
p-Aminophenylarsonic acid	[98-50-0]	482
2-Aminotoluene	[95-53-4]	425
Amitrole	[61-82-5]	463
Amizine	[122-34-9]	205
Ammonia	[7664-41-7]	482
sec-Amyl acetate	[626-38-0]	483
n-Amyl acetate	[628-63-7]	483
Anilazine	[101-05-3]	205
Aniline	[62-53-3]	295,481
Anisidine	[29191-52-4]	482
o-Anisidine	[90-04-0	482,483,501
p-Anisidine	[104-94-9]	482
Anisole	[100-66-3]	469
Anthon	[52-68-6]	218
Anthracene	[120-12-7]	166,486,491,501
Anthraquinone	[84-65-1]	463
Antimony hydride	[7803-52-3]	405
Aprocarb	[114-26-1]	201
Aquacide	[85-00-7]	517
Aramite	[140-57-8]	501
Arelon	[34123-59-6]	203
Aroclor 1016	[12674-11-2]	244,501
Aroclor 1221	[11104-28-2]	244,501
Aroclor 1232	[11141-16-5]	244,502
Aroclor 1242	[53469-21-9]	244,502
Aroclor 1248	[12672-29-6]	244,502
Aroclor 1254	[11097-68-1]	244,502–503
Aroclor 1260	[11096-82-5]	244,503
Aroclor 1262	[37324-23-5]	244
Aroclor 1268	[11100-14-4]	244
Arsenic	[7440-38-2]	482
Arsenic hydride	[7784-42-1]	297
Arsenic trihydride	[7784-42-1]	297
Arsenic trioxide	[1237-53-3]	482
Arsine	[7784-42-1]	297,482
Asbestos, bulk	—	482
Asbestos, fibers	—	482
Aspon	[3244-90-4]	217
Atratol	[1912-24-9]	205
Atraton	[1610-17-9]	205
Atrazine	[1912-24-9]	205,463,503
Avadex	[2303-16-4]	503
Azabenzene	[110-86-1]	397

Azelaic acid	[123-99-9]	482
Azimethylene	[334-88-3]	331
Azine	[110-86-1]	397
Azinphos ethyl	[2642-71-9]	217,219
Azinphos methyl	[86-50-0]	217,219,503
Azobenzene	[103-33-3]	463
Azodrin	[6923-22-4]	217
Azofene	[2310-17-0]	217
Azulene	[275-51-4]	463
Banol	[671-04-5]	201
Barban	[101-27-9]	463
Basalin	[33245-39-5]	503
Basudin	[333-41-5]	217,511
Baycarb	[3766-81-2]	201
Baygon	[114-26-1]	201
Baytex	[55-38-9]	520
Bendiocarb	[22781-23-3]	201,463
Benfluralin	[1861-40-1]	463
Benlate	[17804-35-2]	201
Benomyl	[17804-35-2]	201,463
Benzal chloride	[98-87-3]	503
Benzaldehdye	[100-52-7]	108,468,490
Benz[*a*]anthracene	[56-55-3]	166,486,491
Benzene	[71-43-2]	301,484,491,503
Benzeneamine	[62-53-3]	295
Benzenecarboxylic acid	[65-85-0]	505
Benzenechloride	[108-90-7]	319
1,4-Benzenediamine	[106-50-3]	503
1,2-Benzenediol	[120-80-9]	399
1,3-Benzenediol	[108-46-3]	403,529
1,4-Benzenediol	[123-31-9]	367,485
Benzenethiol	[108-98-5]	530
1,2,3-Benzenetriol	[87-66-1]	401
Benzidine	[92-87-5]	303,482,483,503
Benzo[*b*]fluoranthene	[205-99-2]	166,486,491
Benzo[*k*]fluoranthene	[207-08-9]	166,486,491
Benzoic acid	[65-85-0]	463,505
Benzoic aldehyde	[100-52-7]	108
Benzol	[71-43-2]	301
Benzonitrile	[100-47-0]	463
Benzo[*g,h,i*]perylene	[191-24-2]	166,486,491,504
Benzophenone	[119-61-9]	463
Benzo[*a*]pyrene	[50-32-8]	166,486,491
Benzo[*e*]pyrene	[192-97-2]	166,486
Benzoyl chloride	[98-88-4]	463
Benzoyl peroxide	[94-36-0]	482
Benzyl alcohol	[100-51-6]	505
Benzyl bromide	[100-39-0]	147
Benzyl chloride	[100-44-7]	147,305,463,484,505
Beryllium	[7440-41-7]	482
BHC	[608-73-1]	505
α -BHC	[319-84-6]	208,505
β-BHC	[319-85-7]	208,505
δ-BHC	[309-86-8]	505
g-BHC	[58-89-9]	523

Bidrin	[141-66-2]	515
Biethylene	[106-99-0]	307
Biphenyl	[92-52-4]	482
4,4′-Biphenyldiamine	[92-87-5]	303
Birlane	[470-90-6]	217
bis(2-*n*-butoxyethyl) phthalate	[117-83-9]	241
bis(2-chloroethoxy) methane	[111-91-1]	509
bis(2-chloroisopropyl)ether	[108-60-1]	509
bis(2-ethylhexyl) adipate	[103-23-1]	520
bis(2-ethylhexyl)phthalate	[117-81-7]	241,520
bis(2-ethoxyethyl) phthalate	[605-54-9]	241
bis(2-methoxyethyl) phthalate	[117-82-8]	241
bis(4-methyl-2-pentyl)phthalate	[146-50-9]	241
Bivinyl	[106-99-0]	307
Bladex	[21725-46-2]	205
Bolstar	[35400-43-2]	219
Bomyl	[122-10-1]	217
Boroethane	[19287-45-7]	333
Boron carbide	[12069-32-8]	482
Boron hydride	[19287-45-7]	333
Bromacil	[314-40-9]	463,505
Bromobenzene	[108-86-1]	505
1-Bromobutane	[109-65-9]	468
1-Bromo-2-butene	[29576-14-5]	468
Bromochloromethane	[74-97-5]	146,468,484,506
Bromodichloromethane	[75-27-4]	146
4-Bromofluorobenzene	[460-00-4]	146,468
1-Bromo-2-fluoroethane	[762-49-2]	468
Bromoform	[75-25-2]	146,484,506
Bromomethane	[74-83-9]	375,524
1-Bromo-3-methylbutane	[107-82-4]	468
1-Bromonaphthalene	[90-11-9]	463
1-Bromo-4-phenoxybenzene	[101-55-3]	506
4-Bromophenylphenylether	[101-55-3]	506
1-Bromopropane	[106-94-5]	468
2-Bromopropane	[75-26-3]	468
Bromopropylate	[18181-80-1]	463
Bromotrichloromethane	[75-62-7]	468
Bromotrifluoromethane	[75-63-8]	482,484
Bromoxynil	[1689-84-5]	482,506
Bromoxynil octanoate	[1689-99-2]	482
Butachlor	[23184-66-9]	464
1,3-Butadiene	[106-99-0]	307,482
Butanal	[123-72-8]	108
1-Butanethiol	[109-79-5]	482
2-Butanone	[78-93-3]	108,468,482,524
2-Butenal	[123-73-9]	108
2-Butoxyethanol	[111-76-2]	483
Butyl acetate	[123-86-4]	463,468
n-Butyl acetate	[123-86-4]	483
tert-Butyl acetate	[540-88-5]	483
tert-Butylamine	[75-64-9]	463
n-Butyl alcohol	[71-36-3]	468,481
sec-Butyl alcohol	[78-92-2]	468,481
tert-Butyl alcohol	[75-65-0]	468,481

n-Butyl benzene	[104-51-8]	506
sec-Butyl benzene	[135-98-8]	506
tert-Butyl benzene	[98-06-6]	506
Butyl benzyl phthalate	[85-68-7]	241,506
Butyl cellosolve	[111-76-2]	463
n-Butyl chloride	[109-69-3]	468
sec-Butyl chloride	[78-86-4]	468
tert-Butyl chloride	[507-20-0]	468
n-Butyl ether	[142-96-1]	468
Butyl ethyl ether	[628-81-9]	468
Butyllate	[15271-41-7]	463
n-Butylmercaptan	[109-79-5]	468
p-*tert*-Butyltoluene	[98-82-8]	484
Butyraldehyde	[123-72-8]	108,468,490
Butyryl chloride	[141-75-3]	468
Cadmium	[7440-43-9]	482
Calcium	[7440-70-2]	482
Camphechlor	[8001-35-2]	530
Camphor	[76-22-2]	463,485
Caparol	[7287-19-6]	205
Captafol	[2425-06-1]	507
Captan	[133-06-2]	208,315,491,507
Carbamult	[2631-37-0]	201
Carbanolate	[671-04-5]	201
Carbaryl	[63-25-2]	201,463,482,507
Carbazole	[86-74-8]	507
9H-Carbazole	[86-74-8]	463
Carbicron	[141-66-2]	217
Carbofenothion	[786-19-6]	217,218,532
Carbofuran	[1563-66-2]	201,463,507
Carbon black	[1333-86-4]	482
Carbon disulfide	[75-15-0]	309,482,507
Carbon monoxide	[630-08-0]	311
Carbon oxychloride	[75-44-5]	395
Carbon tetrachloride	[56-23-5]	146,313,484,492,507
Carbonyl chloride	[75-44-5]	395
CDEC	[95-06-7]	507
CFC-11	[75-69-4]	531
CFC-12	[75-71-8]	513
Chloral	[75-87-6]	468
Chloranil	[118-75-2]	463
Chlorbromuron	[13360-45-7]	203
Chlordane	[57-74-9]	208,491,507
α -Chlordane	[5103-71-9]	208,507
γ-Chlordane	[5103-74-2]	507–508
Chlordecone	[143-50-0]	523
Chlorfenvinphos	[470-90-6]	217,218,508
Chlorinated diphenyl ether	[55720-99-5]	482
Chlorinated terphenyl	—	482
Chlorneb	[2675-77-6]	208
Chloro IPC	[101-21-3]	201
Chloroacetaldehdye	[107-20-0]	468
Chloroacetic acid	[79-11-8]	317,482
Chloroacetone	[78-95-5]	464,468
Chloroacetonitrile	[107-14-2]	468

Chloroacetyl chloride	[79-04-9]	468
4-Chloroaniline	[106-47-8]	508
Chlorobenzene	[108-90-7]	146,319,484,492,508
m-Chlorobenzene	[541-73-1]	335
o-Chlorobenzene	[95-50-1]	335
p-Chlorobenzene	[106-46-7]	335
Chlorobenzilate	[510-15-6]	208,464,508
2-Chlorobenzoic acid	[118-91-2]	464
2-Chloro-1,3-butadiene	[126-99-8]	146
Chlorobromomethane	[74-97-5]	484,506
2-Chlorobutane	[78-86-4]	468
Chlorocyclobutane	[1120-57-6]	468
4-Chloro-1-butanol	[928-51-8]	468
3-Chloro-2-butanone	[4091-39-8]	468
3-Chloro-1-butene	[563-52-0]	468
Chlorocyclohexane	[542-18-7]	468
Chlorocyclopentane	[930-28-9]	468
Chlorodibromomethane	[124-48-1]	146
Chloroethane	[75-00-3]	351,519
Chloroethanoic acid	[79-11-8]	317
2-Chloroethanol	[107-07-3]	353
Chloroethene	[75-01-4]	532
2-Chloroethyl alcohol	[107-07-3]	353
2-Chloroethylvinyl ether	[110-75-8]	508
Chloroform	[67-66-3]	146,321,484,492,508
Chloroformyl chloride	[75-44-5]	395
Chlorofos	[52-68-6]	218
1-Chlorohexane	[544-10-5]	146,468
Chloromethane	[74-87-3]	377,524
(Chloromethyl)benzene	[100-44-7]	305,505
4-Chloro-3-methylphenol	[59-50-7]	508
(4-Chloro-2-methylphenoxy)acetic acid	[94-74-6]	153
2-Chloro-3-methylpropene	[563-47-3]	464
3-Chloro-2-methylpropene	[563-47-3]	146
3-Chloromethyl-pyridine hydrochloride	[6959-48-4]	509
2-(Chloromethyl)oxirane	[106-89-8]	347
Chloronaphthalene	[90-13-1]	464
1-Chloronaphthalene	[90-13-1]	147,509
2-Chloronaphthalene	[91-58-7]	507,509
4-Chloronitrobenzene	[100-00-5]	485
1-Chloropentane	[543-59-9]	468
2-Chlorophenol	[95-57-8]	323,509
1-Chloro-4-phenoxybenzene	[7005-72-3]	509
4-Chlorophenyl phenyl ether	[7005-72-3]	509
Chloroprene	[126-99-8]	146,482,492
1-Chloropropane	[540-54-5]	468
2-Chloropropane	[75-29-6]	468
1-Chloropropene	[590-21-6]	468
2-Chloropropene	[557-98-2]	468
3-Chloropropene	[107-05-1]	146,467
Chlorothalonil	[1897-45-6]	464,491,509
2-Chlorotoluene	[95-49-8]	146,468
4-Chlorotoluene	[106-43-4]	147
p-Chlorotoluene	[106-43-4]	509
α -Chlorotoluene	[100-44-7]	305

o-Chlorotoluene	[95-49-8]	509
5-Chloro-*o*-toluidine	[95-79-4]	509
Chloroxuron	[1982-47-4]	203
Chlorpropham	[101-21-3]	201
Chlorpyrifos	[2921-88-2]	217,218,491,517
Chlorsulfuron	[64902-72-3]	468
Chromium	[7440-47-3]	482
Chromium, hexavalent	—	482
Chrysene	[218-01-9]	166,486,491,509
Cinnamaldehyde	[104-55-2]	108
Ciodrin	[7700-17-6]	217
Coal tar naphtha	—	485
Coal tar pitch volatiles	[65996-93-2]	482
Cobalt	[7440-48-4]	482
Copper	[7440-50-8]	482
Coronene	[191-07-1]	166,464
Cotoran	[2164-17-2]	203
Coumaphos	[56-72-4]	217,219,510
Counter	[13071-79-9]	218,529
Cresol	[1319-77-3]	482
m-Cresol	[108-39-4]	490,510
o-Cresol	[95-48-7]	490,510
p-Cresol	[106-44-5]	490,510
Cresylic acid	[1319-77-3]	510
Crotonaldehdye	[123-73-9]	108,464,468,490
Crotoxyphos	[7700-17-6]	217,218,510,
Crufomate	[299-86-5]	217
Cumene	[98-82-8]	325,464,469,484,492,510
Curacron	[41198-08-7]	218
Cyanazone	[21725-46-2]	205
Cyanogen	[460-19-5]	327
Cyanomethane	[75-05-8]	289
Cyanuric acid	[108-80-5]	329,483
Cycle	[32889-48-8]	205
Cyclohexane	[110-82-7]	464,484
Cyclohexanol	[108-93-0]	464,468,481
Cyclohexanone	[108-94-1]	108,485
Cyclohexatriene	[71-43-2]	301
Cyclohexene	[110-83-8]	484
Cyclohexylamine	[108-91-8]	464
Cyclohexylketone	[108-94-1]	108
1,3-Cyclopentadiene	[542-92-7]	483
Cyclopentanone	[120-92-3]	108
Cygon	[60-51-5]	217
Cylan	[947-02-4]	217
p-Cymene	[99-87-6]	510
Cythion	[121-75-5]	217
2,4-D	[94-75-7]	153,464,486,510
Dacthal	[1861-32-1]	510
Dalapon	[75-99-0]	464,510
2,4-DB	[94-82-6]	153,464
DBCP	[96-12-8]	464
DCPA	[1861-32-1]	464,510
4,4′-DDD	[72-54-8]	208,510
4,4′-DDE	[72-55-9]	208,491,510

o,*p*-DDT	[789-02-6]	511
4,4′-DDT	[50-29-3]	208,491,511
Decanal	[112-31-2]	108
Dechlorane	[2385-85-5]	511
Decyl aldehyde	[112-31-2]	108
Demeton	[8065-48-3]	483
Demeton-O	[298-03-3]	219
Demeton-O,S	[8065-48-3]	217
Demeton-S	[126-75-0]	219
Diacetone alcohol	[123-51-3]	481
Diallate	[2303-16-4]	464
4,4′-Diamino-1,1-biphenyl	[92-87-5]	303
4,4′-Diaminodiphenyl ether	[101-80-4]	527
2,4-Diaminotoluene	[95-80-7]	511
Diamyl phthalate	[131-18-0]	241
p,p′-Dianiline	[92-87-5]	303
Diazinon	[333-41-5]	217,219,511
Diazomethane	[334-88-3]	331,483
Dibenz[*a,j*]acridine	[224-42-0]	166,511
Dibenz[*a,h*]anthracene	[53-70-3]	166,486,491,511
Dibenzofuran	[132-64-9]	511
Dibenzo[*a,e*]pyrene	[192-65-4]	166,511
Diborane	[19287-45-7]	333,483
Dibrom	[300-76-5]	217
Dibromochloromethane	[124-48-1]	511
Dibromochloropropane	[96-12-8]	512
1,2-Dibromo-3-chloropropane	[96-12-8]	147
Dibromodifluoromethane	[75-61-6]	483
1,2-Dibromoethane	[106-93-4]	468,492,519
Dibromofifluoromethane	[75-61-6]	483
Dibromomethane	[74-95-3]	524
1,2-Dibromopropane	[78-75-1]	468
1,3-Dibromopropane	[109-64-8]	468
2,2-Dibromopropane	[594-16-1]	468
2-(Dibutyl)aminoethanol	[102-81-8]	482
Dibutyl phosphate	[107-66-4]	483
Dibutyl phthalate	[84-74-2]	512
Dicamba	[1918-00-9]	153,464,512
Dicapthon	[2463-84-5]	217
Dichlofenthion	[97-17-6]	217,218
Dichlone	[117-80-6]	464,512
Dichloran	[99-30-9]	208
Dichloroacetaldehdye	[79-02-7]	468
1,1-Dichloroacetone	[513-88-2]	468
Dichloroacetyl chloride	[79-36-7]	467
1,1-Dichlorobenzene	[75-34-3]	484
1,2-Dichlorobenzene	[95-50-1]	146,147,335,492,512
1,3-Dichlorobenzene	[541-73-1]	147,335,492,513
1,4-Dichlorobenzene	[106-46-7]	146,147,335,492,513
o-Dichlorobenzene	[90-50-1]	484
para-Dichlorobenzene	[106-46-1]	484
3,3′-Dichlorobenzidine	[91-94-1]	464,483,513
1,1-Dichlorobutane	[541-33-3]	468
1,2-Dichlorobutane	[616-21-7]	468
1,3-Dichlorobutane	[1190-22-3]	469

2,3-Dichlorobutane	[7581-97-7]	468
cis-1,4-Dichloro-2-butene	[1476-11-5]	146
trans-1,4-Dichloro-2-butene	[110-57-6]	146
Dichlorodifluoromethane	[75-71-8]	146,465,468,483,513
1,1-Dichloroethane	[75-34-3]	146,468,492,505,514
1,2-Dichloroethane	[107-06-2]	146,468,493,519
1,1-Dichloroethene	[75-35-4]	146,505,514
cis-1,2-Dichloroethene	[156-59-4]	146
trans-1,2-Dichloroethene	[156-60-5]	146
Dichloroethyl ether	[111-44-4]	514
1,1-Dichloroethylene	[75-35-4]	337
1,2-Dichloroethylene	[107-06-2]	484,492
cis-1,2-Dichloroethylene	[156-59-2]	514
trans-1,2-Dichloroethylene	[156-60-5]	512
Dichlorofluoromethane	[75-43-4]	483
Dichloromethane	[75-09-2]	379,493,524
3,6-Dichloro-2-methoxybenzoic acid	[1918-00-9]	153,468
Dichloromethyl benzene	[98-87-3]	503
1,2-Dichloro-2-methylpropane	[594-37-6]	469
1,1-Dichloro-1-nitroethane	[594-72-9]	483
Dichlorophene	[97-23-4]	464
2,4-Dichlorophenol	[120-83-2]	339,468,506,514
2,6-Dichlorophenol	[87-65-0]	468,506,514
(2,4-Dichlorophenoxy)acetic acid	[94-75-7]	153
4-(2,4-Dichlorophenoxy)butyric acid	[94-82-6]	153
2-(2,4-Dichlorophenoxy)propionic acid	[120-36-5]	153
Dichloroprop	[120-36-5]	153,465
1,1-Dichloropropane	[78-99-9]	469
1,2-Dichloropropane	[78-87-5]	146,468,484,492,514
1,3-Dichloropropane	[142-28-9]	146,468,493
2,2-Dichloropropane	[594-20-7]	146,468,506
sec-Dichloropropane	[594-20-7]	515
2,2-Dichloropropanoic acid	[75-99-0]	510
cis-1,3-Dichloropropene	[10061-01-5]	146
trans-1,3-Dichloropropene	[10061-02-6]	146,468
1,1-Dichloropropene	[563-58-6]	468
1,3-Dichloropropene	[542-75-6]	146,465,468
1,3-Dichloro-1-propene	[542-75-6]	468
2,3-Dichloro-1-propene	[78-88-6]	468
1,2-Dichlorotetrafluoroethane	[76-14-2]	483
Dichlorvos	[62-73-7]	217,218,468,491
Dicofol	[115-32-2]	465,491
Dicrotophos	[141-66-2]	217,515
Dicyan	[460-19-5]	327
Dicyclohexyl phthalate	[84-61-7]	468
Dieldrin	[60-57-1]	468,491
Diethanolamine	[111-42-2]	465,482
Diethylamine	[109-89-7]	465,481
2-Diethylaminoethanol	[100-37-8]	482
Diethylenetriamine	[111-40-0]	483
Diethyl ether	[60-29-7]	341,468
Diethyl ketone	[96-22-0]	108
Diethyl oxide	[60-29-7]	341
Diethyl phthalate	[84-66-2]	241
Diethylstilbestrol	[56-53-1]	516

Diethyl sulfate	[64-67-5]	516
Diethyl sulfide	[352-93-2]	469
1,4-Difluorobenzene	[540-36-3]	146,468
Difolatan	[2425-06-1]	507
Dihexyl phthalate	[84-75-3]	241
1,3-Dihydroxybenzene	[108-46-3]	403
o-Dihydroxybenzene	[120-80-9]	399
p-Dihydroxybenzene	[123-31-9]	367
Diisobutyl ketone	[108-83-8]	108,469,485
Diisobutyl phthalate	[84-69-5]	241
Diisopropyl ether	[108-20-3]	469
2,4-Diisocyanatotoluene	[584-84-9]	423
Dilantin	[57-41-0]	516
Dimecron	[13171-21-6]	468,528
Dimethoate	[60-51-5]	217,219
3,3-Dimethoxybenzidine	[119-90-4]	516
Dimethyl benzene	[1330-20-7]	435
Dimethyl ketone	[67-64-1]	287
Dimethyl phthalate	[131-11-3]	241
Dimethyl sulfate	[77-78-1]	483
Dimethyl sulfide	[75-18-3]	469
Dimethylacetamide	[127-19-5]	483
Dimethylamine	[124-40-3]	481
N,N-Dimethylaniline	[121-69-7]	482
Dimethylarsonic acid	—	482
2,5-Dimethylbenzaldehyde	[5779-94-2]	108,491
7,12-Dimethylbenz[a]anthracene	[57-97-6]	516
Dimethylformamide	[68-12-2]	483
2,4-Dimethylphenol	[105-67-9]	468,506
1,1-Dimethyl-2-phenyl enanamine	[122-09-8]	528
N,N-Dimethyl-4-(phenylazo)benzenamine	[131-11-3]	516
N,N-Dimethyl-*p*-toluidine	[99-97-8]	482
Dinex	[131-89-5]	516
1,2-Dinitrobenzene	[512-56-1]	517
1,3-Dinitrobenzene	[99-65-0]	517
1,4-Dinitrobenzene	[100-25-4]	517
2,4-Dinitrophenol	[51-28-5]	468,506,517
2,6-Dinitrotoluene	[606-20-2]	345,506,517
2,4-Dinitrotoluene	[121-14-2]	343,506,517
Dinonyl phthalate	[84-76-0]	241
Dinoseb	[88-85-7]	465
Dioctyl phthalate	[117-84-0]	517
Dioxacarb	[6988-21-2]	201
Dioxane	[123-91-1]	483
Dioxathion	[78-34-2]	468
Dioxin	[1746-01-6]	517
o-Diphenol	[120-80-9]	399
1,2-Diphenyl hydrazine	[122-66-7]	517
Diphenylamine	[122-39-4]	465,517
Dipropetryne	[4147-51-7]	205
Dipropyl ketone	[123-19-3]	108
Diquat dibromide	[85-00-7]	517
Disulfoton	[298-04-4]	468
Disulfoton sulfoxide	[2497-07-6]	517
Disyston	[298-04-4]	517

Dithiocarbonic anhydride	[75-15-0]	309
Diuron	[330-54-1]	203,464
Di-*n*-butyl phthalate	[84-74-2]	241
Di-*n*-octyl phthalate	[117-84-0]	241
Dowicide 6	[58-90-2]	217,218,517
Dursban	[2921-88-2]	217,218,517
Dybar	[101-42-8]	203
Dyfonate	[994-22-9]	217
Dyrene	[101-05-3]	205
Ebuzin	[64529-56-2]	205
EDB	[106-93-4]	519
Elocron	[6988-21-1]	201
EL-103	[34014-18-1]	203
Endosulfa*n*-1	[959-98-8]	208,518
Endosulfa*n*-11	[33213-65-9]	208,518
β-Endosulfan	[33213-65-9]	518
Endosulfan sulfate	[1031-07-8]	208,518
Endothall	[145-73-3]	464
Endrin	[72-20-8]	208,486,491,518
Endrin aldehyde	[7421-93-4]	208,491,518
Endrin ketone	[53494-70-5]	208,519
Entex	[55-98-9]	217
Epichlorohydrin	[106-89-8]	347
EPN	[2104-6-5]	217,218,483,529
1,4-Eoxybutane	[109-99-9]	217,419
1,2-Epoxy-3-chloropropane	[106-89-8]	347
1,2-Epoxyethane	[75-21-8]	359
2,3-Epoxy-propyl chloride	[106-89-8]	347
Ethanal	[75-07-0]	108,285
Ethanedial	[107-22-2]	108
Ethanedinitrile	[460-19-5]	327
1,2-Ethanediol	[107-21-1]	357
Ethanethiol	[75-08-1]	464
Ethanol	[78-83-1]	481
Ethenylbenzene	[100-42-5]	409
Ethephon	[16672-87-0]	217,218
Ethrel	[16672-87-0]	217,218
Ethion	[563-12-2]	217,219,519
Ethiozin	[64529-56-2]	205
Ethoprop	[13194-48-4]	519
Ethoprophos	[13194-48-4]	519
2-Ethoxyethanol	[110-80-5]	483
2-Ethoxyethyl acetate	[111-15-9]	483
Ethyl acetate	[141-78-6]	469
Ethyl acrylate	[140-88-5]	469,483
Ethyl amyl ketone	[106-68-3]	108,485
Ethyl benzene	[100-41-4]	349,464,484,519
Ethyl bromide	[74-96-4]	146,484
2-Ethyl-1-butanol	[97-95-0]	469
Ethyl butyl ketone	[106-35-4]	108,485
Ethyl butyrate	[105-54-4]	469
Ethyl carbamate	[51-79-6]	519
Ethyl chloride	[75-00-3]	146,351,484,519
Ethyl ether	[60-29-7]	341,483
Ethyl formate	[109-94-4]	469

Ethyl mercaptan	[75-08-1]	469
Ethyl methacrylate	[97-63-2]	469
Ethyl methanesulfonate	[62-50-0]	519
Ethylbenzene	[100-41-4]	349,464,519
Ethyledene dichloride	[75-34-3]	514
Ethylene chlorohydrin	[107-07-3]	353,469,483
Ethylenediamine	[107-15-3]	483
Ethylene dibromide	[106-93-4]	146,147,484,519
Ethylene dichloride	[107-06-2]	484,492,519
Ethylene glycol	[107-21-1]	357,483
Ethylene oxide	[75-21-8]	359,483
Ethylene tetrachloride	[127-18-4]	413
Ethylene thiourea	[96-45-7]	483
Ethylene trichloride	[79-01-6]	429
2-(Ethylthio)ethanol	[110-77-0]	464
4-Ethyltoluene	[622-96-8]	492
Etridiazole	[2593-15-9]	208
Etrolene	[299-84-3]	218
Famophos	[52-85-7]	217,520
Famphur	[52-85-7]	217,520
Fenamiphos	[22224-92-6]	217,520
Fenchlorphos	[299-84-3]	218
Fenitrothion	[122-14-5]	217
Fenobucarb	[3766-81-2]	201
Fensulfothion	[115-90-2]	217,219,520
Fenthion	[55-38-9]	217,219,520
Fenuron	[101-42-8]	207,464
Fenuron TCA	[4482-55-7]	207
Ferbam	[14484-64-1]	464
Ficam	[22781-23-3]	201
Fluometuron	[2164-17-2]	203
Fluoranthene	[206-44-0]	166,486,491,520
Fluorene	[86-73-7]	166,486,491,520
9H-Fluorene	[86-73-7]	520
Fluorobenzene	[462-06-6]	146
Fluorocarbo*n*-113	[76-13-1]	363
o-Fluorophenol	[367-12-4]	520
Folithion	[122-14-5]	217
Folpet	[133-07-3]	491
Fonofos	[944-22-9]	217,464
Formaldehyde	[50-00-0]	108,361,483,490
Fosthietan	[21548-32-3]	217
Freon-11	[75-69-4]	490
Freon-12	[75-75-8]	490
Freon 113	[76-13-1]	363,490,531
Freon-114	[76-14-1]	490
Furadan	[1563-66-2]	201,507
Furfural	[98-01-1]	464,483
Furfuryl alcohol	[98-00-0]	483
Gardoprim	[5915-41-3]	205
Geofos	[21548-32-3]	217
Gesamil	[139-40-2]	205
Gesapax	[834-12-8]	468
Gesaprim	[1912-24-9]	205
Gesapun	[122-34-9]	205

Gestamin	[1610-17-9]	205
Glean	[64902-72-3]	205
Glutaraldehyde	[111-30-8]	108,483
Glycidol	[556-52-5]	483
Glyoxal	[107-22-2]	108
Glyphosate	[1071-83-6]	520
Guthion	[86-50-0]	217,503
Guthion ethyl	[2642-71-9]	217
Gybon	[1014-70-6]	205
1,2,3,4,7,8-H CDD	[57653-85-7]	492
Heptachlor	[76-44-8]	208,491,521
Heptachlor epoxide	[1024-57-3]	208,491
1,2,3,4,6,7,8-Heptachlorodibenzo-*p*-dioxin	[35822-46-9]	521
Heptaldehyde	[111-71-7]	108
Heptanal	[111-71-7]	108
n-Heptane	[142-82-5]	484
4-Heptanone	[123-19-3]	108,469
3-Heptanone	[106-35-4]	108,469
2-Heptanone	[110-43-0]	469
Hexachlorobenzene	[118-74-1]	147,208,464,484,491
Hexachlorobutadiene	[87-68-3]	146,147,490
Hexachlorocyclohexane	[608-73-1]	505
α-Hexachlorocyclohexane	[319-84-6]	490
Hexachlorocyclopentadiene	[77-47-4]	208,490,522
1,2,3,4,7,8-Hexachlorodibenzofuran	[70648-26-9]	147
1,2,3,6,7,8-Hexachlorodibenzo-*p*-dioxin	[57653-85-7]	522
Hexachloroethane	[67-72-1]	146,147,464,484
Hexachlorophene	[70-30-4]	147,522
Hexachloropropene	[1888-71-7]	147,522
Hexaldehyde	[66-25-1]	108
Hexamethylene diisocyanate	[822-06-0]	485
Hexamethylphosphoramide	[680-31-9]	522
Hexanal	[66-25-1]	108,490
n-Hexane	[110-54-3]	484
2-Hexanone	[591-78-6]	108,485,522
Hexazinone	[51235-04-2]	205
Hexone	[108-10-1]	383
Hexyl 2-ethylhexyl phthalate	[75673-16-4]	241
Hydrobromic acid	[10035-10-6]	481
Hydrochloric acid	[7647-01-0]	481
Hydrofluoric acid	[7664-38-2]	481
Hydrogen antimonide	[7803-52-3]	405
Hydrogen arsenide	[7784-42-1]	297
Hydrogen cyanide	[74-09-8]	365,485
Hydrogen sulfide	[7783-06-4]	369
Hydroquinol	[123-31-9]	367
Hydroquinone	[123-31-9]	367,485
2-Hydroxyphenol	[120-80-9]	399
3-Hydroxyphenol	[108-46-3]	403
m-Hydroxyphenol	[108-46-3]	403
Imidan	[732-11-6]	217,528
Indenol[1,2,3-*cd*]pyrene	[193-39-5]	166,468,486,491,522
Iodine	[7553-56-2]	485
Iodomethane	[74-88-4]	381
IPC	[122-42-9]	201

Isazophos	[42509-80-8]	217
Isoacetophorone	[78-59-1]	371
Isoamyl acetate	[123-92-2]	483
Isoamyl alcohol	[67-63-0]	481
Isobutanal	[78-84-2]	108
Isobutenal	[78-85-3]	108
Isobutyl acetate	[110-19-0]	464,483
Isobutyl alcohol	[105-30-6]	481
Isobutyl chloride	[513-36-0]	469
Isobutyraldehyde	[78-84-2]	108,469,490
Isocil	[314-42-1]	464
Isocyanatomethane	[624-83-9]	385
Isocyanic acid methyl ester	[624-83-9]	385
Isocyanic acid 4-methyl-*m*-phenylene ester	[584-84-9]	423
Isodrin	[465-73-6]	464,522
Isofenphos	[25311-71-1]	217
Isophorone	[78-59-1]	371,485,522
Isopropyl acetate	[108-21-4]	469
Isopropyl acetone	[108-10-1]	383
Isopropyl alcohol	[71-23-8]	481
Isopropylbenzene	[98-82-8]	325,469,510
Isopropyl chloride	[75-29-6]	469
p-Isopropyl toluene	[99-87-6]	510
Isoproturon	[34123-59-6]	203
Isoquinoline	[148-24-3]	464
Isosafrole	[120-58-1]	523
Isothiocyanatomethane	[566-61-6]	464
Isovaleraldehyde	[590-86-3]	108,469,490
Isovalerone	[108-83-8]	108,469
Karmex	[330-54-1]	203
Kemate	[101-05-3]	205
Kepone	[143-50-0]	464,485,523
Kerb	[23950-58-5]	523
Kerosene	[8008-20-6]	485
Ketocyclopentane	[120-92-3]	108
Kloben	[555-37-3]	203
Lannate	[16752-77-5]	201,523
Lead	[7439-92-1]	485
Lead sulfide	[1314-87-0]	485
Lead tetraethyl	[78-00-2]	487
Lepton	[21609-90-5]	217
Leptophos	[21609-90-5]	217,218,219,523
Lexone	[21087-64-9]	205
Lindane	[58-89-9]	208,486,491,523
Linuron	[330-55-2]	203,464
Lorox	[330-55-2]	203
Malathion	[121-75-5]	217,219,485,523,
Maleic anhydride	[108-31-6]	523
Maloran	[13360-45-7]	203
Marsh gas	[74-82-8]	373
Matacil	[2032-59-9]	201
MCPA	[94-74-6]	153
Mercaptodimethur	[2032-65-7]	201
Mercury	[7439-97-6]	485
Merphos	[150-50-5]	217,219

Mesitylene	[108-67-8]	523
Mesityl oxide	[141-79-7]	108,485
Mestranol	[72-33-3]	523
Mesurol	[2032-65-7]	201,523
Metaphos	[298-00-0]	523
Meta Systox-R	[301-12-2]	217
Methacrolein	[78-85-3]	108
Methacrylic acid methyl ester	[80-62-6]	387
Methacrylonitrile	[126-98-7]	469
Methamidophos	[10265-92-6]	217
Methanal	[50-00-0]	108,361
Methane	[74-82-8]	373
Methanol	[67-56-1]	485
Methapyrilene	[91-80-5]	523
Methenyl trichloride	[67-66-3]	321
Methidathion	[950-37-8]	217
Methiocarb	[2032-65-7]	201,464,523
Methomyl	[16752-77-5]	201,465,523
Methoxybenzene	[100-66-3]	469
Methoxychlor	[72-43-5]	208,491,523
2-Methoxyethanol	[109-86-4]	484
Methyl acrylate	[96-33-3]	469
Methylal	[109-87-5]	485
Methyl-*n*-amyl ketone	[110-43-0]	485
2-Methylaniline	[95-53-4]	425
Methylarsonic acid	[124-58-3]	482
2-Methylbenzaldehyde	[529-20-4]	108
3-Methylbenzaldehyde	[620-23-5]	108
4-Methylbenzaldehyde	[104-87-0]	108
2-Methylbenzenamine	[95-53-4]	425
Methylbenzene	[108-88-3]	421
Methyl bromide	[74-83-9]	146,375,484,524,
3-Methylbutanal	[590-86-3]	108
3-Methyl-2-butanone	[563-80-4]	108
Methyl butyl ketone	[591-78-6]	108
3-Methylcholanthrene	[56-49-51]	166,524
Methyl chloride	[74-87-3]	146,377,484,492,524
Methyl chloroacetate	[96-34-4]	469
Methyl chloroform	[71-55-6]	427,484,492,531
Methyl chlorpyrifos	[5598-13-0]	218
Methyl cyanide	[75-05-8]	289
2-Methyl-4,6-dinitrophenol	[534-52-1]	228,524
Methyl dursban	[5598-13-0]	218
1,2-Methylendioxy-4-allyl benzene	[94-59-7]	529
4,4′-Methylene-*bis*-(2-chloroaniline)	[101-14-4]	525
Methylene bromide	[74-95-3]	379,469,524
Methylene chloride	[75-09-2]	147,379,484,492,524
Methylene oxide	[50-00-0]	361
4,4′-Methylenediphenyl isocyanate	[101-68-8]	485
(1-Methyl ethyl) benzene	[98-82-8]	325
Methyl ethyl ketone	[78-93-3]	108,469,524
Methyl ethyl ketone peroxide	[1338-23-4]	485
5-Methyl-3-heptanone	[541-85-5]	228,524
5-Methyl-2-hexanone	[110-12-3]	108
Methyl iodide	[74-88-4]	147,381,469,484

Methyl isoamyl acetate	[108-84-9]	147,483
Methyl isoamyl ketone	[110-12-3]	108
Methyl isobutenyl ketone	[141-79-7]	108
Methyl isobutyl carbinol	[105-30-6]	481
Methyl isobutyl ketone	[108-10-1]	108,383,469,524
2-Methyl isobutyl ketone	[583-60-8]	485
Methyl isocyanate	[624-83-9]	385
Methyl isopropyl ketone	[563-80-4]	108,469
Methyl methacrylate	[80-62-6]	387,469,485
Methyl methanesulfonate	[66-27-3]	524
Methyl parathion	[298-00-0]	219,523
2-Methylnaphthalene	[91-57-6]	166,524
Methylparaben	[99-76-3]	465
4-Methyl-2-pentanone	[108-10-1]	108,383
4-Methylphenol	[106-44-5]	228,510
2-Methylphenol	[95-48-7]	228,510
3-Methylphenol	[108-39-4]	228,510
Methyl propyl ether	[557-17-5]	469
Methyl propyl ketone	[107-87-9]	108
2-Methylpyridine	[109-06-8]	528
N-Methyl-*N*-nitrosoethanamine	[10595-95-6]	524
N-Methyl-*N*-nitrosomethanamine	[62-75-9]	526
α-Methylstyrene	[98-83-9]	484
Metobromuron	[3060-89-7]	203
Metolachlor	[51218-45-2]	465
Metribuzin	[21087-64-9]	205,465
Mevinphos	[7786-34-7]	217,219,485
Mexacarbate	[315-18-4]	465,491
MIBK	[108-10-1]	383,524
MIC	[624-83-9]	385
Milogard	[139-40-2]	205
Mineral oil mist	[8012-95-1]	485
Mineral spirits	—	485
Miral	[42509-80-8]	217
Mirbane oil	[98-95-3]	389
Mirex	[2385-85-5]	208,465,491
MOCA	[101-14-4]	525
Monceren	[66063-05-6]	203
Monochloroacetic acid	[79-11-8]	317
Monocrotophos	[6923-22-4]	217,219
Monoethanolamine	[141-43-5]	482
Montrel	[299-86-5]	217
Monuron	[150-68-5]	203,465
Monuron TCA	[140-41-0]	203
Morpholine	[110-91-8]	465
Muscatox	[56-72-4]	217
Nabac	[70-30-4]	522
Naled	[300-76-5]	217,218
Namafos	[297-97-2]	218
Naphthalene	[91-20-3]	166,484,486,491
1,4-Naphthalenedione	[130-15-4]	525
Naphthas	—	485
1,4-Naphthoquinone	[130-15-4]	525
1-Naphthylamine	[134-32-7]	465,525
2-Naphthylamine	[91-59-8]	465,525
α -Naphthylamine	[134-32-7]	485

β-Naphthylamine	[91-59-8]	485
Navadel	[78-34-2]	217
Neburon	[555-37-3]	203,465
Nemacide-VC-13	[97-17-6]	217
Nemacur	[22224-92-6]	217,520
Neocidol	[333-41-5]	217
Nerkol	[62-73-7]	217
Nialate	[563-12-2]	217,519
Nickel carbonyl	[13463-39-3]	485
Nicotine	[54-11-5]	465
Niran	[56-38-2]	527
Nitran	[298-00-0]	217
Nitric acid	[7697-37-2]	481
5-Nitroacenaphthene	[602-87-9]	166,468
2-Nitroaniline	[88-74-4]	525
3-Nitroaniline	[99-09-2]	525–526
p-Nitroaniline	[100-01-6]	526
5-Nitro-*o*-anisidine	[99-59-2]	525
2-Nitrobenzenamine	[88-74-4]	525
Nitrobenzene	[98-95-3]	389,485,490,526
Nitrobenzol	[98-95-3]	389
4-Nitrobiphenyl	[92-93-3]	526
Nitroethane	[79-24-3]	485
Nitrofen	[1836-75-5]	465,527
Nitrogen dioxide	[10102-44-0]	391,485
Nitrogen peroxide	[10102-44-0]	391
Nitroglycerine	[55-63-0]	485
Nitromethane	[75-52-5]	486
2-Nitrophenol	[88-75-5]	288,526
4-Nitrophenol	[100-02-7]	288,526
2-Nitropropane	[79-46-9]	486
4-Nitroquinoline-1-oxide	[56-57-5]	526
N-Nitrosodibutylamine	[924-16-3]	183,486
N-Nitrosodi-*n*-butylamine	[924-16-3]	183,526
N-Nitrosodiethylamine	[55-18-5]	183,486,526
N-Nitrosodimethylamine	[62-75-9]	183,486,490,526
N-Nitrosodiphenylamine	[86-30-6]	183,526
N-Nitrosodipropylamine	[621-64-7]	486
N-Nitrosodi-*n*-propylamine	[621-64-7]	183,526
N-Nitrosomethylethylamine	[10595-95-6]	183
N-Nitrosomorpholine	[59-89-2]	183,486
N-Nitrosopiperidine	[100-75-4]	183,486
N-Nitrosopyrrolidine	[930-55-2]	183,465,486
2-Nitrotoluene	[88-72-2]	485
3-Nitrotoluene	[99-08-1]	485
4-Nitrotoluene	[99-99-0]	485
5-Nitro-*o*-toluidine	[99-55-8]	525
Nitrous oxide	[10024-97-2]	486
cis-Nonachlor	[5103-73-1]	527
trans-Nonachlor	[39765-80-5]	208,491,527
2-Nonenal	[18829-56-6]	108
Nonenaldehyde	[18829-56-6]	108
Nonyl phenol	[25154-52-3]	465
N-(trichloromethylmercapto)-4-cyclohexene-1,2-dicarboximide	[133-06-2]	315
Nuisance dust	—	486

OCDD	[3268-87-9]	527
Octachlorodibenzo-*p*-dioxin	[3268-87-9]	491,527
Octaldehyde	[124-13-0]	108
Octamethylpyrophosphoramide	[152-16-9]	527
Octanal	[124-13-0]	108
1-Octanethiol	[111-88-6]	486
Oftanol	[25311-71-1]	217
Orthene	[30560-19-1]	217
Ortholide-406	[133-06-2]	507
Oxacyclopentane	[109-99-9]	419
Oxalonitrile	[460-19-5]	327
Oxalyl cyanide	[460-19-5]	327
Oxamyl	[23135-22-0]	201,465
Oxirane	[75-21-8]	359
Oxolane	[109-99-9]	419
Oxychlordane	—	491
Oxydemeton-methyl	[301-12-2]	217
4,4′-Oxydianiline	[101-80-4]	527
Oxydisultoton	[2497-07-6]	517
Oxygen	[7782-44-7]	486
Oxymethane	[50-00-0]	361
Paraoxon	[311-45-5]	217
Paraphos	[56-38-2]	217
Paraquat	[1910-42-5]	527
Paraquat tetrahydrate	[4685-14-7]	486
Parathion	[56-38-2]	219,486,527
Parathion-ethyl	[56-38-2]	217
Parathion-methyl	[298-00-0]	217
Patoran	[3060-89-7]	203
PCB-1016	[12674-11-2]	244,501
PCB-1221	[11104-28-2]	244,501
PCB-1232	[11141-16-5]	244,502
PCB-1242	[53469-21-9]	244,502
PCB-1248	[12672-29-6]	244,502
PCB-1254	[11097-69-1]	244,502
PCB-1260	[11096-82-5]	244,502
Pencycuron	[66063-05-6]	203
Pentachlorobenzene	[608-93-5]	486,491,527
1,2,3,7,8-Pentachlorodibenzofuran	[57117-41-6]	527
1,2,3,7,8-Pentachlorodibenzo-*p*-dioxin	[40321-76-4]	527
Pentachloroethane	[76-01-7]	147,484
Pentachloronitrobenzene	[82-68-8]	527
Pentachlorophenate	[87-86-5]	393
Pentachlorophenol	[87-86-5]	228,393,486,491,527
Pentafluorobenzene	[363-72-4]	147
Pentanal	[110-62-3]	108,490
n-Pentane	[111-65-9]	484
1,5-Pentanedial	[111-30-8]	108
2,4-Pentanedione	[123-54-6]	108,463
2-Pentanone	[107-87-9]	108,469,485
3-Pentanone	[96-22-0]	108,469
Perchloroethylene	[127-18-4]	413,530
cis-Permethrin	[54774-45-7]	208
trans-Permethrin	[1877-74-8]	208
Perylene	[198-55-0]	166

Petroleum ether	[8032-32-4]	485
Petroleum naphtha	[8030-30-6]	485
Phenanthrene	[85-01-8]	166,486,491,527
Phenobarbital	[50-06-6]	528
Phenol	[108-95-2]	228,486,490,528
Phentermine	[122-09-8]	528
Phenylamine	[62-53-3]	295
p-(Phenylazo)aniline	[60-09-3]	528
N-Phenylbenzeneamine	[122-39-4]	517
Phenyl chloride	[108-90-7]	319
p-Phenylenediamine	[106-50-3]	503
Phenylethane	[100-41-4]	349
1-Phenylethanone	[98-86-2]	500
Phenyl ether	[101-84-8]	465
Phenylethylene	[100-42-5]	409
Phenyl mercaptan	[108-98-5]	530
Phenylmethane	[108-88-3]	421
Phenyl phthalate	[84-62-8]	465
2-Phenylpropane	[98-82-8]	325
3-Phenyl-2-propenal	[104-55-2]	108
Phorate	[298-02-2]	217,219,528
Phosalone	[2310-17-0]	217,219
Phosdrin	[7786-34-7]	217
Phosethoprop	[13194-48-4]	218
Phosfolan	[947-02-4]	217
Phosgene	[75-44-5]	395,490
Phosmet	[732-11-6]	217,219,528
Phosphacol	[311-45-5]	217
Phosphamidon	[13171-21-6]	217,219,528
Phosphan	[115-78-6]	218
Phosphoric acid	[7664-38-2]	481
Phosphorus	[7723-14-0]	486
Phosphorus trichloride	[7719-12-2]	486
Phosphothion	[121-75-5]	523
Phosvel	[21609-90-5]	217,523
Phthalic anhydride	[85-44-9]	528
Phygon	[117-80-6]	512
Picene	[213-46-7]	166
Picloram	[1918-02-1]	465,528
2-Picoline	[109-06-8]	528
Piperonyl sulfoxide	[120-62-7]	528
Pirimicarb	[23135-22-0]	201
Pirimor	[23135-22-0]	201
Prebane	[886-50-0]	205
Primatol	[1610-18-0]	205
Procyazine	[32889-48-8]	205
Profenofos	[41198-08-7]	218
Prolate	[732-11-6]	217
Promecarb	[2631-37-0]	201,465
Prometon	[1610-18-0]	205,528
Prometryne	[7287-19-6]	205,465
Pronamide	[25057-89-0]	528
Propachlor	[1918-16-7]	208,465
Propanal	[123-38-6]	108,490
Propanil	[122-42-9]	465

2-Propanone	[67-64-1]	108,287
Propargyl alcohol	[107-19-7]	469
Propazine	[139-40-2]	205
2-Propenal	[107-02-8]	108,291
2-Propenenitrile	[107-13-1]	293
Propetamphos	[31218-83-4]	218
Propham	[122-42-9]	201,465
Prophos	[13194-48-4]	218,219
Propionaldehyde	[123-38-6]	108,469
Propionitrile	[107-12-0]	469
Propionyl chloride	[79-03-8]	469
Propoxur	[114-26-1]	201,465
n-Propyl acetate	[109-60-4]	469,483
n-Propyl alcohol	[71-23-8]	481
n-Propyl benzene	[103-65-1]	469,528
Propylene dichloride	[78-87-5]	514
Propylene oxide	[75-56-9]	486
n-Propyl mercaptan	[107-03-9]	469
Propylthiouracil	[51-52-5]	528
Protothiophos	[34643-46-4]	218
Pyrene	[129-00-0]	166,486,491,528
Pyrethrum	[8003-34-7]	486
Pyridine	[110-86-1]	397,486,529
Pyrocatechol	[120-80-9]	399
Pyrogallic acid	[87-66-1]	401
Pyrogallol	[87-66-1]	401
Pyrophosphoric acid tetraethyl ester	[107-49-3]	417
Quinoline	[83-79-4]	465
Quinone	[106-51-4]	529
Rabon	[22248-79-9]	529
Reglone	[85-00-7]	517
Resorcinol	[108-46-3]	403,529
Ribaviron	—	487
Ronnel	[299-84-3]	218,491
Rotenone	[83-79-4]	465,487
Rubber solvent	—	485
Rubitox	[2310-17-0]	217
Ruelene	[299-86-5]	217
Safrole	[94-59-7]	529
Safrotin	[31218-83-4]	218
Sancap	[4147-51-7]	205
Santox	[2104-64-5]	529
Sencor,Lexone	[21087-64-9]	205
Sevin,Arylam	[63-25-2]	201
Siduron	[1982-49-6]	203,465
Silica	—	487
Silvex	[93-72-1]	153,465,529
Simazine	[122-34-9]	205,208,465,529
Simetryne	[1014-70-6]	205,465,468
Solvent ether	[60-29-7]	341
Standak	[1646-88-4]	201
Stibine	[7803-52-3]	405,487
Stirophos	[22248-79-9]	218,529
Stoddart solvent	[8052-41-3]	485
Strobane	[71-55-6]	427

Strychnidin-10-one	[57-24-9]	407
Strychnine	[57-24-9]	407,465,487
Strychnine sulfate	[60-41-3]	529
Styrene	[100-42-5]	409,465,484,492,529
Sulfallate	[95-06-7]	465,507
Sulfotepp	[3689-24-5]	218,219
Sulfur dioxide	[7446-09-5]	411,487
Sulfuric acid	[7664-93-9]	481
Sulprophos	[35400-43-2]	217
Supracide	[950-37-8]	217
Swat	[122-10-1]	217
Swep	[1918-18-9]	201
sym-Dichloroethyl ether	[111-44-4]	483
Systox	[8065-48-3]	217
2,4,5-T	[93-76-5]	153,465,486,529
Tamaron	—	217
2,3,7,8-TCDD	[1746-01-6]	249,491,517
1,2,3,4-TCDD	—	491
2,3,7,8-TCDF	[51207-31-9]	529
TDI	[584-84-9]	423
Tebuthiuron	[34014-18-1]	203
Telvar	[150-68-5]	203
Temephos	[3383-96-8]	218
Temik	[116-06-3]	201,500
Tenoran	[1982-47-4]	203
TEPP	[107-49-3]	218,417,530
Terbacil	[5902-51-2]	465
Terbucarb	[1918-11-2]	201
Terbufos	[13071-79-9]	218,219,529
Terbuthylazine	[5915-41-3]	205
Terbutol	[1918-11-2]	201
Terbutryne	[886-50-0]	205
α -Terphenyl	[84-15-1]	487
Terracur P	[115-90-2]	217
1,1,2,2-Tetrabromoethane	[79-27-6]	484
1,2,3,4-Tetrachlorobenzene	[634-66-2]	486
1,2,4,5-Tetrachlorobenzene	[95-94-3]	147,468,529
Tetrachlorocarbon	[56-23-5]	313
2,3,7,8-Tetrachlorodibenzofuran	[51207-31-9]	529
Tetrachloroethane	[127-18-4]	413
1,1,1,2-Tetrachloroethane	[630-20-6]	147,468,529
1,1,2,2-Tetrachloroethane	[79-34-5]	468,484
Tetrachloroethylene	[127-18-4]	147,413,484,493,530
Tetrachloromethane	[56-23-5]	313
Tetrachloronaphthalene	[53555-64-9]	465
2,3,4,6-Tetrachlorophenol	[58-90-2]	517
Tetrachlorophenol	[25167-83-3]	530
Tetrachlorvinphos	[22248-79-9]	218,219,529
Tetraethyl lead	[78-00-2]	415,487
Tetraethyl pyrophosphate	[107-49-3]	218,219,417,487,530
Tetraethylplumbane	[78-00-2]	415
Tetrafenphos	[3383-96-8]	218
Tetrahydrofuran	[109-99-9]	419,487
Tetralin	[119-64-2]	465
Tetramethyl lead	[75-74-1]	487

Tetramethyl thiourea	[2782-91-4]	487
Tetron	[107-49-3]	217,218
Thimet	[298-02-2]	217,528
Thiodan-11	[33213-65-9]	518
Thiodemeton	[298-04-4]	217
Thionazin	[297-97-2]	218,219,533
Thiophenol	[108-98-5]	530
Thiram	[137-26-8]	465,487
TOK	[1836-75-5]	527
Tokuthion	[34643-46-4]	218,219
2-Tolidine	[119-93-7]	465
α-Tolidine	[119-93-7]	483
m-Tolualdehyde	[620-23-5]	490
o-Tolualdehyde	[529-20-4]	108,490
p-Tolualdehyde	[104-87-0]	108,490
Toluene	[108-88-3]	421,484,492,530
2,4-Toluenediamine	[95-80-7]	423,487,530
2,6-Toluenediamine	[823-40-5]	487
Toluene-2,4-diisocyanate	[584-84-9]	485,487,530
α-Toluidine	[95-68-1]	482
o-Toluidine	[95-53-4]	425,530
o-Tolylamine	[95-53-4]	425
Tolylchloride	[100-44-7]	305
α-Tolylchloride	[95-49-8]	468
Toxaphene	[8001-35-2]	208,530
2,4,5-TP	[93-72-1]	529
Treflan	[1582-09-8]	532
Tribromomethane	[75-25-2]	492
Trichlorfon	[52-68-6]	218
Trichloroacetaldehyde	[75-87-6]	468
Trichloroacetic acid	[76-03-9]	465
1,1,1-Trichloroacetone	[918-00-3]	469
Trichloroacetyl chloride	[76-02-8]	469
Trichlorobenzene	[87-61-6]	465
1,2,3-Trichlorobenzene	[87-61-6]	147,531
1,2,4-Trichlorobenzene	[120-82-1]	147,487,531
2,2,3-Trichlorobutane	[10403-60-8]	469
Trichloroethane	[79-01-6]	429,531
1,1,1-Trichloroethane	[71-55-6]	147,427,492,531
1,1,2-Trichloroethane	[79-00-5]	147,484,492,531
Trichloroethylene	[79-01-5]	147,429,484,492
Trichlorofluoromethane	[75-69-4]	147,465,484,531
Trichloromethane	[67-66-3]	321
(Trichloromethyl) benzene	[98-07-7]	532
Trichloronaphthalene	[55720-37-1]	465
Trichloronate	[327-98-0]	218,532
Trichlorophenol	[25167-82-2]	532
2,4,5-Trichlorophenol	[95-95-4]	228,532
2,4,6-Trichlorophenol	[88-06-2]	228,431,532
(2,4,5-Trichlorophenoxy) acetic acid	[93-76-5]	153
1,2,3-Trichloropropane	[96-18-4]	147,469,484,532
1,1,2-Trichloropropylene	[2567-14-8]	470
1,1,2-Trichloro-1,2,2-trifluoroethane	[76-13-1]	363,470,532
1,1,1-Trichloro-2,2,2-trifluoroethane	[354-58-5]	470
Trichlorpyrphos	[2921-88-2]	217

Triethanolamine	[102-71-6]	465,482
Triethylamine	[121-44-8]	465
Triethylenetetramine	[280-57-9]	483
O-O-O-Triethylphosphorothioate	[126-68-1]	532
Trifluralin	[1582-09-8]	208,465,532
1,2,3-Trihydroxybenzene	[87-66-1]	401
Trihydroxycyanidine	[108-80-5]	329
2,4,6-Trihydroxy-1,3,5-triazine	[108-80-5]	329
Trimethylamine	[75-50-3]	465
2,4,5-Trimethylaniline	[137-17-7]	532
1,2,4-Trimethylbenzene	[95-63-6]	532
1,1,3-Trimethyl-3-cyclohexene-5-one	[78-59-1]	371
Trimethylphosphate	[512-56-1]	219,532
1,3,5-Trinitrobenzene	[99-35-4]	532
2,4,7-Trinitrofluoren-9-one	[129-79-3]	487
tris-(2,3-dibromopropyl)phosphate	[126-72-7]	219,512
Trithion	[786-19-6]	217,218,532
Tri-*p*-tolyl phosphate	[78-32-0]	219
Tungsten	[7440-33-7]	487
Tupersan	[1982-49-6]	203
Turpentine	[8006-64-2]	487
Tycor	[64529-56-2]	205
Urab	[4482-55-7]	203
Urethan	[51-79-6]	465
Urox	[140-41-0]	203
Valeraldehdye	[110-62-3]	108,470,487,490
Vancide-89	[133-90-4]	532
Vandium oxides	—	487
Vapona	[62-73-7]	515
Velpar	[51235-04-2]	205
Vinyl acetate	[108-05-4]	465,470
Vinyl aldehyde	[107-02-8]	291
Vinyl benzene	[100-42-5]	409,492,529
Vinyl bromide	[593-60-2]	470,485
Vinyl chloride	[75-01-4]	147,433,485,492,532
Vinyl cyanide	[107-13-1]	293
Vinyl ether	[109-93-3]	470
Vinylethylene	[106-99-0]	307
Vinylidene chloride	[75-35-4]	337,485,492,514
Vinyltoluene	[25013-15-4]	484
Vydate	[23135-22-0]	533
Warfarin	[81-81-2]	487
Xylene	[1330-20-7]	435,465,484,533
m-Xylene	[108-38-3]	435,492,533
o-Xylene	[95-47-6]	435,492,533
p-Xylene	[106-42-3]	435,492,533
2,4-Xylidine	[95-68-1]	482
Zinc	[7440-66-6]	487
Zinc oxide	[1314-13-2]	487
Zinophos	[297-97-2]	218,533

CAS REGISTRY INDEX

CAS Registry Number	Chemical Compound	Page Number
[50-00-0]	Methanal	108
[50-00-0]	Formaldehyde	108,361,483,490
[50-00-0]	Methanal	361
[50-00-0]	Oxymethane	361
[50-06-6]	Phenobarbital	528
[50-29-3]	4,4'-DDT	208,491,511
[50-32-8]	Benzo[*a*]pyrene	166,486,491
[51-28-5]	2,4-Dinitrophenol	506,517
[51-52-5]	Propylthiouracil	528
[51-79-6]	Urethan	465
[51-79-6]	Ethyl carbamate	519
[52-68-6]	Trichlorfon	218
[52-68-6]	Anthon	218
[52-68-6]	Chlorofos	218
[52-85-7]	Famophos	217,520
[52-85-7]	Famphur	217,520
[53-70-3]	Dibenz[*a,h*]anthracene	166,486,491,511
[53-96-3]	2-Acetylaminofluorene	500
[53-96-3]	Acetamidofluorene	500
[54-11-5]	Nicotine	465
[55-18-5]	*N*-Nitrosodiethylamine	183,486,526
[55-38-9]	Fenthion	219,520
[55-38-9]	Baytex	520
[55-63-0]	Nitroglycerine	485
[55-98-9]	Entex	217
[55-98-9]	Fenthion	217
[56-23-5]	Tetrachlorocarbon	313
[56-23-5]	Carbon tetrachloride	146,313,484,492,507
[56-23-5]	Tetrachloromethane	313
[56-38-2]	Parathion	219,486,527
[56-38-2]	Niran	527
[56-38-2]	Paraphos	217
[56-38-2]	Parathion-ethyl	217
[56-49-5]	3-Methylcholanthrene	166,524
[56-55-3]	Benzo[*a*]anthracene	166,486,491

[56-57-5]	4-Nitroquinoline-1-oxide	526
[56-72-4]	Coumaphos	217,219,510
[56-72-4]	Muscatox	217
[57-24-9]	Strychnidin-10-one	407
[57-24-9]	Strychnine	407,465,487
[57-41-0]	Dilantin	516
[57-74-9]	Chlordane	208,491,507
[57-97-6]	7,12-Dimethylbenz[a]anthracene	468
[58-89-9]	Lindane	208,486,491,523
[58-89-9]	γ-BHC	208,523
[58-90-2]	Dowicide 6	517
[58-90-2]	2,3,4,6-Tetrachlorophenol	517
[59-50-7]	4-Chloro-3-methylphenol	508
[59-89-2]	*N*-Nitrosomorpholine	183,486
[60-09-3]	Aminoazobenzene	528
[60-09-3]	*p*-(Phenylazo)aniline	528
[60-11-7]	3,3-Dimethoxybenzidine	516
[60-29-7]	Solvent ether	341
[60-29-7]	Diethyl ether	341,468
[60-29-7]	Diethyl oxide	341
[60-29-7]	Ethyl ether	341,483
[60-41-3]	Strychnine sulfate	529
[60-51-5]	Cygon	217
[60-51-5]	Dimethoate	217,219
[60-57-1]	Dieldrin	468,491
[61-82-5]	Amitrole	463
[62-50-0]	Ethyl methanesulfonate	519
[62-53-3]	Aminobenzene	295
[62-53-3]	Aniline	295,481
[62-53-3]	Benzeneamine	295
[62-53-3]	Phenylamine	295
[62-53-3]	Aniline	295
[62-53-3]	Aminobenzene	295
[62-73-7]	Dichlorovos	491
[62-73-7]	Vapona	515
[62-75-9]	Triethanolamine	466
[62-75-9]	*N*-Nitrosodimethylamine	183,486,490,526
[62-75-9]	*N*-methyl-*N*-nitrosomethanamine	526
[63-25-2]	Arylam	201
[63-25-2]	Sevin	201
[63-25-2]	Carbaryl	201,463,482,507
[64-19-7]	Acetic acid	481
[64-67-5]	Diethyl sulfate	516
[65-85-0]	Benzoic acid	463,505
[65-85-0]	Benzenecarboxylic acid	505
[66-25-1]	Hexanal	108,409
[66-25-1]	Hexaldehyde	108
[66-25-1]	Hexanal	491
[66-27-3]	Methyl methanesulfonate	524
[67-63-0]	Isoamyl alcohol	481
[67-64-1]	Acetone	108,287,467,485,490
[67-64-1]	2-Propanone	287,108
[67-64-1]	Dimethyl ketone	287
[67-66-3]	Chloroform	146,321,468,484,492,508
[67-66-3]	Trichloromethane	321

[67-66-3]	Chloroform	146,321,484,492,508
[67-66-3]	Methenyl trichloride	321
[67-72-1]	Hexachloroethane	146,147,464,484
[70-30-4]	Nabac	522
[70-30-4]	Hexachlorophene	147,522
[71-23-8]	Isopropyl alcohol	481
[71-23-8]	*n*-Propyl alcohol	481
[71-36-3]	*n*-Butyl alcohol	468,481
[71-43-2]	Benzene	301
[71-43-2]	Cyclohexatriene	301
[71-43-2]	Benzol	301
[71-43-2]	Benzene	484,492
[71-55-6]	Strobane	427
[71-55-6]	Methyl chloroform	493
[71-55-6]	1,1,1-Trichloroethane	147,427,492,531
[71-55-6]	Methylchloroform	427,484,531
[72-20-8]	Endrin	208,486,491,518
[72-33-3]	Mestranol	523
[72-43-5]	Methoxychlor	208,491,523
[72-54-8]	4,4-DDD	208,510
[72-55-9]	4,4'-DDE	208,491,510
[74-09-2]	Chloromethane	493
[74-09-8]	Hydrogen cyanide	365,485
[74-82-8]	Marsh gas	373
[74-82-8]	Methane	373
[74-83-9]	Methyl bromide	146,375,484,524
[74-83-9]	Ethyl bromide	146,468
[74-83-9]	Bromomethane	375,524
[74-87-3]	Chloromethane	377,524
[74-87-3]	Methyl chloride	146,377,484,492,524
[74-87-9]	Methyl bromide	484
[74-88-4]	Methyl iodide	147,381,469,484
[74-88-4]	Iodomethane	381
[74-95-3]	Dibromomethane	524
[74-95-3]	Methylene bromide	379,469,524
[74-96-4]	Ethyl bromide	146,484
[74-97-5]	Chlorobromomethane	146,468,484,506
[74-97-5]	Bromochloromethane	468,506
[75-00-3]	Chloroethane	351
[75-00-3]	Ethyl chloride	146,351,484,519
[75-00-3]	Chloroethane	519
[75-01-4]	Chloroethene	532
[75-01-4]	Vinyl chloride	147,433,485,492,532
[75-01-4]	Zvinyl chloride	433
[75-05-8]	Methtyl cyanide	289
[75-05-8]	Acetonitrile	289,481
[75-05-8]	Cyanomethane	289
[75-07-0]	Acetaldehdye	108,285,481,490
[75-07-0]	Acetic aldehyde	285
[75-07-0]	Ethanal	108
[75-08-1]	Ethyl mercaptan	469
[75-08-1]	Ethanethiol	464
[75-09-2]	Methylene chloride	147,379,484,4921,524
[75-09-2]	Dichloromethane	379,492
[75-15-0]	Carbon disulfide	309,482,507

[75-15-0]	Dithiocarbonic anhydride	309
[75-18-3]	Dimethyl sulfide	469
[75-21-8]	Ethy!ene oxide	359,483
[75-21-8]	Oxirane	359
[75-21-8]	1,2-Epoxyethane	359
[75-25-2]	Bromoform	146,484,492,506
[75-25-2]	Tribromomethane	492
[75-26-3]	2-Bromopropane	468
[75-27-4]	Bromodichloromethane	146
[75-29-6]	2-Chloropropane	468
[75-29-6]	Isopropyl chloride	469
[75-34-3]	1,1-Dichlorobenzene	484
[75-34-3]	1,1-Dichloroethane	146,468,492,505,514
[75-34-3]	Ethyledene dichloride	514
[75-35-4]	1,1-Dichloroethene	146,505,514
[75-35-4]	Vinylidene chloride	337,485,492,514
[75-35-4]	1,1-Dichloroethylene	337,493
[75-36-5]	Acetyl chloride	467
[75-43-4]	Dichlorofluoromethane	483
[75-44-5]	Carbon oxychloride	395
[75-44-5]	Chloroformyl chloride	395
[75-44-5]	Carbonyl chloride	395
[75-44-5]	Phosgene	395,490
[75-50-3]	Trimethylamine	465
[75-52-5]	Nitromethane	486
[75-56-9]	Propylene oxide	486
[75-61-6]	Dibromodifluoromethane	483,484
[75-62-7]	Bromotrichloromethane	468
[75-63-8]	Bromotrifluoromethane	482,484,
[75-64-9]	*tert*-Butylamine	463
[75-65-0]	*tert*-Butyl alcohol	468,481
[75-69-4]	Trichlorofluromethane	147,465,484,531
[75-69-4]	Freon-11	490
[75-69-4]	CFC-11	531
[75-71-8]	Dichlorodifluoromethane	146,465,468,483
[75-71-8]	CFC-12	513
[75-71-8]	Dichlorofluoromethane	513
[75-74-1]	Tetramethyl lead	487
[75-75-8]	Freon-12	490
[75-86-5]	Acetone cyanohydrin	463,481
[75-87-6]	Chloral	468
[75-87-6]	Trichloroacetaldehyde	468
[75-99-0]	2,2-Dichloropropanoic acid	510
[75-99-0]	Dalapon	464,510
[76-01-7]	Pentachloroethane	147,484
[76-02-8]	Triachloroacetyl chloride	469
[76-03-9]	Tichloroacetic acid	465
[76-13-1]	Fluorocarbon-113	363
[76-13-1]	Freon-113	363,490,531
[76-13-1]	1,1,2-Trichloro-1,2,2-trifluoroethane	363,470
[76-14-1]	Freon-114	490
[76-14-2]	1,2-Dichlorotetrafluoroethane	483
[76-22-2]	Camphor	463,485
[76-44-8]	Heptachlor	208,491,521
[77-74-4]	Hexachlorocyclopentadiene	147,208

[77-78-1]	Dimethyl sulfate	483
[78-00-2]	Tetraethyllead	415,487
[78-00-2]	Tetraethylplumbane	415
[78-00-2]	Lead tetraethyl	415
[78-32-0]	Tri-*p*-tolyl phosphate	219
[78-34-2]	Navadel	217
[78-34-2]	Dioxathion	468
[78-59-1]	Isoacetophorone	371
[78-59-1]	1,1,3-Trimethyl-3-cyclohexene-5-one	371
[78-59-1]	Isophorone	371,485,522
[78-75-1]	1,2-Dibromopropane	468
[78-83-1]	Ethanol	481
[78-84-2]	Isobutanal	108
[78-84-2]	Isobutyraldehyde	108,469,490
[78-85-3]	Isobutenal	108
[78-85-3]	Methacrolein	108
[78-86-4]	2-Chlorobutane	468
[78-86-4]	*sec*-Butyl chloride	468
[78-87-5]	1,2-Dichloropropane	146,468
[78-87-5]	Propylene dichloride	514
[78-87-5]	1,2-Dichloropropane	468,484,492,514
[78-87-5]	Propylene dichloride	514
[78-88-6]	2,3-Dichloro-1-propene	468
[78-92-2]	*sec*-Butyl alcohol	468,481
[78-93-3]	2-Butanone	108,482,524
[78-93-3]	Methyl ethyl ketone	108,469,524
[78-95-5]	Chloroacetone	464,468
[78-99-9]	1,1-Dichloropropane	469
[79-00-5]	1,1,2-Trichloroethane	147,484,492,531
[79-01-5]	Trichloroethylene	493
[79-01-6]	Trichloroethene	531
[79-01-6]	Trichloroethylene	147,429,484,492
[79-01-6]	Trichloroethane	429,531
[79-01-6]	Ethylene trichloride	429
[79-02-7]	Dichloroacetaldehyde	468
[79-03-8]	Propionyl chloride	469
[79-04-9]	Chloroacetyl chloride	468
[79-06-1]	Acrylamide	463
[79-11-8]	Chloroacetic acid	317,482
[79-11-8]	Chloroethanoic acid	317
[79-11-8]	Monochloroacetic acid	317
[79-24-3]	Nitroethane	485
[79-27-6]	1,1,2,2-Tetrabromoethane	484
[79-34-5]	1,1,2,2-Tetrachlorethane	147,484,529
[79-34-5]	1,1,2,2,-Tetrachloromethane	493
[79-36-7]	Dichloroacetyl chloride	469
[79-46-9]	2-Nitropropane	486
[80-62-6]	Methyl methacrylate	485
[80-62-6]	Methacrylic acid methyl ester	387
[80-62-6]	Methylmethacrylate	387,469
[81-81-2]	Warfarin	487
[82-68-8]	Pentachloronitrobenzene	527
[83-32-9]	Acenaphthene	166,486,491
[83-79-4]	Rotenone	465,487
[83-79-4]	Quinoline	465

[84-15-1]	α-Terphenyl	487
[84-61-7]	Dicyclohexyl phthalate	241
[84-62-8]	Phenyl phthalate	465
[84-65-1]	Anthraquinone	463
[84-66-2]	Diethyl phthalate	241
[84-69-5]	Diisobutyl phthalate	241
[84-74-2]	Di-*n*-butyl phthalate	241
[84-74-2]	Dibutyl phthalate	512
[84-75-3]	Dihexyl phthalate	241
[84-76-0]	Dinonyl phthalate	468
[85-00-7]	Reglone	517
[85-00-7]	Aquacide	517
[85-00-7]	Diaquat dibromide	517
[85-01-8]	Phenanthrene	166,486,491,527
[85-44-9]	Phthalic anhydride	528
[85-68-7]	Butyl benzyl phthalate	241,506
[86-30-6]	*N*-Nitrosodiphenylamine	183,526
[86-30-7]	Fluorene	486
[86-50-0]	Azinphos-methyl	217,219,503
[86-50-0]	Guthion	217,219,503
[86-73-7]	Fluorene	166,486,491,520
[86-73-7]	9H-Fluorene	520
[86-74-8]	9H-Carbazole	463
[86-74-8]	Carbazole	507
[87-61-6]	1,2,3-Trichlorobenzene	147,531
[87-61-6]	Trichlorobenzene	465
[87-65-0]	2,6-Dichlorophenol	468,506,514
[87-66-1]	Pyrogallol	401
[87-66-1]	1,2,3-Trihydroxybenzene	401
[87-66-1]	1,2,3-Benzenetriol	401
[87-66-1]	Pyrogallic acid	401
[87-68-3]	Hexachlorobutadiene	146,147,490
[87-86-5]	Pentachlorophenol	228,393,486,491,527
[87-86-5]	Pentachlorophenate	393
[88-06-2]	2,4,6-Trichlorophenol	228,431,532
[88-72-2]	2-Nitrotoluene	485
[88-74-4]	2-Nitrobenzenamine	525
[88-74-4]	2-Nitroaniline	525
[88-75-5]	2-Nitrophenol	228,526
[88-85-7]	Dinoseb	464,517
[90-04-0]	*o*-Anisidine	482,483
[90-11-9]	1-Bromonaphthalene	463
[90-13-1]	1-Chloronaphthalene	147,509
[90-50-1]	*o*-Dichlorobenzene	484
[91-08-7]	2,4-Toluene diisocyanate	485
[91-20-3]	Naphthalene	166,484,486,491
[91-57-6]	2-Methylnaphthalene	166,525
[91-58-7]	2-Chloronaphthalene	507,509
[91-59-8]	2-Naphthylamine	465,525
[91-59-8]	β-Naphthylamine	485
[91-80-5]	Methapyrilene	523
[91-94-1]	3,3'-Dichlorobenzidine	464,483,513
[92-52-4]	Biphenyl	482
[92-67-1]	4-Aminobiphenyl	501
[92-87-5]	Benzidine	303,483,503

[92-87-5]	4,4′-Diamino-1,1-biphenyl	303
[92-87-5]	Benzidine	482,483
[92-87-5]	*p,p′*-Dianiline	303
[92-87-5]	4,4′-Biphenyldiamine	303
[92-93-3]	4-Nitrobiphenyl	526
[93-72-1]	Rotenone	466
[93-72-1]	2,4,5-TP	529
[93-72-1]	Silvex	153,465,529
[93-76-5]	2,4,5-T	153,465,486,529
[93-76-5]	(2,4,5-Trichlorophenoxy) acetic acid	153
[94-36-0]	Benzoyl peroxide	482
[94-59-7]	1,2-Methylendioxy-4-allyl benzene	529
[94-59-7]	Safrole	529
[94-74-6]	(4-Chloro-2-methylphenoxy)acetic acid	153
[94-74-6]	MCPA	153
[94-75-7]	2,4-D	153,464,486,510
[94-75-7]	(2,4-Dichlorophenoxy)acetic acid	153
[94-82-6]	2,4-DB	153,464
[94-82-6]	4-(2,4-Dichlorophenoxy)butyric acid	153
[95-06-7]	CDEC	507
[95-06-7]	Sulfallate	465,507
[95-47-6]	*o*-Xylene	435,492,533
[95-48-7]	*o*-Cresol	490,510
[95-48-7]	2-Methylphenol	228,510
[95-49-8]	α-Tolylchloride	468
[95-49-8]	*o*-Chlorotoluene	509
[95-49-8]	2-Chlorotoluene	146,468
[95-50-1]	1,2-Dichlorobenzene	147,335,492,512
[95-50-1]	*o*-Chlorobenzene	335
[95-53-4]	*o*-Toluidine	425,530
[95-53-4]	2-Methylaniline	425
[95-53-4]	*o*-Tolylamine	425
[95-53-4]	2-Methylbenzenamine	425
[95-53-4]	2-Aminotoluene	425
[95-57-8]	2-Chlorophenol	323,509
[95-63-6]	1,2,4-Trimethylbenzene	532
[95-68-1]	α-Toluidine	482
[95-68-1]	2,4-Xylidine	482
[95-79-4]	5-Chloro-*o*-toluidine	509
[95-80-7]	2,4-Diaminotoluene	511
[95-80-7]	2,4-Toluenediamine	487
[95-94-3]	1,2,4,5-Tetrachlorobenzene	147,529
[95-95-4]	2,4,5-Trichlorophenol	228,532
[96-12-8]	DBCP	464
[96-12-8]	Dibromochloropropane	512
[96-12-8]	1,2-Dibromo-3-chloropropane	147
[96-18-4]	1,2,3-Trichloropropane	147,469,484,532
[96-22-0]	3-Pentanone	108
[96-22-0]	Diethyl ketone	108
[96-22-0]	3-Pentanone	108,469
[96-33-3]	Methyl acrylate	469
[96-34-4]	Methyl chloroacetate	469
[96-45-7]	Ethylene thiourea	483
[97-17-6]	Nemacide-VC-13	217
[97-17-6]	Dichlofenthion	217,218

[97-23-4]	Dichlorophene	464
[97-63-2]	Ethyl methacrylate	469
[97-95-0]	2-Ethyl-1-butanol	469
[98-00-0]	Furfuryl alcohol	483
[98-01-1]	Furfural	464,483
[98-06-6]	*tert*-Butylbenzene	506
[98-07-7]	(Trichloromethyl) benzene	532
[98-50-0]	*p*-Aminophenylarsonic acid	482
[98-82-8]	2-Phenylpropane	325
[98-82-8]	Cumene	325,464,469,484,492,510
[98-82-8]	(1-Methyl ethyl) benzene	325
[98-82-8]	Isopropylbenzene	325
[98-82-8]	*p-tert*-Butyltoluene	484
[98-82-8]	Isopropylbenzene	469,510
[98-83-9]	α-Methylstyrene	484
[98-86-2]	1-Phenylethanone	500
[98-86-2]	Acetophenone	300
[98-87-3]	Dichloromethyl benzene	503
[98-87-3]	Benzal chloride	463
[98-88-4]	Benzoyl chloride	463
[98-95-3]	Nitrobenzene	389,485,490,526
[98-95-3]	Nitrobenzol	389
[98-95-3]	Mirbane oil	389
[99-08-1]	3-Nitrotoluene	485
[99-09-2]	3-Notroaniline	526
[99-30-9]	Dichloran	208
[99-35-4]	1,3,5-Trinitrobenzene	532
[99-55-8]	5-Nitro-*o*-toluidine	525
[99-59-2]	5-Nitro-*o*-anisidine	525
[99-65-0]	1,3-Dinitrobenzene	517
[99-76-3]	Methylparaben	465
[99-87-6]	*p*-Isopropyl toluene	510
[99-87-6]	*p*-Cymene	510
[99-97-8]	*N,N*-Dimethyl-*p*-toluidine	482
[99-99-0]	4-Nitrotoluene	485
[100-00-5]	4-Chloronitrobenzene	485
[100-01-6]	*p*-Nitroaniline	526
[100-02-7]	4-Nitrophenol	228,526
[100-25-4]	1,4-Dinitrobenzene	517
[100-37-8]	2-Diethylaminoethanol	482
[100-39-0]	Benzyl bromide	147
[100-41-4]	Ethylbenzene	349,464,519
[100-41-4]	Phenylethane	349
[100-42-5]	Styrene	465,409,484,492,529
[100-42-5]	Phenylethylene	409
[100-42-5]	Vinylbenzene	529
[100-42-5]	Styrene	529
[100-42-5]	Vinylbenzene	409,492,529
[100-42-5]	Ethenylbenzene	409
[100-44-7]	(Chloromethyl)benzene	305
[100-44-7]	α-Chlorotoluene	305
[100-44-7]	Benzyl chloride	146,147,305,484,505
[100-44-7]	Chloromethyl benzene	505
[100-44-7]	Tolyl chloride	305
[100-47-0]	Benzonitrile	463

[100-51-6]	Benzyl alcohol	505
[100-52-7]	Benzaldehyde	108,468,490
[100-52-7]	Benzoic aldehyde	108
[100-66-3]	Methoxybenzene	469
[100-66-3]	Anisole	469
[100-75-4]	*N*-Nitrosopiperidine	183,486
[101-05-3]	Anilazine	205
[101-05-3]	Dyrene	205
[101-05-3]	Kemate	205
[101-14-4]	MOCA	525
[104-14-4]	4,4'-Methylene-*bis*(2-chloroaniline)	525
[101-21-3]	Chlorpropham	201
[101-21-3]	Chloro IPC	201
[101-27-9]	Barban	463
[101-42-8]	Fenuron	203,464
[101-42-8]	Dybar	203
[101-55-3]	4-Bromophenylphenylether	506
[101-55-3]	1-Bromo-4-phenoxybenzene	506
[101-55-3]	4-Bromophneylphenylether	506
[101-68-8]	4,4'-Methylenediphenyl iocyanate	485
[101-80-4]	4,4'-Oxydianiline	527
[101-80-4]	4,4-Diaminodiphenyl ether	527
[101-84-8]	Phenyl ether	465
[102-71-6]	Triethanolamine	465,482
[102-81-8]	2-(Dibutyl)aminoethanol	482
[103-23-1]	*bis*(2-ethylhexyl) adipate	520
[103-33-3]	Azobenzene	463
[103-65-1]	*n*-Propylbenzene	469,528
[104-55-2]	Cinnamaldehyde	108
[104-55-2]	3-Phenyl-2-propenal	108
[104-87-0]	4-Methylbenzaldehyde	108
[104-87-0]	*p*-Tolualdehyde	108,490
[104-94-9]	*p*-Anisidine	482
[105-30-6]	Methyl isobutyl carbinol	481
[105-30-6]	Isobutyl alcohol	481
[105-46-4]	*sec*-Butyl acetate	483
[105-54-4]	Ethyl butyrate	469
[105-67-9]	2,4-Dimethylphenol	468,506
[106-35-4]	Ethyl butyl ketone	108,485
[106-35-4]	3-Heptanone	108,469
[106-42-3]	*p*-Xylene	435,492,533
[106-43-4]	*p*-Chlorotoluene	509
[106-43-4]	4-Chlorotoluene	146,147,468
[106-44-5]	4-Methylphenol	228,510
[106-44-5]	*p*-Cresol	490,510
[106-46-1]	*para*-Dichlorobenzene	484
[106-46-7]	1,4-Dichlorobenzene	146,147,335,492,513
[106-46-7]	*p*-Chlorobenzene	335
[106-47-8]	4-Chloroaniline	508
[106-50-3]	1,4-Benzenediamine	503
[106-50-3]	*p*-Phenylenediamine	503
[106-51-4]	Quinone	529
[106-68-3]	Ethyl amyl ketone	485
[106-89-8]	2,3-Epoxy-propyl chloride	347
[106-89-8]	Epichlorohydrin	347,483

[106-89-8]	2-(Chloromethyl)oxirane	347
[106-89-8]	1,2-Epoxy-3-chloropropane	347
[106-93-4]	1,2-Dibromoethane	146,492
[106-93-4]	Ethylene dibromide	146,147,484,519
[106-93-4]	EDB	519
[106-93-4]	1,2-Dibromoethane	468,519
[106-94-5]	1-Bromopropane	468
[106-95-6]	Allylbromide	467
[106-99-0]	Binvinyl	307
[106-99-0]	1,3-Butadiene	482
[106-99-0]	Vinylethylene	307
[106-99-0]	Biethylene	307
[106-99-0]	1,3-Butadiene	307
[107-02-8]	2-Propenal	108,291
[107-02-8]	Acrolein	108,481,490
[107-02-8]	Vinyl aldehyde	291
[107-02-8]	Acraldehyde	291
[107-03-9]	*n*-Propyl mercaptan	469
[107-05-1]	3-Chloropropene	467
[107-05-1]	Allyl chloride	146,467,481,484,491
[107-06-2]	Ethylene dichloride	484,492,519
[107-06-2]	1,2-Dichloroethylene	484,492
[107-06-2]	1,2-Dichloroethane	146,468,519
[107-07-3]	2-Chloroethanol	353
[107-07-3]	Ethylene chlorohydrin	353,469,483
[107-07-3]	2-Chloroethyl alcohol	353
[107-12-0]	Propionitrile	469
[107-13-1]	2-Propenenitrile	293
[107-13-1]	Acrylonitrile	481
[107-13-1]	Vinyl cyanide	293
[107-13-1]	Acrylonitrile	293,491
[107-13-1]	Vinyl chloride	147,433,485,492,532
[107-14-2]	Chloroacetonitrile	468
[107-15-3]	Ethylenediamine	483
[107-18-6]	Allyl alcohol	467,481
[107-19-7]	Propargyl alcohol	469
[107-20-0]	Chloroacetaldehdye	468
[107-21-1]	1,2-Ethanediol	357
[107-21-1]	Ethylene glycol	357,483
[107-22-2]	Ethanedial	108
[107-22-2]	Glyoxal	108
[107-49-3]	Tetraethyl pyrophosphate	218,219,417,487,530
[107-49-3]	TEPP	218,417,530
[107-49-3]	Pyrophosphoric acid tetraethyl ester	417
[107-49-3]	Tetron	218
[107-49-3]	TEPP	468,530
[107-66-4]	Dibutyl phosphate	483
[107-82-4]	1-Bromo-3-methylbutane	468
[107-87-9]	2-Pentanone	108,469,485
[107-87-9]	Methyl propyl ketone	108
[108-05-4]	Vinyl acetate	465,470
[108-10-1]	4-methyl-2-pentanone	108,383
[108-10-1]	Methyl isobutyl ketone	108,383,469,524
[108-10-1]	Hexone	383
[108-10-1]	MIBK	383

[108-10-1]	Methyl isobutyl ketone	383,468,469
[108-10-1]	Isopropyl acetone	383
[108-10-1]	4-Methyl-2-pentanone	108,383
[108-10-1]	MIBK	524
[108-20-3]	Diisopropyl ehter	469
[108-24-5]	Acetic anhydride	481
[108-21-4]	Isopropyl acetate	469
[108-31-6]	Maleic anhydride	523
[108-38-3]	*m*-Xylene	435,492,533
[108-39-4]	*m*-Cresol	510
[108-39-4]	3-Methylphenol	228,510
[108-46-3]	1,3-Dihydroxybenzene	403
[108-46-3]	Resorcinol	403,529
[108-46-3]	1,3-Benzenediol	403,529
[108-46-3]	3-Hydroxyphenol	403
[108-46-3]	*m*-Hydroxyphenol	403
[108-60-1]	*bis*(2-Chloroisopropyl)ether	509
[108-67-8]	Mesitylene	523
[108-80-5]	2,4,6-Trihydroxy-1,3,5-triazine	329
[108-80-5]	Trihydroxycyanidine	329
[108-80-5]	Cyanuric acid	329,483
[108-83-8]	Diisobutylketone	108,469,485
[108-83-8]	Isovalerone	108,469
[108-84-9]	Methyl isoamyl acetate	147,483
[108-86-1]	Bromobenzene	146,147,505
[108-88-3]	Toluene	421,484,492,530
[108-88-3]	Methylbenzene	421
[108-88-3]	Phenylmethane	421
[108-90-7]	Chlorobenzene	146,319,484,492,508
[108-90-7]	Phenyl chloride	319
[108-90-7]	Benzenechloride	319
[108-91-8]	Cyclohexylamine	464
[108-93-0]	Cyclohexanol	464,468,481
[108-94-1]	Cyclohexanone	108,485
[108-94-1]	Cyclohexylketone	108
[108-95-2]	Phenol	228,486,490,528
[108-98-5]	Phenyl mercaptan	530
[108-98-5]	Benzenethiol	530
[108-98-5]	Thiophenol	530
[109-06-8]	2-Picoline	528
[109-06-8]	2-Methylpyridine	528
[109-60-4]	*n*-Propyl acetate	469,483
[109-64-8]	1,3-Dibromopropane	468
[109-65-9]	1-Bromobutane	468
[109-69-3]	*n*-Butyl chloride	468
[109-79-5]	*n*-Butylmercaptan	468
[109-79-5]	1-Butanethiol	482
[109-86-4]	2-Methoxyethanol	484
[109-87-5]	Methylal	485
[109-89-7]	Diethylamine	465,481
[109-93-3]	Vinyl ether	470
[109-94-4]	Ethyl formate	469
[109-99-9]	1,4-Epoxybutane	217,419
[109-99-9]	Tetrahydrofuran	419,487
[109-99-9]	Oxolane	419

[109-99-9]	Oxacyclopentane	419
[109-99-9]	Tetrahydrofuran	419
[110-12-3]	5-Methyl-2-hexanone	108
[110-12-3]	Methyl isoamyl ketone	108
[110-19-0]	Isobutyl acetate	464,483
[110-43-0]	Methyl-*n*-amyl ketone	485
[110-43-0]	2-Heptanone	469
[110-54-3]	*n*-Hexane	484
[110-57-6]	*trans*-1,4-Dichloro-2-butene	146,468
[110-62-3]	Valeraldehyde	108,470,487,490
[110-62-3]	Pentanal	108,490
[110-75-8]	2-Chloroethylvinyl ether	508
[110-77-0]	2-(Ethylthio)ethanol	464
[110-80-5]	2-Ethoxyethanol	483
[110-82-7]	Cyclohexane	464,484
[110-83-8]	Cyclohexene	484
[110-86-1]	Pyridine	397,486,529
[110-86-1]	Azine	397
[110-86-1]	Azabenzene	397
[110-91-8]	Morpholine	465
[111-15-9]	2-Ethoyethyl acetate	483
[111-30-8]	Glutaraldehyde	108,483
[111-30-8]	1,5-Pentanedial	108
[111-40-0]	Diethylenetriamine	483
[111-42-2]	Diethanolamine	465,482
[111-42-2]	Diethanolamine	482
[111-44-4]	*sym*-Dichloroethyl ether	483
[111-44-4]	Dichloroethyl ether	514
[111-65-9]	*n*-Pentane	484
[111-71-7]	Heptaldehyde	108
[111-71-7]	Heptanal	108
[111-76-2]	2-Butoxyethanol	483
[111-76-2]	Butyl cellosolve	463
[111-88-6]	1-Octanethiol	486
[111-91-1]	*bis*(2-chloroethoxy)methane	508
[112-31-2]	Decanal	108
[112-31-2]	Decyl aldehyde	108
[114-26-1]	Baygon	201
[114-26-1]	Aprocarb	201
[114-26-1]	Propoxur	201,465
[115-32-2]	Dicofol	465
[115-78-6]	Phosphan	218
[115-90-2]	Fensulfothion	217,520
[115-90-2]	Terracur P	217
[116-06-3]	Aldicarb	201,463,500
[116-06-3]	Temik	201,500
[117-79-3]	2-Aminoanthraquinone	501
[117-80-6]	Dichlone	464,512
[117-80-6]	Phygon	512
[117-81-7]	*bis*(2-ethylexyl)phthalate	241,520
[117-82-8]	*bis*(2-methoxyethyl) phthalate	241
[117-83-9]	*bis*(2-*n*-butoxyethyl) phthalate	241
[117-84-0]	Dioctyl phthalate	517
[117-84-0]	Di-*n*-octyl phthalate	241
[118-74-1]	Hexachlorobenzene	147,208,464,484,491

[118-75-2]	Chloranil	464
[118-91-2]	2-Chlorobenzoic acid	463
[119-61-9]	Benzophenone	463
[119-64-2]	Tetralin	465
[119-90-4]	3,3-Dimethoxybenzidine	516
[119-93-7]	2-Tolidine	465
[119-93--7]	α-Tolidine	483
[120-12-7]	Anthracene	166,486,491,501
[120-36-5]	Dichloroprop	153,465
[120-36-5]	2-(2,4-Dichlorophenoxy)propionic acid	153
[120-58-1]	Isosafrole	523
[120-62-7]	Piperonyl sulfoxide	528
[120-71-8]	*o*-Anisidine	501
[120-80-9]	*o*-Dihydroxybenzene	399
[120-80-9]	1,2-Benzediol	399
[120-80-9]	*o*-Diphenol	399
[120-80-9]	2-Hydroxyphenol	399
[120-80-9]	Pyrocatechol	399
[120-82-1]	1,2,4-Trichlorobenzene	147,487,531
[120-83-2]	2,4-Dichlorophenol	339,468,506,514
[120-92-3]	Cyclopentanone	108
[120-92-3]	Ketocyclopentane	108
[121-14-2]	2,4-Dinitrotoluene	343,506,517
[121-44-8]	Triethylamine	465
[121-69-7]	*N*,*N*-Dimethylaniline	482
[121-75-5]	Cythion	217
[121-75-5]	Malathion	217,219,485,523
[121-75-5]	Phosphothion	523
[122-09-8]	1,1-Dimethyl-2-phenyl enanamine	528
[122-09-8]	Phentermine	528
[122-10-1]	Bomyl	217
[122-10-1]	Swat	217
[122-14-5]	Fenitrothion	217
[122-14-5]	Folithion	217
[122-34-9]	Simazine	205,208,465,529
[122-34-9]	Amizine	205
[122-34-9]	Gesapun	205
[122-39-4]	Diphenylamine	465,517
[122-39-4]	*N*-Phenylbenzeneamine	517
[122-42-9]	IPC	201
[122-42-9]	Propanil	465
[122-42-9]	Propham	201,465
[122-66-7]	1,2-Diphenyl hydrazine	517
[123-19-3]	4-Heptanone	108
[123-19-3]	Dipropyl ketone	108
[123-19-3]	4-Heptanone	469
[123-31-9]	1,4-Benzenediol	367,485
[123-31-9]	Hydroquinone	367,485
[123-31-9]	Hydroquinol	367
[123-31-9]	*p*-Dihydroxybenzene	367
[123-38-6]	Propionaldehyde	108,469
[123-38-6]	Propanal	108,490
[123-51-3]	Diacetone alcohol	481
[123-54-6]	Acetylacetone	108,463
[123-54-6]	2,4-Pentanedione	108,463

[123-54-6]	Acetyl acetone	108,463
[123-72-8]	Butyraldehyde	108,468,490
[123-72-8]	Butanal	108
[123-73-9]	2-Butenal	108
[123-73-9]	Crotonaldehyde	108
[123-86-4]	Butyl acetate	463,468
[123-86-4]	*n*-Butyl acetate	483
[123-91-1]	Dioxane	483
[123-92-2]	Isoamyl acetate	483
[123-99-9]	Azelaic acid	482
[124-13-0]	Octanal	108
[124-13-0]	Octaldehyde	108
[124-40-3]	Dimethylamine	481
[124-48-1]	Chlorodibromomethane	146
[124-48-1]	Dibromochloromethane	511
[124-58-3]	Methylarsonic acid	482
[126-68-1]	*O-O-O*-Triethylphosphorothioate	532
[126-72-7]	*tris*-(2,3-dibromopropyl)phosphate	219,512
[126-75-0]	Demeton-S	219
[126-98-7]	Methacrylonitrile	469
[126-99-8]	Chloroprene	146,482,492
[126-99-8]	2-Chloro-1,3-butadiene	146
[127-18-4]	Tetrachloroethylene	147,413,484,493,530
[127-18-4]	Perchloroethylene	413,530
[127-18-4]	Ethylene tetrachloride	413
[127-18-4]	Tetrachloroethane	413
[127-18-4]	Perchloroethylene	530
[127-19-5]	Dimethylacetamide	483
[129-00-0]	Pyrene	166,486,491,528
[129-79-3]	2,4,7-Trinitrofluoren-9-one	487
[130-15-4]	1,4-Naphothoquinone	525
[130-15-4]	1,4-Naphthalenedione	525
[131-11-3]	*N,N*-Dimethyl-4-(phenylazo)benzenamine	516
[131-13-3]	Dimethyl phthalate	241
[131-18-0]	Diamyl phthalate	241
[131-89-5]	Dinex	516
[131-89-5]	2-Cyclohexyl-4,6-dinitrophenol	516
[132-64-9]	Dibenzofuran	511
[133-06-2]	Ortholide-406	507
[133-06-2]	*N*-(trichloromethylmercapto)-4-cyclohexene-1,2-dicarboximide	315
[133-06-2]	Captan	208,315,491,507
[133-07-3]	Folpet	491
[133-90-4]	Vancide-89	532
[134-32-7]	1-Naphthylamine	465,525
[134-32-7]	α-Naphthylamine	485
[135-98-8]	*sec*-Butylbenzene	506
[137-17-7]	2,4,5-Trimethylaniline	532
[137-26-8]	Thiram	465,487
[139-40-2]	Gesamil	205
[139-40-2]	Milogard	205
[139-40-2]	Propazine	205
[140-41-0]	Monuron TCA	203
[140-41-0]	Urox	203
[140-57-8]	Aramite	501

[140-88-5]	Ethyl acrylate	469,483
[141-43-5]	Monoethanolamine	482
[141-43-5]	2-Aminoethanol	482
[141-66-2]	Dicrotophos	217,515
[141-66-2]	Bidrin	515
[141-66-2]	Carbicron	217
[141-75-3]	Butyryl chloride	468
[141-78-6]	Ethyl acetate	469
[141-79-7]	Mesityl oxide	108,485
[141-79-7]	Methyl isobutenyl ketone	468
[142-28-9]	1,3-Dichloropropane	146,468,493
[142-82-5]	*n*-Heptane	484
[142-96-1]	*n*-Butyl ether	468
[143-50-0]	Kepone	464,485,523
[143-50-0]	Chlordecone	523
[145-73-3]	Endothall	464
[146-50-9]	*bis*(4-methyl-2-pentyl)phthalate	241
[148-24-3]	Isoquinoline	464
[150-50-5]	Merphos	217,219
[150-68-5]	Telvar	203
[150-68-5]	Monuron	203,465
[152-16-9]	Octamethylpyrophosphoramide	527
[156-59-2]	*cis*-1,2-Dichloroethylene	468,514
[156-59-4]	*cis*-1,2-Dichloroethene	146
[156-60-5]	*trans*-1,2-Dichloroethene	146
[156-60-5]	*trans*-1,2-Dichloro ethylene	512
[191-07-1]	Coronene	166,464
[191-24-2]	Benzo[*g,h,i*]perylene	166,486,491
[192-65-4]	Dibenzo[*a*,e]pyrene	166,511
[192-97-2]	Benzo[*e*]pyrene	166,468
[193-39-5]	Indeno[1,2,3-*cd*]pyrene	166,486,491,522
[198-55-0]	Perylene	166
[205-99-2]	Benzo[*b*]fluoranthene	166,486,491
[206-44-0]	Fluoranthene	166,486,491,520
[207-08-9]	Benzo[*k*]fluoranthene	166,486,491
[208-96-8]	Acenaphthylene	166,486,491
[2104-6-5]	EPN	219,483,529
[213-46-7]	Picene	166
[218-01-9]	Chrysene	166,491,509
[224-42-0]	Dibenz[*a,j*]acridine	166,511
[260-94-6]	Acridine	463
[275-51-4]	Azulene	463
[280-57-9]	Triethylenetetramine	483
[297-97-2]	Zinophos	218,533
[297-97-2]	Namafos	218
[297-97-2]	Thionazin	218,219,533
[298-00-0]	Metaphos	523
[298-00-0]	Nitran	217
[298-00-0]	Methyl parathion	219,523
[298-00-0]	Parathion-methyl	217
[298-02-2]	Phorate	217,219,528
[298-02-2]	Thimet	217,528
[298-03-3]	Demeton-O	468
[298-04-4]	Disyston	517
[298-04-4]	Disulfoton	468

[298-04-4]	Thiodemeton	217
[299-84-3]	Ronnel	218,491
[299-84-3]	Etrolene	218
[299-84-3]	Fenchlorphos	218
[299-84-3]	Ronnel	468,492
[299-86-5]	Crufomate	217
[299-86-5]	Montrel	217
[299-86-5]	Ruelene	217
[300-57-2]	Allyl benzene	467
[300-76-5]	Naled	217,219
[300-76-5]	Dibrom	217
[300-76-5]	Naled	217,218
[301-12-2]	Meta Systox-R	217
[301-12-2]	Oxydemeton-methyl	217
[309-00-2]	Aldrin	208,486,490
[311-45-5]	Paraoxon	217
[311-45-5]	Phosphacol	217
[314-40-9]	Bromacil	463,505
[314-42-1]	Isocil	464
[315-18-4]	Mexacarbate	465,491
[319-84-6]	α-BHC	208,505
[319-84-6]	Hexachlorocyclohexane	490
[319-85-7]	β-BHC	208,505
[319-86-8]	δ-BHC	208,505
[327-98-0]	Trichloranate	218,532
[330-54-1]	Siduron	465
[330-54-1]	Diuron	203,464
[330-54-1]	Karmex	203
[330-55-2]	Linuron	203,464
[330-55-2]	Lorox	203
[333-41-5]	Diazinon	217,219,511
[333-41-5]	Basudin	217,511
[333-41-5]	Neocidol	217
[334-88-3]	Azimethylene	331
[334-88-3]	Diazomethane	331,483
[352-93-2]	Diethyl sulfide	469
[354-58-5]	1,1,1-Trichloro-2,2,2-trifluoroethane	470
[363-72-4]	Pentafluorobenzene	147
[367-12-4]	*o*-Fluorophenol	520
[460-00-4]	4-Bromofluorobenzene	146
[460-19-5]	Oxalyl cyanide	327
[460-19-5]	Cyanogen	327
[460-19-5]	Ethanedinitrile	327
[460-19-5]	Oxalonitrile	327
[460-19-5]	Dicyan	327
[462-06-6]	Fluorobenzene	146
[465-73-6]	Isodrin	464,522
[470-90-6]	Birlane	217
[470-90-6]	Chlorfenvinphos	217,218,508
[501-15-6]	Chlorobenzilate	468
[506-96-7]	Acetyl bromide	467
[507-20-0]	*tert*-Butyl chloride	468
[510-15-6]	Chlorobenzilate	208,464,508
[510-15-6]	Acaraben	508
[512-56-1]	1,2-Dinitrobenzene	517

[512-56-1]	Trimethylphosphate	219,532
[513-36-0]	Isobutyl chloride	469
[513-88-2]	1,1-Dichloroacetone	468
[529-20-4]	2-Methylbenzaldehyde	108
[529-20-4]	*o*-Tolualdehyde	108,490
[534-52-1]	2-Methyl-4,6-dinitrophenol	228,524
[540-36-3]	1,4-Difluorobenzene	146,468
[540-54-5]	1-Chloropropane	468
[540-88-5]	*tert*-Butyl acetate	483
[541-33-3]	1,1-Dichlorobutane	468
[541-73-1]	1,2-Dichlorobenzene	146,468
[541-73-1]	1,3-Dichlorobenzene	147,335,492,513
[541-73-1]	*m*-Chlorobenzene	335
[541-85-5]	5-Methyl-3-heptanone	108
[541-85-5]	Ethyl amyl ketone	108
[542-18-7]	Chlorocyclohexane	468
[542-75-6]	1,3-Dichloropropene	146,465,468
[542-75-6]	1,3-Dichloro-1-propene	468
[542-92-7]	1,3-Cyclopentadiene	483
[543-59-9]	1-Chloropentane	468
[544-10-5]	1-Chlorohexane	146,468
[555-37-3]	Kloben	203
[555-37-3]	Neburon	203,465
[556-52-5]	Glycidol	483
[557-17-5]	Methyl propyl ether	469
[557-31-3]	Allyl ethyl ether	467
[557-40-4]	Allyl ether	467
[557-98-2]	2-Chloropropene	468
[563-12-2]	Nialate	217,519
[563-12-2]	Ethion	217,219,519
[563-47-3]	3-Chloro-2-methylpropene	146
[563-47-3]	2-Chloro-3-methylpropene	464
[563-52-0]	3-Chloro-1-butene	146,468
[563-58-6]	1,1-Dichloropropene	468
[563-80-4]	Methyl isopropyl ketone	108,469
[563-80-4]	3-Methyl-2-butanone	108
[566-61-6]	Isothiocyanatomethane	464
[583-60-8]	2-Methyl isobutyl ketone	485
[584-84-9]	TDI	423
[584-84-9]	Toluene-2,4-diisocyanate	423,485,487,530
[584-84-9]	Isocyanic acid 4-methyl- -*m*-phenylene ester	423
[590-20-7]	2,2-Dichloropropane	146,468
[590-21-6]	1-Cholorpropane	468
[590-86-3]	3-Methylbutanal	108
[590-86-3]	Isovaleraldehyde	108,469,490
[591-08-2]	1-Acetyl-2-thiourea	505
[591-78-6]	2-Hexanone	108,485,522
[591-78-6]	Methyl butyl ketone	108,522
[593-60-2]	Vinyl bromide	470,485
[594-16-1]	2,2-Dibromopropane	468
[594-20-7]	*sec*-Dichloropropane	515
[594-20-7]	2,2-Dichloropropane	468,506
[594-37-6]	1,2-Dichloro-2-methylpropane	468
[602-87-9]	5-Nitroacenaphthene	166

[605-54-9]	*bis*(2-ethoxyethyl) phthalate	241
[606-20-2]	2,6-Dinitrotoluene	506,517
[606-20-2]	2,6-Dinitrotoluene	345
[608-73-1]	Hexachlorocyclohexane	505
[608-73-1]	BHC	505
[608-93-5]	Pentachlorobenzene	147,486,491,526
[616-21-7]	1,2-Dichlorobutane	468
[620-23-5]	3-Methylbenzaldehyde	108
[620-23-5]	*m*-Tolualdehyde	108,490
[621-64-7]	*N*-Nitrosodipropylamine	486
[621-64-7]	*N*-Nitrosodi-*n*-propylamine	183,527
[622-96-8]	4-Ethyltoluene	492
[624-83-9]	Isocyanic acid methyl ester	385
[624-83-9]	Isocyanatomethane	385
[624-83-9]	Methyl isocyanate	385
[624-83-9]	MIC	385
[626-38-0]	*sec*-Amyl acetate	483
[627-40-7]	Allyl methy ether	467
[628-63-7]	*n*-Amyl acetate	483
[628-81-9]	Butyl ethyl ether	468
[630-08-0]	Carbon monoxide	311
[630-20-6]	1,1,1,2-Tetrachloroethane	147,468,529
[634-66-2]	1,2,3,4-Tetrachlorobenzene	487
[671-04-5]	Carbanolate	201
[671-04-5]	Banol	201
[680-31-9]	Hexamethylphosphoramide	522
[709-98-8]	Propachlor	466
[732-11-6]	Phosmet	217,219,528
[732-11-6]	Imidan	217,528
[732-11-6]	Prolate	217
[762-49-2]	1-Bromo-2-fluoroethane	468
[786-19-6]	Carbofenothion	217,218,532
[786-19-6]	Trithion	217,218,532
[786-19-6]	Carbofenothion	217,218,532
[789-02-6]	*o,p*-DDT	511
[822-06-0]	Hexamethylene diisocyanate	485
[823-40-5]	2,6-Toluenediamine	487
[834-12-8]	Gesapax	205
[834-12-8]	Ametryne	205,463
[870-23-5]	Allyl mercaptan	467
[886-50-0]	Terbutryne	205
[886-50-0]	Prebane	205
[918-00-3]	1,1,1-Trichloroacetone	469
[924-16-3]	*N*-Nitrosodi-*n*-butylamine	183,526
[924-16-3]	*N*-Nitrosodibutylamine	183,486
[928-51-8]	4-Chloro-1-butanol	468
[930-28-9]	Chlorocyclopentane	468
[930-55-2]	*N*-Nitrosopyrrolidine	183,465,486
[944-22-9]	Fonofos	217,464
[947-02-4]	Cylan	217
[947-02-4]	Phosfolan	217
[950-37-8]	Methidathion	468
[950-37-8]	Supracide	217
[959-98-8]	Endosulfan-1	208,518
[961-11-5]	Tetrachlorvinphos	468

[961-11-5]	Stirifos	468
[994-22-9]	Dyfonate	217
[994-22-9]	Fonofos	205
[1014-70-6]	Gybon	205
[1014-70-6]	Simetryne	205,465
[1024-57-3]	Heptachlor epoxide	208,491
[1031-07-8]	Endosulfan sulfate	208,518
[1071-83-6]	Glyphosate	520
[1120-57-6]	Chlorocyclobutane	468
[1190-22-3]	1,3-Dichlorobutane	469
[1237-53-3]	Arsenic trioxide	482
[1314-13-2]	Zinc oxide	487
[1314-87-0]	Lead sulfide	485
[1319-77-3]	Cresol	482
[1319-77-3]	Cresylic acid	510
[1330-20-7]	Xylene	435,466,484,533
[1330-20-7]	Dimethyl benzene	435
[1333-86-4]	Carbon black	482
[1338-23-4]	Methyl ethyl ketone peroxide	485
[1471-03-0]	Allyl propyl ether	467
[1476-11-5]	*cis*-1,4-Dichloro-2-butene	146
[1563-66-2]	Carbofuran	201,463,507
[1563-66-2]	Furadan	201,507
[1582-09-8]	Trifluralin	208,465,532
[1582-09-8]	Treflan	532
[1610-17-9]	Atraton	205
[1610-17-9]	Gestamin	205
[1610-18-0]	Prometon	205,528
[1610-18-0]	Primatol	205
[1610-18-0]	Prometon	468
[1646-87-3]	Aldicarb sulfoxide	201,500
[1646-88-4]	Aldicarb sulfone	201,500
[1646-88-4]	Aldoxycarb	201,500
[1646-88-4]	Standak	201
[1689-84-5]	Bromoxynil	482,506
[1689-99-2]	Bromoxynil octanoate	482
[1746-01-6]	2,3,7,8-TCDD	249,491,517
[1746-01-6]	Dioxin	517
[1757-18-1]	Akton	217,218
[1836-75-5]	Nitrofen	465,527
[1836-75-5]	TOK	527
[1861-32-1]	DCPA	464,510
[1861-32-1]	Dacthal	510
[1861-40-1]	Benfluralin	463
[1888-71-7]	Hexachloropropene	147,522
[1897-45-6]	Chlorothalonil	464,491,509
[1910-42-5]	Paraquat	527
[1912-24-9]	Atratol	205
[1912-24-9]	Gesaprim	205
[1912-24-9]	Atrazine	205,463
[1918-00-9]	Dicamba	512
[1918-00-9]	3,6-Dichloro-2-methoxybenzoic acid	153,468
[1918-00-9]	Dicamba	153,464
[1918-02-1]	Picloram	465,528
[1918-11-2]	Terbucarb	201

[1918-11-2]	Terbutol	201,468
[1918-16-7]	Propachlor	208,465
[1918-18-9]	Swep	201
[1929-82-4]	Metribuzin	465
[1982-47-4]	Chloroxuron	203
[1982-47-4]	Tenoran	203
[1982-49-6]	Siduron	203,465
[1982-49-6]	Tupersan	203
[2032-59-9]	Matacil	201
[2032-59-9]	Aminocarb	201,463
[2032-65-7]	Methiocarb	201,464,523
[2032-65-7]	Mesurol	201,523
[2032-65-7]	Mercaptodimethur	201
[2104-64-5]	EPN	217,219,483,529
[2104-64-5]	Santox	529
[2164-17-2]	Cotoran	203
[2164-17-2]	Fluometuron	203
[2303-16-4]	Diallate	463
[2303-16-4]	Avadex	503
[2310-17-0]	Azofene	503
[2310-17-0]	Rubitox	217
[2310-17-0]	Phosalone	217,219
[2385-85-5]	Dechlorane	511
[2385-85-5]	Mirex	208,465,491
[2425-06-1]	Captafol	507
[2425-06-1]	Difolatan	507
[2463-84-5]	Dicapthon	217
[2497-07-6]	Oxydisulfoton	517
[2497-07-6]	Disulfoton sulfoxide	517
[2567-14-8]	1,1,2-Trichloropropylene	470
[2593-15-9]	Etridiazole	208
[2631-37-0]	Promecarb	201,465
[2631-37-0]	Carbamult	201
[2642-71-9]	Guthion ethyl	217
[2642-71-9]	Azinphos ethyl	217,219
[2675-77-6]	Chlorneb	208
[2782-91-4]	Tetramethyl thiourea	487
[2921-88-2]	Dursban	217,517
[2921-88-2]	Chlorothalonil	464,491,509
[2921-88-2]	Dursban	468,517
[2921-88-2]	Chlorpyrifos	217,218,491,517
[2921-88-2]	Trichlorpyrphos	217
[3060-89-7]	Patoran	203
[3060-89-7]	Metobromuron	203
[3244-90-4]	Aspon	217
[3268-87-9]	Octachlorodibenzo-*p*-dioxin	491,527
[3268-87-9]	OCDD	527
[3383-96-8]	Tetrafenphos	218
[3383-96-8]	Temephos	218
[3383-96-8]	Abate	218
[3689-24-5]	Sulfotepp	218,219
[3766-81-2]	Baycarb	201
[3766-81-2]	Fenobucarb	201
[4091-39-8]	3-Chloro-2-butanone	468
[4147-51-7]	Dipropetryne	205

[4147-51-7]	Sancap	205
[4170-30-3]	Crotonaldehyde	108.464,468,490
[4170-30-3]	2-Butenal	468
[4482-55-7]	Urab	203
[4482-55-7]	Fenuron TCA	203
[4685-14-7]	Paraquat	486
[5103-71-9]	α-Chlordane	208,507
[5103-73-1]	*cis*-Nonachlor	527
[5103-74-2]	γ-Chlordane	507
[5598-13-0]	Methyl dursban	218
[5598-13-0]	Methyl chlorpyrifos	218
[5779-94-2]	2,5-Dimethylbenzaldehyde	108,491
[5902-51-2]	Terbacil	465
[5915-41-3]	Terbuthylazine	205
[5915-41-3]	Gardoprim	205
[6923-22-4]	Monocrotophos	217,219
[6923-22-4]	Azodrin	217
[6959-48-4]	3-Chloromethyl-pyridine hydrochloride	510
[6988-21-1]	Elocron	201
[6988-21-2]	Dioxacarb	201
[7005-72-3]	4-Chlorophenyl phenyl ether	509
[7005-72-3]	1-Chloro-4-phenoxybenzene	509
[7287-19-6]	Caparol	205
[7287-19-6]	Prometryne	205,465
[7421-93-4]	Endrin aldehdye	208,491,518
[7429-90-5]	Aluminum	481
[7439-92-1]	Lead	485
[7439-97-6]	Mercury	485
[7440-33-7]	Tungsten	487
[7440-38-2]	Arsenic	482
[7440-41-7]	Beryllium	482
[7440-43-9]	Cadmium	482
[7440-47-3]	Chromium	482
[7440-48-4]	Cobalt	482
[7440-50-8]	Copper	482
[7440-66-6]	Zinc	487
[7440-70-2]	Calcium	482
[7446-09-5]	Sulfur dioxide	411,487
[7553-56-2]	Iodine	485
[7581-97-7]	2,3-Dichlorobutane	468
[7647-01-0]	Hydrochloric acid	481
[7664-38-2]	Hydrofluoric acid	481
[7664-38-2]	Phosphoric acid	481
[7664-41-7]	Ammonia	482
[7664-93-9]	Sulfuric acid	481
[7697-37-2]	Nitric acid	481
[7700-17-6]	Crotoxyphos	217,219,510
[7700-17-6]	Ciodrin	217
[7719-12-2]	Phosphorus trichloride	486
[7723-14-0]	Phosphorus	486
[7782-44-7]	Oxygen	486
[7783-06-4]	Hydrogen sulfide	369
[7784-42-1]	Arsine	297
[7784-42-1]	Arsenic hydride	297
[7784-42-1]	Hydrogen arsenide	297

[7784-42-1]	Arsenic trihydride	297
[7784-42-1]	Arsine	482
[7786-34-7]	Mevinphos	217,219,485
[7786-34-7]	Phosdrin	217
[7803-52-3]	Stibine	405,487
[7803-52-3]	Hydrogen antimonide	405
[7803-52-3]	Stibine	487
[7803-52-3]	Antimony hydride	405
[8001-35-2]	Toxaphene	208,530
[8001-35-2]	Camphechlor	530
[8003-34-7]	Pyrethrum	486
[8006-64-2]	Turpentine	487
[8008-20-6]	Kerosene	485
[8012-95-1]	Mineral oil mist	485
[8030-30-6}	Petroleum naphthe	485
[8032-32-4]	Petroleum ether	485
[8052-41-3]	Stoddart solvent	485
[8065-48-3]	Demeton	483
[8065-48-3]	Systox	217
[8065-48-3]	Demeton-O,S	217
[10024-97-2]	Nitrous oxide	486
[10035-10-6]	Hydrobromic acid	481
[10061-01-5]	*cis*-1,3-Dichloropropene	146
[10061-02-6]	*trans*-1,3-Dichloropropene	146,468
[10102-44-0]	Nitrogen dioxide	391,485
[10102-44-0]	Nitrogen peroxide	391
[10265-92-6]	Methamidophos	217
[10265-92-6]	Tamaron	217
[10403-60-8]	2,2,3-Trichlorobutane	469
[10595-95-6]	*N*-Methyl-*N*-nitrosoethanamine	524
[10595-95-6]	*N*-Nitrosomethylethylamine	183
[11096-82-5]	PCB-1260	244,503
[11096-82-5]	Arochlor 1260	244,503
[11097-68-1]	Arochlor 1254	244,502
[11097-69-1]	PCB-1254	244,502
[11100-14-4]	Arochlor 1268	244
[11104-28-2]	PCB-1221	244,501
[11104-28-2]	Arochlor 1221	244,501
[11141-16-5]	Arochlor 1232	244,502
[11141-16-5]	PCB-1232	244,502
[12069-32-8]	Boron carbide	482
[12672-29-6]	PCB-1248	244,502
[12672-29-6]	Aroclor-1248	244,502
[12674-11-2]	Arochlor 1016	244,501
[12674-11-2]	PCB-1016	244,501
[13071-79-9]	Counter	529
[13071-79-9]	Terbufos	218,219,529
[13071-79-9]	Counter	218,529
[13171-21-6]	Phosphamidon	217,219,528
[13171-21-6]	Dimecron	468,528
[13171-21-6]	Phosphamidon	468
[13194-48-4]	Ethoprophos	519
[13194-48-4]	Phosethoprop	218
[13194-48-4]	Ethoprop	519
[13194-48-4]	Prophos	218,219

[13360-45-7]	Chlorbromuron	203
[13360-45-7]	Maloran	203
[13463-39-3]	Nickel carbonyl	485
[14484-64-1]	Ferbam	464
[15271-41-7]	Butyllate	463
[15972-60-8]	Alachlor	208,463
[16672-87-0]	Ethephon	217,218
[16672-87-0]	Ethrel	217,218
[16752-77-5]	Lannate	201,523
[16752-77-5]	Methomyl	201,465,523
[17804-35-2]	Benlate	201
[17804-35-2]	Benomyl	201,463
[18181-80-1]	Bromopropylate	463
[18829-56-6]	2-Nonenal	108
[18829-56-6]	Nonenaldehyde	108
[19287-45-7]	Diborane	333,483
[19287-45-7]	Boroethane	333
[19287-45-7]	Boron hydride	333
[21087-64-9]	Sencor	205
[21087-64-9]	Metribuzin	205
[21087-64-9]	Lexone	205
[21548-32-3]	Acconame	205,465
[21548-32-3]	Fosthietan	217
[21548-32-3]	Geofos	217
[21609-90-5]	Lepton	217
[21609-90-5]	Leptophos	217,218,219,523
[21609-90-5]	Phosvel	217,523
[21725-46-2]	Cyanazone	205
[21725-46-2]	Bladex	205
[22224-92-6]	Fenamiphos	520
[22224-92-6]	Phenamiphos	468
[22224-92-6]	Nemacur	468,520
[22248-79-9]	Rabon	529
[22248-79-9]	Tetrachlorvinphos	218,219,529
[22248-79-9]	Stirophos	218,529
[22781-23-3]	Bendiocarb	201,463
[22781-23-3]	Ficam	201
[23135-22-0]	Vydate	533
[23135-22-0]	Oxamyl	201,465
[23135-22-0]	Pirimor	201
[23135-22-0]	Pirimicarb	201
[23184-66-9]	Butachlor	464
[23950-58-5]	Kerb	523
[25013-15-4]	Vinyltoluene	484
[25057-89-0]	Pronamide	528
[25154-52-3]	Nonyl phenol	465
[25167-82-2]	Trichlorophenol	532
[25167-83-3]	Tetrachlorophenol	530
[25311-71-1]	Isofenphos	217
[25311-71-1]	Oftanol	217
[29191-52-4]	Anisidine	482
[29576-14-5]	1-Bromo-2-butene	468
[30560-19-1]	Orthene	217
[30560-19-1]	Acephate	217
[31218-83-4]	Propetamphos	218

[31218-83-4]	Safrotin	218
[32889-48-8]	Procyazine	205
[32889-48-8]	Cycle	205
[33213-65-9]	β-Endoulfan	518
[33213-65-9]	Endosulfan-II	208,518
[33213-65-9]	Thiodan-II	518
[33245-39-5]	Basalin	503
[34014-18-1]	EL-103	203,468
[34014-18-1]	Tebuthiuron	203
[34123-59-6]	Arelon	203
[34123-59-6]	Isoproturon	203
[34643-46-4]	Protothiophos	218
[34643-46-4]	Tokuthion	218,219
[35400-43-2]	Sulprophos	217
[35400-43-2]	Bolstar	219
[35822-46-9]	1,2,3,4,6,7,8-Heptachlorodi -benzo-*p*-dioxin	521
[37324-23-5]	Arochlor 1262	244
[39765-80-5]	*trans*-Nonachlor	208,491,527
[40321-76-4]	1,2,3,7,8-Pentachlorodibenzo-*p*-dioxin	527
[41198-08-7]	Profenofos	218
[41198-08-7]	Curacron	218
[42509-80-8]	Isazophos	217
[42509-80-8]	Miral	217
[50594-66-6]	Aciflurofen	500
[51207-31-9]	2,3,7,8-TCDF	529
[51207-31-9]	2,3,7,8-Tetrachlorodibenzofuran	529
[51218-45-2]	Metolachlor	465
[51235-04-2]	Hexazinone	205
[51235-04-2]	Velpar	205
[51877-74-8]	*trans*-Permethrin	208
[53469-21-9]	Arochlor 1242	244,502
[53469-21-9]	PCB-1242	244,502
[53494-70-5]	Endrin ketone	208,519
[53555-64-9]	Tetrachloronaphthalene	465
[54774-45-7]	*cis*-Permethrin	208
[55720-37-1]	Trichloronaphthalene	465
[55720-99-5]	Hexachlorocyclopentadiene	491
[55720-99-5]	Chlorinated diphenyl ether	482
[57117-41-6]	1,2,3,7,8-Pentachlorodibenzofuran	527
[57653-85-7]	1,2,3,6,7,8-Hexachlorodibenzo-*p*-dioxin	512
[57653-85-7]	1,2,3,4,7,8-HCDD	492
[64529-56-2]	Tycor	205
[64529-56-2]	Ethiozin	205
[64529-56-2]	Ebuzin	205
[64902-72-3]	Glean	205
[64902-72-3]	Chlorsulfuron	205
[65996-93-2]	Coal tar pitch volatiles	482
[66063-05-6]	Pencycuron	203
[66063-05-6]	Monceren	203
[70648-26-9]	1,2,3,4,7,8-Hexachlorodibenzofuran	147
[75673-16-4]	Hexyl 2-ethylhexyl phthalate	241